MOZAMBIQUE
a nation in crisis

First published 1989

by The Claridge Press
43 Queen's Gardens
London W2
and Box 420
Lexington
Georgia 30648

ISBN 1-870626-95-8

Hoile, David
Mozambique, A Nation In Crisis
1. Political Science and Economics

CONTENTS

Foreword 9

Introduction 11

Is Frelimo Marxist? 15

Frelimo in Power 25

The Mozambican Economy Under Frelimo 51

The Mozambique National Resistance:
The Origins of Renamo 66

Structure and Control of Renamo 85

The Internationalisation of the Mozambican Conflict 104

Mozambique: Dialogue or Disaster? 125

Notes 130

Appendices 137

Acknowledgements

I would like to thank Roger Scruton, Barbara Day and all the staff at Claridge Press for their advice, assistance and encouragement in facilitating the publishing of my original text.

My appreciation also goes to Fiona Ellis for her time and effort in proof reading the final text.

About The Author

David Hoile has been a member of the Conservative Party for eight years. He has held local, regional and national office, serving on the Conservative Party National Union Executive Committee 1985-86 and representing conservative students on the Conservative Foreign and Commonwealth Council 1984-86. He is currently deputy chairman of his constituency association.

David Hoile holds degrees in politics and International Studies from the University of Warwick, and has worked as a parliamentary research assistant. He presently runs his own political and design consultancies. A particular research interest of his is political and economic conflict, and his previous publications have included *Nicaragua for Beginners* and *Understanding Sanctions.*

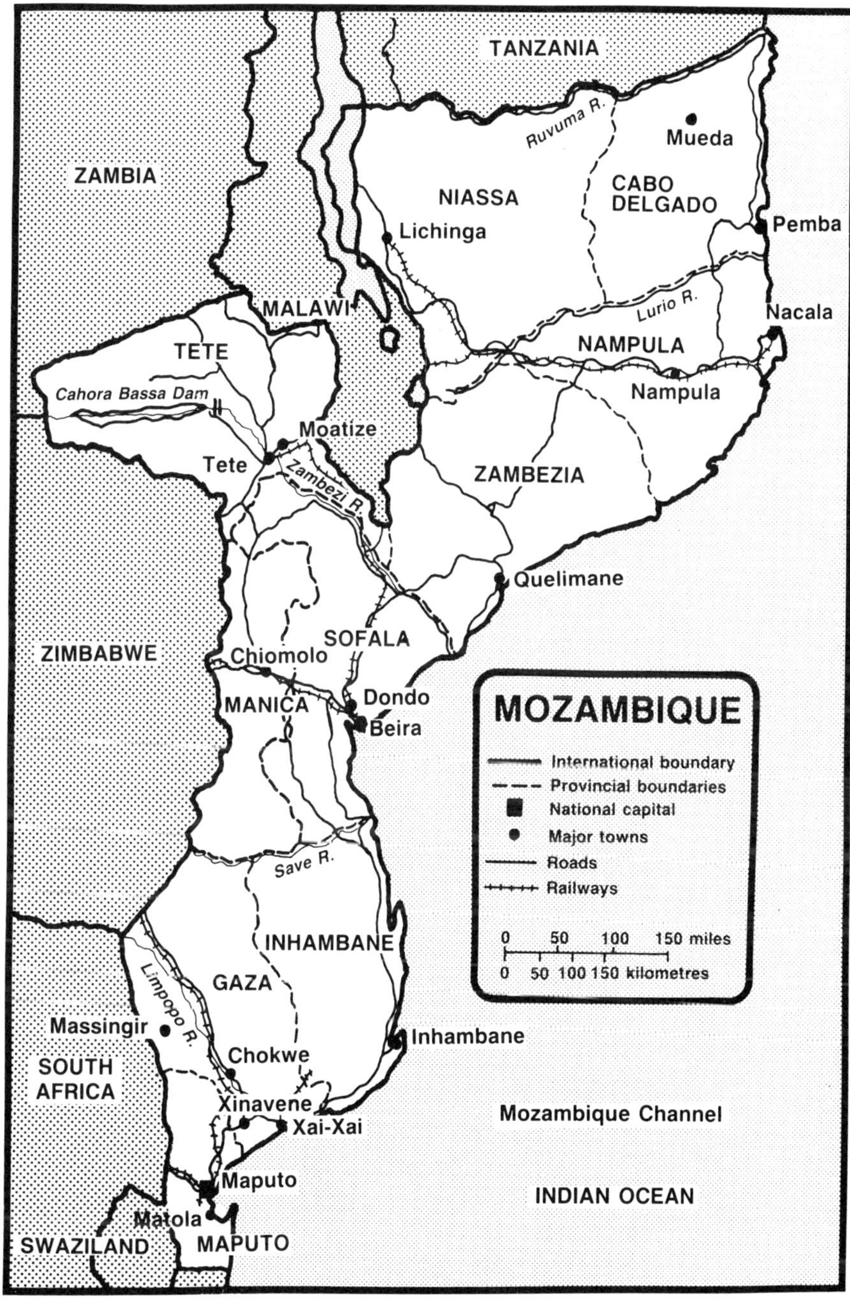
TANZANIA
ZAMBIA
Ruvuma R.
Mueda
NIASSA
CABO DELGADO
Pemba
Lichinga
MALAWI
Lurio R.
Nacala
TETE
NAMPULA
Cahora Bassa Dam
Nampula
Moatize
Tete
Zambezi R.
ZAMBEZIA
Quelimane
SOFALA
ZIMBABWE
Chiomolo
MANICA
Dondo
Beira
MOZAMBIQUE
International boundary
Provincial boundaries
National capital
Major towns
Roads
Railways
0 50 100 150 miles
0 50 100 150 kilometres
Save R.
INHAMBANE
GAZA
Limpopo R.
Massingir
Inhambane
SOUTH AFRICA
Chokwe
Xinavene
Mozambique Channel
Xai-Xai
Maputo
INDIAN OCEAN
Matola
SWAZILAND
MAPUTO

FOREWORD

Mozambique offers not only one of the most interesting challenges to British foreign policy of the late eighties but also presents some of its most striking contradictions. In many ways the conflict which has engulfed this nation presents a test case for the West, and Britain in particular, of future foreign policy strategies.

The former Portuguese colony is today ravaged by civil war, famine of massive dimensions and political repression.

As with so much of the communist bloc, Mozambique is torn between Marxist-Leninist loyalties and economic realities – desperate to retain its revolutionary fervour, whilst seeking to attract as much Western, including South African, capital as possible. Yet, as with so many other 'reforming' communist states there is a marked unwillingness to match economic liberalisation with the extension of political freedom. In issues central to the retention of power, such as the seperation of the Party (Frelimo) from the State and the retention of Party control over the military, compromise does not appear to be an acceptable path.

For the West to accept, and materially encourage, economic liberalisation and restructuring without demanding political reform, through national reconciliation, is an unrealistic and unwise strategy. As with the current situation in the Soviet Union and Poland, the West owes it to the people of Mozambique to encourage national reconciliation between the opposing forces, Frelimo and Renamo.

Perhaps the greatest irony is that Mozambique remains rich in natural resources, with the potential for mining and agriculture outstripping that of almost all her neighbours. With this potential conflict and famine can be replaced by development and prosperity.

'*Mozambique – A Nation in Crisis*' offers a new perspective on a much reported civil war. Whatever the merits of Renamo, or Frelimo for that matter, its conclusion, that political reconciliation must accompany any economic restructuring cannot be doubted.

Chapter One

INTRODUCTION

"Throughout Mozambique these primary resistance movements had common raisons d'etre - to drive out the imperialist forces, to protect the indigenous homelands and historic way of life, and to avoid harsh taxes and the expropiation of land and labor. These goals explain the popular support the anticolonial campaigns generally enjoyed." [1]

These words, written to describe the political unrest and armed conflict in the Nineteenth century and early Twentieth century between black Mozambicans, Portuguese and Portuguese-aligned black local leaders, aimed at overthrowing Portuguese authority, are as true today as they were a century or so ago. The Mozambique people, innately conservative, culturally and historically land-centred, and deeply resentful of externally-imposed authority are today caught up in a conflict with military and political forces intent on destroying their way of life, on demolishing villages and challenging property-rights as well as on re-introducing detested forced labour practices. The Mozambican people have met a relentlessly undemocratic project head-on, a project dictated by an alien Marxist-Leninist ideology and introduced by Mozambique's Frelimo government.

Unilaterally installed in power in 1974, despite the existence of several democratic opposition groupings enjoying similar political support, Frelimo's ascendancy was assisted by the sympathetic leftist Portuguese administration in Lisbon at the time. As only one of several Marxist regimes which came to power as the result of violent political change in the mid-and-late 1970s, Frelimo exploited the détente honeymoon atmosphere prevalent in Western political attitudes towards not only the Soviet Union but also with regard to pro-Soviet 'national liberation movements'. A combination of the détente mood, incomplete and often inaccurate information, and a capable propaganda projection of these Marxist insurgencies produced an attitude willing to give regimes, such as that of Frelimo in Mozambique, the benefit of the doubt: the image that groups such as the MPLA in Angola, ZAPU/ZANU in Rhodesia, Kampuchea's *Khmer Rouge*, Ethiopia's Dergue and Frelimo were liberal nationalists, imagined to be temporarily embracing Marxist terminology in order to secure the means necessary to liberate their respective countries.

What became very clear all too soon was that far from being liberal nationalists exploiting a veneer of Marxism in order to gain power, these movements were as often as not of Marxists dogmatically committed to their own agenda, sheltering behind a veneer of nationalism. It is no coincidence that most, if not all, of these countries were politically and socially devastated within months of these ideological regimes coming to power. The tragedy of Kampuchea has been graphically illustrated in studies such as the film *The Killing Fields*, and several books by survivors of the *Khmer Rouge*'s Marxist programme. Similar television and documentary evidence, together with books such as Myles Harris' *Breakfast in Hell*, have clearly underlined the Ethiopian state's responsiblity for war, famine and disease. That the Addis Ababa regime's explicitly Marxist agenda is responsible for the Ethiopian holocaust is beyond question, disputed only by ideologues. That Mozambique has experienced its own tragic holocaust is similarly obvious to any observer. Several international aid agencies and charities have given clear warnings about the present situation in Mozambique, as well as alarming projections of rapidly impending famine and disease. Pictures rivalling the horror of the Ethiopian famine have been screened on our television. What is also evident is a clear confusion over the causes of Mozambique's intense difficulties.

There is, seemingly, a distinct attempt to lay all of Mozambique's many problems at the feet of South Africa, arguing that Mozambique's famine, pestilence and conflict all have their roots in South African foreign policy towards the 'Frontline States' of Southern Africa, of which Mozambique is a strategic member. This blends in with similar efforts to deny the political responsiblity for human and material holocaust of comparable Marxist regimes, in Kampuchea and Ethiopia for instance. In Kampuchea, successive apologists for the fact that the *Khmer Rouge* regime destroyed two million people, have claimed that it was the result of American involvement in the Indo-Chinese conflict, specifically because of the United States bombing of strategic targets. In Ethiopia, the weather and separatist insurgencies were cited as the reasons for famine and death on an unprecedented scale, it was ignored that the regime's ruthless policy of human relocation into state villages had destroyed both agricultural production and incentive. Added to this, the systematic abuse of overseas aid, aimed at alleviating the resultant state-created refugee problem, heightened an already scarcely believable situation.

That the Frelimo regime has made as many disastrous policy decisions as the Ethiopian state, with much the same effect in human misery, is now beyond question. Attempts to centralise control over Mozambique's population and economy, a key component of which was the creation of state villages and farms following the Ethiopian model, have failed dismally. And the price of failure has been famine, disease, escalating levels of political and military conflict, and the loss of whatever claim Frelimo may have had to legitimacy. What is also beyond question is the degree of Western confusion over the Mozambican situation. United States and British attitudes towards Frelimo and the Mozambican civil war have, at best, been naive and, at worst, have shown a tendency towards a cynical regional *realpolitik*. Support for the Marxist Frelimo regime has been used as a counter-balance, for example by Britain, to opposition to mandatory economic sanctions against South Africa. In so doing British and American foreign policy has ignored the legitimate origins of the economic, political and ultimately military opposition of growing numbers of the Mozambican people to the Maputo regime. Far from condemning the foreign military intervention in Mozambique of tens of thousands of Zimbabwean, Tanzanian and Soviet-bloc military personnel, British and other European Community military advisers openly and actively train, assist and equip such occupation and intervention forces. British 'civilian' military advisers, many of them former British special forces personnel, serve inside Mozambique, with at least tacit British government approval: several have been killed in action against anti-Frelimo guerillas. The rise and consolidation of the anti-Marxist democratic resistance in Mozambique is a further example of not only confusion, but deliberate disinformation surrounding any opposition to the Maputo regime.

Against the backdrop of what can only be described as Frelimo-instigated political holocaust and economic chaos, some degree of resistance could only be expected, especially from a people with a long history of opposition to oppression. Inevitably, within a totalitarian system, an obvious way of fighting to secure freedom is through the armed struggle. The Mozambican civil war has been waged for more than a decade, growing from conflict in one or two districts within a province, to armed insurrection throughout all of Mozambique's provinces and cities. The strength and success of the Mozambique National Resistance (Renamo or MNR) in the people's war against Frelimo is clear. In a direct parallel with Afghanistan, vast portions of the countryside have been conceded to the resistance, with government forces desperately attempting to hold on to key towns and cities. In a further parallel with the equally unpopular Soviet-aligned Kabul regime, there is no doubt that without massive foreign military intervention and assistance, Frelimo would have collapsed years ago. Governments and organisations, either politically sympathetic to Frelimo, or motivated by consensus or 'pragmatism' in their Mozambique policies, have deliberately challenged the legitimacy of the pro-Western Renamo movement through propaganda and disinformation. In so doing, these institutions, including the British Foreign Office, only serve to prolong Mozambique's tragedy. The Mozambican civil war cannot be brought to an end without addressing the legitimate aspirations of substantial portions of the Mozambican people for political and economic pluralism, democracy and religious tolerance. Repeated calls by the Pope, the Mozambican Catholic Bishops and even portions of

Frelimo's military forces for negotiation between Frelimo and the Mozambique National Resistance have so far gone unheeded.

It is vital to examine the mainstay of the propaganda case against the Mozambican rebels – the claim that Renamo is a South African creation, reliant on South African military support and finance. There has been incredibly little evidence of what would have to be a massive South African logistical effort to maintain ongoing offensives in Mozambique's ten provinces. What there is ample evidence of, however, is an ever-increasing South African financial, logistical and military commitment to the Frelimo government. Following the rationale that South African connections delegitimise those in receipt, then Frelimo by its own claims is illegitimate. There is a need, therefore, to examine the Mozambican situation to attempt to contribute towards a realistic British foreign policy assessment of options.

Chapter Two

IS FRELIMO MARXIST?

"To understand the place of research in Mozambique today, one must begin with its context: Mozambique is a socialist state, with a Marxist-Leninist party, engaged in a struggle which encompasses many fronts to build a socialist society, to develop a socialist economy."[1]

"In our country Marxism is the product of the War of National Liberation. We did not proclaim Marxism after independence. The war turned itself, in the process of its development, into a people's revolutionary war. It was this which allowed the Mozambique Liberation Front to change itself into a Marxist-Leninist party."[2]

Throughout Frelimo's political history, both before and after the movement's coming to power in Mozambique, there has been considerable speculation as to whether or not Frelimo was a Marxist organisation. Certainly for several years before 1975, Frelimo downplayed any such overt political orientation while not denying direct Chinese or Soviet bloc diplomatic, moral and logistical support. This led those sympathetic to the cause of African black nationalism to advance the theory that 'national liberation movements', such as Frelimo, Robert Mugabe's ZANU or the South African ANC, were merely black nationalists temporarily and half-heartedly embracing Marxist rhetoric to secure assistance in their bids for power. The theory concluded that, once in government, these movements would drop the pretence and assume their true black nationalist identity. This certainly had been the image portrayed of Frelimo in Mozambique. Typical of this somewhat naïve school of thought was Colin Legum's 1975 view that:

"Although the Marxist intellectuals managed to acquire a strong position in the exile movement during the period of armed struggle, their political strength is likely to decline when Frelimo confronts the realities of independent Mozambique. One may predict that, once Frelimo has assumed power, it will draw more upon the perspective of Tanzania's President, Julius Nyerere, than upon any non-African political influences. Nyerere served as the patron of the Mozambican struggle, and his marked aversion to foreign ideo-

> **logies and his championing of a non-Marxist African socialism have strongly influenced Samora Machel."[3]**

The reality of the situation was very different. Far from distancing themselves from Marxist-Leninism, Frelimo politicians have ruthlessly institutionalised a Marxist one-party state in Mozambique.

The Origins of Frelimo

The *Frente de Libertacao de Mocambique* was formed on 23-28 September 1962 at a congress in Dar-es-Salaam, Tanzania. The Frelimo organisation was created out of a merger of three Mozambican nationalist groups – the Malawi-based *Uniao Nacional Africana de Mocambique Independente* (UNAMI), the Rhodesian-based *Uniao Democratica Nacional de Mocambique* (Udenamo) based on Joshua Nkomo's National Democratic Party and led by the Rev Uria Simango and Adelino Guambe; and the Tanzanian-based *Mozambique African National Union* (MANU), modelled on both the Tanzanian African National Union and the Kenya African National Union. Frelimo's first president was US-educated Dr Eduardo Chivane Mondlane, a United Nations Trusteeship Department research officer, and associate professor of anthropology at Syracuse University in New York State. Elected as vice-president was Udenamo's Rev Uria Simango. Other important executive members were Cape Verde-born Marcelino dos Santos (UNAMI), who was elected Secretary for External Relations, Lazaro Nkavandame (MANU) and Fanuel Mahluza (Udenamo).

The history of Frelimo from 1962 until 1969/70 was a clear case study of the struggle within a genuine nationalist liberation movement between nationalists and pro-Soviet Marxists, and of how the Marxist grouping gained ascendancy. John Saul, formerly a lecturer at University College in Dar-es-Salaam and keen observer of the Frelimo movement, lecturing in Marxism-Leninism at Frelimo institutions in the 1980s, described the tensions which had existed within Frelimo for most of the sixties:

> **"This advanced stage is not reached easily. For a movement like Frelimo is two entities for much of the early period of its existence; a conventional nationalist movement frustrated in achieving any easy transition to power, and, a revolutionary movement struggling to be born."[4]**

The tensions between the nationalist and Marxist factions spilt over into conflict on several occasions. The conflict centred around, amongst other things, the issue of education. Uneasy at the influx of American-educated graduates back into Frelimo, the Marxist faction managed to cut off, at source, US scholarships for Mozambicans wanting to undertake post graduate studies – on the pretext of instructing them instead to enter the armed struggle. They would thereby come under the political control of Marxist hardliners such as Samora Machel, a militant who had left Mozambique in 1963 and who had been ideologically and militarily trained in both Algeria and the Soviet Union. Conflict nearer to home also soon emerged. The Mozambique Institute, based in Dar-es-Salaam, served as Frelimo's high school education centre. A black Mozambican educator at the Institute, Roman Catholic priest Father Mateus Gwenjere, clashed with the radical faction in their attempt to gain control of the Institute and use it as a power base in their attack on the moderate leadership of Mondlane. Gwenjere complained of the Marxist influence and called for the dismissal of far left foreign teachers from the Institute. The conflict was so intense that the Institute actually closed for a while. Father Gwenjere, however, was forced out of Frelimo. Two months after the school confrontation, a group of dissident black Mozambican nationalists attacked the Frelimo headquarters building in Dar-es-Salaam, in the course of which a central committee member was killed.

Concern at the strength and machinations of the pro-Soviet faction emerged as early as 1964, when Frelimo executive member Fanuel Mahluza broke away in protest, re-establishing Udenamo. Intra-Frelimo division had reached such a level that, in June 1965, the Zambian government convened a conference aimed at reconciling the differing factions and personalities. Mondlane withdrew when it became apparent that Frelimo would not be rejoined en masse. The groups remaining at the conference formed the *Comite Revolucionario de Mocambique* (Coremo), led by Adelino Guambe, an organisation which was to launch a significant guerilla campaign against the Portuguese authorities.

While Eduardo Mondlane had undoubtedly been an asset in presenting a respectable, moderate image for Frelimo internationally, thereby cloaking the consolidation of the pro-Soviet factions within the movement, he did become an obstacle to their aspiration of ultimate total control. His re-election as president, and that of fellow moderate Uria Simango as vice-president at Frelimo's second congress in July 1968, was to be for the last time. The Congress and elections took place against the backdrop of distinct manoeuvring for power within the organisation by the political hardliners. It had become increasingly apparent that the radicals had succeeded in gaining control of the military wing of the movement. This grip was tightened by Samora Machel's appointment as Frelimo's military commander, in the wake of the unexplained death of the founding guerilla chief, Filipe Samuel Magaia. A protégé of Marcelino dos Santos, Samora Machel was a hard-line radical cadre. Moderates soon became very critical of, and concerned about, the guerilla wing's concentration on political activities instead of fighting the Portuguese – activities aimed rather at further consolidating control within both Frelimo-dominated areas of the northern provinces and the organisation itself.

Dr Mondlane was killed in early February 1969 by a parcel bomb at his home in Dar-es-Salaam. Although Frelimo automatically blamed Portuguese intelligence for the assassination, there has always been considerable speculation surrounding the circumstances. There are strong suspicions that the pro-Soviet faction was responsible. Dr Edward Cain, writing in Charles Moser's excellent study *Combat on Communist Territory*, touched on this theory:

> "It was widely rumoured that (Machel) was responsible for Mondlane's death, and not the Portuguese secret police. Certainly Machel's advent accentuated Frelimo's swing to the left, and marked the beginning of an escalation of the armed struggle with increased supplies of communist arms."[5]

Mondlane's death also saw the implementation of obviously clearly thought-out moves to isolate Uria Simango. Rather than automatically acceding to the presidency, vice-president Simango was effectively marginalised by the pro-Soviet faction. Simango's adherence to a black nationalist line and his rejection of a class orientated Marxist perspective was clear:

> "Our struggle today is not principally an ideological or class one; it is a struggle of masses of people against foreign domination, Portuguese colonialism, for freedom and independence of these masses. The question of scientific socialism and capitalism in Mozambique should not be allowed to divide us..."[6]

The presidency of Frelimo was temporarily assumed by Samora Machel, Marcelino dos Santos and Uria Simango. Within ten months, Simango had been politically outmanoeuvred and expelled from Frelimo. With him went the last genuinely nationalist leader within the organisation. Ordered out of Tanzania in February 1970, Simango moved to Egypt and joined Coremo. Another key moderate leader within Frelimo, Lazaro Nkavandame, had earlier shared much the same fate in what had clearly been pre-arranged purges in the wake of Mondlane's assassination. Frelimo Secretary within his Cabo Delgado province, (a key operational area for the movement), and a powerful Central Committee member, Nkavandame was a political leader among the large Makonde ethnic group in northern Mozambique. Very much a father figure of Mozambican nationalism, Nkavandame had been the founding president of the *African Voluntary Cotton Society of Mozambique* (SAAVM), one of the first manifestations of an organised black Mozambican voice. Two of his early SAAVM colleagues were the future Frelimo Mozambique defence minister and Politburo member, Alberto Chipande, and Mozambican state inspector and central committee member, Raimundo Pachinuapa. Imprisoned by the Portuguese for his nationalist activities, Nkavandame became active in the Mozambique African National Union, and subsequently with Frelimo. A political moderate in the mould of Jomo Kenyatta, Nkavandame clashed openly with the radical faction over the administration of the 'liberated zones' within Mozambique at that time and was expelled from Frelimo in late 1969. His political control over the areas was usurped by military cadres answerable to military commander Samora Machel. After the death of Mondlane, the expulsion or marginalisation of moderate elements within the Frelimo leadership and the May 1970 central committee meeting (which replaced the triumvirate and appointed Samora Machel as President and Marcelino dos Santos as vice-President of Frelimo), great satisfaction was felt by the victorious Marxist grouping within Frelimo. This satisfaction was reflected in an editorial of the movement's official journal *Mozambique Revolution*:

> *"The spirit which prevailed at the latest meeting of the Central Committee revealed that we have already reached an advanced phase in that process of purifying our ranks. Frankness reigned — there was friendship and revolutionary fraternity among all members. The climate that we felt was the result of the ideological unity that existed among us. For the first time in the history of Frelimo there were no discordant voices on the Central Committee which were opposed to the revolutionary positions; it constituted a solid and united block.*
>
> *. . .The enemy has two faces: the principal and direct enemy, i.e. Portuguese colonialism and imperialism . . . The other face is that of the indirect or secondary enemy, who presents himself under the cover of a nationalist and even a revolutionary. . .The Central Committee reaffirmed that the characterization of the enemy for us will never be derived from colour, nationality, race or religion. . .*
>
> *Thus, a new period is being opened in the life of Frelimo. We took an important step forward in the consolidation of our unity, we elected a truly revolutionary leadership, we clarified our political line. . ."*[7]

Samora Machel, during a speech made in Maputo at the end of 1979 some four and a half years after Frelimo gained power, analysed the earlier struggle within the movement:

> "The very term 'comrade' bears with it the history of the struggle between the revolutionaries and the reactionaries, between the old and the new, between the dialectic and a static conservatism, within Frelimo...

The designation 'comrade' is the outcome of the defeat of those who wanted to make independent Mozambique a puppet state, a preserve of the capitalist bourgeoisie and of international imperialism."[8]

Significantly, of Frelimo's original central committee only three members remain – dos Santos, Mariano Matsinhe and Pascoal Mocumbi. Marina Ottaway's study of post-independence Mozambique in *Africa and International Communism* perhaps sums up the origins of Frelimo's Marxist ideology:

"The proclamation of Marxism-Leninism as Mozambique's ideology at the third party congress in February 1977 was. . .simply the logical extension of a process began well before independence"[9]

The Third Congress of Frelimo, meeting less than two years after independence on 3-7 February 1977, saw the formal declaration of Frelimo's commitment to Marxism-Leninism, transforming the organisation into an official vanguard party of "workers and peasants":

"The hard class battles demand that the working class, closely allied with the peasantry, its fundamental ally, and with the progressive elements of the other labouring classes, organise itself into a vanguard party, guided by the scientific ideology of the proletariat - Marxism-Leninism"[10]

Mozambique's new masters now made no secret of their ideological standpoint. Frelimo ideologue, Marcelino dos Santos, speaking on state radio in August 1977, confirmed the party's clear Marxist analysis:

"The fight of the workers of our country is a fight against the capitalists. . .Our objectives are clear; destroy capitalism in our country and create socialism."[11]

A 1978 Frelimo Politburo report took Dos Santos' line even further:

"For us, Marxism-Leninism was not something we chose out of a book. It was in the process of the struggle that the people's interest asserted themselves and became more and more clearly demarcated from the interests of the colonial exploiter and the would-be national exploiter. . .Marxism-Leninism is a living science which is continually developing the process of revolutionary struggles. It is not a sclerotic, dogmatic catechism; not a complete and finalized pharmacopoeia."[12]

It is also clear that the socialism envisaged by Frelimo is very far removed from the democratic socialist image created of Frelimo by the movement's Western sympathisers. Frelimo's concept of socialism coincides with that practised in the Soviet Union:

> "We reject the idea that there can be an 'African socialism' or a 'Mozambican socialism.' We consciously affirm that there can be no socialism other than scientific socialism."[13]

Frelimo's political programme as set out in the wake of the 1977 Congress, furthermore defined the "...international workers movement and Marxist-Leninist parties" as the party's "natural allies".

Samora Machel made it very clear that the belief that, somehow, Frelimo had been fighting for a nationalist revolution, rather than a Marxist solution, was totally misguided:

> "Our revolution is not a bourgeois revolution; is not the triumph of the bourgeois conceptions of the world and society...
>
> Our revolution is a proletarian revolution, a socialist revolution. It is power of the worker-peasant alliance, and its exercise through dictatorship of the proletariat..."[14]

One of the litmus tests of how pro-Soviet certain countries and regimes are, was the rise of the independent *Solidarity* trade union in Poland. Frelimo condemned *Solidarity* and supported the Polish state in its attempt to crush all opposition to the country's Marxist regime.

By far the most clear indication of both Samora Machel's and Frelimo's commitment to Marxism was given in Machel's key-note address, entitled "Every Revolution is a contribution to Marxism", delivered in Berlin on 11 April 1983, at international celebrations marking the centenary of Marx's death:

> **"Dear comrades, with deep feeling we greet this August assembly where representatives of the forces of progress and socialism, from all the continents and peoples, from all races and nations, are together celebrating the centenary of the death of one of mankind's most beloved sons, Karl Marx.**
>
> **...In four continents, workers, taking control of their destiny, are building a happy future, are building socialism, Communism.**

> **. . .Marxism is a shining path, a sun of hope and certainty that never sets, a sun that is always at its zenith.**
>
> **Marxism, the science of revolution, is the fruit of practice, of mankind's struggle for a better future and so is renewed and developed through human practice. The experience of revolutionary struggle of the Mozambican people provides an illustration of this principle.**
>
> **The dictatorship of the proletariat is as alive today as at the time of the Commune."**

The Mozambican president then went on to affirm that:

> **". . .(A)t the Third Congress in 1977, our Marxist-Leninist Party was born from its embryo in the Front."**

Samora Machel paid tribute to the East German leader and his party:

> **"We salute through Comrade Erich Honecker, General Secretary of the Central Committee of the Socialist Unity Party of Germany, the workers – labourers, peasants, revolutionary intellectuals – who have been able to transform the genius of Marx into an invincible material force, for the triumph of socialism and peace."**

East Germany, a state where there is extensive political and religious persecution, which seeks to keep its citizens in by means of walls, minefields and barbed wire entanglements, and where the price for even attempting to escape is so often death, is lauded by Machel:

> **"We are celebrating this centenary in the German Democratic Republic, a Germany in which socialism has triumphed. A Germany where working man continually manifests the breadth of his intelligence, his energy, his ability, by creating prosperity, progress and happiness."**[15]

Frelimo's international outlook is one further indication of its ideological perspective. The Frelimo central committee report, *Out of Underdevelopment to Socialism*, published at the 1983 Fourth Congress, gives a clear insight into the Party's standpoint. The Party's alignment with the Soviet Union is manifest. In March 1977, Machel signed a 20-year Treaty of Friendship between the USSR and Mozambique, during Soviet President Podgorny's state visit – a treaty following on from trade and economic agreements signed in February 1976. In Moscow, in 1980, Frelimo opened only its second embassy outside

Africa, the other being to Portugal. An acid test of Soviet alignment is the Afghanistan issue. Frelimo lauded the consolidation of "progressive power" in Afghanistan, describing the Kabul regime as a "legitimate revolutionary government", and voted against United Nations resolutions deploring the Soviet invasion of Afghanistan in December 1979. Mozambique's relationships with other Soviet bloc states are similarly close. During the transition to power in 1974/5, Machel made formal visits to East Germany, Bulgaria, Rumania, China and North Korea. Cuban leader Fidel Castro visited Mozambique in March 1977, and a few months afterwards, in October, Machel signed a twenty year Treaty of Friendship between Cuba and Mozambique while himself on a state visit to Cuba.

Despite the projection of a political re-alignment by Frelimo away from Soviet-style Marxism-Leninism and towards Western values, an image carefully cultivated by the Maputo government from 1983 onwards, the reality is somewhat different. Mozambique became the first Southern African government to become associated with the Soviet Bloc Council of Mutual Economic Assistance, an economic and development co-operation agreement covering agricultural, industrial, health and educational projects.

That the warmth of Frelimo's relationship with the Soviet Union has not been lost is clear from Soviet-Mozambique declarations. The following four sample quotations by Chissano are typical:

On the occasion of the tenth anniversary of trade and economic co-operation between the Soviet Union and Mozambique, in February 1986: ". . .we highly value the Soviet people's heroic solidarity and are convinced that Lenin's compatriots, working hand in hand with us will continue to be our reliable bulwark"[16].

The ninth anniversary of the Treaty of Friendship and Co-operation between the Soviet Union and Mozambique, in March 1986: "Our ties of friendship and co-operation grow and consolidate on a basis of solid and common principles – Marxism-Leninism and proletarian internationalism"[17].

On the occasion of the 69th anniversary of the October revolution, Chissano's message to General Secretary Gorbachev said: ". . .we would like to take this opportunity to reaffirm our sincere desire to still further strengthen the bonds of friendship which unite our peoples, parties and governments"[18]. On Soviet ambassador Dybenko's presentation of credentials in November 1986, Chissano was cited as stating that: "Mozambique desires to continue developing and consolidating its fraternal ties, friendship and varied co-operation with the Soviet Union"[19].

That, for all the economic liberalism and stated non-alignment, Frelimo was still committed to a Marxist-Leninist policy, was put on the record once again to Joaquim Chissano in February 1987. Speaking to several hundred Frelimo members in Maputo, Chissano declared:

> "The rehabilitation program states that we are going to support the private sector. We have been holding talks with the IMF and the World Bank. One might feel that we are having doubts about the socialist option. No, that is not the case. We must not confuse the issue. We chose socialism as a result of our experiences. We have combined these experiences with the scientific teachings of Marxism-Leninism."

Frelimo's ideological solidarity extends very much into the Third World. In 1982, Mozambique established diplomatic relations at ambassadorial level with the 'Palestinian Liberation Organisation' terrorist grouping, a link "reflecting the strengthening bonds of friendship and solidarity between the Palestinian and Mozambican peoples". A Frelimo Party central committee report spelt out Frelimo's political relationship with the PLO in black and white terms: "The Frelimo Party unconditionally supports the just struggle of the Palestinian Arab people and their legitimate representative, the PLO."[20]

Furthermore, Frelimo is supportive of the FMLN/FDR terrorist movement in El Salvador, the Sandinista regime in Nicaragua, the Timorese FRETILIN insurgents, Polisario in the Western Sahara, and the Vietnamese-installed Heng Samrin regime in Kampuchea.

Chapter Three

FRELIMO IN POWER

"People say we are a one-party state. We are not a one-party state, we are a one-party democracy."[1]

There is, and always will be, a conflict when political movements, Marxist or Fascist, attempt to impose an all-encompassing totalitarian one-party state upon a country. There is a self-evident contradiction between such a political project and human rights and civil liberties as understood by Western democrats – one or the other can be secured, but certainly not both. Frelimo's attempt to create such a Marxist one-party state in Mozambique has been no exception to this rule.

Frelimo's concerted attempts to dictate and regulate virtually every facet of society within Mozambique, particularly against the backdrop of disastrous political and economic misjudgments, have led to opposition and criticism – manifestations of civil and political resistance which in turn have led to increasing state repression. In addition to Frelimo's political and economic mismanagement, widespread corruption has further distanced the Frelimo government from the people.

In order to both impose its Marxist agenda and overcome post-independence problems of political legitimacy and organisation, Frelimo initiated a dual strategy of extending its political authority to those areas in which it had not been active, and by clamping down on any opposition. Limited as it was to areas within the three northern-most provinces, Frelimo had no organisation or presence in any of the others. Frelimo was a movement consisting of 250 trained cadres in 1964, according to pro-Frelimo sources, such as Father Cesare Bertulli. There were only some one thousand Frelimo activists in the country by September 1974, and these were supported by several hundred Tanzanians and Zambians. The repressive ferocity with which Frelimo sought to impose itself on the Mozambican masses, carried with it the seeds of the civil war which was to follow.

The Fusion of State and Party

On Frelimo's attainment of power in 1975, Mozambique became a Marxist-Leninist one-party state. Article 3 of the Constitution of 25 June 1975 clearly outlined Frelimo's status:

> "(T)he People's Republic of Mozambique will be guided by the political line defined by Frelimo, which is the leader of the state and of society. Frelimo will establish the basic political policy of the state, so as to ensure the conformity of the policy of the State with the interests of the people".

That the Mozambican state had become nothing more than a vehicle for the Marxist Frelimo party was also clearly articulated by Samora Machel:

> "The state exercises above all the political and administrative power...
>
> The state is an instrument of the party."[2]

Several clear definitions exist of Frelimo's role. Frelimo's own statutes, as approved at the party's Fourth Congress in 1983, state that:

> "The Frelimo Party is the vanguard organisation of the worker-peasant alliance in Mozambique. Under the leadership of the working class the Frelimo party unites in a voluntary and militant alliance based on scientific socialism the workers, peasants, soldiers, revolutionary intellectuals and other working people, forged and tempered in the struggle against exploitation of man by man, and for the victory of the people's interests."[3]

That Frelimo adheres to a Soviet-style 'democratic centralism' is also evident, as outlined by the Frelimo central committee at the 1977 Third Congress. 'Democratic centralism' in essence:

> "Lies above all in the existence of a single centre which enjoys the confidence of all militants and of the masses in general, to lead the Party organisations in their entirety. Our Party represents the interests of the labouring classes as a whole, and not the private interests of individuals or isolated groups. Therefore the Party has to carry out the gigantic task of bringing together all the forces existing within the masses, and orienting them in the direction of a single objective, thus lending unity to disparate actions. The Party can only realise these tasks if the leadership is centralised."[4]

Frelimo's official role is of setting the political line and strategy, and to secure their implementation by the state. This distinction of function, tenuous at best, has vanished in Mozambique. All the Frelimo politburo are usually ministers, and well over half of the then Frelimo central committee elected in 1977 were appointed ministers, provincial governors or military commanders, or diplomats. Furthermore, provincial governors and district administrators are also party secretaries in their areas.

Frelimo lost no time in using its new authority to start creating a Marxist Mozambique. Moves towards nationalisation started even before the formal hand-over of power by the Portuguese. The provisional Frelimo administration enacted legislation in February 1975, empowering the state to dismiss the management of private businesses and to replace them with state officials. Large-scale nationalisations started one month after independence, including all schools and colleges, hospitals and clinics, legal practices and funeral businesses. In early 1976, the nationalisation, without compensation, of all land, privately-owned buildings and property was enacted – all rents being collected from then on by Frelimo officials.

The creation of 'dynamising groups' of young ideological Frelimo cadres was designed to extend the movement's political structure into every province and district of Mozambique and enforce state authority within Mozambican society. This tangible, and often bloody, extension of the Marxist one-party state exercised considerable new political powers. Hand-in-hand with these 'dynamising groups' went the ruthless use of police state repression by Frelimo to secure its often tenuous authority in both urban and rural areas. That Frelimo desired an all-encompassing presence became evident:

> "We shall form people's brigades in each neighbourhood to go from house to house finding out who lives there and registering them. Each neighbourhood must have a complete register of its residents. Any newcomers to the neighbourhood must report to the Dynamising Group for registration.
>
> This campaign will be a fundamental step in the process of organising the cities and communal neighbourhood.
>
> The party structure, the organs of people's power, the People's Assemblies, the democratic mass organisations (for women, youth, journalists, production councils), neighbourhood Dynamising Groups, the *FPLM* and the defence and security forces, must all take part in this campaign, incorporating the people..."[5]

These 'dynamising groups' exercised considerable power. Together with the Frelimo-designated block committees, they controlled and enforced a system of residency cards needed for food rationing – a means of politically controlling movement into the urban areas.

The deliberate intrusion of young Marxists into the political, economic and sociological structure of what was still a largely conservative rural Mozambican society was traumatic and inevitably resulted in conflict. Samora Machel was very conscious of the obstacles to radical change. Addressing a women's meeting in Maputo in April 1976, less than a year after independence, he admitted:

> "And we find that there is a great deal of conservatism in our country."[6]

Frelimo's concerted attempted to destroy, very often by assassination, the traditional Mozambican power structures of tribal chieftains and headmen, was bitterly resented. Resentment of these moves is just as strong today as it was in 1975, as a Frelimo official speaking in February 1988 confirmed:

> "In some areas, this government is seen as an intruder into a world of age-old traditions. It breeds a very vicious reaction, a desire to destroy any vestige of central... authority."[7]

President Machel also went so far as to claim that Frelimo had destroyed the very essence of Mozambican society – ethnic structures and consciousness: "We killed the tribe to give birth to the nation"[8]. In the event, his claim was a premature one, but Frelimo's heavy-handedness was to create ever increasing degrees of resistance to the regime.

Resistance to this alien ideology resulted inevitably in both death and imprisonment for opponents. In order to consolidate its control of Mozambican society, Frelimo built up large defence and security forces, including a 30,000 strong army, the *Forcas Populares de Libercao de Mocambique* (FPLM), together with police and secret police.

Frelimo's political control over these security forces is total. Frelimo central committee member, and government minister, Sergio Vieira wrote in a 1981 *World Marxist Review* article:

> **"Our organs of defence and security are the armed hand of the dictatorship of the proletariat. It is mandatory for officers to be party members, communists."**[9]

He was echoing Machel's September 1980 statement that: **"All officers will have to be Frelimo members. To be an officer in the Frelimo army, it is first necessary to be a communist."**[10]

This Marxist political orientation and control also applied to the police, the *Policia Popular de Mocambique*: "Our people's police must speed up the political education of its members... This requires strengthening their Party structures and constant political and ideological work by the base structures."[11]

The Mozambican secret police, *Servicio Nacional de Seguranca Popular* (Snasp), was brought into being in early October 1976. Modelled on the Soviet KGB, and set up with the technical assistance and training of the East German and Cuban state secret police, Snasp is answerable only to the president. Mozambican president Joachim Chissano, a former minister of security, served as its head. *Snasp* has draconian powers of arrest and indefinite detention, search and seizure. Law No. 21/75 of October 1975 dictates that *SNASP* detainees cannot contest the legality of their detention.

Within two weeks of its existence, the organisation had arrested more than three thousand people across the country.

Several international human rights organisations have strongly criticised Frelimo's human rights record. Amnesty International has recorded killings, torture, and mutilation by government forces over the years. It has also expressed concern about the disappearance of political prisoners, Frelimo's unfair judicial process, appalling prison conditions and detention without trial. The 1987 Amnesty Report on Mozambique cites several examples of Frelimo human rights violations. One case of unfair conviction was secondary school teacher Leonardo Mabunda. Accused of writing an essay critical of government policy and submitting it as the work of one of his students, Mabunda was sentenced in April 1983 to eight years imprisonment. Amnesty was "concerned that those tried. . .were not given an adequate opportunity to defend themselves." Prisoners such as Mabunda had not been informed of the charges against them before the trial took place, and had not been able to call witnesses in their defence[12]. Leonardo Mabunda was released in December 1987.

The West German-based International Society for Human Rights 1987 review of Mozambique similarly recorded political murder, disappearances, unlawful imprisonment, denial of due process and torture, as well as heavy censorship and hinderance of religion by Frelimo.[13]

The "brutal behaviour" of Frelimo police units was even cited in the government-controlled daily *Noticias*. An editorial in early 1987 stated that police behaviour had been "characterised by arrogance and easy resort to violence in cases in which this is completely unjustified." [14]

The *Economist* 's *World Human Rights Guide* cites Mozambique as having a 25% human rights rating. The *Guide* places on record Frelimo extrajudicial killings, torture and coercion, conscription of labour, indefinite detention without charge, Frelimo-controlled courts, trade unions, television and radio networks, political indoctrination in schools, political censorship of the press, mail censorship and telephone-tapping. The *Guide* also confirms that there is no freedom of movement either within, or to leave the country, nor is there the right of free assembly or association. [15]

The Frelimo Electoral System

Article 37 of the Constitution states that "the People's Assembly is the supreme organ of the state". Within any one-party state system, legislative institutions are, at best, token. Mozambique is no exception. For the first two years after independence the People's Assembly was directly appointed by the Frelimo central committee. Only towards the end of 1977 was there any pretence of elections.

The delay was a tactical one – to gain time for Frelimo to have tightened its political control nationwide. Limited to areas within Niassa, Cabo Delgado and Tete provinces during the war, Frelimo had to extend its one-party state framework to those other provinces in which it had been denied a presence either by black political hostility or by the Portuguese counter-insurgency effort.

A Frelimo central committee, meeting in late August 1977, evaluated the progress made in enforcing the party line in these areas and proposed electoral legislation – legislation rubber-stamped a few days later by the Frelimo-appointed People's assembly as Act No 1/77.

A Legislative Rubber Stamp

The legislation provided for elections at four levels – local, district and city, provincial, and for the national People's Assembly – to be staggered between late September and December 1977. In keeping with the Soviet Union's electoral system, Frelimo's one party state system ensured that these, and all of Mozambique's subsequent elections, have been parodies of democracy as understood by Western democrats with, for example, the Mozambique electoral law stating: "The list of candidates will be proposed by the Frelimo committee at the appropriate level". Articles 14 and 15 of the electoral law also stipulated that even if elected, the delegate's tenure was at all times dependent on Frelimo being satisfied that he/she had "never identified with colonialism, with the schemes and intrigues of imperialism and of reactionaries."

In addition to the nomination by Frelimo of all the candidates, the electoral system brought into being by Frelimo denied universal suffrage to the Mozambican people in all but the local elections. Representatives elected at local level would nominate district assembly members, who in their turn would nominate delegates for the provincial assemblies. The 226-member national assembly is then elected when the ten provincial assemblies, themselves Frelimo approved and all made up of Frelimo members, rubber-stamp the Frelimo central committee's selection of members. Even after all the façade of elections, the role of the People's Assembly is a very token one indeed. Meeting once or twice a year for a few days, it merely approves political lines and decisions already finalised by Frelimo. Interestingly, the official national election report, issued on 24 December 1977, indicated that restrictive clauses in the electoral legislation were used by Frelimo to dismiss 2,182 of the 22,230 elected at local level for holding "erroneous attitudes".

Pro-Frelimo Propaganda cannot hide totalitarianism

Typical of attempts to play down the totalitarian one-party state within Mozambique was Oxfam's 1987 *Chicualacuala: Life on the Frontline*. Written by Ian Bray, Oxfam's assistant field director in Mozambique, this piece of work attempted to record "in words and photos some of the features and incidents which make up day-to-day life for the people on Mozambique's frontline". In it, Bray describes village-level elections within Chicualacuala district, an area of Mozambique bordering on South Africa and Zimbabwe.

> "Today is election day for Chicualacuala and the surrounding villages. It is the first of three phases in the National elections. First delegates are elected at village level – as today – then at provincial level, finally going on to National level.
>
> Eventually, the beaten up Landrover of the District Administration arrives, bringing the election officers...
>
> The elections begin...
>
> A government official addresses the crowd, explaining the elections... The delegates must be exemplary citizens and they must represent the interests of the people...
>
> Once they have all been presented, the first candidate, an elderly genial-looking man is called forward. But there is some argument

about whether he is an official of the church. Someone asks a party official to explain the ruling on religious candidates for the elections. He replies that since Mozambique is a secular state, religious officials like priests cannot stand as it might lead to contradictions between their religious and state duties... so he cannot stand.

The next candidate is called forward...

Proceedings end with the party official appealing for voluntary contributions to cover the election costs, and then closing the meeting with the customary singing." 16

What is conveniently ignored, but still paradoxically evident, in Ian Bray's idyllic description of this local election, is the fact that Mozambique is a one-party state. The Landrover brings not only the Frelimo district administration officials, but also the Frelimo election officers. The government officials who open, explain, conduct and close the election are Frelimo cadres; it is the Frelimo official who is the arbiter of what is in the interests of the people and who gives rulings on the eligibility of candidates. Bray's account of the first candidate being prevented from standing for office hints at the true nature of Mozambican democracy. No-one not approved by, or a member of, Frelimo is allowed to stand for any election, whether local district, provincial or national in Mozambique. This point is brought out by *Guardian* journalist Joseph Hanlon:

"Elections of all sorts are tightly controlled by the Party... No matter what the election-officials of a Party cell, villagers electing a council, Party members selecting delegates to a Congress... it is Party officials who nominate a list of candidates."17

It is as democratic a process as that of Nazi Germany under the Third Reich or as in the Soviet Union today.

The Frelimo Legal System

All law practices were nationalised in 1975. The Frelimo party is *de facto* and *de jure* the legal force within Mozambique. The role for the party is immense. Where Portugese law is still on the statute books, Frelimo directives can over-ride it: where it does not exist, Frelimo directives apply.

The 1983 Frelimo Fourth Congress saw the introduction of further draconian legislative measures. These included the death penalty for "economic crimes", public executions, public flogging for a wide variety of crimes, wide-ranging pass laws and the forced relocation of thousands of Mozambicans living in urban areas. Hanlon dismisses criticism of Frelimo's imposition of the death sentence as somehow invalid because the majority of people in Britain are in favour of it. He misses a key point. If the death penalty were reintroduced in Britain, there would be a very sophisticated legal system in operation to ensure that any such sentence would be fair and just. The Mozambican legal system's capacity to judge any crime, let alone capital ones, must be challenged.

Senior Frelimo leader Sergio Vieira summed up Frelimo's total political control over Mozambique's legal system: "the composition of the courts embodies the power of the proletariat".

Allen and Barbara Isaacman's *Mozambique From Colonialism to Revolution* gives a graphic account of the unbelievably rudimentary and essentially political nature of Frelimo's judicial system. The Third Party Congress in 1977 ordered the reorganisation of Mozambique's legal system "in accordance with the Constitution and Frelimo's broad socialist principles". This new legal system came into being in late 1978, fitting into Frelimo's provincial, district and local administrative structure. Provincial courts dealt with serious crimes, district courts assessed crimes carrying a maximum penalty of two years, and local/communal village courts heared minor offences. Within three years some three hundred 'peoples tribunals' had been held.

Isaacman makes clear the political sham that passes for justice in Mozambique:

> "The court system is guided by the broad principles of popular justice that Frelimo first formulated in the liberated zones."

The checks and balances of any legal system were done away with:

> "The formalistic procedures of the colonial past were drastically simplified because the lay judges could not comprehend them and because there were no private attorneys to capitalise on procedural errors. Thus, procedures have become more intelligible even without a new procedural code."[18]

What must be of considerable interest to anyone concerned with justice is the basis on which judgement is made:

> "Judges reach decisions based on common sense and general notions of justice outlined in the Constitution. Similarly, much of the formalism and paperwork has disappeared, as litigants do not know what paper must be filed by what date or in what form."[19]

Serious offences are dealt with at district or provincial level by a state-appointed judge. The requirements, apart from adherence to Marxist political viewpoint, for appointment at district level are a sixth-grade education and the completion of a six-month legal course: judges appointed by the legislative assembly are excused both criteria. The political input into the whole system is hall-marked by the sentencing procedure:

> "Periodic meetings between Justice Ministry officials, their provincial counterparts, and judges have produced a consensus that adapts the range of penalties for each crime to the new priorities and values of the socialist society in the making."[20]

In other words, essentially political justice.

Amnesty International has persistently criticised the Frelimo legal system and the resultant abuse of human and civil rights. The Frelimo 'Revolutionary Military Tribunal', an extension of Snasp and designed to process some of the thousands of political detainees, was singled out by Amnesty as a particularly unfair and biased pseudo-judicial instrument.

The Reintroduction Of Flogging

Flogging had been one of the most detested of Portuguese colonial punishments, featuring heavily in Frelimo propaganda as being typical of colonial repression. Journalist Joseph Hanlon pointed out the obvious comparisons with South Africa:

> "Comparison with South Africa was inevitable. Frelimo has always claimed. . .that it was building a fundamentally different society. And then it introduced what the Mozambicans themselves called passes and the sjambok (the name for the whip used in South Africa)." [21]

Dr Joseph Hanlon's *Mozambique: The Revolution Under Fire* provides a unique insight into Frelimo in power. Hanlon, the Mozambique correspondent for the BBC, *Guardian* and *Observer* from 1974-84 provides documentation of Frelimo excesses and brutality in an account which cannot be dismissed simply as anti-Frelimo propaganda.

Flogging became very widespread. Police and security forces, as well as local and regional party and district administrators use it for minor violations and 'disrespect'. Among the first people flogged was Raul Cumbe, lashed for selling 15 small cakes for 75MT (instead of the government set price of 60MT). Hanlon records that:

"High officials were forced repeatedly to assert that Frelimo flogging is different from colonial flogging." 22

It soon became clear that Frelimo officials were using flogging as a punishment. Manica Governor Manuel Antonio was forced to publically order the halt of "the practice of flogging people who criticise incorrect actions of officials."23

Frelimo's Pass Laws And Forced Removals

Frelimo's introduction of pass laws went hand-in-hand with the forcible relocation of people from urban areas. In legislation which made South Africa's long abolished pass system look simple, people living in urban centres were always to carry four passes: a residents card, an identity card, a work card, and – for young people – a national service card. Any person caught without his full documentation was to be "detained immediately and returned to his place of origin, or some other location if the authorities consider it more convenient"24. In a mirror image of the forced relocation of urban dwellers in South Africa, Frelimo's apartheid-esque 'Operation Production' in 1984, directed by the Justice, Defence, Security and Interior ministries, expelled unemployed people from the built-up areas.

Thousands of people without the correct documentation were detained at 'verification centres'. Those declared to be 'marginals', unemployed, prostitutes, etc., were then detained further at 'evacuation centres' and transported to northern provinces such as Niassa and Cabo Delgado. The old people, the crippled, the sick and thousands of innocent single women (unjustly dubbed prostitutes) were also transported. Hanlon was intellectually honest enough to concede the South African parallel:

> "Those without the right documents were being removed from Mozambican cities at the same time as South Africa was removing people from so-called 'black spots'." 25

The planning of 'Operation Production' was dismal. 'Evacuation' and 'verification'centres were set up without any administrative or logistical back-up. Very few had sanitation services. In Maputo, inhabitants were given 14 days to ensure that all of their documentation was in order, or would run the risk of falling victim to the pass laws.

Further restrictions on movement were imposed in January 1984. Frelimo Minister of the Interior, Armando Guebeza, decreed that no-one could move from one district to another without a travel pass – a *guia de marcha* – issued by Frelimo officials. Any visits longer than two days needed the authority of the Interior ministry and the local Frelimo body. Manica Governor, Antonio, further announced that anyone absent from home without the knowledge and consent of the local Frelimo authorities should be detained and interrogated by security forces on his return.

The Mozambican Gulag

Thousands of opponents, real or imaginary, of the Frelimo regime have been detained in the state's chain of 're-education centres'. To all intents and purposes concentration camps, centres such as Rulumba, Mozambique-D, Chibembe, Ruarua, Chaimite, Codzo, Nhangau, Inchope and Donde served initially to immobilize the vanguard of any democratic opposition within the country as well as to intimidate the general population. The camps' catchment was extended as Frelimo's totalitarian project grew. It is estimated that over 75,000 people may have died in these camps since 1975: upwards of 200,000 may have been imprisoned.[26] The International Society for Human Rights currently estimates that there are somewhere between sixteen and twenty-one thousand political prisoners in Mozambique.[27]

Jorge Costa, the former head of Snasp, revealed the existence of between fourteen and twenty camps, each one holding up to 4,000 people. He has also described the structure of the camps:

> "The idea was to make them (the camps) self-sufficient and ultimately provide food for 'export' to the rest of the country. It could have worked if it were run properly, but it dissolved into feudal kingdoms run by the soldiers who controlled the camps. They lived on the people in the camps. They raped the women, forced the others to work, and ate the produce. A disaster."

Da Costa described one of these camps in detail:

> "When I arrived at Machava prison there were over one thousand five hundred people in the cells. It only takes four hundred. Many of them were there without trial and even without having been charged. I found people in Machava who didn't even have a police file opened in their name. They had been put in by the army or the police on the spur of the moment and when the policeman or soldier responsible were transferred elsewhere he left behind these kinds of people, innocent but in jail."

It was clear that torture and abuse of Frelimo's political prisoners was common.

> "The army only had one way of interrogation and punishment. Total beating and whatever the security police were supposed to have done

was like a drop in the ocean compared with what I saw the soldiers doing. At night the screams of the prisoners could be heard a long way off. They seemed to specialise in the early morning as the best time. Sometimes it was the screams of men – sometimes it was women." 28

The savage conditions of the Mozambique Gulag were also described by the Roman Catholic deacon of Sao Leonardo parish at Mussorize in Manica province. Arrested for complaining at the detention of several of his church-goers, he was able to describe conditions after he escaped:

"When the security men come to inspect the underground cells, they usually point the muzzles of their guns to the cells and just open fire against anybody they choose to kill. Fellow prisoners are not supposed to remove the body without the order to do so. The body remains in the cell for two or three days when the order comes from the command post to remove the dead body. The body is not buried. It is taken and thrown away near the camp. The decomposing flesh is eaten by hyenas and other wild animals at night."29

Lisbon-born Antonio Malaquias de Lemos, a Maputo-based lawyer detained for over four months in a Frelimo gaol, provides a further insight into Mozambique's inhumane jails. His enthusiasm for the Frelimo regime makes him a particularly reliable source. While claiming, incongruously, that Frelimo is not aware of the "sub-human" conditions, his account of incarceration was revealing. His backdrop to the Frelimo police state was that:

"Nationals and foreigners are arrested without any charges brought against them simply on reports of anonymous informers, often for puerile reasons. This creates an atmosphere of instability, worry and fear."30

Lemos described widespread mistreatment of Frelimo detainees. Jailors would flood cells with water, keeping the occupants wet for days at a time. They would also flood prison toilets, force prisoners to undress and keep them in the water for long periods. Lemos cited further examples of the abuse, including rape, of foreign female detainees by Frelimo warders. One Swazi woman was locked away in a toilet – even having her meals inside. 31

The experience of Nigerian citizen Peter Ighofose showed that the situatuion for both civil and political detainees and prisoners has not changed, even though this is denied by the British Foreign Office. Ighofose, a Scottish-educated 52 year-old businessman who was running a deep-sea fishing company out of

Mozambique, was detained for alleged commercial irregularities. Held initially in Maputo's civil jail, he was transferred to the notorious Cadea Central gaol. Ighofose estimated that Cadea, which was built by the Portuguese for five hundred inmates, now holds about eight thousand prisoners – many of them are mad; one thousand are children. He revealed both that these children were prostituted in homosexual brothels within the prison, and that the prison authorities were party to a slave labour system, whereby the authorities allowed local farmers to choose prisoners as labourers for payments of £2 a month to prison guards. The prison regime was extremely harsh: "We were given one meal a day only, usually between 1am and 3am. It was a bowl of maizemeal porridge." In addition to forced labour, child abuse and poor food, the Nigerian businessman also accused Frelimo prison authorities of flogging prisoners, of allowing guard dogs to kill and eat detainees who were attempting to escape and of raping, at whim, female prisoners and youths.

After four months detention without trial, Peter Ighofose managed to escape from Frelimo's Cadea prison in April 1988. His account of the nightmarish reality of the Mozambican Gulag was brought to the attention of *The Sunday Times* by a well-known British author, with whom he had had dealings in the past [32].

The treatment of citizens as property of the state is a standard facet of Marxist-Leninist states. Frelimo's Mozambique has been no exception. Antonio Rocha, a senior Frelimo diplomat who defected to the West, speaking at a Washington-DC press conference in October 1987, revealed details of the forced overseas political indoctrination of young Mozambicans:

> "One of the first shocks I experienced in the 'new order' was the mass deportation of Mozambican youth to various parts of the East-bloc, particularly the Soviet Union... Children as young as seven years of age were soon trooped aboard Soviet aircraft and ships. Their heart-broken and bewildered parents were powerless to object as they were told 'children are property of the state'."[33]

Rocha stated that thousands of Mozambican children had been sent to Cuba and East Germany, as cheap agricultural and industrial labour. Those transported to the Soviet Union have been forced to undergo extensive political and ideological indoctrination. Many have never returned.

The London *Sunday Telegraph* has also reported on Frelimo's extensive international forced labour policies. Chissano's administration has sent eight thousand Mozambican miners, many against their will, to work in East German coalfields, as a result of a 1987 employment agreement between the Frelimo government and East Germany. According to Western diplomatic sources, there is extensive dissatisfaction amongst the Mozambican miners, often resulting in violence between the Mozambicans and local communities. They are ill at ease in the cold East European climate, subject to anti-black racism and are earning considerably less in comparison with those in similar mining employment in South Africa. Frelimo's current involvement in what amounts to the supply of forced, cheap labour for one of the most totalitarian police states in Europe is yet one more contradiction with the image of a reforming Marxist regime in Maputo.[34]

There has been continuous evidence of the large-scale degree of Frelimo abuse of power, corruption and maladministration. Frelimo personnel and government officials have been implicated in everything from murder to petty-theft. Chapter 6 of this study touches on the banditry issue, giving an account of the abuses openly admitted by Samora Machel – killings, rape, child abuse, armed robbery and so on.

One clearly authentic account of some other excesses came again from Machel himself. At a speech given in Maputo on 5 November 1981, President Machel revealed the extent of corruption within Frelimo's security forces:

> "We find elements who adopt the methods of capitalist secret police, methods conceived to oppress the people, to repress and humiliate.
>
> Another factor... is that the rot has reached people in positions of responsibility. The failing of a soldier or policeman often goes unpunished because this soldier or that policeman knows too much about the life of the commander. And when the big fish are corrupt, the small fry follow their example. Obligations, favouritism and nepotism are established. A state of godfathers is created, a state of personal favours, a state of accomplices. Networks are established with the aim of robbery, corruption, the violation of legality."

Machel singled out the Frelimo security forces as especially blameworthy:

> "A particularly serious example of this situation is that these violations are in many cases committed by members of the defence and security forces. They are committed by members of the Mozambique Armed Forces (FPLM), by members of the police force and militia, and by personnel of the Ministry of Security (Snasp)."

The spectre of intra-black consciousness, and the depth of corruption, was also touched on by Samora Machel:

> "There is also the tribal problem among the Blacks who are found today in the police or militias. Someone from the south arrests people from the north or in the centre, just to show his tribal superiority. But he lets off the real criminal, because he comes from the same tribe, speaks the same mother tongue. Even within the same tribe, there is rivalry between clans and families. Disagreements between parents, uncles, or god-fathers become motives for arrest, beating, ill treatment."

President Samora Machel perhaps summed up the national feeling towards his regime, when he mentioned complaints he had received from the people:

> "Frelimo liberated us... But the government is fooling about. It is ill-treating us. We face arrest even when we do nothing wrong. Frelimo... does not govern well. It has handed government over to children. They arrest us even when we do nothing wrong."[35]

When overseas observers search around trying to explain why the Frelimo regime is on the verge of collapse, many deliberately ignore the nature of the state as a contributory factor. Joseph Hanlon indicates that the opposition to the government of Mozambique is home based. He believes that the grievous Frelimo abuse of human rights and civil liberties is reponsible, and that the resistance is not merely the creation of an external power:

> "Frelimo's own high-handedness seemed to be creating an opposition where none had existed before." [36]

Frelimo, Health and Education in Mozambique

Health in Mozambique

Even the health service, together with education and literacy, is used misleadingly by Frelimo and pro-Frelimo propagandists to exemplify a shining achievement of the revolution.

The health service is clearly not succeeding. The government estimates that 40% of children die before five years of age. Joseph Hanlon believes that to be an underestimate. He cites, for example, a visit to a maternity ward in a rural health centre. All the women there had lost at least half of their previous children.[37] More than half of all Mozambique's inhabitants suffer from malaria and bilharzia. Plans to inoculate all babies against tuberculosis, measles, whooping cough, tetanus, diptheria and polio simply did not materialise. Hanlon also cites a World Health Organisation report which found that only some eight percent of one year-old children in rural areas had been immunised. In Maputo, only one-third of this age-group had been inoculated, which resulted in an outbreak of measles in the capital in late 1981: 1,283 children were hospitalised; over 150 subsequently died. [38]

Although considerable rhetoric has been projected, both at home and abroad, about the health programmes which have been designed to help the rural population, the provision of services for Mozambique's peasants, who form a vital part of Frelimo's worker-peasant alliance, has not been a priority issue. In 1981, half of Mozambique's doctors practised in the capital: only 15% were based in the rural areas. Once again, Joseph Hanlon provided several key statistics. Maputo Central Hospital employs just under 2000 staff - 15% of Mozambique's national health workforce, more than any one of the country's ten provinces. As health personnel accounts for more than half of the national health budget, the government's priority becomes clear:

> "In 1979, the only year such a calculation was made, expenditure on staff was £7 for each resident of the city of Maputo, and only 40p for each rural resident." [39]

A joint 1981 World Health Organisation and UNICEF report on 'National Decision-making for Primary Health Care', stated that Mozambique's "concentration on developed urban health centres slowed down the expansion of rural coverage."

It is important to remember that most of the statistics above were from years during which the Mozambique National Resistance was still a relatively small organisation, limited to several districts within one or two provinces. The excuse, therefore, that war had disrupted the provision of health care does not ring true. Once again the reality of Frelimo's government planning differed vastly from theory. This did not, however, prevent the empty claims, that Frelimo was concentrating on the provision of health services to peasants in rural Mozambique, from being projected internationally as established achievements.

Even apologists for the Maputo government have documented the grave shortcomings of the health service in Mozambique. Several contributors to *Mozambique: Towards a People's Health Service* (a publication written by predominantly British doctors, nurses, health planners and social scientists who had spent between two and four years working as "cooperantes", foreign specialists, within Frelimo's health service) afforded the reader a unique insight into the system's bureaucratic incompetence and confusion. Dr David Bell served as the only doctor in a rural part of Inhambane province from 1976-78. Governmental inefficiency was evident:

> "Accompanying this bureaucracy was a bureaucratic mentality that knew no bounds. Sick peasants were often interrogated about all their personal details, name of mother, name of father, number of identity card, etc., before they were seen. I remember a young peasant who had just witnessed his wife's death, being told abusively that he had been stupid not to have brought his various documents with him so that the proper forms could be filled in. Needless to say the higher people were in the local class structure the less this attitude was taken." 40

Bell described a public meeting convened by the authorities to discuss several of his district's health problems. Three hundred people arrived. After some hesitance, the crowd began to articulate what were obviously deep-felt criticisms:

> "Then a woman got up and said she would like to make her criticisms but she could fall ill tomorrow and was worried about what would happen if she put herself in bad favour with the hospital. But then another woman got up and said that she had once come to the hospital and had been turned away by the midwife. Another complained of bad treatment, being pushed about and shouted at, 'We are not animals' she said. Slowly more people got up and made their criticisms. . ." 41

It was clear that this behaviour wasn't isolated. Julie Cliff worked as an epidemiologist in Maputo from 1976-79, and with the Ministry of Health from 1980 onwards. She wrote of attitude problems amongst nursing staff:

> "Many nurses treated patients in a hostile and degrading manner. Almost every day I saw patients shouted at, bundled along corridors on stretchers and herded into queues like animals. Little attempt was made to ensure privacy when procedures were carried out and screens to shield patients were rarely used. On one occasion I saw nursing staff abusing a young man for complaining of pain as he lay dying of blood loss after a car accident. . Clearly not all staff treated patients in this way but to my eyes it did seem depressingly like the norm." 42

Education in Mozambique

The educational system was one of the first areas of society in Mozambique targetted for politicisation by Frelimo. The political indoctrination of children through state-controlled educational institutions has always been a particular goal of totalitarianism, and Frelimo has been no exception, widening the definition of education to encompass all aspects of the attempt to consolidate Marxism in Mozambique. Frelimo's 1977 Third Congress, the first after the regime came to power, dictated the overt political orientation of education in Mozambique. It included consolidation at the behest of the "worker-peasant alliance", ideological instruction for both teachers and pupils and, in the words of a leading foreign 'cooperante', the "transformation of the schools into revolutionary bases in the midst of society."

Judith Marshall worked in the Frelimo Ministry of Education from 1978 to 1984, and, as a foreign voluntary worker, witnessed most of the government's educational programmes. One of the first actions of the Maputo regime was to invalidate all existing textbooks. Frelimo introduced in their place: new texts for junior schools; text hailing the "New Mozambican"; history, English and geography texts; a standard political education textbook and teacher's manual. Even history books were clearly political in theme. Marshall records that "the new history program...tried to present history as an ongoing struggle of opposing class forces".43 She also describes the contents of general texts, including one aimed at junior school pupils:

> "Political studies were introduced for the first time into the school program. The textbook prepared for the lower grades combined elements of civics with a brief history of colonialism, Frelimo, the armed struggle, the history of women's emancipation, and other important themes."

High schools also had their mandatory set books:

> "The secondary schools and institutes drew on basic Marxist texts and current policy documents. With the paucity of both texts and teachers who felt themselves capable of teaching political studies, the experiences in many of the schools were somewhat less than positive, with a tendency to use dictated texts and rigid teaching methods rather than create a real dialogue with the students that would begin to develop a capacity to read Marxism as Mozambicans and read Mozambique as Marxists."

The role of the teacher is central to political indoctrination. To this end, Frelimo concentrated considerable time and effort in training teachers for the Mozambican educational system:

> "Thus a main task of all the teacher training programs was ideological transformation, to enable the future teachers to recognise and surmount the distorted views of themselves and the world which the colonial education system had engendered."[44]

Special in-house training for teachers heightens the Frelimo political line:

> "Higher education is to give cadres a solid political foundation and technical expertise, to equip them to do research and investigation and to apply their knowledge to solving the concrete problems of revolutionary transformation."[45]

Little wonder that far left British educationalists, such as Chris Searle, a teacher in Nampula for two years in the 1970s, saw Frelimo's Marxist educational indoctrination as an inspiration. What Marxist teachers have desperately and cunningly attempted to introduce into the British education system, is a *fait accompli* in Mozambique:

> "To effect these changes in the minds and lives of their students, the schools in Mozambique have generated structures that are still only dreams in the minds of progressive teachers in western Europe."[46]

Frelimo And Religion

Vladimir Lenin, one of Frelimo's philosophical mentors, who summed up the ideological Marxist position on religion:

> "Every religious idea, every idea of a God, is unutterable, vileness of the most dangerous kind, 'contagion' of the most abominable kind. Millions of sins, filthy deeds, acts of violence, and physical contagion are less dangerous than the subtle spiritual idea of a god."[47]

In a classic Orwellian theme, Mozambique's Frelimo constitution guarantees, in common with the Soviet constitution, the right of religious belief:

> "(T)he individual liberties are guaranteed by the state to all the citizens of the People's Republic of Mozambique... In the People's Republic of Mozambique the state guarantees to its citizens the freedom of practising or of not practising their religion."

The simple fact of the matter is that Frelimo has tried its best to marginalise, and even destroy, spiritual faith and religious institutions in Mozambique. The Roman Catholic church, as the dominant manifestation of religion within the country, has suffered most. Frelimo legislation, in 1979, forbade children under the age of eighteen to attend church, and subjected them to Marxist indoctrination instead. All Church property and contents were also nationalised. Church schools and hospitals also came within this ambit. Action against the Catholic church has been sustained and ruthless. In 1979, Mozambique bishops from the country's nine dioceses drew attention to an increase in Frelimo interference with their functions. In a pastoral letter smuggled out of Mozambique they stated that:

> "Christians are facing grave difficulties in holding religious services and in instructing their children in the basics of their religion. Bishops and priests are likewise having problems in their pastoral work and several missionaries have been expelled."[48]

Father Onorino Venturini, an Italian Roman Catholic priest who has lived in Mozambique for thirty-seven years, gave a typical account of Frelimo behaviour towards the church:

> "There was always (religious) oppression by Frelimo. They saw religion as a block to their programme. Our religious work was controlled and we were supervised when we travelled." [49]

Church-state relations had reached such a low ebb by the end of 1978 and beginning of 1979, that the Vatican's chief representative in Mozambique was detained while attending a church meeting in the north of the country. In their last two collective pastoral letters in 1984 and 1987, the Roman Catholic bishops expressed concern at Frelimo excesses. The 1987 pastoral letter, signed by Mozambique's Roman Catholic bishops, also called for dialogue and reconciliation between Frelimo and Renamo. Excerpts included:

> "THOSE WHO ARE RESPONSIBLE FOR PEACE. The peace in our country depends, in the first place, on the Frelimo party, the government of the country, and Renamo since these are the ones directly involved in the conflict and can make decisions about it. Those who are equally responsible are all who finance, feed, and help direct the war in Mozambique.
>
> THE ONLY WAY FOR PEACE. In this dramatic hour in which we are living in search of solutions for the conflict, the only solution for peace and honor is dialogue and national reconciliation. This is the

feeling of the Mozambique people and individuals of good will who have an interest in the welfare of the population.

APPEALS. We are appealing to all people, organisations and governments, to help the country achieve peace through peaceful means mainly political dialogue, and national reconciliation to help our people build a society of brotherhood and prosperity."[50]

Frelimo had closely identified with the church's concern over human rights violations and government violence during its war against the Portuguese. Now that the church merely continued to express its concern over continued abuse – this time by Frelimo – it was condemned. President Chissano, for example, labelled the Catholic Bishops – because of their impassioned call for national reconciliation and dialogue between Frelimo and Renamo – as "apostles of treason".

Whereas it can be argued that the Roman Catholic Church in Mozambique had close institutional connections with the Portugese administration, this was not the case with the Protestant churches. Nevertheless, the latter have been equally persecuted by Frelimo authorities (despite the fact that several Protestant churches were openly accused of pro-Frelimo sympathies during the pre-independence struggle, and numerous ministers and missionaries were imprisoned by the Portuguese under security legislation). The respected London-based 'Minority Rights Group', dedicated to securing human rights throughout the world, has cited Frelimo conflict with the following churches in Mozambique, in addition to the Roman Catholic church: Seventh Day Adventists, Church of Christ the Saviour, Church of the Good Father in Mozambique, Dutch Reformed Church, Complete Church of God, Ethiopian Church of Portuguese Africa, United Apostolic Church of Zion, the South Church, the Church of the Apostles in Mozambique, Church of the Nazarene and the Jehovah's Witness church.[51]

Frelimo's savage treatment of the Jehovah's Witnesses is a prime example of state-sanctioned persecution of Mozambique's protestants. Jehovah's Witnesses have had a history of persecution, and harassment by totalitarian regimes. In Nazi Germany, the sect was virtually run to ground, and its remnants consigned to concentration camps. From the outset of Frelimo administration in Mozambique, the Jehovah's Witnesses were targetted for harassment. In 1975, there were over twenty-two thousand Jehovah's Witnesses in Mozambique, gathered together in some 422 congregations. Typical of initial Frelimo intolerance of the sect was their treatment of the Choupal congregation. On 13 September 1975, Frelimo soldiers forced their way into the Choupal Kingdom Hall, on the outskirts of Maputo, and demanded that the 300 or so Jehovah's

Witnesses present declare their allegiance to Frelimo and shout *Viva Frelimo!*. When the congregation refused, many of the Witnesses were severely beaten with clubs, batons and rifle butts.

The official party line was summed up only months after Frelimo's accession to power. When he addressed a meeting at Massingir in Gaza province, on 10 October 1975, Samora Machel termed Jehovah's Witnesses agents of imperialism, and threatened them with destruction if they did not change their ways. One day later, a Frelimo provincial governor, Fernando Matavelas, addressing a rally in Gaza, threatened members of the sect with incarceration in Frelimo concentration camps if they did not participate in helping to consolidate Frelimo's grip on Mozambique. Joao Chauke, a Frelimo leader in Zambesia province, writing a fortnight later in the state-controlled *Noticias da Beira* clearly articulated growing political and religious intolerance:

> "There are found in Mozambique the 'Jehovah's Witnesses' and other religious organisations which are allied to imperialism and try, by subversion, to struggle against the ideological line of Frelimo...
>
> We do not tolerate this and it is our concern to pinpoint those agents of the enemy, where they have their bases. We will not accept anyone using religion against us, against our sovereignty. . .We did not sacrifice ourselves to establish a religion in Mozambique, nor to create religious arrangements... We come up against contradictions, when we encounter 'Jehovahs' who do not want to recognise Frelimo."[52]

Within weeks of these ominous statements, the Frelimo central committee ordered the arrest and detention of all Jehovah's Witnesses in the country. In the words of the sect's official publication:

> "This order was carried out systematically and heartlessly. Frelimo followers went from house to house demanding that residents say *Viva Frelimo*. Those who would not were assumed to be Jehovah's Witnesses and were taken to prison. Whole families, children included, were mercilessly dragged off." [53]

The vast majority of those detained have been sent to 're-education' camps in Zambezia province. As part of an international movement, the plight of the Jehovah's witnesses in Mozambique has inevitably been highlighted. The fate of indigenous black Mozambique churches and their tens of thousands of followers is a cause for concern, given that there is no external monitor to document Frelimo excesses. An indication of the situation emerged in the wake of a routine evangelical mission to northern Mozambique by the South African-based 'Frontline Fellowship' organisation. Missionary, Peter Hammond, travelled many hundreds of kilometres within Mozambique in mid-1986, and visited over fifty villages in nine districts within Tete, Zambezia and Niassa provinces, where he interviewed over one hundred black church leaders. Recording only eye-witness testimonies sworn on oath and confirmed by further first-hand accounts by separately-interviewed witnesses, Hammond documented the razing to the ground by Frelimo military units of forty-two villages, the destruction of seventy-four churches, and over sixty instances of Bibles being burnt. The testimony of one protestant pastor and six of his church members in Milanji district in Zambesia Province is typical:

> "At the beginning of July 1986, Frelimo came during our church service and arrested everyone – all 134 of us. 'Why are you praying?' they asked. 'We never liberated Mozambique by prayer – but by guns.' They then took Brother Robinson, our choir leader, and slit his throat with a knife. Brother Manuel, our church secretary, they chopped up with a machete. He leaves a wife and three children. Brother Paul – they tied up with wire and beat until his back was broken and both his arms also broken... He can only walk with great difficulty."

The Frelimo unit who disrupted this service were particularly incensed to find children present:

> "The Frelimo were very angry because we had children in the church. 'You know that children may not be in church – we want all the children for Frelimo. Your old people can be Christians but the young must be Marxist.' All our possessions – even our clothes and dishes – were stolen by Frelimo. Frelimo also took all our Bibles (16) and all our hymn books (8) and tore them up and burnt them... So there are no more Bibles or hymn books in our village. We are not allowed to bury the bodies of our brethren who have been killed."[54]

Further eye-witness accounts present a very clear picture of a systematic Frelimo attempt to destroy any religious infrastructure or manifestation. A minister in Niassa province gave sworn evidence of state intolerance:

> "In June 1986, an Apostolic Faith Mission minister was killed for preaching the Gospel. Frelimo told him: 'You should obey us – forget about Christ.' When the pastor answered: 'I cannot forget about Him for He is the living Christ', the Frelimo said: 'Then you

must pray for we will certainly kill you.' He knelt and prayed and was then shot by the communists."[55]

A pastor and thirteen of his congregation in the Posto Shire district of Zambezia province survived the Frelimo destruction of both their village and the churches within it:

> "On the night of 15th November 1985, Frelimo soldiers came and arrested me, asking: 'Why are you preaching to the people about God?' They bayoneted me in the head. They shot Joao, Mr Domingo, Brother Nasho, Sister Maria and Dorga. Then they burnt down the churches of Hebrew Christian Church, the Pentecostal Church of God and African Assemblies of God. The whole village was burnt and at least 70 people killed by Frelimo as the houses were burning and the people running in fear. A baby was burnt to death in a hut during the shooting." [56]

Professor Andre Thomashausen documents the themes of several posters, proposed for distribution in parishes by the Mozambique Secretariat for Ecumenical Coordination as part of the Churches' *Justice and Peace* Campaign: *No to Assassinations,* (the) *death penalty* (and) *abortion*, *No to whipping, beating,* (being) *ordered to crawl*, and *No to accusations or arrest without proof*.[57]

Frelimo has sought not only to stifle the Christian faith – Muslims in Mozambique have fared no better. With some two million adherents, centred on the Makua ethnic group, the Islamic faith is particularly strong in the north of the country. The extent of the pre-Portuguese Islamic grip on the eastern coastline of Africa is not widely known. Vasco da Gama's expedition in 1498 recorded brisk business between resident Arab traders and both coastal and inland communities. Despite the officially Roman Catholic nature of the Portuguese administration, the authorities respected the Islamic community, and funded the building and maintenance of mosques and religious schools. One of Mozambique's last Governors-General, Dr Baltazar Robello de Souza, was even granted the title of Defender of Islam by Mozambique Island Muslim leaders.

The contrast between Portuguese and Frelimo attitudes towards Mozambique's Islamic community could not have been sharper. In a premeditated act of contempt, Frelimo leader Samora Machel defiled the sacred Mozambique Island mosque by entering with his shoes on. Machel's continued contempt for his country's Muslims could also be gauged from comments made three years after independence. At a meeting in the largely Islamic northern coastal town of Angoche, Machel's analogy between God and a pig, and sneering dismissal of Muslim taboos were a telling indication of Frelimo attitudes:

> "It seems that here God is a pig, isn't that so? But when you are drunk, you eat pork and don't tell anybody about it. You eat it in the bars as a snack when you are drinking beer. You have all eaten pork. Have you died? Have you caught any disease? No. You have grown fatter and stronger... This thing, the mosque, does not do anybody any good."[58]

Mozambique Muslum Iussufo Moamed served as a *Forcas Populares de Liberacao de Mocambique* (*FPLM*) officer until 1978. He left the military as a result of witnessing Frelimo persecution of adherents of the Islamic faith. Moamed documented the destruction of several mosques in northern Mozambique, including the Great Mosque at Namapa and the amapa Madressa – a mosque school for children. The largest religious association in the north, 'Mahafil Islam' was closed down by a Frelimo decree. Moamed identified a distinct ideological motive:

> "Now everything is Marxism and only Marxism. No one is allowed to be true to himself or to follow his own traditions."

Iussufo Moamed was also able to give an insight into the treatment of Muslim conscripts in the Frelimo military:

> "Islam is the religion of my parents and grandparents. . I am a practicing Muslim myself. In the Forcas Populares I was not only forbidden to say my prayers but they tried to force me to eat pork. . .When I was stationed at Dondo, monkey meat was served as there was no other food. Muslim soldiers who refused to eat it were beaten up and one of them died in hospital."[59]

His views on Frelimo treatment were echoed by Dr Suleiman Valy Mamede, leader of the Mozambique Islamic community in exile in Portugal:

> "One-third of the population of Mozambique is Muslim, but they are not represented in all the levels of the community. I openly say that Frelimo continues to treat Muslim Mozambicans as second-class citizens."[60]

Chapter Four

THE MOZAMBICAN ECONOMY UNDER FRELIMO

"But they are nowadays fated to live under conditions of deepening economic crisis and extreme tension, with city-dwellers unable to live on official salaries, and forced to take extra jobs in order to cope with rates of inflation of over 500%. After more than a decade of peace, living conditions... have shown few signs of improvement. In the over-crowded rural areas of the north, the ever-rising population confronts stagnating food output, threatening a deep crisis. Much state industry operates uneconomically, sucking in valuable resources better used elsewhere. Life for most is still hard, and people lack the peace of mind they seek. After a short period of positive economic growth in the early 1980s, there are now fears of an abrupt slowdown if effective reforms are blocked by vested interests and political stalemate."[1]

Introduction

The above passage from *Marxism Today*, the journal of the Communist Party of Great Britain, referring as it does to post-1975 Vietnam also helps to put the economic chaos that is Mozambique into some perspective. The crucial background is that Marxist-Leninist movements gained power in both Mozambique and Vietnam in the course of 1975, establishing one-party state structures – to totally control, amongst other sectors, the country's economic life. The intellectual and practical bankruptcy of Marxist economics, and the resultant misery, shortages and ultimate starvation is self-evident. Both Mozambique and Vietnam are prime examples. With the case of Mozambique, however, both the Frelimo authorities and their apologists internationally have attempted to explain away the economic chaos of Mozambique by reference to South Africa and the on-going military conflict. The simple fact is that the political and military conflict in Mozambique is at least in part a direct consequence of the totalitarian economic conditions that are to be found in the country. Frelimo's disastrous economic centralism, together with the inevitable political repression which hallmarks all Marxist administrations provided the background for the armed struggle as waged by the Renamo national liberation movement.

The Pre-independence Economy

Mozambique, as a country, is endowed with incredible economic potential. Tremendous hydro-electric power, natural gas in abundance and quantities of oil and uranium. There are, in addition, very large reserves of iron, coal, copper, phosphates, asbestos, and graphite: Mozambique's reserves of tantalite are amongst the largest in the world. Its iron and coal deposits could provide the basis for a flourishing industry.

There are also off-shore marine resources such as shrimp, prawns, lobster and tuna. The economic potential was increased by three major harbours and a strong pre-1975 agricultural sector made up of cotton, sugar, tea, sisal, copra, vegetable oils and cashew nuts. Despite fluctuating weather conditions, two of Southern Africa's largest rivers, the Limpopo and the Zambezi flow through Mozambique providing immense scope for irrigation schemes.

Frelimo and the Economy

Frelimo's comprehensive nationalisation programme, as well as dealing with education, and the health and legal services, concentrated on the economy. The central role accorded to a Marxist-Leninist economic framework as dictated by Frelimo is clear. The conclusions of the eighth central committee meeting in December 1980 speak for themselves:

> "(T)here can be no socialist plan without the party being implanted in all production units. Only the party can direct the workers in the process of elaboration of the plan. Only the party can create the conditions for a conscious, active and creative participation of the workers in the process of realisation (of the Plan)... Only with the dynamic presence of the party, its militants in all sectors of economic and social life, can we guarantee the force required for the fulfilment of the plan." [2]

The previously diverse agricultural sector was a particular target of the new regime for drastic re-structuring in line with its Marxist economics. The result of this was that abstract political ideology dictated the day-to-day running of an agricultural sector upon whom some 85% of the Mozambican population was reliant.

Frelimo's Forced Collectivisation

Frelimo's collectivisation programme, with its introduction of communal villages and huge state farms, resulted in the uprooting and forced relocation of tens of thousands of peasants into state-mandated population centres in the rural areas – several of which had served as 'aldeamentos', or 'strategic hamlets' during the Portuguese counter-insurgency campaign. Frelimo's adoption, at least in part, of Portuguese methods which it had itself so strongly criticised during the first guerilla war, further undermined its legitimacy with the rural population of Mozambique. By mid-1979, Frelimo could boast of over 1000 communal villages, concentrating some one and a half million people. This had increased to just under two million people, in some 1350 state villages, by 1983: 14% of Mozambique's total population. The parallels between Stalin's collectivisation programme and Frelimo's were clear. Myles Harris, a doctor who has worked in Ethiopia and author of *Breakfast in Hell*, a harrowing account of Ethiopia's equally savage peasant collectivisation, has concluded that in Mozambique:

> "As in the Soviet Union of the Thirties, peasants are being forced to relive the horrible consequences of a failed socialist dream – that the peasants of the world can only prosper if they hold their land in common.
>
> In Mozambique, such a policy has led to a desolated capital where the only things on sale in the shops are wooden buttons, a horror that has driven away many of its peasants into murderous anarchy." [3]

Even publications ideologically sympathetic to the Frelimo administration have commented on the blame directly attributable to government policies. Philip Raikes' comprehensive article 'Food Policy and Production in Mozambique since Independence', published in the *Review of African Political Economy*, for example, conceded that while there were several factors explaining Frelimo's disastrous economic and industrial factor:

> "(M)any of the problems have been exacerbated by government policies, both in the agricultural sector and elsewhere. Probably the most important of these has been the almost exclusive emphasis to date on the modern sectors in both agriculture and industry, implying heavy concentration of funds, materials and skills upon large-scale, complex and 'system-dependent' units of production, whose gross product has grown slowly while their net product is probably not much above zero. This in turn has starved the peasant sector of resources, markets for products and infrastructure."[4]

It is evident that Frelimo's state farms were an unmitigated disaster in production and production target terms. Raikes made clear that the government's agricultural planning was essentially political in nature and had "strong roots in the policies of the USSR and Eastern Europe." He takes as one starting point the 'dominant and dynamising' role of the state-farm in socialist agricultural developments[5]. Raikes provides a "guesstimate" that ninety percent of all government agricultural investment was channelled into the state-farm sector. In one of the few studies of Frelimo's state-farm sector production, published by the *Centro de Estudos Africanos* at Maputo's Eduardo Mondlane University, and cited by Raikes, Ole Stage concluded that the net-product of the state-farm sector was zero: the value of gross output did not even cover production costs. Raike's further interpretation of Stage's results is even more telling:

> "If amortisation is included, it seems likely that the net product would be negative, and given the rapid expansion of the sector and the high level of new investment in large projects, this is almost inevitable. . .(S)tudies of specific state-farms tend to show that it is quite unusual for them even to cover operating costs and administration overheads, let along amortisation."[6]

He further cites one case of a state-farm he had studied personally, whose costs of production, excluding amortisation, appeared to be 3-4 times the value of output.

The generally unrealistic approach used by Frelimo in enforcing these collectivisations was exemplified by state policy in Inhambane province. Despite a desperate lack of organisation, the Politburo ordered the province's total collectivisation within one year regardless of the effect on its population. The sites for the new villages were picked at random, without any reference to adequate land or water supplies. One indication of the irresponsibility of the state edict was that in order to relocate everyone, the state villages would require some 450 new wells; this worked out, however, as double the number drilled throughout the entire country the previous year. Joseph Hanlon provides an epitaph on the sheer impracticability of the Inhambane collectivisation process:

> "In the end, Inhambane's villagization target was quietly scaled down, but only after considerable protest and lobbying by technicians who warned that it was not only impossible, but also dangerous: not only would the villages fail but the land would be turned into a desert. But they had to overcome fierce resistance from officials who argued that they had been given targets 'from the highest level' and that their duty was to fulfil them, come what may."[7]

The following factors all helped to grind the Mozambican economy effectively to a standstill: Frelimo's alienation of white commercial farmers; dislocation of black farms from traditional farming areas; the peasantry's alienation from the land by Frelimo's nationalisation; and the replacement of commercial farming by state-regulated quotas and prices. This economic predicament was made worse by the drought conditions and the guerilla-based protests of the Mozambicans against Frelimo's brutal dictatorship. The crucial family sector disintegrated. The production of the five main crops, cashew nuts, sugar, tea, cotton and copra either virtually ceased or was more than halved in the first ten years of independence. Philip Raikes has summed up the impact of Frelimo's economic policies upon Mozambique's rural population:

> "(F)irstly... peasants are pushed out of the door by poor terms of trade or sheer lack of goods to exchange for their produce, and secondly that their flight, if such it may be termed, is less to subsistence than to the black market. In Mozambique... it is often hard to see what the state can do to 'develop' the peasants, but sometimes easier to see how it can stop hindering them." [8]

The Introduction of Forced Labour

Frelimo's state-dictated agricultural policies, such as monocropping, led inevitably to an increased demand for seasonal work forces. They attempted to destroy all economic initiative, so that the incentive to work on badly-administered, unproductive state farms ruled by fear was understandably low. From 1983 onwards, because of these shortages, Frelimo introduced a system of forced labour – a mirror image of the colonial system of compulsory labour of which Frelimo had been so critical in the past. Tens of thousands of unemployed urban Mozambicans were rounded up and transported to state farms and plantations. Some Frelimo apologists have argued that these practices were just "slipped back into" on old former colonial estates. Joseph Hanlon correctly points out that this is simply not the case. These practices were consciously copied by Frelimo cadres, and forced labour methods were introduced to brand new state farms such as Lioma in Zambezia province and N'Guri in Cabo Delgado.

Frelimo also introduced the euphemistically-designated days of 'voluntary work', which are nothing less than compulsory labour days. Faced with both managerial and economic bottlenecks, Frelimo turned to even more coercion. Hanlon draws a further parallel with the Portuguese administration:

> "It seems as if Frelimo has reverted to the Portuguese tradition of increasing exploitation to increase production."[9]

The Industrial Sector

The pre-1975 Mozambican economy was one of the strongest in Africa. It was the eighth most industrialised country on the continent, and its 100 000 strong industrial workforce processed local products for both export and for local consumption. The introduction of comprehensive state planning into all facets of Mozambican industry resulted in a bureaucratic nightmare for both Mozambican workers and citizens. Industry had already been badly affected by the loss of white skilled manpower and management. A Frelimo statement one month after the June 1975 independence provides an insight into the chaos that was inevitably to compound this difficulty within both industry and governmental bureaucratic institutions:

> "We are not concerned with the lack of technical training of our personnel. Skills are secondary in relation to the political orientation and class consciousness... We shall acquire our skills as we work, in the course of the revolution, just as we learned how to win the war while fighting the war." [10]

Frelimo subsequently appointed thousands of unqualified political cadres to various levels of management within all the sectors of Mozambique's nationalised industries. This, coupled with their attempts to institutionalise doctrinaire Marxist state planning, can be seen as a crucial contributory factor to the economic chaos which has resulted in widescale starvation for hundreds of thousands of Mozambicans. These cadres were political administrators rather than economic or industrial managers and, as Frelimo appointees within a rigid totalitarian society, were inflexible and beyond reproach. Samora Machel, speaking in March 1980, confirmed that Frelimo's concentration of power in the hands of state-appointed bureaucrats was total:

> "Let us say in clear terms: at the level of each firm, the power is exercised by the Director. It is the Director who organises, directs and controls the production. It is the Director who has the authority to punish. He is the one responsible for discipline in the firm." [11]

It became clear that these managers and bureaucrats simply could not use the equipment which was either at their disposal or hastily imported from abroad. Senior Frelimo agronomist, Arnaldo Ribeiro, speaking in early 1981 candidly admitted: "we have not sufficiently dominated the technologies we use."

Frelimo's Intervention at the Sena Sugar Company

The state intervention at Sena, Mozambique's premier sugar production company, provided a key example of Frelimo priorities. The giant Sena Sugar Estates, which employed some eight hundred skilled workers and over twenty thousand labourers, was a jewel in the pre-Independence Mozambique economy, realising some $5 million profit in the 1974 financial year. However, the company was subject to a form of Frelimo nationalisation after chronic absenteeism, the departure of most of the management and the inability to secure enough state-sanctioned foreign currency to purchase vital equipment. The theme of the intervention was readily evident in the new state-appointed management's first edict: "The wearing of shorts is embarrassing to women and in future all personnel will wear long trousers." Sena's former general manager, Michael Tinne, a British tropical agriculture and sugar expert, was recruited post-Independence to take over the Sena works. He left Mozambique totally disillusioned with the Frelimo system. The new Frelimo management were more interested in politics than sugar production:

> "Production of sugar, running of the factories, continuation of good husbandry on the company's huge ranches, did not matter - far more important were petty matters such as the order about what must be worn and the painting of Frelimo's slogans all over the walls of the company's installations."[12]

Tinne went on further to point out that Frelimo closed down both the field laboratory and the agronomists department as "not essential to production".

Frelimo's industrial bureaucrats, in addition to alienating Mozambican society by their technical incompetence which resulted in severe food and economic shortages, caused further resentment by widespread abuses of power and corruption. These bureaucrats would have better housing, servants, the use of state vehicles, access to better food and to special shops supplying foreign consumer goods, and private hospital rooms – all state-sanctioned extras. Attempting to play down the wide criticism of such privilege, Machel could only argue that: "The people like to see their officials well treated"[13]. As with any bureaucracy, the Frelimo economic and industrial managers created more and more jobs within the state bureaucracy and institutions – culminating in the creation of the National Planning Commission. This centralised institution gave bureaucrats even more power. As Joseph Hanlon pointed out, the bureaucrats favoured big state institutions above local peasant aspirations – machinery at the expense of hoes and other basic farming equipment. And inevitably with all bureaucracies, the focus was very much on the cities – rural needs and problems taking very much a back-seat. The Third Frelimo Congress, for example, decided to make the national production of white potatoes, favouring urban consumption, a priority at the expense of peasant staples such as cassava and sweet potatoes. Hanlon provides further examples of urban bias. While visiting Mocuba district in Zambezia, he points out that each rural dweller is only able to buy one-twentieth of the soap and sugar guaranteed to each Maputo inhabitant. The supply of 'capulanas', a multi-purpose cloth, was another gripe. Each woman needs two per year: in Mocuba only one 'capulana' was available per three women.

Hanlon also provides one further example of the emphasis Frelimo bureaucrats placed upon urban projects and large schemes: a vital drilling rig, which would have brought water to villagers in the drought-stricken Inhambane province, was diverted to provide water for fruit trees to beautify Maputo airport.

An insight into the insulated and inherently corrupt nature of this Frelimo bureaucracy was given in Frelimo's introduction of wage legislation in Mozambique. This stipulated a minimum wage of some 65.5 Meticals (£1) per day for farm workers and some 2100 Meticals (£33) per month for other workers. Because of the large number of Frelimo personnel who employed servants, this minimum wage was declared to be non-applicable to house servants.

The degree of alienation between Frelimo and the Mozambican people was summed up by Joseph Hanlon in a 1983 *Guardian* article. He described one of the many meetings called by Frelimo, within the areas it then still controlled, in the lead up to the Fourth Party Congress:

> "A peasant stood up in a meeting and said: 'Do you want to know where the problem is? It's under the table'. The party leaders sitting at a table in the front looked, but found nothing. Finally the peasant said: 'There are shoes under that table. You are wearing new shoes. We produce the wealth of this country but we are still barefoot.'"[14]

Conditions for ordinary Mozambicans, particularly with regard to food products, were dire. Hanlon visited Mozambique's richest farming province, Zambezia:

> "Yet when I visited it in 1982, the shops were empty except for salt, tea, curry powder, a few men's clothes and the ubiquitous tinned beans. It had been three years since anyone had seen bicycle parts."

The plight of the ordinary Mozambican was further updated in yet another *Guardian* account. Mike Saunders, reporting in mid-1986, four years later, gives very much the same description of the stagnation of Mozambique society and production:

> "Life for the average person is appallingly difficult. Shops are virtually empty and food is a major preoccupation. In one supermarket in Beira, the second city, I counted 43 brown paper packets of curry powder, 54 bottles of hair oil, a dozen childrens garments, and a quantity of shaving brushes. A second in the city centre had only 33 packets of coarse tea. A third, half a shelf of toilet paper and again

shaving brushes. No bread or flour was available in the town at all. Rice was short and fish almost unobtainable.

In the northern capital Nampula, no food shops had bothered to open at all." 16

The entrenched neglect of rural areas, and especially the northern provinces of Mozambique was also touched on during a 1978 interview by Michael Tinne, Sena's former general manager:

"The rural areas continue to survive but the lot of the local population becomes worse and worse. There is a total lack of organisation of all administrative services. Road and coastal shipping distribution to the remote north has almost come to a standstill with the inevitable lack of essentials. Train services might run, depending on whether the line has collapsed or not through lack of regular maintenance or whether one of the few remaining serviceable locos is available."17

This transport chaos existed while the Renamo insurgency was limited to two districts within one province.

Acknowledgement of Frelimo Mistakes

The scale of Frelimo's economic mismanagement was soon apparent. The causes – Marxist economics and the sheer incompetence of, and abuse of power by, Frelimo economic administrators – were at least, in part, identified by Samora Machel. Speaking as early as 1980, in a speech entitled 'Tranform the State Apparatus into an Instrument of Victory':

"The widespread failures in management and discipline have reached such a pitch..."18

These problems had not changed very much by the 1983 Congress. Addressing Frelimo's Fourth Congress he stated that:

"Our state apparatus is corrupted... It is not directly linked to the enemy, but to comfort. We are slaves to our house and furniture. Even ministers start to think: 'If I go out to the countryside, I won't have my chauffered car with its flag flying on the front.' Little by little, these things corrupt."19

Joseph Hanlon's frank examination of the Frelimo regime in power specifically pinpoints Frelimo's lack of ability to manage effectively:

> "Many of the mistakes in the National Planning Commission and on state farms were pure incompetence."[20]

Far from analysing the root cause of Mozambique's economic malaise, doctrinaire Marxist state planning, the Fourth Congress demonstrated its commitment to measures which had already resulted, at the very least partially, in the deaths from starvation of hundreds of thousands of Mozambicans:

> "We must ensure that by 1985 the state controls wholesale trade at national and provincial level and guarantees effective control of the circulation of goods."[21]

The root cause of the famine affecting hundreds of thousands of Mozambicans was not Renamo or South African involvement but Frelimo itself. As Hanlon states:

> "External factors may have made Mozambique's agricultural crisis worse, but they did not cause it. The cause seems to lie in a fundamental misunderstanding of three related issues: first, the intimate linkages between the various sectors and the extent to which rural people have one foot firmly planted in the cash economy; second, the need to take account of labour supply and demand; and third, the time required for a radical transformation of agriculture."[22]

By its very nature, the Frelimo administration was either directly responsible for, or involved with, all three of these factors. External factors were also downplayed in a 1986 *Guardian* report:

> "The government attributes all problems to the guerilla war with the MNR. In fact many other factors are also to blame. There is widespread lack of real organisation coupled with a proliferation of paper work and regulations in a top heavy bureaucracy beyond the imaginings of a Brussels technocrat. That and endemic corruption contributes to an apathy that drains the population like a disease. Inept policies are applied in a country that is effectually bankrupt."[23]

Frelimo Forced to Introduce Reform

Domestic production in Mozambique is estimated to have fallen by a third between 1981 and 1984. The resultant balance of payments deficit and massive foreign debt continues to severely handicap the Mozambican economy. The $2.3 billion foreign debt, for example, exceeding Mozambique's Gross Domestic Product in 1984. It was, therefore, against a backdrop of a desperately deteriorating economy that Frelimo was forced to introduce some measure of economic liberalisation. As early as 1980-81, Machel began to appeal to the previously persecuted private sector to help reverse Mozambique's economic decline. The combination, however, of small-scale denationalisation of one or two outlets was undermined totally by increased rationing within the economy. In 1985, Frelimo set out to de-nationalise several state-run industrial and energy plants. Internationally, in an attempt to attract foreign investment and aid, Frelimo courted the International Monetary Fund and World Bank, as well as becoming a signatory to the Lome Convention. Frelimo also rescheduled some $259 million of its Paris Club debt. Courting international investors, the regime signed an agreement with the Overseas Private Investment Corporation in July 1984. The net result of this desperate attempt to attract foreign investment was that Frelimo was pledged to introduce new, more liberal investment legislation, as well as to accept IMF restructuring of its economy. This imposed restructuring meant the further easing of state controls on prices and rationing.

Samora Machel decorating Joaquim Chissano at the 10th anniversary of Mozambican Independence 1985

Government sloganising, Maputo 1989

Decaying installations in Beira Harbour

Removing Samora Machel's body from the wreckage of the plane crash which killed him

Afonso Dhlakama, President of Renamo

A young Renamo guerilla

A Renamo school in one of the movement's 'Liberated zones'

Chapter Five

THE MOZAMBIQUE NATIONAL RESISTANCE: THE ORIGINS OF RENAMO

In order to understand the circumstances leading up to the emergence of the Mozambique National Resistance and civil war within Mozambique, it is crucial to examine the roots of Frelimo's political illegitimacy and intolerance.

Indications of Frelimo's disdain for the democratic process emerged soon after the overthrow of the Portugese dictatorship in April 1974. During talks in Lusaka in early June 1974, between the newly appointed Portuguese foreign minister, Mario Soares, and Frelimo, Samora Machel totally rejected any idea of either a ceasefire or referendum, and laid down three conditions for peace in Mozambique: total unconditional independence for Mozambique; the recognition of Frelimo as the sole representative of the Mozambican people; and the total transfer of power to Frelimo. The guerilla war continued as Frelimo went all out to weaken the Portuguese will to fight for the Lisbon government's solution. Confined to the three northern provinces, on 1 July Frelimo desperately initiated action in a further province, Zambezia, and began using new sophisticated weapons such as surface-to-air missiles.

The end of the authoritarian Caetano dictatorship, and the subsequent relaxation of restrictions in Mozambique, saw the creation of several political parties, associations and trade unions. Prominent amongst these was the *Grupo Unido de Mocambique* (Gumo) led by Maximo Dias and Joana Simiao. Formerly a Coremo guerilla, a prominent leader of the two million-strong Makua tribe in Northern Mozambique and a Portuguese detainee, Simaio later went on to merge Gumo with the African National Congress of Mozambique (CNAM) to form the *Frente Comum de Mocambique* (Frecomo) political party. Frecomo sought an end to the guerilla war, and had as its two main objectives a referendum to decide Mozambique's future, together with a 'Congress of the Mozambique People' to map out the result.

Other black political parties which emerged were the Union of the Peoples of Mozambique (Unipomo) led by veteran black nationalist, and Frelimo founder, Lazaro Nkavandame and Father Mateus Gwenjere's *Frente Independente Afri-*

cana (Freina). Much of the political organisation and running was made, however, by Joana Simiao and Frecomo. Several weeks of intensive negotiation between various black nationalist political groupings, led by Frecomo, and the externally-based Coremo nationalist guerilla movement, which was a clear Frelimo rival, resulted in the creation of the *Partido de Coligacao Nacional* (PCN). PCN served as a broad democratic coalition, bringing together several strands of legitimate black nationalism. Its leaders included Simiao, Gwenjere, Paulo Gumane and Uria Simango. Their political agenda called for the Mozambican people to be granted independence by means of a democratic process in which the electorate were able to register their choice in free and fair elections. PCN was also adamant that Mozambique's future should not be the subject of deals made between the Portuguese authorities and Frelimo, the latter of which had never been elected by the Mozambican people.

Even in these first manifestations of political democracy in Mozambique, Frelimo's political intolerance was soon in evidence. Early Gumo political rallies in both Beira and Lourenco Marques were either broken up or disrupted by Frelimo militants. Similar attempts by Frelimo were made to harass and deter many other rival political meetings and manifestations.

Portuguese resolve to seek a democratic solution in Mozambique was weakened by a Lisbon administration which became increasingly sympathetic to Frelimo as radicals consolidated their control. Matters were worsened by a wave of industrial and rural unrest, and the severe morale problems in the Portuguese army which were due both to war fatigue and involvement in a political 'no win' situation. At a September meeting in Lusaka between Frelimo and a Portuguese delegation led by minister without portfolio, Melo Antunes, an Armed Forces Movement radical, the Lisbon administration agreed to all of Frelimo's conditions. This Lusaka agreement, signed on 7 September 1974, and negotiated without foreign minister Soares, set the date for Mozambique's independence as 25 June 1975. In the interim, the country would be administered by a transitional government. The prime minister and six of the nine ministers would be appointed by Frelimo; the Portuguese High Commissioner would nominate the other three. Frelimo Chief of Security Joaquim Chissano was appointed Prime Minister. An analogy, while bearing in mind that no analogy is perfect, would be that of a radical Labour government in Britain desperately seeking a solution to the political and security conflict in Northern Ireland by handing over power to the Provisional IRA, despite calls for a referendum and elections, and then washing its hands of the consequences.

Within two weeks of the signing of the Lusaka Accords, all political parties opposing Frelimo were banned, and their leaders, organisers and supporters incarcerated in Frelimo concentration camps. PCN was made a particular target by Frelimo's secret police. Several hundred non-Marxist black political leaders were arbitrarily imprisoned. Frelimo put two hundred and thirty-nine of these prisoners on show on 16 March 1975, followed by some three hundred and sixty people, also detained without trial, who were displayed in April 1975. Despite a decade of enquiries and investigation by human rights groups such as Amnesty International, into the whereabouts and circumstances of these prisoners, Frelimo still refuses to release any details. These prisoners of conscience have, to use Amnesty's term, "disappeared". There can be very little doubt that the vast majority of these detainees will never be seen again, having died in Frelimo's system of concentration and death camps. On his defection, *SNASP* director Jorge da Costa brought out with him the original Frelimo documentation ordering the execution of democratic Mozambican nationalists such as Uria Simango, Lazaro Nkavandame, Mateus Gwengere, Joana Simiao and Paulo Gumane[1]. A frequent criticism of the Mozambique National Resistance is that it has been essentially military in nature. If this is the case, it may well be because Frelimo cold-bloodedly destroyed those non-Frelimo political leaders who emerged at the end of the Caetano regime.

Opposition to Frelimo Consolidates

Within weeks of Frelimo's coming to power within Mozambique, and the subsequent purge of democratic opposition, several underground anti-Frelimo groupings emerged. The *'Magaia'* movement and its publication, *Magaia's Voice* - the 'free voice of the Mozambique people' – named after Frelimo's first military commander, who died in unexplained circumstances making way for Machel's crucial command of Frelimo's military forces, supported the aims and objectives of Frelimo under Mondlane, and denounced the Frelimo of Dos Santos and Machel:

> "Frelimo, which represents our struggle for freedom and independence should not be confused with Samora Machel, symbol of tyranny and slavery." [2]

Magaia's Voice catalogued examples of Frelimo's web of corruption and of Machel's personal excesses, and exposed both Frelimo's systematic repression of any opposition and the creation of a network of concentration and forced labour camps. The Magaia movement specifically cited the example of the Massacaze concentration camp in Gorongosa district. *Magaia's Voice* compared Massacaze to both Dachau and the Soviet Gulag:

> "Thousands of prisoners are taken by lorry or train to isolated areas without any lodging or feeding facilities where, besides the brutal treatment, they are used as forced labour under the threat of bayonets."

In its first communique, issued in January 1976, the *Frente Unido Democratica de Mocambique* (FUMO) declared its total opposition to Frelimo. The difference between what may have been an authoritarian Portuguese administration and Frelimo's totalitarian regime had become very apparent:

> "The dictatorship installed by Frelimo in Mozambique is so atrocious that it has made people wish that they were back under Portuguese rule." 3

FUMO strongly criticised the wholesale nationalisations in Mozambique. Frelimo's nationalisation of health services, it complained, had whittled down the number of doctors left in Mozambique to twenty: these doctors were forced to attend party militants and their dependants "while other people die on the streets, in the villages and at the doors of the hospitals for lack of medical assistance". The nationalisation of the legal system had also destroyed any chances for an impartial judicial process – those lawyers left were all employed by the state.

The first anniversary of independence and of the Frelimo administration, 25 June 1976, saw FUMO's clandestine first congress at Mocuba in Zambesia province. Fumo was led by a former Frelimo activist Dr Domingos Arouca. Arouca had been Mozambique's only fully trained black lawyer. Imprisoned for several years by the Portuguese for his involvement in anti-colonial activities, he was subsequently released and placed under house arrest in Inhambane. The Portuguese withdrawal from Mozambique saw Arouca's reinvolvement in nationalist politics. Fumo's political agenda included calls for multi-party democracy and free enterprise in Mozambique. Arouca and his political supporters were severely handicapped by being an illegal organisation within Mozambique, and were increasingly forced into exile politics.

The Cabo Delgado Front came into being soon after independence in 1975 – and most of the information about its existence came from Frelimo sources and communiques. Operating in northern Mozambique, it has obviously created a number of problems for Frelimo. Thirty-five Cabo Delgado Front members were arrested in Nangada in February 1976, and a few weeks later Frelimo announced the arrest of a senior Frelimo political commissar based in Montepuez, south of Nangada, who was detained because of his involvement in the Front.

Patently, therefore, the degree of opposition to the Frelimo regime was intense and focused on both the doctrinaire Marxist perspective and the repression associated with Frelimo. A strong factor in the emerging opposition to Frelimo was the resilient Mozambican nationalism, a nationalism born, in the words of

Mondlane, "like practically all African nationalism... of direct European colonialism". For all its sloganising about national liberation and nationalism, Frelimo's political direction dictated hostility to genuine Mozambican nationalism. Marxism and genuine nationalism, as articulated by patriots such as Kenyatta and Mondlane, are incompatible. This is made clear by Frelimo's political mentors:

> "Nationalism is hostile to socialism and the Marxist-Leninist world outlook in its nature; it contradicts the objective evolution and rapprochement of the socialist nations...
>
> No Communist party can remain a genuine Marxist-Leninist party and fulfil its mission as vanguard in the building of a new society, unless it creates an atmosphere of intolerance towards nationalism and overcomes it first of all within its own ranks."[4]

As a self-avowed Soviet-aligned and Marxist-Leninist vanguard movement, Frelimo's renunciation of, and its virtually mandatory 'atmosphere of intolerance' towards nationalism, together with the increasing repression and state-terrorism alienated significant cross-sections of Mozambique's population – not least of all were the considerable numbers of Frelimo combatants. Many of these men and women had fought a bitter ten-year war to secure freedom and democracy within a nationalist Mozambique, and were very disappointed to see Portuguese colonialism merely replaced by a pro-Soviet hegemony: a hegemony characterised by what was essentially a rejection of the very nationalism which had provided the motivation for the vast majority of Frelimo forces during the armed struggle against the Portuguese state.

If there had been objective conditions explaining both the necessity and legitimacy of Frelimo's commencement of the armed struggle against the Portuguese in 1964, then these conditions, if applied to post-1975 Frelimo-administered Mozambique, would have more than justified the emergence of a renewed armed struggle for freedom – a second war of liberation. The spontaneous emergence of deep opposition to Frelimo's ruthless political and social repression, disastrous economic mismanagement, rejection of true Mozambican nationalism, and enthusiastic alignment with the Soviet bloc was manifested by the underground political movements, the emergence of an active black market, a continuing adherence to the values of traditional Mozambican society, and as the only last resort against totalitarian dictatorship, the armed struggle.

A people's ultimate right to take up arms against a totalitarian state, Marxist or fascist, should be a clear one. It was the last resort of the people of Western Europe, who were helped by Britain, during the Second World War. There is also no challenge, for example, to the armed Afghan resistance to both the pro-Soviet Kabul regime, or to the Soviet soldiers and gunships that once sustained it. Similarly the Mozambican peoples' rejection of their régime, as led by the Mozambique National Resistance, is just as valid as that of the French maquis or the Afghan mujaheddin in their struggles with equally fascistic states.

It is clear that Renamo inherited the legacy of several political groupings antipathetic to Frelimo's Marxist ideology. These included several tendencies within Frelimo itself, organisations Frelimo had attempted to destroy physically in their quest for a uniform Marxist Mozambique. A common indictment of Renamo is the claim that they are essentially a military and not a political organisation. The reason why there is no Mozambican Nelson Mandela or Walter Sisulu is because Frelimo secret police, under direct orders from Maputo, executed all Mozambican opposition political leaders of note in the years immediately after independence. This goes at least some of the way in explaining the initially military orientation of the Mozambique National Resistance.

The immediate background to Renamo can be seen as the establishment of the *Voz da Africa Livre*, the Voice of Free Africa, radio station which started broadcasting from Gwelo, Rhodesia in mid-1976. It was directed by a white Mozambican, Orlando Cristina, together with several former Frelimo members. A disillusioned member of Frelimo in Tanzania in the early 1960s, Orlando Cristina had been imprisoned by the Portuguese authorities on his return to Mozambique in 1964. He was freed through the personal intervention of Beira superbusinessman, and Malawi's consul-general Jorge Jardim. Cristina became his personal assistant, and spent several years building up indigenous black units within the Portuguese army and training President Hastings Banda's black security units in Malawi. Frelimo's accession to power in Mozambique saw Cristina relocate to Rhodesia. Frelimo's closure of Mozambique's border with Rhodesia in early March 1976, together with Samora Machel's open sanctuary and logistical support for Robert Mugabe's ZANLA guerillas saw the go-ahead for Cristina's idea of an anti-Frelimo radio station, funded at least in part by Jardim.

The *Voz da Africa Livre* radio broadcasts served as a rallying point for those tens of thousands of Mozambicans who were either politically or socially hostile to Frelimo's Marxism or themselves victims of the Maputo regime's totalitarian excesses. Thousands of these Mozambicans, many of whom were former Frelimo combatants disillusioned with the new society for which they had fought, left the country as refugees, and settled in Rhodesia. By the end of 1976, therefore, the nucleus of what was to become the Mozambique National Resistance had established itself in Rhodesia.

The one man who served as a focus for what were several quite disparate strands of opposition to Frelimo was André Matesangaissa. He had been a Frelimo combatant since 1972, serving as a platoon commander in the Gorongosa district of Sofala province. After independence, Matesangaissa was the commander of the Donde base near Beira. He was concerned, as were hundreds of other Frelimo soldiers, with the widening gap between his vision of what Frelimo had been fighting for, and that being implemented by Frelimo's ideology. His vocal opposition to the Maputo regime's behaviour caused him, and several of his colleagues to be sent to the Massacaze concentration camp in Sofala province. He escaped in October 1976 and made his way to Rhodesia. In an interview in October 1977, Matesangaissa outlined the reason for his having denounced Frelimo:

> "During the long period of the war with the Portuguese we believed in freedom and that this freedom would have allowed the people to choose its own form of government and elect the President of its choice. This never happened. Samora Machel and the Central Committee lied to us. They took over power and rule us as they wished... The people have not been liberated... So our fight goes on until we achieve the real liberation of the people of Mozambique."[5]

Matesangaissa specifically mentioned Frelimo's ideology as a major factor explaining the continuing conflict:

> "Do you know that all Mozambicans who belong to the party can only work for the state. He has no rights whatsoever, is unable to choose his employment, run his own farm, he can do nothing. With Machel's Communism he is from the time of birth until he is a dead servant of the Party. This is pure slavery. Here you have a Government of 20 to 30,000 living the good life while the other 8,500,000 are just slaves... The Colonialism was bad but 'Machelism' is worse. We all have the right to say that we have been betrayed."[6]

André Matesangaissa was joined some months later by Afonso Dhlakama. Dhlakama was the son of a chief of the Shangaan-Ndau ethnic group, and had been educated at a Catholic seminary until conscripted into the Portuguese army. He was sympathetic to Mozambican nationalism, and he deserted to join Frelimo in March 1972. Trained at the Nachingwa base in southern Tanzania, Dhlakama served as a Frelimo commander in Niassa province, and after independence was made logistical commander for Sofala province. He was similarly disillusioned, as Matesangaissa had been, with the society being created by Machel, sought sanctuary in Rhodesia in 1977. There he came into contact with André Matesangaissa and soon emerged as an efficient deputy to 'Commander André', as Renamo's first commander was to become known.

The first training camp, with six guerillas, was set up near the Rhodesian border town of Umtali in March 1977. André Matesangaissa declared the "Second Struggle of National Liberation" to have started 7. The first operation staged by Matesangaissa was to raid the Massacaze concentration camp in April 1977. The lightly-armed unit penetrated deep into Sofala province, attacked the camp and liberated its inmates. Some three hundred of the five hundred men and women freed from the camp opted to return with Matesangaissa to Rhodesia and carry on the struggle for freedom in Mozambique, this time against Frelimo. The Mozambique National Resistance was operational.

Another typical Renamo recruit was the Soviet-trained former Frelimo cadre Daniel Lacerda Caetano. Caetano joined Frelimo in 1973 at the age of sixteen, seeing active service in the Milange area. In 1976 he was sent to the Soviet Union to specialise in anti-aircraft weapons and tactics. He explained that his alienation from Frelimo was a result of disatisfaction with its ideology and methods:

> "I saw how people lived oppressed and unhappy and realised I could not go on fighting for a communist Frelimo... The so-called *Forcas Populares* (*FPLM*) are really the party's own military forces, sworn to defend the Machel regime. And Machel's Frelimo is a minority party that does not represent the will of the Mozambique people." 8

Caetano also described how Frelimo policies, particularly those which enforced collective farms and villages, were causing wholesale discontent amongst Mozambique's workers:

> "People are forced to work on them but the crops go to the Frelimo barns. The workers have not the rights to keep the maize or the rice they grow. To avoid dying of hunger, they have to work on their own plots hidden in the bush... People have no freedom or rights. They are treated as dogs and the party members are the masters."9

Despite their commitment to an ever-expanding conflict of their own, Rhodesia granted aid to Renamo through the Central Intelligence Organisation. The assistance of the Rhodesian Special Air Service, particularly towards the end of the Rhodesian Civil War, is also documented. Small units of the Rhodesian SAS, who were largely concerned with training the insurgents also occasionally accompanied Renamo teams during incursions inside the country. From 1977 until early 1979, Renamo operated largely from bases within Rhodesia. From then onwards, the organisation was headquartered deep inside Mozambique, in the Serra da Gorongosa mountains of Sofala province, some 130 kilometres north west of Beira. An operational staging point situated in the Sitatonga mountains, at Gogol, some 35 kilometres from the Rhodesian border. Rhodesian assistance was always restricted. As Professor Andre Thomashausen points out:

> "(T)here were limits to the Rhodesian willingness to support the resistance and the supply of material; the group's movements, and its numbers were strictly limited by the Rhodesian authorities."[10]

The Rhodesian reluctance was, at least in part, based on hopes that a compromise or accommodation of some sort could be made with Machel. In the event, Frelimo was intransigent and an open state of war existed from 1977 onwards, with hundreds if not thousands of Frelimo soldiers serving in an 'internationalist' role alongside ZANLA guerillas inside Rhodesia[11]. By the time Frelimo's intentions were clear, the Rhodesians were stretched to capacity with their own war.

Nonetheless, despite minimalist aid in its crucial first few months of existence, the Mozambique National Resistance grew out of all proportion. Renamo was primarily concerned with establishing a basis of political support among Mozambique's rural population, particularly in Manica province, but also engaged Frelimo military and state targets. They gained more and more confidence and augmented their supply of weapons. Several towns and Frelimo concentration camps were also overrun by Renamo fighters, who freed their inhabitants – many of whom subsequently joined the resistance. From a nucleus of six men in 1977, Renamo had an operational strength of 917 trained guerillas by December 1978 [12]. Towards the end of 1979 there were between one and two thousand Renamo cadres in the field.

Mozambican regular forces launched a huge offensive against Renamo in October 1979, which was aimed primarily at the movement's Gorongosa mountain headquarters. Armour, heavy artillery, infantry and air forces were concentrated against the base. Tragedy struck when, in a bold attack on Frelimo infantry and armour in Gorongosa town on 17 October, André Matsangaissa was killed in action. Assaulting Frelimo positions across four hundred metres of open ground, he died leading his men from the front. After some initial confusion, Afonso Dhlakama succeeded Matsangaissa as leader of Renamo.

By late 1979 the Mozambique National Resistance was extending its operational areas both north and south, and units were becoming operational in the northern province of Tete and as far south as the Save river. The December 1979 Lancaster House peace agreement between Zimbabwean nationalists and Salisbury's multiracial government ended the Rhodesian Civil War, restored British administration and led the way to the general election of February 1980, with Robert Mugabe's ZANU forming the new government of Zimbabwe. Elite force units such as the Special Air Service were disbanded. This all meant a dramatic reorientation for Renamo. Rhodesian intelligence and military officers warned Renamo of possible adverse circumstances days before the election. The *Voz da Africa Livre* stopped broadcasting on 20 February and Mugabe's election victory, when announced on 4 March, severed all connections with Zimbabwe. By this stage all Renamo personnel were operational within Mozambique with a build-up near the Sitatonga base. The leadership skills of Dhlakama, who was confirmed as commander-in-chief of Renamo during mass meetings of Renamo members at the Sitatonga base in June 1980, were severely tested in the months which followed. Frelimo military forces assaulted and captured the Sitatonga base, killing well over two hundred guerillas and capturing others. Dhlakama, and what remained of his movement, regrouped at a small transit base near Garagua on the Save river, on the border of Manica and Gaza provinces.

Renamo and South Africa

It is now clear that as the Rhodesian door closed on the Mozambique National Resistance in early 1980, assistance was offered by elements of the South African Defence Force. Orlando Cristina was now living in South Africa, and in addition to any South African committment, was acting again as a channel for private sector investment in Renamo. Renamo's principle financier, Jorge Jardim, had died in Libreville, Gabon in 1979, and was replaced by the millionaire industrialist Manuel Bulhosa, who was the richest man in southern Europe and a past victim of Frelimo's industrial nationalisation programme in Mozambique. Dr Evo Fernandes, a Mozambican of Goan extraction, and former editor of the *Noticias de Beira* newspaper, also began to emerge as a key link between Renamo and its external, largely Portuguese, supporters and financiers, as well as helping to build up the movement's political profile. The *Voz da Africa Livre* recommenced broadcasting to Mozambique in mid-1980. Renamo was still based within Mozambique, and South African assistance included supplies of weaponry, and other war material, and the air-dropping of training teams of South African Special Forces personnel into Renamo-held areas of Mozambique.

Aided by the complacence into which Machel and Frelimo slipped after Zimbabwe's April 1980 independence and the June 1980 destruction of Renamo's Sitatonga headquarters, Renamo consolidated its existing bases, retrained considerable numbers of its personnel and exploited the ever-growing peasant discontent with Frelimo. This alienation ensured a continuing and increasing supply of recruits to the movement. Professor Thomashausen estimates that by

December 1980 Renamo had some 10,000 men in the field. [13]. This is partially corroborated by a Zimbabwean military intelligence briefing dated 26 June 1981, which put Renamo strength in December 1980 at "between 6,000 and 7,000 fully armed men with an estimated 2,000 recruits in the pipeline" [14]. Circumstances in Mozambique were so stable for Renamo that President Dhlakama was able to absent himself from the war zones in order to visit Western Europe. Much of November was spent overseas, meeting politicians, businessmen and journalists in Portugal, France and West Germany. In West Germany, for example, he met with representatives of both the Christian Democratic Union and Christian Social Union. As a result of this international excursion, Renamo offices were set up in both West Germany and Portugal, the latter headed by Dr Evo Fernandes. John Saul in *A Difficult Road: The Transition to Socialism in Mozambique* has also touched on the considerable international support for Renamo that emerged in the early 1980s:

> "(O)ther MNR backers have taken on higher profiles than previously, most importantly various right-wing Portugese elements, many of them with a history of previous involvement in the colonial economy and/or security system. Apparently such countries as Israel, the Comoros, and Oman have also joined more actively in the game."[15]

Renamo's reorganisation in the latter half of 1980 had dramatic results in 1981. The movement became operational in extensive areas of both Manica and Sofala province. Mozambique's communications, infrastructure and energy system became obvious targets. The Beira-Mutare (formerly Umtali) oil pipeline and railway were both damaged in Renamo attacks. The Maputo-Zimbabwe rail-link was also disrupted in October 1981, as were several national highways. The Cahora Bassa powerline, supplying both Mozambique and South Africa with electricity, was cut for the first time in November 1981, causing power cuts in the northern Transvaal. Frelimo's major offensive in 1981 resulted in the capture of the by-then empty Garagua base: Renamo's headquarters had already been relocated back to the Gorongosa mountains.

1982 saw the Mozambique National Resistance in action as far south as Gaza and Maputo province, and as far north as Tete province, with a front also opened up in Zambezia. In Zambezia province Renamo linked up, and merged, with *Africa Livre*, an independent grouping of anti-Frelimo guerillas identified with Fanuel Mahluza. A founding member of Frelimo, Mahluza had become disillusioned with Frelimo's political drift, and had helped to form Coremo, serving as its first secretary for defence. He was detained during Frelimo's accession to power, and escaped from the Ruarwa concentration camp in August 1977 to help build up the *Africa Livre* organisation [16]. Despite Renamo's expanding base, Sofala and Manica provinces were still the centre of military activity. Renamo recorded some 1500 military engagements in 1982, which were co-ordinated from more than fifty base camps throughout the length of the country. The capital's electricity supply was cut four times in the course of 1982; the water supply was also disrupted. By the end of the year, the Mozambique National Resistance was able to host an overseas journalist in one of their liberated zones [17].

Joseph Hanlon provides evidence of the ease of Renamo movement and activity in the operational areas. He gives the example of a Renamo unit which entered the sawmill town of Savane, some 77 kilometres north of Beira, in July 1982. The guerillas stayed in Savane for seven days and then moved towards Mafambisse, a large population centre on a national highway. Frelimo eventually tried to challenge the Renamo presence two weeks later. Renamo withdrew, but were back in Savane on 9 August [18]. The ease with which the organisation was able to move around Mozambique was also documented by John Burlison, a British wildlife specialist captured by Renamo guerillas in Gorongosa in December 1981. Freed in May 1982, Burlison had covered some eight hundred kilometres in central Mozambique before being released in Zimbabwe. The unit accompanying him was able to move "fairly easily", without encountering any Frelimo presence whatsoever. Burlison personally came into contact with about one thousand Renamo guerillas, who "seemed to be popular" with the local populations. Eduardo Ribeiro, a Portuguese trader taken into custody in December 1982, travelled over 1300 kilometres in the company of Renamo units before being released. He too saw no Frelimo presence.

The northern province of Zambezia, a key earner of much-needed foreign currency, was the scene of considerable Renamo activity from August onwards. Guruc, the centre of the tea industry and situated less than sixty kilometres from the provisional Capital of Quelimaine, was easily taken by the guerillas. Frelimo military and economic activity ground to a virtual halt in the province. Several foreign technicians, including six Bulgarians, were captured[19]. The 9 December demolition, by Renamo saboteurs, of thirty-four oil storage tanks in Beira harbour, caused $20 million worth of damage, an acute fuel crisis in Zimbabwe, and brought home the growing capacity of Renamo to Frelimo.

Frelimo prepared for a series of offensives against Renamo in 1983. These saw a number of successes in both Gaza and Maputo province, and were combined with huge Frelimo propaganda rallies during which Renamo prisoners of war were paraded and denigrated. On at least one occasion, seven captured insurgents were publicly executed. At one such rally in Chibuto, Machel clearly indicated that Frelimo was suspending the rules of war:

"Those who supply the bandits with information must die with the bandits. Those who feed the bandits must die with the bandits. Our mission is not to wound, but rather to kill!" [20]

The *Economist* journal, commenting on Renamo's growth from a "small, ragged band" in 1980, described the organisation in early 1983 as a ten thousand-strong "disciplined guerilla army" [21]. It was a manifestation of opposition which Frelimo tried desperately to root out. However, despite serious losses in the south, Renamo was active in nine of Mozambique's eleven provinces by mid-1983 [22].. The guerilla high command launched first its 'Red August' offensive and then its 'Black September' campaign: the first aimed at bleeding Frelimo and the Maputo regime's foreign 'mercenaries' and the latter pledged to make September a month of mourning for Frelimo. 'Red August''s most publicised result was the capture of 24 Soviet technicians at Morrua, in Zambezia province[23]. The freedom fighters continued to focus on disrupting the national transport network, and cut the Beira-Zimbabwe rail-link, even though there was an increasing level of Zimbabwean military commitment to its protection and that of the Beira corridor itself. Despite setbacks for the Mozambique National Resistance in Mozambique's southern provinces, and a concerted Frelimo offensive in Inhambane province which saw the capture of the important Tome base in August 1983, Renamo was still present. Frelimo's Inhambane provincial military commander admitted to journalists, that month, that some 3500 Renamo insurgents were still operational in his province alone. His point was brought home to accompanying journalists when a Frelimo public relations visit to a supposedly Renamo-free district, which had been secured some days earlier by Frelimo, was ambushed by guerillas [24].

The assassination of Renamo secretary-general Orlando Cristina at a farm near Pretoria on 16 April 1983 was a blow to the Resistance. Several early Renamo members such as Artur Vilankulu, Gideaon Macheluze, Zeca Caliate and Armando Kembo Santos left the organisation, after the subsequent infighting, and later formed several Mozambican splinter groups in Western Europe. Evo Fernandes became the movement's new secretary-general in December 1983. The year ended with Renamo's involvement in over two and a half thousand military operations against the Maputo regime, which were directed and carried out from 127 guerilla bases throughout Mozambique [25]. Frelimo's end of year status was summed up by the London *Times*:

"The anti-Frelimo guerilla movement has gradually advanced southwards so that now it even controls some elements of Maputo province itself. North of the Zambezi Frelimo forces are confined to garrison duties, rather like the Soviet army in Afghanistan. South of the Zambezi the Frelimo effort is concentrating on the security of a narrow strip of land which runs along the rail and pipeline link from Beira to Zimbabwe."[26]

The Nkomati Accord

The intensity of the peoples' war against Frelimo was clearly defined by Frelimo's massive propaganda and misinformation offensive which was launched towards the end of 1983. As the Soviet Union's commitment to a losing cause started to falter, Frelimo desperately sought other avenues of aid. President Machel visited six Western European states, including Britain, with the message that Frelimo's commitment to the Soviet Union and Marxism-Leninism could well be loosened if substantial Western economic and military aid for Frelimo's war was forthcoming. It is short-sighted to suppose that a Marxist state, which is temporarily relaxing economic controls, providing incentives to farmers and moving towards a more Western-orientated economy, is undergoing a political reorientation – Lenin's New Economic Policy in the 1920s, did not lead to a liberal democratic state, but merely served to consolidate his, and subsequent Soviet leaders' fascistic grip on their country (and increasingly on other nations). Nevertheless, this somewhat unlikely scenario of a moderating Marxist state was built upon by the United States Assistant Secretary of State for African Affairs, Chester Crocker, in his negotiations with South African Foreign Minister Pik Botha, notably during meetings in February 1984.

This new phase of Frelimo's last ditch attempt to stay in existence, aided by American and South African diplomacy, culminated in the signing of the treaty of 'Non-Aggression and Good Neighbourliness' between South Africa and Mozambique on 16 March 1984. The Nkomati Accord, as it was subsequently termed, committed both governments to ensure that their countries were not used as a base to "plan or prepare to commit acts of violence, terrorism or aggression" against each other. South Africa would end support for Renamo in return for Frelimo's denial of assistance and bases to the anti-Pretoria African National Congress organisation.

The end of official South African aid to the Mozambique National Resistance revealed deep divisions within government circles. Anticipating the end of logistical assistance, elements of the South African military, in direct violation of cabinet-level decisions, arranged a massive pre-emptive resupply of war material. Apart from this pre-Nkomati assistance there is no evidence of further officially sanctioned South African logistical help to date. The 'Gorongosa diaries', Renamo papers subsequently captured during an assault on the movement's headquarters in August 1985, reveal only that there was unauthorised help from sympathetic military sources – assistance which ended within months of Nkomati.

There is every indication that the Nkomati Accord came as bitter and unexpected news to the Renamo leadership. In an interview with a Western defence correspondent in April 1987, President Dhlakama "showed some hostility to his former mentors". "Slighted" by the Accord, the guerilla leader stated: "We were told about it, just a week before it was signed" 27. Machel's ideology dictated that unrest in Mozambique was the direct result of external factors, and that Renamo was merely an "extension of the South African armed forces". Nkomati had cut the cord. The Frelimo president declared that:

> "We have turned off the tap. All that is left is the water which is already in the pipe". 28

Reality, however, ran contrary to Marxist dialectic. The peoples' war against Maputo did not de-escalate. One day after the Accord a heavily-protected Zimbabwean convoy was attacked along the main Zimbabwe-Malawi highway in Tete province: several Mozambican soldiers were killed and $3 million of damage was inflicted by guerillas 29. Insurgent activity was still nation-wide and within a month of Nkomati insurgents had hit the main highway between Maputo and South Africa. Renamo activity in southern Mozambique was so intense that both the American and Portuguese embassies warned their nationals against night-time use of the road network in southern Maputo province. The destruction of key power installations to the south of Maputo in April left the capital without electricity for more than a week. The strategic town of Inhaminga, situated on the key Beira-Malawi railway was occupied on 6 April and held by Renamo insurgents for several days. By the end of the year of Nkomati, the peoples' insurgency had consolidated its presence and support in all ten of Mozambique's provinces, and Renamo was now also operational in Niassa and Cabo Delgado. With an estimated fifteen thousand guerillas and a national network of some ten thousand active supporters, Renamo had convincingly dispelled the myth of an umbilical relationship with South Africa.

Despite, and maybe because of the Nkomati Accord, the conflict in Mozambique reached unprecendented levels in the years that followed. Renamo was able easily to adopt classical guerilla tactics of concentrating when least expected and vanishing when the military correlation of forces was unfavourable. An example of the ability of the Renamo guerillas was the ease with which they eluded a massive combined Zimbabwean/Mozambican offensive which was centred on the movement's Gorongosa headquarters. Several thousand Zimbabwean soldiers, spearheaded by special forces units, together with 8,000 Frelimo personnel and extensive air support, including fighter-bombers and Mi-24 helicopter gunships, captured the *Casa Banana* base on 28 August 1985. Eighteen days earlier, however, Renamo President Afonso Dhlakama had issued the following statement:

Message Of The Commander-in-chief To The National Forces

General, Commanders and Fighters!

The Commander-in-chief has reliable information that a major Zimbabwean/Tanzanian offensive against the headquarters of Gorongosa is being prepared. Powerful air support will be deployed in this attack, including MiG-21s manned by Ethiopians and armoured helicopters Mi-24 flown by Soviets, thousands of Zimbabwean parachutists trained by British officers, and intensive napalm bombing. We also have knowledge that this operation will be directed by the British military mission in Harare as regards the bombing raids, the air reconnaissance and the overall co-ordination, while the financing is provided by the British and the North American governments. The attacking forces have orders to kill the civilian population and to use defoliation agents in the thicket.

In view of the high concentration of enemy forces which are going to be launched against the headquarters and the high risk of heavy casualties amongst the civilian population which is under our protection, we ordered the immediate civilian and military evacuation of the area subject to the attack. The Commander-in-chief and RENAMO forces will nevertheless remain in Gorongosa with the mission of circling the invading foreign troops and in order to destroy the logistic lines of the enemy until its complete defeat and the withdrawal of the Zimbabweans.

FRELIMO will most certainly exploit this foreign offensive for a major propaganda campaign, and in order to create the impression abroad, on the eve of the visit of the dictator Machel to the Vatican, to Italy, to the United Kingdom, to the USA and the UN, that the course of the war has turned. But the mere fact that this operation against our headquarters is being financed, planned and carried out by foreigners shows the degree of debility of the Maputo government.

This fact assures us that militarily, the war is practically won and that Final Victory is in sight.

Consequently, the military planning established for execution as from November 15th, 1985 remains unaltered.

That everyone may accomplish this task, so that the Commander-in-chief may accomplish his.

THE STRUGGLE CONTINUES (A LUTA CONTINUA!)
Gorongosa Headquarters on 10 August 1985

Afonso Dhlakama (President and Commander in Chief) 30

The Zimbabwean/Frelimo forces, led ironically by several former Rhodesian special force trainers of Renamo, captured an evacuated base, constantly harassed by highly mobile guerilla units who killed numerous attackers including two Soviet security advisers 31. The Sofala province offensive soon ground to a halt while Renamo forces reached new levels of activity throughout the rest of Mozambique. Whatever propaganda victory Frelimo had achieved by the seizure of *Casa Banana* in August 1985, was destroyed when the Mozambique National Resistance recaptured their Gorongosa headquarters on February 14 1986.

Frelimo's inability to come to terms with Renamo at a military level, even with extensive and unprecedented external assistance, was commented on internationally. The BBC reported that "Mozambique Frelimo government forces have been suffering serious setbacks in its war against Renamo" 32. It can be argued that since these early 1986 defeats, and the death of President Samora Machel in a plane crash in October of that year, Frelimo has virtually ceased to exercise authority in up to eighty per cent of Mozambique, limited as it is to the centre of Maputo and a few cities, and has kept in power only through external military forces, investment and assistance. The destruction, by Renamo special forces, of the massive Maputo arms depot in September 1985, showed that Renamo could also operate in an urban environment 33. This action caused some $400 million of damages, and was quickly followed later that year by the destruction of the Maputo-based national aluminium factory and Olsa Citrus factory 34.

It became evident that Frelimo was beginning to abdicate its role in the war against Renamo. The London *Sunday Times* correspondent, Peter Godwin, reporting from the war zone in northern Mozambique in April 1987 demonstrated how the Zimbabwean military had taken over from Frelimo in parts of the country, in what was becoming an increasingly unpopular war for Zimbabwean soldiers. One Zimbabwean non-commissioned officer interviewed by Godwin summed up what seemed to be a common complaint:

> "We are very far from home. Why should we be killed in someone else's war? First we were told just to guard the pipeline. Now we seem to be fighting the whole war."

The article described Zimbabwean military morale as "flagging", and revealed Zimbabwean awareness of both Renamo ability and support:

> "All behind us is Renamo. As far as we're concerned they have already won this war. All that is left is for them to announce it. These Renamo are good. They control the countryside. The local people support them." 35

The increasing use of Zimbabwean military units was also confirmed by Lord Michael Cecil, a British journalist who travelled extensively throughout Mozambique's war zones in early 1987:

> "The government relies for its own defence on Zimbabwean troops who supports its demoralised army. It cannot even protect the remaining road and rail links effectively."

Writing in July 1986, *Guardian* journalist Mike Saunders observed that the Mozambique National Resistance had:

> "...gradually gained strength until today the country is all but crippled by their activities. Their main strategy is to mount hit and run attacks on communications. About 80 per cent of the country is dominated by the MNR".

Frelimo's authority in Mozambique, Saunders stated, was minimal:

> "Apart from the towns and their environs, a few transport corridors and the Maconda's area around Cabo Delgado up by the Tanzanian border, one can say that most of the country is a no-go area for the government."37

One year later Renamo's reach had extended even further, including deep into Cabo Delgado province and amongst the Makonde ethnic group itself: Frelimo's authority was being challenged everywhere. Peter Tygesen, a pro-Frelimo Danish Radio News journalist who spent August/September 1987 visiting Mozambique, recorded that in September 1987 alone, 215 of Mozambique's 219 national districts registered Renamo activity:

> "On only one of the country's main roads – that from Beira to Zimbabwe – can cars travel without the risk of daily armed assaults; at nights no road is safe. All the traffic between the major towns takes place in the form of flights, and in order to get beyond these towns and travel to the district capitals... you have to go by taxi aircraft or in convoys... protected by soldiers." 38

This reading of the situation in September 1987 was reinforced by Richard Grenier. Writing in the *Sunday Telegraph* he states that:

> "...the simple, observable fact is that nothing moves in Mozambique outside the main cities except by military convoy, escorted by armoured cars and troop carriers. Zimbabwean troops... patrol timidly

> in large units, stick to main roads, and scurry back to their bases before dark. Renamo owns the night. This is not a winning strategy." [39]

It was only a matter of time before confident and well-trained Renamo guerilla units began to attack military convoys. In December, the insurgents destroyed a convoy on the outskirts of Maputo itself [40]. By February 1988, every single road and rail link to Maputo had been interdicted by the Mozambique National Resistance. Between October 1987 and February 1988 over one hundred and fifty vehicles under military escort were destroyed on the main highway north from Maputo [41].

Renamo was increasingly coming to operate in large numbers. The Frelimo news agency, *AIM*, repeatedly reported activity involving two or more battalions of Renamo soldiers: a unit of more than 500 insurgents in Gaza province in August 1987 [42], and a February 1988 account of 500 guerillas in Sofala province [43].

It would be accurate to state that the security position by 1989 is a military stalemate in which Frelimo, even with massive foreign backing, is unlikely ever to win. It is quite possible that Renamo could seize power within weeks, should they wish to, merely consolidating what they already control and pushing into the few cities still administered by Frelimo. The correlation of political and international forces, however, is still not favourable to this end. The Mozambique National Resistance is obviously content with a military stalemate which is advantageous to the guerillas and is waiting to negotiate for peace and national reconciliation to solve the country's dire political problems.

Chapter Six

STRUCTURE AND CONTROL OF RENAMO

The Mozambique National Resistance has at its head Afonso Dhlakama, who is both President of Renamo and the Supreme Military Commander of the organisation. The liberation movement has a political and military structure. The National Council heads Renamo's political wing, serving as a cabinet. Chaired by Afonso Dhlakama, there are ten departments headed by senior officials: Administration, Education, Information, Health and Welfare, Foreign Affairs, Agriculture, Youth Affairs, Defence and Ideology: Dhlakama is both Secretary of Defence and Ideology. The General Secretariat has been vacant since 1986. All the National Council, with the exception of Renamo Secretary for Foreign Affairs Artur Fonseca (resident in West Germany) and Information Secretary Francisco Nota Moises (Canada), are based at Gorongosa.

Renamo's fifteen-man Military Council is also chaired by Dhlakama. The military structure comes under three chiefs-of-staff for the North, Centre and South military sectors of Mozambique. Niassa, Cabo Delgado, Nampula, Zambesia and Tete provinces come under the Northern chief-of-staff, General Dick; Sofala and Manica provinces form the Central Sector under General Ismail: and General Domingos controls operations within the southern sector, which includes Gaza, Inhambane and Maputo province. Altogether, they control an estimated twenty thousand Renamo guerillas. Contrary to earlier Frelimo propaganda images of former Portuguese secret policemen or special forces, Renamo military leaders are usually former Frelimo soldiers. The propaganda assault has now changed from portraying Renamo as former colonial servicemen, to highlighting how young many of the insurgents are.

Each province has a general who is responsible for the prosecution of the war within his area. There is usually a central military base, which provides a political/administrative presence as well as initiating and controlling military operations within each province. The provincial structure incorporates several regions, and within each region are sector commanders at a battalion level, who control 250 men. Several 100-strong companies usually operate within each sector. Below the companies are either sections or platoons of ten and thirty guerillas respectively.

Renamo soldiers are equipped at the section level with the standard Soviet AK-47 or AKM assault rifle. Company sized units have access to both light and heavy machine-guns, usually Soviet RPK or RPD models, as well as RPG-7 rocket launchers and mortars; companies may also have a demolitions specialist. As with any guerilla army and conflict, structures are very fluid, and sections of ten fighters often operate independently while still retaining the capacity to come together in platoon, company, battalion or even brigade-size operations. There are also upwards of three special 'strike' battalions, each consisting of several hundred seasoned guerillas, who are deployed as and when Gorongosa sees the necessity. In addition to this special intervention force, which has smaller versions in several provinces, Renamo also deploys several hundred 'special forces', many of whom were probably trained by the South African military in pre-Nkomati days: these are used on special sabotage and reconnaissance operations, including, for example, the 1982 destruction of the Beira fuel storage installations.

Control of this vast political and military organisation is facilitated by a sophisticated radio network, which keeps the Gorongosa headquarters of Renamo in touch with its provincial and regional base camps, as well as with private supporters, and possibly unofficial government sources, in South Africa and Malawi. This pre-Nkomati radio system, which works by portable generator, has been improved by new equipment taken from Frelimo and Zimbabwean units, and by lap-top computers – a cost-effective and comparatively inexpensive gift from private Western supporters which is the latest in high-technology communications. Several visitors to the liberated zones have also commented on Renamo's widespread use of motorcycle couriers as a communication system between various regional heaquarters and bases inside Mozambique.

Strategy and Tactics

The Mozambique National Resistance, in its military struggle against the Maputo-regime, has consciously adopted a variant of Mao Tse-Tung's "People's War" model. This method of insurgency is to secure control of the countryside, temporarily concede but ultimately cut off and win the cities. The basis of this theory is the guerilla movement's identification with, and the crucial support of, the masses. The guerillas must be as "fish in the sea", with the sea being the rural population.

Renamo, therefore, has painstakingly fostered and built up close relationships with the people in its operational areas, manifesting the organisation's ideology by removing the political and physical shackles of Frelimo's totalitarian experiment. Tens of thousands of Mozambicans have been freed from enforced incarceration in collective villages and state farms. They have been able to live in freedom and comparative safety in Renamo liberated zones throughout most of central and northern Mozambique. Frelimo accusations that Renamo was responsible for the spate of massacres such as Homoine patently run against the entire nature, ideology and tactics of the organisation; it is suspected that either Frelimo or Zimbabwean units perpetrated these actions in an attempt to loosen local support for Renamo, as well as to create a propaganda event which the Maputo regime could exploit.

It was pointed out by the *Independent's* Lisbon correspondent, Sharon Behn, in an account of her four week visit to Mozambique's operational zones in 1987 that areas controlled by the Mozambique National Resistance are prosperous and capable of feeding their inhabitants, whereas large portions of Frelimo territory have been struck by wide-scale famine. Her account, "The Unknown Side", appeared in *Africa Events:*

> "An unexpected sight was that of sweeping fields of corn and manioc machambas or farms which contrasted dramatically with international aid organisation warnings that Mozambique is swiftly becoming a second Ethiopia. Private property is one of Renamo's keystone principles. Local communities have their own well-tended and productive machambas, and they also grow rice and peanuts."[1]

The liberated zones have seen the re-establishment of traditional authority and social structures: systems which are more in keeping with Mozambican and African heritage than Frelimo's alien Marxist ideology. Sharon Behn quotes Commander Alface, head of Renamo's administration department in Zambezia province: "Frelimo imposed a political regime alien to the Mozambican understanding of life. It is our objective to win over the people."[2]

Most independent observers who have visited the liberated zones of Mozambique, areas under direct or indirect Renamo control, have commented on the movement's popularity with the population. Miss Kindra Bryan, a missionary taken into protective custody by Renamo forces, after trekking several hundred kilometres through Mozambique, concluded that:

> "(T)he soldiers seemed disciplined and well organised and seemed to have a good relationship with the villagers, who, in turn, seemed to recognise them as the government."[3]

Sharon Behn's 1987 report echoed that of Kindra Bryan:

> "In the areas of Zambezia I visited at least, Renamo seems to have gained strong grass-roots support and is well-entrenched militarily.

> The rebels' military strength is backed up by considerable local support in what Renamo calls 'liberated zones'."

Her view is that Renamo's popularity is based, at least in part, on its political orientation. Behn cites 20 year-old Azevedo Jackson's comments as typical of several stories she heard during her four weeks trips:

> "Under Frelimo it was like living in a prison. We were forced to work in communal villages, and we were whipped when we didn't produce enough... Now with Renamo we have our own farms and we're free. I'm all for them and I know they will win."[4]

Behn also interviewed Father Onorino Venturini, an Italian priest with 37 years of missionary work in Mozambique.

> "The people supported Frelimo when it won the war of independence. But then Frelimo slowly began to squeeze the country, controlling people's movements, taking away their freedom, and using forced labour on the communal farms... The people are suffering; they are oppressed. They want a change, and that's what they see in Renamo."[5]

Dr Jack Wheeler, a leading authority on anti-Soviet insurgencies visited Renamo-controlled liberated areas of Mozambique in mid-1985. In addition to noting that in stark contrast to Frelimo-controlled zones, the liberated zones had enough to eat, he also perceived the development of a political consciousness amongst Mozambican peasants:

> "At a fishing village on a marsh near the border I asked two villagers who spoke some English how things were. 'Things are much better with Renamo controlling the border and this entire area... Frelimo tried to control everything, nobody could have private business. With Renamo, we have freedom, confidence – Renamo likes us to have private business.'
>
> With that, a Renamo soldier near me who knew a few English words smiled, and cried out: 'Renamo – Capitalist! Frelimo – Communist!' Everyone laughed. He raised his Kalashnikov high and yelled: *'Abassos communismo!'* [Down with communism] *'Capitalismo muito bom!'* [Capitalism very good.]"[6]

Defence correspondents of *Jane's Defence Weekly*, in a July 1986 report from Mozambique formed much the same impression of the support enjoyed by the Mozambique National Resistance:

> "The people appear to live in the resistance's 'liberated areas', where they have their own fields and can grow crops as they wish.
>
> Three nuns, who had been prisoners of the rebels and travelled extensively with them in Nampula, Niassa and Zambesia provinces, told JDW they crossed vast areas where the rebels apparently enjoy total control, getting food and help from the farmers."[7]

In keeping with the Maoist theory of insurgency, military operations can only commence once a base of support and co-operation amongst the population has been established. Rhodesian SAS observers documented Renamo's policy of close identification with the local population as early as 1978, which partly explains why the guerillas have enjoyed consistent support from Mozambique's rural population. The Mozambique National Resistance's military strategy is a long-term one, and Renamo's level of activity has increased immensely over the past ten years. President Dhlakama summed up the strategy as the maintenance of:

> "(P)ressure all over the country, launching a lot of small-scale attacks to force the enemy to the country, launching a lot of small-scale attacks to force the enemy to disperse his forces."[8]

Two prime targets of this strategy have been firstly any manifestation of the Frelimo party, and secondly the Mozambican infrastructure. The aim was to paralyse the state politically and economically. *Guardian* journalist Mike Saunders summed up Renamo's tactics quite succinctly:

> "The idea seems to be that if you chop at a man's windpipe he will begin to suffocate. If you chop often enough and hard enough he will eventually collapse. This gives you time to mull over what you want to do with the body."[9]

In line with the first objective, all government instalations and projections are deemed to be legitimate targets, just as in occupied France, for example, any manifestations of the Nazi party or their Vichy collaborators – military or civil – were targets for the French maquis. This policy has of necessity meant the capture of dozens of towns, some of them several times, throughout Mozambique. In keeping with the guerillas' present strategy, Renamo forces then withdraw, often after several days presence. Criticism has often been levelled at Renamo for destroying collective villages, state farms, government shops, educational establishments and health centres. Christian Aid has revealed the following details about Renamo activity up until the end of 1986 in Zambezia province alone: 44 Frelimo administration buildings, 503 state shops, 68 primary and eight secondary schools, and 144 out of 280 health posts were destroyed by guerillas[10]. What must be understood is that the Mozambican people are engaged in a war with a ruthless state, a state which has been responsible for tens, if not hundreds of thousands of deaths. All the targets mentioned above are seen by the resistance as projections of the Frelimo state, and are therefore legitimate targets. To take the example of health posts, we observe that Frelimo sees and institutes a very clear relationship between its politics and the health services. Samora Machel defined this inter-relationship:

> "The hospital is a state institution where Party policy touches the most sensitive points of the population... It is often in the hospital that the people see reflected the organisation of our State. It is there that the people feel directly the fruits of independence and of the construction of socialism."[11]

Machel also underlined Frelimo's political abuse of the health service, perceiving it as a political unit first and foremost:

> "(O)ur hospital is a center of national unity, a center of class unity, a center of clarification of ideas, a center of revolutionary and organisational propaganda, a combat unit."[12]

Several foreign workers in Frelimo's health service have also provided evidence of how parts of the service became extensions of Frelimo's ideology, with medical assistants, for example, being trained as political cadres and political sloganising preceding health visits[13]. As Machel correctly pointed out, the health service may well have been a point of contact between the state and people: at the same time it provided an obvious outlet for the political frustration of the masses. The *Independent's* Karl Maier, reporting from Mozambique, has also recorded Frelimo's military use of lorries from the National Disasters Office, the organisation charged with distributing emergency aid to famine-blighted areas[14]. Such cynical use of aid vehicles may well have been extended to ambulances.

The question which must be asked is whether a Nazi teacher, actively propagating a fascist system, or a village chief appointed by the pro-Soviet Kabul régime in Afghanistan could be deemed legitimate targets? In comparison with, for example, the massive and devastating fire-bombing of Dresden which killed over one hundred thousand German civilians, mainly women and children,

Renamo's selective elimination of Frelimo institutions and personnel, all representative of an equally hated and fascistic ideology, is surgical.

The infrastructure has also been severely damaged by extensive guerilla activity. Every single major or minor road and rail link in Mozambique has been cut at some point in the last ten years. Road traffic on all national highways is subject to military escort, and even that has not prevented constant attacks. A particular target has been the so-called Beira Corridor of road, rail and oil pipeline links which connects Zimbabwe to the vital port of Beira. Despite the commitment of an escalating number of Zimbabwean and other foreign military personnel to its security, this transport and fuel lifeline continues to be disrupted. Mozambique's power grid, particularly that supplied by the Cahora Bassa project, has been another priority target for the insurgents, who have, on several occasions, caused power blackouts of up to one week in Maputo itself. The *Guardian* summed up Renamo activity:

> "They are, in Mozambican terms relatively strong yet make no concerted effort to secure major military objectives. They emerge from the bush to attack vehicles, cut roads and railways, cut the occasional power line and then retreat."

Mike Saunders described the cutting of Beira's electricity supply in mid-1986, commenting on how a number of population centres such as Beira and Nampula were being 'partially strangled'

> "Yet the MNR have made no effort to secure a town, such as Chimoio which they effectively surround. Objectives like these are certainly within their capability..."[15]

Renamo's Political Programme

Contrary to standard Frelimo propaganda claims, repeated by numerous media outlets, that the Mozambique National Resistance 'lacks political direction', Renamo has a well-defined political manifesto. It consists of seven chapters which deal with politics, economy, justice, constitutional matters, health and education, public services and international policy, and was adopted by the National Council in August 1981. Renamo's political programme is considerably more advanced, coherent and unified than, for example, the various political and theological strands symbolising the uneasy Afghan Mujaheddin alliance.

The programme calls for the end to totalitarianism in Mozambique without vindictiveness, and guarantees "the people's right to choose and freely vote on the country's political, social and economic system." It envisages a Mozambique governed by respect for the "traditions, customs and personality" of the people. The programme includes a firm commitment to human rights and an independent judiciary. Free enterprise and foreign investment are defined as "Key elements of economic democracy". Jack Wheeler confirmed the fundemental importance of economic freedom to both Renamo and Mozambicans in general:

> "For the Renamo guerillas... to whom I talked, communism and socialism means collectivism: property and possessions owned in common; being treated not as an individual but as a member of a work or social unit. Capitalism is synonymous with individual freedom to them: freedom to own and use things individually, freedom to open your own store, freedom to work at a job you wanted (and not one the state tells you to work at), freedom to quit a job you don't like and find another.
>
> Capitalism also means political freedom to them – democracy. Who chose Samora Machel to be leader of Mozambique? ...He chose himself – he is a dictator, he gives orders, everyone else must obey. Renamo wants democracy, where people can chose the leader they want."[16]

In international affairs, the Mozambique National Resistance is committed to a policy of non-alignment, declaring respect for the principles of the United Nations, the Organisation of African Unity and the Geneva Conventions.

The immediate political goals of Renamo can be summed up as negotiating towards a Government of National Unity, pending the holding of internationally-supervised free and fair elections, the withdrawal of all foreign military forces from Mozambique and the integration of the Mozambican military forces. Renamo's calls for free elections have been consistent. The 1982 Renamo Congress, held near Chimoio, called for a ceasefire pending elections. Frelimo ignored the offer[17]. Renamo's solution to the Mozambican situation is, in essence, a mirror-image of the Lancaster House Rhodesian settlement: a ceasefire, the withdrawal of foreign troops, national reconciliation, and free and fair elections within a country which is governed by the rule of law and respect for human rights.

A key Renamo appeal for national reconciliation, peace and democracy in Mozambique was issued again in March 1988. The communique started by quoting Afonso Dhlakama; "It is time to stop talking with guns, we must now sit down and seriously start talking about peace in our war ravaged land." The communique went on to state:

"On numerous occasions in recent months, Frelimo President Joaquim Chissano has conceded that he cannot win the war. Initially, this admission was made to President Reagan in October and most recently at a Politburo meeting in Maputo this month...

Now is the golden opportunity for true nationalists within Frelimo and Renamo to set the stage for negotiations aimed at national unity and the establishment of peace in Mozambique. We must come together in the spirit of brotherhood, without recrimination, for our mutual benefit and protection.

Renamo's objective has been, and still is, to break the grip of foreign arms and ideology imposed on our people. Our goal is independence for Mozambique, within a democratic framework, with respect for human rights...

(W)e invite Frelimo to a new relationship of national unity and we are prepared to participate. On February 2nd of this year, our President Afonso Dhlakama, stated his willingness to talk with President Chissano...

Peace will not come through Portugese infiltration. Peace will not come through State Department political games. Peace will not come through more weapons. Peace will not come through multi-national business deals. Peace will not come through South African divide and rule. Peace will not come through Commonwealth consolidation. Peace will not come through Zimbabwe's mercenary invasion. Peace will not come through Soviet colonialism...

Peace will come through a respect for religious freedom, private property, human rights, and political tolerance...

Thousands of Mozambicans have died in 12 years of civil war, fueled and perpetuated by foreign interests. Thousands more will die unless Mozambicans come together now... (W)e pray (for) a resurrection of our land and the re-birth of independence."[18]

It is very clear and freely admitted by the organisation itself that the Mozambique National Resistance has a problem with the external projection of its political structure and programme.

This problem has several causes. The imposition of a totalitarian system on the Mozambican people destroyed the political élite of a newly-independent nation.

Because of the Portuguese denial of political ideals and institutions to Mozambique, this élite had in any case barely time to form itself. Portuguese policy had in effect left the future of the country in the hands of those prepared to use violence. Although even Renamo's military successes were based, at least in part, on its political liberation of the largely rural Mozambican masses from Frelimo diktat, a very low priority was placed on Renamo's external explanation and presentation of these accomplishments.

The assassination of Evo Fernandes, the architect of the Mozambique National Resistance's political programme and its international profile, in April 1988, seriously damaged the liberation movement's external projection at a time when a political solution was in sight. Renamo's international image has been a Frelimo target for several years. Factional infighting and the resultant tensions within the organisation, particularly following Orlando Christina's murder, have been exploited by the Maputo regime and external actors. Professor Thomashausen has documented the creation, encouraged by United States and West German sources, of several new organisations by Renamo dissidents. These splinter groups include *Comite para a Unaio Mocambicana* (Cunimo), *Partido Democratico de Libertacao de Mocambique* (Padelimo) and *Frente de Salvacao de Mocambique* (Fresamo): previously existing groups such as Maximo Dias' *Movimento Nacional Mocambicano* (Monamo) and Domingos Arouca's Fumo were similarly encouraged[19]. The existence of these splinter groupings, marginal to the struggle for freedom inside Mozambique, has served to further confuse the issue of Renamo's external political representation. The defection of Paulo Oliveira to Frelimo, a Renamo spokesman in Western Europe, has indicated possible Frelimo infiltration specifically aimed at damaging the movement's international work.

Perhaps the most distinct obstacle to Renamo's international image has been a very well-developed Frelimo propaganda and misinformation campaign. Frelimo's pro-Soviet standpoint, and its ten years of armed struggle against the Portugese, automatically connected the organisation – and subsequently Mozambique's Frelimo regime – with continental, regional and international organisations such as the Organisation of African Unity, the United Nations and the World Council of Churches, as well as with support and religious groupings sympathetic to its political standpoint.

In Britain, for example, several charities such as War on Want, Oxfam, Action Aid and Christian Aid, have all been actively assisting Frelimo in its international propaganda, blaming the Mozambique National Resistance for the famine, labelling the liberation movement as an unrepresentative 'bandit' grouping, and portraying Frelimo as a moderate, reformist Marxist government.

The Death of Samora Machel

The campaign launched after President Samora Machel's death in 1986 was a spin-off from, and very typical of, Frelimo's capacity for propaganda and disinformation. On 19 October 1986, a Soviet Tupolov airliner, which was carrying President Samora Machel and several aides, crashed just inside South Africa as Machel was travelling to Maputo after a visit to Zambia.

A report published on 2 July 1987, by an International Commission of Inquiry, (whose members included former Lord Chief Justice of Appeal Sir Edward Eveleigh of Britain, Mr Geoffrey Wilkinson, a former Chief Inspector of Accidents at the British Department of Transport and a former chairman of British European Airways, as well as Colonel Frank Borman, Apollo 8 Astronaut and former Eastern Airlines president and chief executive officer) found that the disaster was due primarily to crew error. Nevertheless, a well co-ordinated disinformation campaign tried to allege that Machel died because of South African action.

Ninety percent of all the reports on the Machel airplane crash originated from three agencies: the Mozambican News Agency *AIM*; the Zimbabwean government news agency *ZIANA* and the Soviet Unions's *APN/TASS*. Several suggestions were made in these reports: that South Africa arranged the crash to kill President Machel; that South Africa was a regional threat and that this was the start of a series of assassinations of Southern African leaders; that a South African mobile beacon had lured the Tupelov off-course; that the South Africans had mistreated survivors and blocked co-operation with Soviet and Mozambican investigation teams. Copy originated by these three agencies alleging, without any evidence, South African complicity in Machel's death was carried without reservation or comment by much of the Western media.

A similar exercise followed the Homoine massacre. On 18 July 1987, 386 men, women and children in this Mozambican coastal town were murdered by uniformed gunmen in an orgy of destruction which involved pregnant women being bayonetted and bed-ridden patients being killed in hospital. Responsibility for the atrocity has not yet been attributed. Suspicion has pointed to Zimbabwean units, or to a local Frelimo garrison. Within hours, however, the Frelimo information machine had categorically blamed Renamo for the incident, a claim which was carried internationally. A study of the media manipulation of the incident revealed that of the twenty-six accounts of the massacre which appeared in the United States media, seventeen were based solely on Frelimo sources such as Mozambique Television, Mozambican Radio or Frelimo's official news agency, *AIM*. Only one report was from the actual massacre site. Eight of the accounts claimed to have been independently confirmed, but none of the correspondents had visited the area. Renamo denials were limited to one line in most accounts: nine of the twenty-six articles even failed to carry the denial[20].

In any examination of pro-Frelimo propaganda, the recent United States State Department report on Mozambican refugees must be mentioned. Published in April 1988, the 43-page report, prepared by Robert Gersony, a consultant to the State Department's bureau for refugee programmes, alleged systematic violence by Renamo against civilians. Based on interviews with two hundred refugees and fifty relief workers, Gersony's report claims that one hundred thousand civilians have died as a result of this violence. The report was carried internationally within days, adding to the international isolation of Renamo. Its conception, methodology and findings were strongly challenged by Renamo and several conservative authorities. The author could not speak the language of the refugees, and used translators openly hostile to Renamo; all the refugees were interviewed in camps guarded by either Frelimo or Frelimo-allied soldiers, which was an automatic barrier to spontaneity. The figure of one hundred thousand deaths was extrapolated from some 80 interviews, the methodology of which was highly suspect. The contradictions of the report's findings conflict not only with the political/military strategy of the Mozambique National Resistance, a strategy designed to win and keep the support and backing of the rural population, but also run contrary to every independent account of Renamo's relationship with the Mozambican people.

Frelimo's disinformation campaign against the Mozambique National Resistance,which has involved several of the organisations and agencies mentioned above and often carried exactly as originated, has been very effective. From the basic attempts to delegitimise Renamo, by portraying them as South African-controlled and supplied 'armed bandits' without any political constituency or viewpoint, Frelimo propaganda has increased both in intensity and degree as the Mozambique National Resistance moves closer to victory. Allegations of involvement in several massacres of Mozambican civilians, such as those at Homoine and Manjacaze, and of ritual mutilation have even been augmented by bizarre claims that Renamo guerillas have eaten children[21].

'Pseudo-operations' by special Zimbabwean and Frelimo military units, together with wide-scale banditry, have provided a convenient backdrop for Frelimo propaganda efforts. The concept is an established ploy in propaganda and counter-insurgency. The Sandinistas, fighting a similar war against the Nicaraguan people, have, for example, special units specifically trained in the Soviet Union and East Germany for black propaganda and 'pseudo-operations'. Major Alvaro Baldizon, a senior Sandinista security officer who defected in 1985, documented their deployment. Dressed and armed as 'Contra' guerillas, these units attacked villages, killed peasants who showed themselves sympathetic to the anti-Sandinista cause, and committed 'Contra' atrocities, thus attempting to destroy local support for the Nicaraguan democratic resistance within the areas concerned and to provide the Sandinistas with excellent propaganda material. Their Soviet tutors have had a long involvement in such matters. The Katyn Wood massacre of several thousand Polish officers by the Soviet NKVD secret police in 1941 was blamed on the Nazis. The Soviets still have not officially accepted responsibility.

One clear example of a deliberate attempt to blame Renamo fighters for an atrocity they did not commit was the Inhamizua incident on the outskirts of Beira in late July 1988. Fifteen civilians were killed and seventy-nine injured when gunmen attacked a local festival. Frelimo automatically blamed Renamo guerillas for the carnage. Within days, however, both the Zimbabwean *Financial Gazette* and the BBC were able to reveal that the massacre was in fact carried out by a Frelimo security unit. The Frelimo soldiers had tried to gatecrash the festival only to be told that there was an admission charge. They returned with weapons and indiscriminantly opened fire on the civilians[22].

The presence of Zimbabwean units in Mozambique, commanded by former Rhodesian military personnel who are well versed in the very similar 'pseudo' tactics used during the Rhodesian war points to another obvious source of dirty war tactics in the international propaganda war against the Mozambique National Resistance. Other recent suggestions have been that Renamo forces are kidnapping and conscripting ten and twelve year-old children[23], force recruits to kill civilians, as well as using drugs[24].

Another tactic has been to cast the Renamo liberation movement as the Southern African equivalent of the *Khmer Rouge* in Kampuchea, a doctrinaire Marxist 'national liberation movement' in the 1970s which was responsible, through murder and starvation, for the destruction of one third of the Kampuchean nation. The contradiction in this claim is clear. President Samora Machel called Kampuchea, under *Khmer Rouge* control, a 'liberated zone of humanity'. Several facets of Frelimo's socialist experiment mirrored that of the *Khmer Rouge*, including famine, attempts at the destruction of religion, de-urbanisation and concentration camps.

The Spectre of Banditry

In a standard Marxist attempt to delegitimise its political and armed opposition, Frelimo and its apologists, have consistently described the Mozambique National Resistance's fighters as "armed bandits". That Renamo guerillas are a well-disciplined, legitimate military opposition to a repressive regime is evident. To put the issue into context, however, the existence of armed criminals, either serving or former Frelimo soldiers, and numerous acts of banditry (one of Mozambique's current problems) must be explained.

There can be no doubt that within the political and social-economic chaos that Mozambique is today, the phenomena of armed bandits and criminality has taken a firm root. It would be strange for this not to have been the case. In a country which has been racked by military and political conflict for twenty-four years, conditions which have produced thousands of men hardened by years of relentless war, an abundance of sophisticated weapons, a peasantry alienated from the central authorities and economic hardship, the emergence of an aggressive criminal element with access to weapons is not an unexpected occurrence. The situation has been deliberately muddied by the Frelimo authorities in an attempt to associate Renamo with this distinct and unrelated phenomena. That Renamo is not accountable for all the criminal acts it is accused of was conceded by Richard Dowden, the *Independent's* Africa Editor, who stated that "it is conceivable that these gangs have no direct links with the MNR. Some of those that have been caught and questioned seem not to have heard of the MNR."[25]

Banditry has had a long history and even tradition in Mozambique. Peasant opposition to authority is deeply rooted. Allen Isaacman in his *Social Banditry in Zimbabwe (Rhodesia) and Mozambique 1884-1907: An Expression of Early Peasant Protest*[26] documents just such activity in several areas of Mozambique including the Zambezi valley. The example of Mapondera is a case in point. A bandit leader who operated along portions of the Rhodesia-Mozambique border in the 1890's and early 1900s, Mapondera attacked and robbed numerous outposts of colonial authority and commerce, enjoying a certain support from the peasantry. Mapondera was captured and executed in 1904, but Isaacman records that oral traditions of his activities are strong to this day. His successor, Dambakuhamba, continued with attacks on Portugese outposts and businesses. Banditry was an activity which continued through to the 1950s and 1960s. Indeed, Alberto Chipande, whose Frelimo unit commenced the armed struggle against the Portugese authority in September 1964, mentions the presence of such bandits in Cabo Delgado province in the same month as his attack. These armed criminals had been involved in the murder of a local Dutch priest and implicated in the robbery and looting of several Indian shops in the area[27]. There is every reason to believe that just as such criminal elements carried out atrocities, robberies and murders in the mid-1960's, so similar elements may exist in the 1980's – and may if anything, with access to weaponry, have grown in number throughout the country.

'Social Banditry' aimed at attacking 'specific symbols of rural oppression', symbols which 40-50 years earlier would have been colonial authorities, outposts and overseers has found another, more contemporary outlet. Today's symbols of oppression are any manifestation of Frelimo. Renamo cannot be held accountable for banditry or crime in Mozambique. Indeed, it is Frelimo's creation of such socio-economic despondency and oppression which is responsible for the continuing criminality of this kind.

Frelimo's attempts to cloak all opposition and all hostile acts as banditry and the work of criminals, go hand in hand with the labelling of Afghan freedom fighters as 'bandits' by Frelimo's Soviet and Afghan Marxist allies. It is clear by its very definition that the term 'bandit' is merely a propaganda device, masking the legitimate armed struggle of the Mozambican people. Banditry implies a certain monetary or economic motivation. The blowing up of rail-links, bridges and oil pipelines, as well as the ambushing of Frelimo military convoys brings no profit; it is, if anything, a drain on valuable resources. It is clear that the motivation is political, not economic. Renamo's attitude towards bandits is clear; there have been several accounts of shoot-outs between guerillas and bandits, incidents which Frelimo propaganda have on occasion cleverly portrayed as inter-factional tension within the liberation movement.

It is clear that banditry has become a major factor. In the Rhodesian civil war, a conflict quite similar to Mozambique's, there was a strong element of armed banditry, the so-called pseudo-terrorists (not to be confused with Rhodesian pseudo-operations), who were nothing more than armed criminal elements exploiting the confusion and chaos of war. The same element continues to this day. The present Zimbabwean government draws a distinction between the activities of 'dissidents', with legitimate political grievance against President Mugabe's state, and bandits, often former guerillas, with criminal motives. President Mugabe's 1988 amnesty offer, for example, was expressly for political dissident fighters, not ordinary criminals.

No-one can deny the quite often horrific instances of banditry and crime documented over the last ten years in Mozambique. The involvement of Frelimo soldiers and personnel in several of these incidents is clear, confirmed even by Samora Machel himself. Machel's keynote November 1981 speech 'The Enemy Within' promised change and reform from this state banditry. In his speech he gave example after example of Frelimo banditry. The extent of this behaviour was nation-wide:

> "We received information and reports from every province, from every district, and from many individual citizens."

Machel candidly admitted that the Mozambican people had accused its Frelimo masters of:

> "(U)ndue and arbitrary arrests... as a means of settling personal quarrels; as revenge against people who lodge complaints about the abuses committed by these same infiltrators; to intimidate the population; to steal the belongings of detainees; to obtain the houses of detainees..."

The Mozambican head-of-state also revealed that Frelimo security forces had been guilty of:

> **"(R)ape by physical force or by threats with a firearm: rape of minors."**

He furthermore had evidence that Frelimo soldiers:

> **"...steal from the population especially in the following circumstances: armed assault; theft at road and border control points on the pretext of confiscating goods for the state; break-ins at houses around the barracks, with the thieves then hiding in the barracks itself."**

Faced with such extensively documented evidence of Frelimo banditry, Machel promised action which would:

> **"(P)urge the ranks of our defence and security forces. It will remove.. the kidnappers... the corrupt, the despots... the incompetent, the abusers, the thieves, the rapists of women and minors, the murderers, and those who would ride roughshod over other people."**[28]

Yet, despite this rhetoric, as late as June 1984, the ministers of security and internal affairs were dismissed because of the continuing abuse of civilians by Frelimo members and units. A month before these two ministers were dismissed, Machel publically castigated them, and a senior *FPLM* commander, Sebastiao Mabote, at a Maputo rally:

> **"Here is (Armando) Guebuza. He is the one who arrests many people and just leaves them there. And this is Mariano (Matsinhe). He detains many people and forgets them... Mabote's men create secret prisons. They detain people and don't know what to do with them."**[29]

It is very unlikely that if as forceful a personality as Samora Machel was unable to curtail Frelimo state terrorism and banditry, the chaos and the weakening of what little authority Frelimo had, which followed his death, has increased the indiscipline of Frelimo soldiers and militia.

The issue of who was responsible for banditry in Mozambique was also raised by *Guardian* reporter Mike Saunders. Commenting on an act of banditry, allegedly carried out by Renamo, Saunders also states that "the boot also fits neatly on the other foot". Frelimo conditions of service were poor:

> "There is no doubt that many of the Frelimo live periodically by banditry and the outsider cannot distinguish Frelimo from MNR. Both wear the same motley battledress, both speak the same language, both carry the same Russian weapons. Many incidents blamed on the MNR are certainly committed by Frelimo troops."[30]

The *Independent's* Karl Maier describes what may well be a typical day-to-day example of murderous Frelimo attitudes to peasants:

> "As a lorry loaded with 24 Frelimo soldiers singing and blowing whistles crashed through brush on the way to Vavate, one teenage trooper spotted two men walking 50 yards away. 'Stop the truck, I want to kill those mujibas (informers),' he begged his commander. The sergeant calmly shook his head. 'How do you know they are not ours?'"[31]

Although in this instance, which was perhaps tempered by the presence of an overseas journalist, two Mozambican peasants were not slaughtered out of hand, it is not too difficult to imagine that this callous frame of mind is ingrained.

The Nature and Composition of the 'Armed Bandits'

An interesting insight into the difficulty with which the Renamo phenomenon, and the Frelimo attempts to delegitimise national opposition, has been approached can be seen in exchanges such as those between Bertil Egero, author of *Mozambique: A Dream Undone The Political Economy and Democracy 1975-84* and Yussuf Adam, a Mozambican academic at the Eduardo Mondlane University's Centre for African Studies in Maputo. Egero's work, a mildly critical assessment of Frelimo in power, published by the Scandinavian Institute of African studies, was strongly criticised by Adam, an associate of the late Professor Aquino de Braganca, who was one of Samora Machel's mentors.[32]

Egero's belief that the conflict was essentially between Mozambicans and could be seen as a permanent state of affairs, was especially challenged by Adam. Yet even in Adam's criticism, the essentially nationalist nature of Renamo's struggle against the Maputo regime is reinforced. He mentions that despite the purges of nationalists within Frelimo in the 1960's, rationalised as removing 'contradictions' within Frelimo, some of the nationalists still remained. This point is taken up by Egero in a rejoinder to Yussuf Adam's review:

> "I find it surprising that Adam, himself underlining the links between today's enemies to Frelimo and the contradictions of the armed struggle tries to make a point of my simple assertion (that Mozambicans are fighting Mozambicans)."[33]

Bertil Egero, still a Frelimo apologist and by no means a Renamo supporter refers to them as 'armed bandits', and goes on to state quite correctly that despite purges throughout the 1960s, Mozambican black nationalist opposition to Frelimo's Marxism-Leninism and how its post-independence politics was still present:

> "The old enemies of Frelimo did not disappear overnight – nor will these disappear when the external sources of support dry up."[34]

Frelimo, in its propaganda campaign against the Mozambique National Resistance, has consistently denied the essentially Mozambique nationalist nature of the movement. This has been coupled with equally propagandistic allegations that Renamo is a South African creation, an extension of the South African political-military establishment. The inherent contradictions of this blinkered projection lead to exchanges such as those above and a constant underestimation of the movement's local, regional and national support. Yet on occasion Frelimo's propaganda guard was lowered. President Samora Machel's keynote 18 March 1980 speech 'We are Declaring War on the Enemy Within', which launched the internal offensive which destroyed Renamo's Sitatonga bases in June of that year, made it clear that the enemy was Mozambican:

> "We are here solemnly declaring war on the enemy within.
>
> On 25 September 1964 we declared war on the foreign enemy – Portugese colonialism. Here today, 18 March 1980, we declare war on the enemy within...
>
> This is a decisive battle. It is a battle in the class struggle."[35]

Frelimo's enemy was thus clearly defined as a black Mozambican national enemy. Machel also made it clear that there was no racial dimension to this essentially Mozambican conflict, ignoring the usual propagandistic projection of Renamo as an external South African racist manifestation:

> "The enemy is the same! Black, yellow or white, the enemy is the enemy and deserves the same treatment."[36]

That the resistance was internal was also conceded by Frelimo-apologist Joseph Hanlon:

> "(A)ll the fighters are Mozambican, and it is clear from their rapid spread that some peasants feed and at least tacitly support them."[37]

Chapter Seven

THE INTERNATIONALISATION OF THE MOZAMBIQUE CONFLICT

Mozambique's continuing post-independence political and military crisis has become increasingly internationalised. Frelimo's political orientation, and its acceptance by the Soviet Bloc as an approved national liberation movement, provided the régime with an 'internationalist' framework, resulting in immediate foreign political and military assistance from the Soviet Union, Warsaw Pact and other Soviet Bloc nations. The Mozambican civil war, and Frelimo's active involvement in the Rhodesian conflict which was soon to follow, saw an increasing international military commitment to keeping Frelimo in power; a military commitment which came initially from Eastern bloc states but which was augmented by growing military assistance from Tanzania, Zambia and Zimbabwe. Increasing tension between Mozambique and South Africa, before the 1984 Nkomati Accord, added a further dimension. Following Samora Machel's successful 1983 Western European tour, several Western nations, notably Britain, as well as the United States, extended assistance to the Maputo regime.

By far the largest involvement is that of Marxist Zimbabwe. An estimated 15 – 20,000 Zimbabwean soldiers are now committed to the maintenance of Frelimo government. From an initial one thousand men in November 1982, the Zimbabwean military effort has escalated massively. Tripled to 3,000 soldiers in 1984, the Zimbabwean commitment is now between one-third and one-half of the Zimbabwean National Army, with extensive support from Zimbabwean fighter-bombers and helicopter gunships – a war effort, according to the *Guardian*, costing $500,000 per day in 1986 alone[1]. The 1988 bill may well be one million dollars a day. Initially deployed to protect the strategic Beira-Zimbabwe transport corridor, a vital commercial lifeline for Zimbabwean and SADCC exports and imports, Zimbabwean military forces are now operational in several other areas of Mozambique, including Tete and Zambezia provinces in the north, and extensively in the southern third of the country.

It is now clear that the Zimbabweans are doing most of the fighting while Frelimo is based in heavily garrisoned urban centres, either too unreliable or ineffective for serious commitment against Renamo forces. Zimbabweans have ruthlessly spearheaded most of the strikes into Renamo liberated zones, but despite all their well planned offensives, training, aggression, combat experience and absolute air power they have never caught Renamo units within a major base. Units such as the Zimbabwean Sixth Brigade, commanded by officers such as Brigadier Lionel Dyke, notorious for their involvement in pseudo-operations during the Rhodesian civil war and for the wanton killings of thousands of civilians in Matabeleland in the early 1980s, have been deployed. Numerous atrocities have been committed against Mozambican civilians by these units. Brigadier Dyke's own description of the Sixth Brigade under his command as a "pack of wild dogs looking for a fight" certainly boded ill for Mozambican peasants on its deployment in Mozambique[2]. It is also clear that friction exists between the Frelimo government and military command and the Zimbabwean military, who operate either totally or semi-independently from Frelimo control. Some of these units have serious morale problems, factors which often contribute to atrocities and war crimes.

Zimbabwe's military effort in support of Frelimo in Mozambique is clearly a drain on its economy. Military spending accounted for some 270 million of Zimbabwe's 1987 1.8 billion national budget[3].

One consequence of the Zimbabwean military intervention in Mozambique has been the extension of the conflict into Zimbabwe. The Zimbabwean army's increasingly aggressive and atrocity-ridden operations among Mozambique's civilian population have provoked the Mozambique National Resistance into commencing hostilities along Zimbabwe's 1,200 kilometre border with Mozambique. The extent of Renamo co-operation with Zimbabwean political dissident supporters of Nbadaningi Sithole, whose political and ethnic base saddles the border, is unclear. *Radio Truth*, the Zimbabwean dissidents' broadcasting outlet, defined their opposition to Robert Mugabe's military commitment to Frelimo in a July 1986 broadcast:

> "The flood of refugees from Mozambique is caused principally by the fact that we in Zimbabwe are involved in a war which basically, has nothing to do with us. Our stay in Mozambique prolongs the struggle between Frelimo and Renamo and we stand accused of increasing the hardship, which causes the flood of refugees from Mozambique..."[4]

Increasing military activity in Zimbabwe's eastern highlands was one reason cited by the Zimbabwean government for extending the country's state of emergency legislation. In January 1988, the Zimbabwean defence minister admitted one hundred and one Renamo military operations inside Zimbabwe: the Mozambique National Resistance and foreign correspondents indicate that the official Zimbabwean figure was an under-statement. Military sources have confirmed that every district along Zimbabwe's border with Mozambique has seen Renamo activity. Several areas of eastern Zimbabwe, such as the Gona-re-Zhou national park, have been declared closed security zones. Leaflets left behind after Renamo activity inside Zimbabwe explain, at least in part, the reasons for such incursions:

> "...Nobody, today, ignores the war which exists in Mozambique particularly Zimbabwean people whose government is directly involved with an official presence of her troops in our country to help Frelimo... We... do not think that with the introduction of Zimbabwean troops in 1984 was simply a system adopted by Zimbabwe for security or safeguard of her own interest in this country.
>
> Zimbabwean attitudes in Mozambique are of destabilization. Zimbabwe is making the condition too tough for Mozambican people. Zimbabwe is feeding and worsening the war situation prevailing in Mozambique and she is continuing to cause a high cost of more human lives of civilian populations in Mozambique..."[5]

That Mozambican refugees in Zimbabwe have fled Frelimo excesses, and are mostly sympathetic to the Mozambique National Resistance was made clear by measures take by the Harare administration to control them. Christian Aid's pro-Frelimo book, *Mozambique: Caught in the Trap*, and several British newspapers, revealed that the Zimbabwean authorities have been rounding up thousands of Mozambican refugees and concentrating them for 'security reasons' in strictly regulated camps[6].

In addition to the heavy Zimbabwean commitment, between six and ten thousand Tanzanian soldiers are stationed in Mozambique, mainly in the north and north-west. Tanzanians tend not to be as aggressive as the Zimbabwean soldiers, preferring to keep to garrison bases and defensive perimeters. Peter Godwin, writing in the *Sunday Times* early in 1988, recorded Tanzanian unease at their involvement in the conflict. He visited a contested Tanzanian garrison at Mopeia in Zambezia province, where he found the officers, who requested anonymity, pessimistic about ever winning the war for Frelimo. He stated that eighty percent of Mozambique's population was behind the Mozambique National Resistance.

According to the International Institute for Strategic Studies, six hundred Cuban, eight hundred and fifty Soviet and five hundred East German security and military advisers are present in Mozambique[7]. Jorge da Costa, the former director of SNSAP secret police put the 1985 Soviet figures at between 4-5,000 men, including some one thousand East Germans, mainly security specialists, under the command of an East German secret police general. He also revealed the presence of about 4,000 Cuban, personnel, mainly military[8].

Western Involvement and Peace in Mozambique

In any realistic assessment of the Mozambican civil war, the role of Western investment and Western military involvement in prolonging and even intensifying a particularly bitter conflict cannot be understated. Western aid to the Maputo regime is artificially sustaining Frelimo in power, and has in effect taken over that role from the Soviet Union. This transfer of roles is in itself an indication of the Soviet Union's attitude towards Mozambique and the viability of the Frelimo regime.

In order to put Western economic and military assistance to Frelimo into context, it is essential first to examine and analyse the factors influencing Soviet reassessment of the Mozambique conflict, within both a national and international framework – a reassessment which has opened up the way for misguided opportunism by Western states.

Mozambique, Gorbachev and Glasnost

Mikhail Gorbachev's public re-orientation of Soviet goals has also seen a distinct re-analysis of Soviet foreign policy aims, objectives and realities – particularly in relation to the Soviet Union's, and, to a lesser extent, the Soviet bloc's extensive military and related economic commitments within the Third World. It can be speculated that the foreign policy manifestations of 'Glasnost' have included what can be described as, at the very least, a realistic look at the cost effectiveness of Soviet commitments and, at best, a process of either partial or total military disengagement from costly instances of Soviet 'internationalism'. These moves are arguably not because of any deep-rooted concern about world peace, human rights or sovereignty, but through sheer economic expediency and reality.

The Soviet Union has learnt that it is one thing to nurture a pro-Soviet 'national liberation' movement during armed struggles against the status quo, but the price of sustaining the movement once in power is quite another prospect. The cost of supplying small amounts of military hardware, logistical support and small-scale training at Soviet bloc military institutions, flying movement leaders around the world, and of expressing diplomatic solidarity at international conferences during the 'struggle' pales into financial insignificance when compared with the military and economic cost of keeping the unpopular Marxist-Leninist client state in power. Leaving aside the cost of sustaining the Warsaw Pact economically, itself a financial nightmare for any Soviet politician, Soviet subsidy of political interests and client states has escalated virtually out of control since the 1970s. It is this spending which frames recent Soviet attitudes to Mozambique. It is estimated that the Soviet Union is now committed to subsidising Cuba to the tune of six billion dollars annually; Vietnam receives almost two billion dollars per year, Nicaragua up to $600 million annually. The 1987 Soviet-Cuban offensive in Angola saw losses of $2 billion in equipment alone. Other Soviet commitments, for example in Ethiopia, add additional billions to the Soviet treasury. The war in Afghanistan, patently a diplomatic disaster, was, and to an extent still is, an additional economic drain. The key factor explaining what has become a seemingly open-ended escalation in Soviet support and subsidy to several of these client states has been the emergence of a popular, well-rooted anti-Marxist and anti-Soviet insurgencies within these nations, or as in the case of Cuba and Vietnam, the extensive use of their military personnel in an 'internationalist' role – 50,000 Cubans in Angola alone, 180,000 Vietnamese soldiers in Kampuchea. Military conflict of any sort is expensive – counter-insurgency especially so. It may well be possible that the economic bill for continuing to support the MPLA in its 13 year-long war against UNITA, the Addis Ababa regime's decades-long conflict with separatists, the Sandinistas' eight year war with Contra rebels, Frelimo's 13 year war against Renamo, Vietnam's occupation of Kampuchea – all conflicts which have the capacity for continuing and indeed escalating for years to come – has been assessed with a reality now associated with Gorbachev's 'Glasnost', a reality which assesses these commitments against the opportunity cost of transforming the Soviet economy.

The prospect of subsidising inconclusive and intensifying counter-insurgency wars supporting Marxist-Leninist client states against popular anti-Marxist liberation movements for decades to come has obviously been re-evaluated by the Soviets. President Mohammad Najibullah, leader of Afghanistan's pro-Soviet régime, undoubtebly articulated a reason for Gorbachev's new policy, a policy revealing at least a movement towards widespread Soviet military and economic disengagement, when he stated in early 1988:

> "Our experience has shown us that the military way of solving things is futile"[9]

This correct assessment has been accompanied by Soviet Union pressure on its various client states to at least start the process of trying to negotiate political settlements to their military conflicts. This has manifested itself in Nicaragua, Afghanistan, Kampuchea and, most recently, Angola – all situations in which Soviet military/economic aid was sufficiently crucial to secure a certain level, however reluctant, of compliance. Both President Najibullah and Daniel Ortega of Nicaragua were noticeably hostile to suggestions of negotiation, which would in itself legitimise the resistance in their respective states: there is little doubt that Soviet pressure was crucial.

That Mozambique falls into the Soviet analysis of eluding any possibility of a military solution is now self-evident. The Soviet Union's $475 million dollar military investment since 1975 had witnessed, if anything, an increase in anti-Frelimo insurgency[10]. Soviet bloc military involvement has been minimalised for some time, limited seemingly to 'care and maintenance' of the security of Maputo itself and one or two other political or military enclaves. Soviet bloc economic subsidy of the weakening Maputo government is at an all-time low.

The irony is that it is now Western states, having taken it upon themselves the task of maintaining the Frelimo administration in power, that are keeping the civil war going. The United States has supplied Frelimo, for example, with some US$ 300,000,000 since 1975, US$ 85,000,000 in 1987 alone. There has been United States military assistance since 1982, according to William Pascoe, a Heritage Foundation policy analyst, assistance – including military vehicles and weaponry[11]. Far from bringing peace to Mozambique, this military/economic involvement merely serves to exacerbate the conflict, prolonging the civil war and in so doing placing at risk the whole population of Mozambique, six million of whom are already grievously undernourished.

Caught up perhaps in their own propaganda, and now undoubtedly bolstered by unrealistic encouragement, military aid and investment from Western nations, Frelimo has been able to ignore the salient lessons of new realism in Soviet foreign policy. Namely, those pointing to the impossibility of a military solution to political problems such as those faced by similar Soviet client states – lessons in realism vigorously enforced on other pro-Soviet régimes through critical aid linkages. This temporary insulation is due entirely to the unprecedented level of Western support for Frelimo.

There can be no doubt that the Soviet Union is content to let Western military and economic support prop up the visibly weakening Frelimo regime. It is no doubt being dialectically evaluated on a par with Lenin's statement that Western capitalists will sell Marxist-Leninists the very rope with which they themselves will be hanged. In this instance, it is Western states who are literally selling, or in some cases giving away, to Frelimo the weapons with which to kill pro-Western freedom fighters and repress its rapidly starving population, while also propping up the state and the economy sufficiently to allow the regime the breathing space to do so. The London *Times* issued a clear warning about assistance to the Maputo regime:

> "(The) Frelimo government is still fighting a guerilla war against pro-Western Renamo movement; its leaders are still frequent and welcome guests in Moscow.
>
> The Soviet Union prefers not to pay for its satellites if it can avoid it. Britain ought not to take up the burden. Still less should it fool itself that by paying the Mozambique communists it can tempt the country away from Soviet influence. For Britain, and the EEC, to give economic and humanitarian aid to Mozambique is already a controversial matter. Military aid should be stopped entirely."[12]

Having found Western European nations unwilling to finance either the Red army's role in Afghanistan or the Sandinista régime's ideological war against the Nicaraguan people, the Soviet Union must be somewhat surprised that in the wake of their partial disengagement from Mozambique on grounds that it was a lost case, Western takers paradoxically should rush in where the Soviet Union fears to tread. Britain's involvement, as both a key military and economic supporter of Frelimo, must be challenged for the obstacle to peace and reconciliation in Mozambique that it is.

British Involvement in Mozambique

Britain's involvement in the Mozambican civil war is an aspect of the conflict which is inconsistent, paradoxical and directly harmful to prospects of peace and national reconciliation in Mozambique. In the form of extensive military, economic and diplomatic aid, British assistance has been instrumental in propping up the Maputo régime.

That this British help is crucial to Maputo's war against Renamo, was freely admitted by President Chissano in a press conference in Harare in 1987[13]. It is also significant that Chissano singled out the United Kingdom, among all other Western aid donors, as contributing specifically toward's Frelimo's defence.

The tip of the iceberg of British involvement is the work undertaken by the British Military Advisory and Training Team in eastern Zimbabwe. The official British military training assistance to Frelimo started in February 1986 as an instruction programme for Frelimo officers. In 1987, the course changed over to training integrated companies of over one hundred men each, for subsequent deployment as elite units inside Mozambique. The course of instruction lasts sixteen weeks and costs the British taxpayer some £230,000 per company, funding which is guaranteed until at least March 1989, and is to be extended[14]. This training is in addition to officer training for Frelimo nominees at Sandhurst and to overt military equipment supplies, such as the latest Enfield assault rifle and rifle sights. Once trained, British-instructed units are normally stationed in strategic positions along either the Beira or Limpopo rail corridors.

A Renamo communique in May 1988 recorded the destruction of one of these British trained units. In an attack on 15 May, on a combined Frelimo/Zimbabwean garrison at Macossa in Manica province, 80 kilometres east of the Zimbabwean border, 23 Frelimo soldiers and 15 Zimbabwean servicemen were killed by Renamo guerillas. Eight more British-trained Frelimo troops were captured. The Frelimo commander, a graduate of the first BMATT training course in Zimbabwe, was amongst the dead. Three guerillas died in the attack. A report by Karl Maier on the training of Frelimo servicemen, headlined 'Small army, large country', which appeared in *Pretoria News* in June 1988, summed up the ideological contradictions in Britain's military involvement with the Maputo régime with its sub-heading, "British-trained 'green berets' are on the same side as Russian-trained 'red berets' in Mozambique."[15]

In addition to the open British military training schemes inside Zimbabwe, covert British assistance can also be seen in, for example, the sanctioning of a Foreign Office-approved private security firm, Defence Systems Limited, operating within Mozambique, with fifty former British special forces military personnel deployed to defend and police a 'cordon sanitaire' through to the coast, ostensibly in co-operation with the multinational Lonrho concern. Several British personnel have already been killed in action with Renamo guerillas. The further involvement of the Foreign Office approved Hall and Watts weapons and equipment procurement company, acting in support of the DSL operation, adds another level of British connections. Exactly who is paying for the operation is unclear, but with the intention of ultimately deploying one hundred personnel, the financial commitment is immense. The British government is one possible source other than the ostensible Lonrho link. Defence Systems Ltd's exact mandate is still unclear. The Paris based *Lettre D'Afrique*, citing political-diplomatic sources in Britain, speculated that British ex-SAS personnel may have been involved in the assassination of Renamo's former secretary-general Evo Fernandes[16].

The Foreign Office have obviously realised that British military forces' casualties, however minor, would be unacceptable to the British public, however much junior foreign minister Lynda Chalker may be supportive of direct British intervention. The use of technically 'privatised' security personnel for much the same ends is a cynical extension of what may well be covert British foreign policy.

A second area of covert British military assistance to Frelimo has been through aid to Zimbabwe. British aid, military or economic, directly or indirectly helps to sustain the Zimbabwean military effort within Mozambique. Aimed originally at securing the Beira Corridor, this effort soon extended to several other heavily-contested areas and provinces. With perhaps as many as twenty thousand Zimbabwe National Army personnel on active service within Mozambique, just under one half of the Zimbabwean military could now be committed. British military training, hardware and other logistical support for Zimbabwe is used and deployed within Mozambique as a matter of course.

There are those who argue that British and other Western economic aid can be divorced from the military conflict in Mozambique. This thesis falls short of the reality of any war. External economic assistance automatically frees national, and other external resources, for military uses. A gift of several civilian vehicles, for example, frees an equivalent number of vehicles already in the country for military duties. Foreign investment in an essential agricultural project frees local resources which are invariably used to prop up the Frelimo régime. There has also been a careful attempt by Frelimo to draw Western economic investors into a military involvement. This policy was made very clear by the Governor of the Bank of Mozambique, Dr Eneas da Conceicao Comiche, in the course of an interview while attending the 1987 Vancouver Commonwealth Conference as an observer:

> "The Governor sees investment as one way of fighting the war against Renamo... 'We select projects in areas which are a priority for defence. The defence of a country and economic issues are related'".[17]

It is very clear, therefore, that Frelimo is consciously endangering Western investment projects, deliberately placing the lives of both Western specialists and local workers at risk in order to internationalise their war against the Mozambican people.

That this underhand mechanism has been at least partially successful is evident. Britain has already stated that investment in Mozambique must be 'protected'. Five per cent of the $140 million pledged by the European Community, under Lome III in November 1987, is to be defence related. Under the arrangements for a Spanish agricultural investment project, a Frelimo security unit is to be trained in Spain. Mozambican transport minister Lieutenant-General Armando Guebuza has also stated that Portugal and Italy will be implementing similar programmes. Even Nordic nations have declared support for investment and defence linkage.

While there is little which can effectively be done against Western multinationals investing where they see fit, even if it is to perpetuate a bloody conflict, the role of Western European governments, and particularly the British government, in encouraging, and even bankrolling a latterday Nero régime while Mozambique burns and starves, must be challenged. It is shortsighted, immoral and ideologically bankrupt. The one billion dollars of Western investment has replaced, and actually doubled, the Soviet subsidy[18]. In so doing it has temporarily insulated the Frelimo régime from the reality of the Mozambican situation, a reality that dictates a political solution to the conflict. The irony is that without the extensive Western investment Frelimo would, in all probability, already be negotiating with Renamo about the future of Mozambique. Western investment is merely buying the regime time.

The official reasons given for British involvement in Mozambique are two-fold. First, it is in appreciation for Machel's assistance in securing the Lancaster House Rhodesian settlement. It is a further indication of the flawed Foreign Office logic dictating British foreign policy that Machel is to be thanked for helping to transform the authoritarian multi-racial Western administration into a self-admitted Marxist-Leninist one-party state which is responsible for racial and political oppression in Matabeleland and which constantly flouts the rule of law. The second motive cited is that Frelimo can somehow be weaned away from the Soviet Union and Marxist politics by large-scale British and Western economic and military assistance. Richard Grenier gave his opinion on the possibility of such an ideological change in the *Sunday Telegraph*, following a visit to Mozambique:

> "...I would think not. No more than Lenin was enticed from Communism when he introduced his New Economic Policy in 1921 or the Sandinistas by $75 million American aid in 1979."[19]

The Marquis of Salisbury, whose family has had a long-standing connection with Southern Africa, and whose son Lord Michael Cecil has travelled extensively throughout Mozambique has expressed similar reservations:

> **"The suggestion that a communist dictator will respond to a nice, friendly approach, especially if it is accompanied by large hand-outs without strings at the British taxpayer's expense... is nonsense. There is no precedent for this belief. Indeed, I suggest that a communist dictator will regard this as a sign of weakness in the same way that Hitler did when Mr Chamberlain visited him in 1938.**
>
> **It is more likely that the Frelimo Government will accept the hand-outs and military assistance and continue within the communist fold providing whatever facilities its Marxist allies demand."**[20]

It is a matter of record that hundreds of millions of dollars of Western economic and military aid to Zimbabwe merely helped to 'wean' ZANU off a multi-party democracy and into an openly Marxist one-party state.

The British Foreign Office's approach to the complex Mozambique issue is in direct contradiction to both current and previous foreign policy guidelines on other Southern African situations. Throughout the Rhodesian civil war, the Foreign Office consistently called for negotiations between the Salisbury administration, the Patriotic Front of Joshua Nkomo's *Zimbabwe African People's Union* and the *Zimbabwe African National Union* led by Robert Mugabe. Why will the Foreign Office not adopt the same approach to the Mozambican conflict? Why too did the British Foreign Office persistently call for free and fair elections in Rhodesia, while it ignores calls by Mozambican democrats for such elections to achieve legitimacy and reconciliation within Mozambique? Why are free and fair elections acceptable for Zimbabwe but not for Mozambique?

There exists an insurgency, a people's war, considerably more advanced than that of ZIPRA/ZANLA even at the height of the Rhodesian conflict – Frelimo only controlling some fifteen per cent of the Mozambican state, and even that with extensive foreign military manpower and assistance. If the basis of the Patriotic Front's representation at the Lancaster House constitutional talks ending the Rhodesian war was the strength and success of their military forces, why does the Foreign Office not concede, by its own criteria, that the Mozambique National Resistance, is a legitimate force without whose inclusion there can be no settlement of Mozambique's national crisis.

If the argument is that Renamo is beyond the pale because of allegations of atrocities, then it also delegitimises the Frelimo regime, which has been implicated in dozens of atrocities and war crimes – catalogued at least in part by Samora Machel himself. It is also on record that despite extensive and well-documented ZIPRA and ZANLA atrocities, including the shooting down of two civilian airliners, followed in one instance by the bayonetting of unarmed and injured survivors, and numerous cold-blooded massacres of black civilians, the British Foreign Office bent over backwards to accommodate these two Zimbabwe nationalist organisations, even hosting the December 1979 Lancaster House conference. If the argument then advanced is that the international status-quo is hostile to Renamo's inclusion in a peace process, then we have the Foreign Office falling back on an argument which let Nazi Germany assert itself with impunity in the 1930s – appeasement. We also have yet one more example of how the Left still dominates and dictates the foreign affairs agenda, despite eight years of the Thatcher Revolution.

In a more contemporary example of double-standards, the Foreign Office continues to call for negotiations between the South African government and the African National Congress terrorist organisation. The ANC's involvement in indiscriminate bombings and landmining, identification with the horrific 'necklacing' murders involing the burning alive of hundreds of black ANC opponents, have not stopped Sir Geoffrey Howe or his deputy Lynda Chalker, meeting African National Congress leaders in an effort to achieve peace and reconciliation in South Africa. Why can it not exercise a similar realism with regard to the Mozambique National Resistance, a movement which excercises considerably more authority and which enjoys more support in Mozambique than the ANC in South Africa. A movement, moreover, which identifies itself with Western political values, unlike the pro-Soviet orientation of the ANC, in alliance as it is with the South African Communist Party.

An additional contradiction in British government policy can be noted in its moral opposition to sanctions on the one hand, and a commitment to building up the economic and transport infrastructure of the 'Front-line states' on the other. The very essence of the latter project, building up the 'Southern African Development Coordination Conference' (SADCC), is to heighten the effect of international sanctions upon South Africa. Formed in April 1980, SADCC comprises eight regional governments (Zimbabwe, Zambia, Malawi, Botswana, Lesotho, Swaziland, Mozambique and Angola) who are concerned with advancing economic development through economic co-ordination, and reducing dependence on South Africa. Transport and communications were identified as the most urgent SADCC priority. The first SADCC conference was held in November 1980, in Maputo, and gave an outline of transport rehabilitation schemes asking for some $1,000 million from international donors. Between a third and half of the funding was raised and the SADCC transport and communications secretariat was established in Maputo. It is evident that there was as much of a political as an economic motivation behind the SADCC communications upgrading projects which aimed at aiding the imposition of sanctions on South Africa

Dr Simba Makoni, SADCC executive director, interviewed in the SADCC journal, *The Southern African Economist*, made this point clear:

> "Question: What is SADCC's position on economic sanctions against South Africa?
>
> Dr Makoni: Although we recognise that individual member states may not be in a position to impose sanctions, SADCC supports the imposition of sanctions against South Africa, by those within the organisation and outside who are in a position to do so."[21]

The United Kingdom contributes tens of millions of pounds directly to the Southern African Development Co-ordination Conference.

A prime example of this contradiction in action is the British investment in the SADCC upgrading of the Limpopo rail-link from Maputo to Zimbabwe. This 534-kilometre line, potentially more strategic than the Beira corridor, has been seriously neglected over the years, as a result of Frelimo incompetence and the civil war. The British Overseas Development Administration (ODA) and the Frelimo régime are jointly funding the re-laying of a 62 kilometre stretch of this line. As part of this project, the British taxpayer has allocated fourteen million pounds to finance a British management and consultancy firm, Mott, Hay and Anderson, to supervise technical work on the Limpopo line.

This policy of hostility to sanctions coupled with the tightening of the Frontline States attempts to vigorously enforce them, has been quite correctly criticised by Labour leader Neil Kinnock as "shopping on both sides of the street". In the cold light of day, it would seem to be a typical Foreign Office compromise, an attempt to play both sides that could have come straight from a script of the *Yes, Prime Minister* television satire. In its arrogant smugness, however, the game these Whitehall bureaucrats are playing is killing Mozambicans.

A Conservative British Government is keeping an unpopular pro-Soviet Marxist-Leninist régime in power through both economic subsidy and military assistance against a pro-Western national resistance movement (in so doing taking over from the Soviet Union), as well as being directly involved in strengthening the pro-sanctions infrastructure in Southern Africa. It is, in short, something one would expect from a radical Labour government, and not from a doctrinaire New Right government.

The South African Connection

The central argument used by Frelimo, Frelimo apologists and the pro-Maputo propaganda machine in attempting to delegitimise the Mozambique National Resistance as a political actor within the Mozambican situation is that of its South African connection. The argument is that military or economic assistance to Renamo, or allegations thereof, somehow disqualifies the organisation from Mozambique's political spectrum. Any application of these criteria, as dictated by Frelimo, could well delegitimise the Maputo regime.

It is a matter of record that from early 1980 until the signing of the Nkomati Accord in 1984, there was a limited amount of officially sanctioned South African military assistance to the Mozambique National Resistance. This help took the form of arms supplies, training both in and outside Mozambique for selected Renamo personnel, as well as intelligence and communications assistance. South African aid in this period occurred against the backdrop of massive Frelimo military and logistical assistance to the African National Congress, an organisation openly committed to the violent overthrow of the South African government. It is also clear that elements of the South African military, in the lead-up to the Nkomati accord, ensured a pre-emptive resupply of several months material to Renamo. After 1984, apart from limited intelligence and communications help, again probably officially unsanctioned, South African assistance ended. That this aid, if any, is unauthorised is the view of Eddie Cross, a Zimbabwean business expert on Mozambique's security problem. As chief executive of the powerful Beira Corridor Group (a consortium of SADCC-related business and state interests), Cross has access to the latest Zimbabwean intelligence and therefore has unparalleled insight into the Mozambican conflict. Propagandistic Frelimo claims nevertheless persist in projecting an image of Renamo being wholly reliant on South African logistical support, command and communications.

Despite these claims of massive South African logistical support, and apart from one or two shreds of American-made parachutes, Frelimo has not produced any tangible evidence to back up its allegations. The Frelimo propaganda projection of Renamo dictates a massive constant air and sea reinforcement of Renamo's twenty thousand guerillas, a logistical accomplishment on a par with a mini-D-Day, of which there would have to be ample evidence. Typical pro-Frelimo propaganda claims of extensive South African aid were contained in *Dumba Nengue: Run For Your Life*, a collection of 'peasant tales of tragedy in Mozambique', edited by long-time Frelimo activist Lina Magaia, and published in the United States[22]. The introduction cites Frelimo military, claims to have fought against and captured South African military personnel in assaults on Renamo bases. Victoria Brittain in her *Hidden Lives, Hidden Deaths: South Africa's Crippling of a continent* is also liberal in claims of extensive South African support for the Mozambique National Resistance, even alleging a direct military presence[23]. Again and again, however, no such evidence has ever been presented to the Western media. Even *Guardian* journalist Joseph Hanlon has admitted that Frelimo "has produced very little direct evidence of the South African connection"[24]. Despite several years of consistent Frelimo claims of large-scale South African support, no proof has been forthcoming, much to the dissatisfaction of Western diplomats, eager to support Maputo's claims. A *Daily Telegraph* report, in early 1988, cited a Western diplomat: "We have told them, give us the evidence, and we will confront Pretoria with it."

It is also significant that when key Renamo official Paulo Oliveira (who served as Renamo's Western European representative from 1984-7) defected to Frelimo in March 1988, all he could reveal about South African logistical support to Renamo was that they had provided Renamo's Lisbon and West German offices with a telefax machine and an encoder/decoder device in June 1987[25]. French journalist Alain Louyot, reporting in the French magazine *L'Express* in June 1988, not only stated that ninety-five percent of Mozambique was now contested, but admitted that Frelimo had been unable to show a single weapon over the last two years, which it could claim to have been supplied by South Africa. Hostile to Renamo, Louyot nonetheless points out that Renamo's weapons are captured from Frelimo and Zimbabwean military forces[26].

Mrs Lynda Chalker, the deputy British Foreign Minister, and a vociferous critic of both South Africa and the Mozambique National Resistance, visited several Southern African 'Frontline states' towards the end of 1987. Even after extensive discussions regarding Renamo, in Maputo, with the Frelimo government, she had to admit that she had no proof of a 'direct link' between South Africa and Renamo. This did not prevent her, however, from adding: "I very much hope the South African government will stop all their cross-border activities"[27]. If a British Foreign Office minister, extremely sympathetic to Frelimo, with all the intelligence and information sources that her office and government would have at their disposal, cannot provide any evidence of a South African connection, the question must be posed as to whether any exists at all.

It is apparent, therefore, that for all the allegations of extensive South African support for, and involvement with Renamo, little or no evidence has ever been produced. What there is evidence of, however, is a massive, well-documented economic and increasingly military relationship between the Maputo régime and the South African government. The South African and Frelimo transport ministers, for example, signed a 10-point agreement in August 1984 to secure extensive South African assistance in redeveloping and servicing Maputo's port and rail system. Pretoria made ten million rand available to the Frelimo government to assist this project. The South African government also provides large numbers of technicians and specialists who repair and service Mozambican rail locomotives, rolling stock, and the national airline's aircraft, as well as managerial staff to supervise and advise on Mozambique's transport services.

The exact extent of this South African assistance is seldom revealed. Eddie Cross, whose Beira Corridor Group is intimately concerned with, and dependent on, the Mozambican infrastructure, provided an insight during an interview with *Die Suid-Afrikaan* in late 1987:

> "Renamo took out a big bridge – three spans, just outside Maputo, on the Nkomati road – on the eighth of this month; the second major bridge we've lost the last month. The SATS (South African Transport Services) threw everything they had at it. South African troops went in, swept the area and protected the workmen. SATS worked round

the clock, with floodlights, everything, and fixed the bridge in ten days."[28]

Eddie Cross also revealed that South Africa supplies all Maputo's power, one hundred megawatts[29] and one estimate attributes at least a quarter of the country's national power supplies to South African sources[30]. Frelimo and Pretoria are co-operating in attempts to rehabilitate the massive Cahora Bassa hydro-electric scheme. Power supplies from the dam on the Zambezi River ended in late 1983, following repeated Renamo sabotage of power pylons. Agreements between the two governments saw the creation of a joint South African/Frelimo security scheme to protect the repaired and reactivated power lines[31]. Pretoria has committed itself to providing the Maputo régime with $15 million worth of military equipment to help train three battalions of Frelimo soldiers. An '*Independent*' report of May 1988 also reveals that South African military personnel, in addition to providing anti-Renamo security for transport repair teams in Maputo province, may be involved in training other Frelimo personnel in conjunction with the private British Defence Systems Ltd security firm[32].

It is clear, therefore, that overwhelming evidence of South African military, civil and economic involvement in Mozambique exists. The Frelimo régime has signed several treaties and agreements binding them to South Africa. Pretoria runs Maputo's railways, ports and communication system, supplies the Frelimo administration with logistical backup, food, medical treatment and medical supplies, and spare machinery and engineering parts. South Africa also supplies Frelimo's capital with its electricity, and is actively committed to repairing and protecting the Cahora Bassa power scheme and pylons. Furthermore, Pretoria is providing Frelimo with military and security assistance, which includes costly training in the régime's war against Renamo. Pretoria probably spends more on subsidising Frelimo in one year than it has ever spent on Renamo.

All in all, Frelimo's documented dependence on South African assistance should be enough to horrify any anti-apartheid and pro-Frelimo activist. Frelimo apologists, however, explain away the Maputo régime's acceptance of such military and logistical assistance, and the signing of treaties and agreements as sheer political necessity and geo-political realpolitik. Surely then, cannot the Mozambique National Resistance advance essentially the same reasoning for having accepted South African assistance prior to 1984, or indeed if Frelimo claims were proved correct, for continuing to accept their aid? Indeed, if anything, Renamo's options were considerably fewer: unlike Frelimo, they could not expect $475 million in Soviet bloc military assistance alone.

Another specific example cited by Frelimo of South African 'sponsorship' of Renamo, which was revealed as a result of the 'Gorongosa documents' captured during a combined Frelimo/Zimbabwean assault on the movement's headquarters in September 1985, was visits to Renamo's Gorongosa headquarters by South African civil and military officials. It is a matter of record, confirmed by South African foreign minister Pik Botha at a subsequent press conference, that Louis Nel, his deputy foreign minister, flew into the 'Casa Banana' headquarters on 8 June 1985 for discussions with Afonso Dhlakama. It should also be noted that Louis Nel was also involved in several meetings with Frelimo leaders, both in Maputo and South Africa. The fact is that it would be very strange indeed if South Africa, as the regional Superpower, were not to keep channels of communication open to the two major actors in the Mozambican situation, a situation with a direct economic and security bearing on South Africa. As previously mentioned, far from documenting extensive South African military and logistical help to Renamo, all that key Renamo defector Paulo Oliveira could cite as South-African supplied equipment was two telefaxes and other minor communications material for two offices in Western Europe – lines of communication which would prove invaluable in negotiating an end to the conflict in Mozambique.

Whatever Frelimo and its supporters would like to believe, any hope of peace and national reconciliation in Mozambique would need the political involvement of the Mozambique National Resistance. Frelimo has also conceded as crucial the South African government's involvement in any resolution of the South African civil war. Pretoria, for example, hosted the sadly abortive peace negotiations held in late September/early October 1984. During those talks, the Mozambican régime invited the South African government to consider playing a role in implementing any peace accords[33]. There is no doubt that South Africa would have a crucial role to play in any settlement. To that end, as a possible convenor of, or mediator in, any peace talks, it is essential for channels of communication, however covert, to remain open.

Several Western European states, including Britain, maintain covert political and intelligence links with Afghan guerilla leaders: such connections do not imply political or military command and control of the Mujahiddin forces. Several prominent British politicians have also, to take another example, visited the Lusaka Headquarters of the African National Congress. As the former political authority in South Africa, the United Kingdom does take a close interest in political developments in the country, maintaining a relationship with both the government and with dissident elements such as the ANC. Paulo Oliviera, at his press conference, also mentioned extensive Portuguese government and military interest in Mozambique. Renamo's first external office, after all, was opened in Portugal. The Portuguese government would not have tolerated this unless it too perceived the vital need to keep channels of communication open. There have also been reports of at least one senior Portuguese military officer visiting Renamo's 'Casa Banana' headquarters. It is against this backdrop, of maintaining essential communications links with the Mozambique National Resistance, that such South African contacts that exist must be seen. If these visits are deemed proof of South African sponsorship of Renamo, then fingers could also be pointed at several other governments, particularly the Portuguese.

In conclusion, it can clearly be said that the case for claiming continuing, large-scale support from South Africa for the Mozambique National Resistance is unproven. Any support which may have been supplied before the Nkomati Accord must itself be measured against South African intentions. Even against the backdrop of the legitimacy of assisting people in their attempts to gain their freedom from a Marxist state, South African assistance pre-Accord was in all probability within the context of political expediency, at the whim of politicians concerned with the progress of South African foreign policy. Military aid may well have been extended to Renamo in its opposition to the Frelimo régime because South African and Renamo objectives overlapped at the time. Just how expendable the Mozambican freedom fighters were, was graphically manifested in the 1984 Nkomati Accord, when at the whim of South African diplomats and politicians, and without consultation or notice, military assistance to the liberation movement was ended. Politicians had deemed it more expedient to play the Frelimo card, enticing Maputo into a spider's web of regional, economic and military connections with Pretoria.

It is also clear that South Africa is quite reluctant to see Renamo militarily seizing power in Mozambique. A unilateral seizure of power would merely initiate a new phase in the civil war, one in which Mozambique is well and truly destabilised, Frelimo goes back to the bush as insurgents, and, South Africa has a Lebanon on her borders. Zimbabwean and Tanzanian invasion, supported by Soviet/Cuban combat troops, could not be ruled out in such a scenario. Such a Renamo seizure of power would, therefore, inevitably lead to some level of South African military intervention in Mozambique. Any such military intervention would risk being ground down in a war of attrition, and any significant level of South African military casualties, even with immediate, identifiable objectives would fast become politically unacceptable to Pretoria. The lessons of the Israelis' 1985 military intervention in the Lebanon are clear: a medium or long term military presence, without clear goals and fuelled by casualties is bad for political direction and morale. Israel's 1985 campaign in Lebanon virtually split the nation in half, leading to the creation of the 'Peace Now' movement and an unprecedented schism in Israeli society. For this reason alone, if nothing else, significant South Africa sponsorship of the Mozambique National Resistance is unlikely.

Mozambique and Destabilisation

Claims of destabilisation are not limited to the Mozambique issue, but are a common theme dominating Southern African politics in general. The South African government has been accused of co-ordinating a policy of military, political and economic destabilisation of several Southern African states hostile to the Pretoria régime. Methods cited have included direct military incursions, and the destruction of Southern Africa's transport infrastructure, especially when road and rail links are designed to lessen reliance on South African commercial routing as well as external assistance to right-wing guerilla movements struggling against Marxist-Leninist governments. There is undoubtedly a considerable level of destabilisation at play within Southern Africa. Pretoria's responsibility for some of this destabilisation is clear. At the same time, however, several other states are equally guilty of military and political destabilisation in the region. The MPLA government in Angola, for example, actively supports and harbours SWAPO insurgents destabilising Namibia, terrorists whose targets included Namibia's civilian population and infrastructure. Zimbabwe's military intervention in Mozambique continues to deny that country the stability that a peaceful settlement would bring. Both Zimbabwe and Mozambique's governments have plotted the destabilisation of Malawian society. The Republic of South Africa, nonetheless, seems to draw most of the fire regarding destabilisation, particularly with regard to Mozambique. The whole issue of destabilisation is an area which is worthy of study.

An Analogy of Destabilisation

Joseph Hanlon, in discussing destabilisation in Mozambique as allegedly orchestrated by South Africa, uses the following analogy to illustrate his argument:

> "For example, suppose a foreign power trained Welsh guerillas, aeroplanes dropped arms and other supplies into the Welsh mountains, submarines came close to land instructors and pick up men for training, commandos came ashore and blew up British Rail bridges, and a pirate radio station broadcasting in Welsh for an uprising. All this and more has been done in Mozambique."[34]

In outlining this analogy, Hanlon lays his destabilisation strategy open to immediate challenge.

A Case for Destabilisation

What if the state which these hypothetical Welsh insurgents were trying to overthrow was as repressive as the Frelimo régime in Mozambique? What if this state, let us call it the Hanlon British State, had embraced Marxism or Fascism, had unilaterally declared Britain to be a Marxist/Fascist one-party state, had physically destroyed all political opposition, and had created a system of concentration camps for such opponents, and any other anti-totalitarian elements left? What if this targeted state had presided over a forced implementation of doctrinaire Marxist economic ideology which led, as it inevitably does, to massive economic dislocation, food shortages, starvation and, in some areas, open famine? What if this hypothetical Hanlon state had introduced pass laws forcing Britons to carry not one but four identity cards with the penalty of immediate detention, interrogation, incarceration or even death for failure to comply? And what if this state rounded up all the urban unemployed, sick people and thousands of single women, termed them 'marginals' and sent them to forced labour camps in the north of England? All this and much, much more has been done by Frelimo in Mozambique.

Within this context, Hanlon's hypothetical Welsh insurgents, unless they themselves were Marxist or Fascist, seeking to implement a similar totalitarian system, would certainly qualify as freedom fighters, and those external powers supporting them would also qualify as helping the fight for freedom. Had Britain been occupied, for example, by National Socialist Germany in the Second World War, any non-totalitarian state externally assisting anti-Nazi insurgents, be they Welsh or otherwise, would have been party to a just cause. Similarly, to use another hypothetical example, if the National Front ever gained power in Britain and set about creating a fascist one-party state, based, as it must be, on terror and death camps, any external support for democratic guerillas undermining that state would be welcome. Whether the totalitarianism in Britain was externally or internally imposed would be irrelevant to the argument.

Theoretically, any alleged South African involvement, therefore, in supporting, equipping or training Renamo guerillas in the Mozambican peoples' war against its unilaterally imposed, unelected Marxist-Leninist régime should be supported by those people whose values are those of the free and open society. While it could be termed a reformist authoritarian state, South Africa cannot be labelled as a fascist system, and Renamo has made its commitment to a democratic, pro-Western future for a liberated Mozambique quite clear. The destruction or dislocation of the Frelimo régime would see its replacement with an inevitably more free, genuinely black nationalist Renamo administration.

Far from criticising accounts of the possible destabilisation of a repressive totalitarian régime such as that of Frelimo in Mozambique, democrats and others who seek to maximise freedom should actively support such measures. There is no example ever of a Marxist or fascist régime voluntarily relinquishing power. In a direct parallel with external involvement in Mozambique today,

Britain, after all, waged a very consistent destabilisation policy against the Nazi and pro-Nazi régimes in Europe during the Second World War. British aeroplanes, submarines, commandos and radio propaganda broadcasts actively supported resistance fighters in countries such as France, Belgium, Norway, Denmark and the Netherlands throughout the conflict. And just as the various French, Belgian and Dutch resistance fighters were not 'puppets' of Britain, but rather for the most part anti-totalitarian, nationalist democrats, Renamo is similarly far from being a South African 'puppet' creation. Renamo can be seen as a reaction to a vicious totalitarianism, just as the French maquis were in the Second World War.

Western Destabilisation in Afghanistan

A parallel between Mozambique and Afghanistan is clear. In both countries, politically illegitimate Marxist-Leninist régimes have swiftly alienated their population to the point of military resistance, and have needed extensive foreign military backing (140,000 Soviets in Afghanistan and 20,000 Zimbabwean soldiers in Mozambique) to maintain even a fleeting semblance of power in their countries – limited as it is in both to several major cities and a handful of national highways. In *Afghanistan: A Difficult Path to Peace*, published by the Soviet Novosti Press Agency in 1987 and indicative of a clear Soviet policy, there are allegations that the United States, Pakistan and certain Western European nations are helping to train anti-Marxist Afghan guerillas for action against the Soviet-aligned Kabul-régime – action including subversion "aimed at disrupting the normal operation of factories and cooperatives", putting "power transmission lines out of order" and killing "government and Party officials and servicemen".

Does the fact that a previously authoritarian government such as General Zia's Pakistan was helping to destabilise the Afghan client-state in any way detract from the legitimacy or cause of the Afghan guerillas? The simple answer must be no. And similarly, any alleged South African involvement with Renamo cannot be seen as making the anti-Frelimo struggle of the Mozambican masses any less valid.

Chapter Eight

MOZAMBIQUE: DIALOGUE OR DISASTER?

Conclusion

Peace must come to Mozambique. An examination of both British and United States involvement in the Mozambican situation and any such possible peace process is long overdue. Such an examination is vital to resolve the impediment posed by present British Mozambique policy to any such peace process. The simple reality of the Mozambican situation is that if an end does not come to a conflict raging from the Maputo to Rovuma rivers, the country will die. Six million out of Mozambique's thirteen million people are at risk from famine, partly as the consequence of Frelimo's disastrous agricultural policies, partly because of natural weather conditions and partly as the consequence of a twelve year civil war. The country's infrastructure is virtually non-existent, concentrated on keeping a Zimbabwean economic corridor to the sea. Over one hundred thousand people have lost their lives in the war between the Maputo administration and pro-Western guerillas. Provoked by Frelimo's lack of political legitimacy (for it was not elected into office, but was appointed by the departing Portuguese), the conflict was quickly fuelled by Frelimo's ideologically-motivated state collectivisation of agriculture and villages, by its attempts to kill traditional ethnic leaders, by ruthless political intolerance and by the suppression of religion. Between seventy-five and one-hundred thousand people may have been killed or starved to death in Frelimo's gulag, and thousands more have been, or are still, incarcerated.

It is a bloody conflict kept going by substantial foreign involvement and investment. Even pro-Frelimo observers have acknowledged that the civil war is, at best, a military stalemate in Renamo's favour. Most independent observers and journalists see the Maputo régime as an administration in name only. Its thirty-thousand strong army strengthened by over thirty thousand foreign soldiers (including twenty thousand Zimbabweans) is pushed to maintain an ever fleeting hold on major urban centres and several major road and rail links – a direct parallel of the Kabul régime in Afghanistan. The vacuum caused by the Soviet Union's partial disengagement from Mozambique, dictated by the political and military reality of Frelimo's position, has paradoxically been filled by Western governments in a seriously flawed example of political and economic opportunism. Based on the historically unproven belief that economic reforms (such as those Frelimo was forced to adopt after its disastrous economic experiments) are evidence of an ideological realignment *away* from the Soviet Union and Marxism-Leninism and *towards* the West, Western support for Maputo is a crucial factor, together with Zimbabwean military intervention, in keeping Frelimo in power. Western economic and military involvement, in sustaining the Maputo administration against the clearly popular Mozambique National Resistance, is in turn artificially prolonging the Mozambican civil war. Even the Zimbabwean war effort is indirectly subsidised by the British taxpayer.

Present British policy towards Mozambique is a second crucial reason for a closer study of the Mozambican situation. The contradictions of the Thatcher administration's close identification with the maintenance of a doctrinaire Marxist régime are clear[1]. Anyone concerned with securing a peaceful, negotiated settlement to the Mozambican crisis can but only concur with Renamo's reservation concerning British involvement:

> *"For reasons better known to herself, Mrs Thatcher has sadly led Britain to support the Frelimo government in Maputo in its conflict with the Renamo freedom fighters who now control 80% of the Mozambican territory...*
>
> *Nothing can excuse Mrs Thatcher's decision militarily to bail out the tottering Stalinist régime, which has transformed Mozambique into a prison camp and murdered over 100 000 in its concentration camps – not to speak of the reign of terror characterised by such violence as the burning or burying of people alive, drownings, summary executions, religious persecution... violation of human rights and denial of freedoms...*
>
> *We are fighting for freedom and democracy – for the right of the people of Mozambique to be governed under the rule of law through a popularly elected sovereign-legislative assembly and an executive and also an independent judiciary. For such a system to be credible, it has to ensure the people's freedoms and rights: the right to free movement, free speech, free political and ideological thought, free assembly and association and the right for people to reside where they wish. Accordingly people must be free to chose and pursue objectives for their own profit and benefit, own and dispose of property as they deem fit and conduct trade and business unencumbered and in a free market system. Only this, we believe, will bring stability, lasting peace and development to Mozambique. This is the struggle the British Government ought to support."*[2]

The Mozambique National Resistance is incontrovertibly the key to peace in Mozambique. As an organisation left with no option save that of armed struggle in opposition to an increasingly bloody totalitarian model (a model upheld by extensive foreign military intervention), Renamo have, in fighting the war, gained increasing political support from the increasingly political Mozambican masses. Several independent observers taken into protective custody by Renamo have commented on the movement's affinity with the people. Renamo's claim to control eighty-five percent of the country may be exaggerated, but a presence exists in all ten of Mozambique's provinces as well as in Maputo City, with liberated zones in well over sixty percent of the country – zones and presence which could not exist without a local mandate.

Just as the Afghan Mujaheddin guerillas are crucial to any resolution of Afghanistan's crisis, despite tremendous propaganda by both the Soviets and the Kabul government which portrays them as disorganised bandits and fanatics, Renamo is an equally legitimate actor within the Mozambican situation. There can be no settlement of the Mozambican conflict without recognition of Renamo's calls for an end to Frelimo's totalitarianism, for a ceasefire, for the withdrawal of all foreign military forces, for national reconciliation and for the granting of internationally-recognised free and fair elections.

It is clear that the Mozambique National Resistance and its political agenda are valid political forces, essential to the achievement of peace and democracy in Mozambique. Frelimo has already legitimised Renamo once, when it entered into South African-hosted peace talks in 1984. Factions within Frelimo itself still concede the legitimacy of negotiating with Renamo: the influential Frelimo Veteran Soldiers' Association, chaired by General Bonifacio Gruveta, one of Frelimo's 'old guard', in 1987, published a call for negotiations with Renamo to establish a government of national unity. All seven of Mozambique's Roman Catholic bishops have similarly signed a call for dialogue between the two parties, a sentiment echoed by Pope John-Paul II during his visit to Mozambique. Zimbabwe's President Mugabe has reportedly privately expressed support for such a course of events. As early as 1986, even *Guardian* articles began conceding that "In the end, the two sides must negotiate peace"[3]. Reconciliation between Mozambicans is the only solution.

Ideological hardliners, with a vested political interest in continuing the conflict, together with their often equally doctrinaire propagandists in the West, will deny any ground for negotiation. Observers have attributed the assassination of Evo Fernandes as a last, desperate attempt by such elements to prevent negotiation and reconciliation, which would have broken their political hold and altered Maputo's ideological orientation. Those who support the continuation of the conflict also argue that far too much blood has been spilt for such a rapprochement. In doing so, they purposefully ignore the 1979/80 Rhodesian settlement, negotiated between two guerilla forces, hostile to each other and Salisbury, and the Muzorewa administration, a settlement which saw the integration of eighty thousand ZANLA, ZIPRA, and Rhodesian security force combatants into one force, against the background of an equally vicious civil war.

By far the major difficulty in approaching negotiations is the international image of the Mozambique National Resistance. There is a case for arguing that Frelimo's propaganda war against Renamo has been too effective. Having labelled Renamo as a South African institution, as a grouping of "armed bandits" with no political agenda, responsible for countless atrocities, Frelimo can now be seen as backed into a corner, finding it difficult to admit the inaccuracy of its own misinformation. It was a difficulty overcome in parallel instances of propaganda projections such as the Afghan Mujaheddin and the Nicaraguan 'Contra' democratic resistance, where the respective Soviet client states were pressurised by the Soviet Union into negotiating with resistance forces similarly smeared with images of banditry and political confusion. A billion dollars of Western aid is currently insulating Frelimo from having to come to terms with its situation. A further problem is that the commitment of several of Frelimo's international apologists is not to peace and national reconciliation in Mozambique, but rather to the continuation of Mozambique as a Marxist-Leninist state, and they refuse to concede that the two are mutually exclusive. It is interesting that the very organisations calling for the denial of free elections and for Renamo's continued banishment are amongst the first to call for just such elections, and the legalisation of the African National Congress, within South Africa.

The conditions necessary for peace and reconciliation in Mozambique are clear:

1 **A general ceasefire.**
2 **The repatriation of all foreign military forces.**
3 **The release of all political prisoners held by the Frelimo administration.**
4 **Freedom of the press, television and radio.**
5 **Guarantees of full human rights and civil liberties.**
6 **Freedom for trades unions to organise.**
7 **The ending of compulsory military service, together with the abolition of the party-political FLPM armed forces, and its replacement with a genuine Mozambique volunteer army.**
8 **Internationally supervised free and fair elections within a multi-party system.**

The importance to the reconciliation process of the Mozambique National Resistance drastically raising its international profile cannot be understated. Renamo's external representation is now the most important phase in the struggle for peace and democracy in Mozambique. The military struggle has been won, as has the internal dimension of the political struggle. International awareness and recognition of the economic, political, and as a last resort, armed struggle for freedom within the Third World, be it in Afghanistan or Mozambique, is always important. In Mozambique's case the international projection of the basis of Renamo's legitimate armed struggle, and of Frelimo's totalitarian experiment, was ignored at the obvious cost of United States and Western political confusion, and to the benefit of Frelimo propagandists and apologists. It is, essentially, this projection which stands between Mozambique, reconciliation and peace.

NOTES

Chapter 1 'Mozambique in Crisis'

1 Allen and Barbara Isaacman, *Mozambique From Colonialism to Revolution*, Zimbabwe Publishing House, Harare 1983.

Chapter 2 'Is Frelimo Marxist?'

1 'Strategies of Social Research in Mozambique' in *Review of African Political Economy* No 25, September-December 1982 p29. Authored by Eduardo Mondlane University's CEA, to all intents and purposes, Frelimo's think tank.

2 Press Conference by President Machel, 22 March 1980 in Maputo. Cited in ROAPE *op cit.*

3 Colin Legum '"National Liberation" in Southern Africa' in *Problems of Communism* January -February 1975 Vol XXIV p 9.

4 Giovanni Arrighi and John Saul, *Essays on the Political Economy of Africa*, Monthly Review Press, New York 1973 p387.

5 Dr Edward Cain 'Mozambique's Hidden War' in Charles Moser (Editor) *Combat on Communist Territory*, Regnery Gateway 1985 p40.

6 Aquino de Braganca and Immanuel Wallerstein (Editors) An *African Liberation Reader* Zed Books, London 1982 p125.

7 *Mozambique Revolution*, No 43 April/June 1970.

8 Barry Munslow (Editor) *Samora Machel: An African Revolutionary*, The College Press, Harare 1987 p 160/1.

9 Marina Ottaway 'The Theory and Practise of Marxism-Leninism in Mozambique and Ethiopia' in David Albright *Africa and International Communism*, Macmillan, London, 1980 p 122.

10 Cited in Gwendolen Carter and Patrick O'Meara (Editors) *Southern Africa: The Continuing Crisis*, Macmillan, London 1979, p 67.

11 *Ibid* p 67.

12 Cited in Joseph Hanlon, *Mozambique: The Revolution Under Fire*, Zed Books, London 1984, p 28.

13 *Voz da Revolucao* September 1978.

14 Munslow *op cit* p 166.

15 *Ibid* p 104-9.

16 FBIS - Soviet Union 20 February 1986.

17 FBIS - Soviet Union 2 April 1986.

18 FBIS - Soviet Union 12 November 1986.

19 Broadcast on Maputo Domestic Service, 1800 GMT 3 February 1987.

20 *Out of Underdevelopment to Socialism*, Report of the Frelimo Central Committee, June 1983 p152.

Chapter 3 'Frelimo in Power'

1 Jose Luis Cabaco, Frelimo External General Secretary, cited in *New Internationalist*, February 1989.
2 Munslow *op cit.*
3 *Frelimo Party Programme and Statutes*, Maputo 1983.
4 Hanlon *op cit* p27.
5 Munslow *op cit.*
6 *Ibid* p 169.
7 London *Independent* 24 February 1988.
8 Munslow *op cit* 177.
9 *World Marxist Review* Volume 24 No 1, January 1981.
10 *New African Yearbook 1987-88*, IC Publications, London p224.
11 *Out of Underdevelopment to Socialism op cit* p123.
12 Amnesty International 1987 Yearbook.
13 International Society for Human Rights 1987 Yearbook.
14 *Sowetan* 26 February 1987 p16.
15 *The Economist Human Rights Guide*, 1988 p187/9.
16 Ian Bray, *Chicualacuala: Life on the Frontline*, Oxfam 1987.
17 Hanlon *op cit* p137.
18 Isaacman *op cit* p134.
19 *Ibid* p134/5.
20 *Ibid* p137.
21 Hanlon, *op cit* p244.
22 *Ibid.*
23 *Ibid* p262.
24 London *Guardian*, 12 June 1984.
25 Hanlon *op cit* p244.
26 Heritage Foundation.
27 ISHR *op cit.*
28 *Scope* magazine, 18 February 1983.
29 Rev Sithole in testimony on file at the Freedom Research Foundation, La Jolla, California.
30 *To the Point* Magazine, 20 July 1979.
31 *Ibid.*
32 London *Sunday Times* , 29 May 1988.
33 Statement by Antonio Rocha at press conference, Washington DC, 5 October 1987.
34 London *Sunday Telegraph*, 10 July 1988.
35 Munslow *op cit.*
36 Hanlon *op cit* p262.

37 *Ibid* p63.
38 *Ibid* p64.
39 *Ibid* p66.
40 Gillian Walt and Angela Melamed (Editors), *Mozambique Towards A People's Health Service*, Zed Books, London, 1983.
41 *Ibid* p 85.
42 *Ibid* p 93.
43 John Saul (Editor) *A Difficult Road: The Transition to Socialism in Mozambique*, Monthly Review Press, New York 1985, p180.
44 *Ibid* p183.
45 *Ibid* p203.
46 Chris Searle, *We're Building the New School! Diary of a Teacher in Mozambique*, Zed Books, London 1981.
47 Lenin cited in *White House Digest* 29 February 1984.
48 *Carta Pastoral dos Bispos Catolicos de Mozambique* 1979.
49 *Africa Events*, May 1987.
50 *Carta Pastoral dos Bispos Catolicos de Mozambique* 1987.
51 *Jehovah's Witnesses in Africa*, Minority Rights Group Report No 29, Revised Edition January 1985.
52 *Ibid.*
53 *Ibid.*
54 *Mozambique Report - Eyewitness Testimonies of Persecution And Atrocities*, Frontline Fellowship, 1987.
55 *Ibid.*
56 *Ibid.*
57 Professor Andre Thomashausen, 'The Mozambique National Resistance' in C. J. Maritz (Editor) *Weerstandbewegings in Suider-Afrika*, University of Potchefstroom Press, 1987.
58 *To the Point*, 23 March 1979.
59 *Ibid.*
60 *Africa Events*, May 1987.

Chapter 4 'The Mozambique Economy Under Frelimo'.

1 *Marxism Today*, October 1987.
2 *Review of African Political Economy*, No 25 1982, p90.
3 London *Independent*, 4 November 1987.
4 *Review of African Political Economy*, No 29 1984, p95.
5 *Ibid*, p100.
6 *Ibid*, p101/2.
7 Hanlon, *op cit* p131.
8 *Review of African Political Economy*, No 29 1984, p107.
9 Hanlon, *op cit* p178.

10 Preface cited in *Mozambique: Ten Years Independence*, Lecture by Professor Andre Thomashausen to South African Institute of International Relations, Johannesburg, 8 February 1986.
11 Cited in *Review of African Political Economy*, No 25 1982.
12 *To The Point*, 15 December 1978.
13 Hanlon, *op cit* p192.
14 London *Guardian*, 18 March 1983.
15 Hanlon, *op cit*.
16 London *Guardian*, 21 March 1986.
17 *To The Point*, 8 December 1978.
18 Munslow, *op cit* p81.
19 Hanlon, *op cit* p205.
20 *Ibid* p203.
21 *Ibid* p208.
22 *Ibid* p114.
23 London *Guardian*, 21 July 1986.

Chapter 5 'The Mozambique National Resistance: The Origins of Renamo'

1 *Scope*, 11 February 1985.
2 *To The Point*, 5 March 1976.
3 *To The Point*, 5 March 1976.
4 *Dictionary of Scientific Communism*, Progress Publishers, Moscow, 1983 p148.
5 *Armed Forces*, October 1977.
6 *Ibid*.
7 *A Luta Continua!*, journal of Renamo, January 1981.
8 *To The Point*, 19 January 1979.
9 *Ibid*.
10 Pierre Botha 'Mozambique: The Democratic People's Revolution Fails' in *Africa Insight*, journal of the Africa Institute, Volume 13, No 2 1983.
11 *Out of Underdevelopment to Socialism*, *op cit* p146.
12 *Africa Insight*, *op cit*.
13 Thomashausen, *Africa Insight op cit* p126.
14 Zimbabwean military briefing cited in Phyllis Johnson and David Martin (Editors) *Destructive Engagement: Southern Africa at War* Zimbabwe Publishing House, Harare 1986 p19.
15 John Saul (Editor) *op cit* p126.
16 Professor Andre Thomashausen 'The National Resistance of Mozambique' in *Africa Insight*, Volume 13, No 2 1983.
17 Thomashausen, *Africa Insight*, *op cit*.
18 Hanlon, *op cit*.
19 The Johannesburg *Star*, 8 October 1983.
20 The Johannesburg *Star*, 1 March 1983.

21 *The Economist* 26 March 1983.
22 Hanlon, *op cit* p227.
23 The Johannesburg *Citizen*, 24 August 1983.
24 The Johannesburg *Star*, 13 January 1984.
25 Thomashausen, *op cit.*
26 *The Times* London, 18 October 1983.
27 *Army* magazine, May 1987.
28 *Janes Defence Weekly*, 12 July 1986.
29 The Johannesburg *Star*, 22 March 1984.
30 *A Luta Continua!*, June 1985.
31 *Janes Defence Weekly op cit* p32.
32 *Ibid.*
33 Professor Thomashausen *op cit* p33.
34 *Expresso*, Lisbon, 1 November 1985.
35 London *Sunday Times*, 5 April 1987.
36 London *Sunday Times*, 21 June 1987.
37 *London Guardian*, 23 July 1986.
38 *Southern Africa Political and Economic Monthly*, Harare, November 1987.
39 London *Sunday Telegraph*, 20 September 1987.
40 London *Independent*, 1 December 1987.
41 Cited in *Freedom Research*, London, February 1988.
42 Mozambique Information Office, 111/2.
43 AIM, February 1988.

Chapter 6 'The Mozambique National Resistance: Structure and Control'

1 *Africa Events*, May 1987 p24.
2 *Ibid.*
3 *Washington Post*, 23 August 1987.
4 *Africa Events, op cit.*
5 London *Independent*.
6 *Reason Magazine*, December 1985.
7 *Janes Defence Weekly, op cit.*
8 *Ibid.*
9 London *Guardian*, 23 July 1986.
10 Derrick Knight *Mozambique Caught In The Trap*, Christian Aid, London 1988, p25.
11 *Mozambique Towards a Peoples Health Service, op cit.*
12 *Mozambique Revolution* No 58 1974 12-17.
13 *Mozambique Towards a Peoples Health Service op cit.*
14 London *Independent*, 24 November 1987.
15 London *Guardian*, 23 July 1986.
16 *Reason Magazine*, December 1985.
17 The Johannesburg *Star*, 6 September 1982.

18 Renamo communique, 30 March 1988.
19 Professor Thomashausen, *The Mozambique National Resistance, op cit.*
20 *Media Watch*, December 1987.
21 *Newsweek*, 24 August 1986.
22 BBC *Focus on Africa*. Programme in English 1722 GMT 28 July 1988.
23 *The Children of War*, Frelimo Party Central Committee Information Department, July 1987.
24 AIM, 13 August 1987.
25 London *Independent*, 11 April 1987.
26 Allen Isaacman 'Social Banditry in Zimbabwe (Rhodesia) and Mozambique 1884-1907: An expression of Early Peasant Protest' in *Journal of Southern African Studies* 4,1977, 1-30.
27 Cited in Eduardo Mondlane *The Struggle for Mozambique* p136.
28 Munslow *op cit.*
29 John Saul (Editor) *op cit.*
30 London *Guardian* 23 July 1986.
31 London *Independent*, 24 November 1987.
32 'Towards an Understanding of the Mozambique Situation', in *Southern Africa Political and Economic Monthly*, November 1987 p27.
33 'The Mozambique Crisis and the Analysis: Critique of a Critique', in *SAPEM* February 1988 p44.
34 *Ibid.*.
35 Munslow, *op cit* p86.
36 *Ibid.*
37 Hanlon, *op cit.*

Chapter 7 'The Internationalisation of the Mozambican Conflict'

1 London *Guardian*, 15 March 1986.
2 London *Independent*, 15 September 1986.
3 *Ibid.*
4 *Review of African Political Economy*, No 40 December 1987.
5 *Ibid.*
6 Derrick Knight *op cit* p64.
7 *The Military Balance 1988-89*, International Institute for Strategic Studies, London 1988.
8 *Reason* magazine December 1985.
9 *South* magazine, May 1988 p15.
10 *Time*, 30 January 1984.
11 *The Controversial State Department Report on Mozambique* Backgrounder Update No75, 5 April 1988, Heritage Foundation, Washington DC.
12 London *Times*, 11 September 1987.
13 *AfricaAsia*, Paris No 40, April 1987 p10.
14 *London Financial Times*, 16 May 1988.

15 *Pretoria News*, 28 June 1988.
16 *Lettre D'Afrique*, Paris 29 June 1988.
17 *The Southern African Economist*, Harare, March 1988 p11.
18 London *Financial Times*, 16 May 1988.
19 London *Sunday Telegraph*, 22 September 1987.
20 London *Sunday Telegraph*, 7 August 1988.
21 *The Southern African Economist*, Harare, February/March 1988.
22 Lina Magaia *Dumba Nengue: Run For Your Life*, Africa World Press, Trenton 1988.
23 Victoria Brittain *Hidden Lives, Hidden Deaths: South Africa's Crippling of a Continent*, Faber and Faber, London 1988.
24 Hanlon, *op cit* p223.
25 Mozambique Information Office Briefing No 126.
26 *L'Express*, No 1925, 3 June 1988.
27 London *Daily Telegraph*, 2 December 1987.
28 *Die Suid Afrikaan*, December 1987.
29 *Ibid.*
30 *Mozambique in the Balance*, House of Commons Research Note 380, Westminster, December 1987.
31 London *Sunday Times*, 22 November 1987.
32 London *Independent*, 3 May 1988.
33 Phyllis Johnson and David Martin (Editors) *Destructive Engagement: Southern Africa At War*, Zimbabwe Publishing House, Harare, 1986 p33.
34 Joseph Hanlon *Apartheid's Second Front*, Penguin, 1986 p83.

Chapter 8 'Mozambique: Dialogue or Disaster?'

1 London *Independent*, 19 September 1988.
2 Renamo Communique, 25 August 1988.
3 London *Guardian*, 23 July 1986.

APPENDICES

APPENDIX 1.

Political Program of the Mozambique National Resistance

The objective of the Mozambique National Resistance is to democratize, liberalize and create the conditions for overall progress in the Republic of Mozambique in accordance with the following principles:

1. Political
a) Eradication of the system of communist dictatorship and the removal of responsible Marxist elites, without victimisation. Only by forgiving and forgetting will it be possible to build a Mozambique belonging to all Mozambicans.

b) Establish a Government of National Unity to pacify the country. Stabilize the country through a democratic process and build the foundation for social and economic recuperation. The unity of the Mozambique people is essential to be able to face the challenges of modern times.

c) The people have the right to choose and freely vote for the political, social and economic system of the country. Nothing will be imposed on the majority by an aggressive minority.

d) The traditions, customs and personality of the Mozambique national people will be respected by the government. Progress will be made in step with the lessons of history and never against it.

2. Economic
a) The public sector will be defined and serve to optimize the private sector, the generator of the economic prosperity of Mozambique. Private enterprise and foreign investment are the key elements of economic democracy.

b) Development plans will be orientated towards increasing national wealth, harmonizing the development of all regions and improving the quality of life of our people.

c) Monopolies, oligopolies or the economic exploitation of one class or sector by another will not be permitted. Steps will be taken to protect consumers and to safeguard their respective participation in the establishment of economic laws.

d) The money levied from the people shall be used correctly in the public trust for the limited and well defined function of government.

e) The situation of confiscated or appropriated assets will be analysed with regard to their eventual restitution. The process of dismantling the collectivized system in favour of private ownership will be studied and implemented.

3. Justice

a) All citizens shall be equal before the law and shall have the same rights and duties. No-one shall be tried or sentenced without adequate defence and no-one shall be arrested without an order or warrant from judicial authority.

b) Judges shall be professional, permanent and independent. The widest guarantees and safeguards shall be adopted so that judges may conscientiously execute the law.

c) Trials will be public and will be open to public opinion. Offenders will be defended by a lawyer of their own choice and no restrictions will be imposed on that right.

d) Political prisoners will be given their freedom and other prisoners will have their cases reviewed by impartial judges. The prosperity of Mozambique lies in human dignity and justice.

4. Health and Education

a) Robust economy and a free political system are the prerequisites to enable our people to return to their lands and livelihood to produce food and income. The bulk of Mozambique's health crisis will be cured by peace. A healthy economy will stimulate private and government revenue to provide medical facilities.

b) International organisations and institutions, including religious ones, will be encouraged to restart the work they have done in this field. The state will foster the search for new methods and programs in health and education in a free, participatory and progressive society.

c) Maternity, infancy and old age will be especially protected. State services in the field of health will as far as possible be free.

d) Education will be obligatory and free up to the age of twelve years. The most able will be encouraged to continue in their studies.

5. Public Services

a) Devotion to public duty and welfare will be the perquisite of state, local government and public company employees. Public activities will be open to hierarchical and employee scrutiny and will be accountable to the majority of citizens.

b) The Armed Forces and Police will be under civilian control and will be distanced and not interfere in political activity. They will be the guarantors of democratic political activity and their dignity will derive from this.

c) The Armed Forces of Mozambique will be constituted mainly by the Mozambique National Resistance Forces, by nationalists of RENAMO and FRELIMO and by the incorporation of the actual regular units existing in the country, except for those which, during the Second War of Liberation behaved in an unpatriotic manner. The instruments of military power and political corruption used by the present regime in the repression of fundamental citizen's rights will be dismantled.

d) Veterans, ex-combatants and victims of the Wars of Independence, Liberation and National Defence will be particularly well cared for and protected by the State. Independence, freedom and national defence are the permanent and unifying commitments of past, present and future generation of the Mozambique people.

6. International Politics

a) Agreements established previously with foreign countries or international organisations will be fully respected providing they do not compromise National Independence and the basic democratic freedoms of our citizens, and do not contradict the inviolable principles of this program. Mozambique will not discriminate against any country because of its respective internal regime: international relations will be governed by mutual advantage and the national interest.

b) Mozambique will respect and defend the principles governing the United Nations and the Conventions of Geneva.

c) The country will co-operate actively with neighbouring states, especially those which depend on our ports. International co-operation is the funda mental instrument of peace in the region.

7. Constitution
a) Approval of the Constitution will be notified by the National Assembly and the Participation of the People in the form of a referendum.
b) National Independence, personal freedom, justice under the law, will be the basis for the Constitution of the Republic of Mozambique.

Confirmed by the blood of our martyrs, the supreme ideals of the constitution is the national imperative.

APPENDIX 2.

The Mozambique National Resistance's Provisional Economic Manifesto for Free Mozambique

Renamo's economic philosophy is the complete antithesis of the communist Frelimo government. The Frelimo regime is based upon strict Marxist-Leninist ideology.

Alternatively, Renamo believes state power is not ultimate. Government exists to protect the rights of individuals and serve in such functions as are necessary to the public good. Therefore we believe in limited powers of a Constitutional government based on democratic principles. Rule of law must be enshrined by a system of legal justice and predicated on belief in God given human rights, freedoms, and dignity.

Our political reference logically implies economic freedom based on personal property rights, individual initiative and a free market environment. Competition is essential for a healthy and viable economy.

Foreign investment is encouraged and necessary for the development of Mozambique's untapped wealth. Foreign profits and national development are compatible processes within the bounds of respect for national heritage. Both exploitation and expropriation are destructive to investors and government relations.

Renamo takes a dim view of unprincipled interests who have chosen to profiteer on the last gasp offer of the dying Frelimo regime which sells the blood and birthright of our people for a season of power.

Alternatively, Renamo is anxious to develop economic ties with those like-minded investors who believe real growth results from free market system rather than from exploitive profits of communist concessions.

Renamo is already putting to the test its policy of economic freedom. Although our economy is relatively unsophisticated and rural at present. We believe freedom is a reliable foundational principle when applied to more complex models as well.

APPENDIX 3.

Mozambique: Economic Performance Indicators (1973-1985)

cashew nuts:	216 000 (1973)	5 800 (1983)
sugar:	285 581 (1974)	120 000 (1982)
tea:	18 795 (1973)	14 000 (1983)
maize:	400 000 (1972)	200 000 (1983)
rice:	111 000 (1972)	30 000 (1983)
bananas:	280 000 (1972)	65 000 (1983)
fish & shrimps:	32 000 (1980)	7 695 (1983)
number of goats:	600 000 (1972)	350 000 (1983)
number of pigs:	220 000 (1972)	135 000 (1983)
sisal:	22 003 (1972)	12 000 (1982)
cotton:	280 000 (1972)	73 000 (1982)
coal:	574 800 (1975)	39 500 (1984)
copper:	3 446 (1975)	0,0 (1980/81)
cement:	611 000 (1973)	261 000 (1981)
electricity: max cap.	14 000m.kWh (1980)	2 642m.kWh (1983)
freight traffic (sea):	17 000 000 (1974)	1 200 000 (1984)
tourism:	291 574 (1972)	1 000 (1985)
external debt:	(US$): 0,0 (1973)	1 400 000 (1985)
deaths due to famine:	0,0 (1973)	100 000 (1983)
refugees to RSA:	0,0 (1973)	ca.120 000 (1985)
gross social product at constant 1980 prices in bn Meticals:	111,9 (1973)	55,6 (1984)

Sources: Africa Contemporary Record; Africa South of the Sahara; Economist Intelligence Unit Country Reports & Country Profiles; Expresso (Lisbon); Inst. Democracia e Liberdade (ed), A Descolonizacao vol.1 (1979)

APPENDIX 4.

Frelimo Party Programme and Statutes

Part 1 Basic Principles

Article 1.
The Frelimo Party is the vanguard organisation of the worker-peasant alliance in Mozambique. Under the leadership of the working class the Frelimo Party unites in a voluntary and militant alliance based on scientific socialism and the workers, peasants, soldiers, revolutionary intellectuals and other working people, forged and tempered in the struggle against exploitation of man by man, and for the victory of people's interests. The Frelimo Party is the most advanced form of revolutionary organisation of the Mozambican labouring classes.

Article 2.
The Frelimo Party personifies the highest patriotic traditions and the conquests of the people won by the revolutionary fighters in the course of the national liberation struggle, the people's war of liberation, the defence of the country and the fight against old and new exploiters. The Party, forged and tempered in the process of national and social liberation, raises national unity and patriotic values to their highest level. It is the concrete expression of the most noble patriotic sentiments of our people. It is the guarantee of national independence. For this reason, the Party fights intransigently against tribalism, racism and divisiveness as fundamental enemies of our people's unity, of independence and of our national consciousness.

Article 3.
The Frelimo Party is guided by the synthesis of the experience of the Mozambican people's revolutionary struggle with the universal principles of Marxism-Leninism. These foundations and the class composition of the Party guarantee that the Frelimo Party expresses and embodies the highest legitimate aspirations of the people.

Article 4.
The Frelimo Party has as its objective the continuous consolidation of national independence, of the people's democratic order and of socialism

Article 5.
The Frelimo Party is the leading force in Mozambican society. It fights to consolidate the leading role of the working class and to strengthen the alliance of workers and peasants and of all patriotic and democratic forces. The Frelimo Party leads the Mozambican people in building socialism by mean of radical action to transform economic and social structures. The Frelimo Party defines the policy of the state, and guides and controls the activities of the state. The Frelimo Party directs the Mozambican Armed Forces (FPLM), armed wing of the labouring classes, in defence of the country and the revolution. The Frelimo Party directs the Mozambican security forces as a whole. The Frelimo Party establishes the mass democratic organisations which, under the guidance of the party, guarantee the participation of and integrate the broad masses in the process of transforming society, and in building socialism.

Article 6.
The Frelimo Party defends, as integral and fundamental parts of its political line, proletarian internationalism and solidarity with oppressed classes and peoples in struggle against foreign domination and the system of exploitation of man by man and for peace, independence and social progress. The Frelimo Party promotes and deepens relations with the world revolutionary movement and struggles for the consolidation and broadening of the world anti-imperialist front. The Party develops relations of cooperation, mutual help and solidarity with the world revolutionary movement and struggles for the consolidation and broadening of the world anti-imperialist front. The Party develops relations of cooperation, mutual help and solidarity with revolutionary working class parties, with progressive parties and with the world's democratic forces as a whole. The Party gives particular importance to strengthening relations with progressive parties and forces in African countries. The Party supports the movement for national and social liberation in Africa and the rest of the world in the struggle against the system of exploitation of man by man and against imperialism, colonialism, neocolonialism, fascism and racism.

ZINC ENZYMES

METAL IONS IN BIOLOGY

EDITOR Thomas G. Spiro,
Department of Chemistry
Princeton University, Princeton, New Jersey 08540

VOLUME 1 *Nucleic Acid—Metal Ion Interactions*
VOLUME 2 *Metal Ion Activation of Dioxygen*
VOLUME 3 *Copper Proteins*
VOLUME 4 *Iron-Sulfur Proteins*
VOLUME 5 *Zinc Enzymes*

In Preparation
Volume 6 Calcium in Biology
Volume 7 Molybdenum Enzymes

ZINC ENZYMES

Edited by

THOMAS G. SPIRO
Princeton University

A WILEY-INTERSCIENCE PUBLICATION
JOHN WILEY & SONS
New York • Chichester • Brisbane • Toronto • Singapore

Library of Congress Cataloging in Publication Data:

Main entry under title:

Zinc enzymes.

(Metal ions in biology, ISSN 0271-2911; v. 5)
"A Wiley-Interscience publication."
Includes index.
1. Zinc enzymes—Addresses, essays, lectures.
I. Spiro, Thomas G., 1935- . II. Series. [DNLM:
1. Enzymes—Physiology. 2. Zinc—Physiology. W1 ME9611
v. 5 / QU 135 Z77]
QP601.75.Z54Z56 1983 574.19'24 83-5918
ISBN 0-471-89081-2

Printed in the United States of America

10 9 8 7 6 5 4 3 2 1

Series Preface

Metal ions are essential to life as we know it. This fact has long been recognized, and the list of essential "trace elements" has grown steadily over the years, as has the list of biological functions in which metals are known to be involved. Only recently have we begun to understand the structural chemistry operating at the biological sites where metal ions are found. This has come about largely through the application of powerful physical and chemical structure probes, particularly X-ray crystallography, to purified metalloproteins. From such studies we have learned that nature has evolved highly sophisticated ways of controlling the relatively flexible stereochemistry of metal ions. In one case after another, the structure and the reactivity of a metalloprotein active site have turned out to be different from anything previously encountered in simple compounds of the metals. Indeed, many a reasonable inference about active-site structure, based on the known properties of metal complexes in solution, have turned out to be erroneous. These surprises have inspired inorganic chemists to expand their vision of metal ion reactivity. The biological studies have spurred much fruitful synthetic and mechanistic work in inorganic chemistry, aimed at elucidating the means whereby nature achieves its stereochemical ends. The terra incognita of the biochemical functions of metal ions has become familiar territory to an increasing number of inorganic chemists and biochemists, and several of the more imposing mountains have been scaled. Vast stretches remain uncharted, and the field is alive with a sense of both accomplishment and new opportunities.

The purpose of this series is to convey some of this excitement, as well as the emerging intellectual shape of the field, to a wide audience of nonspecialists. Individual volumes will cover topics that are current and exciting—the recently scaled mountains that are still under active exploration. The chapters are not intended to be exhaustive reviews of the subject matter. Rather, they are intended to be readable accounts of the insights and directions that

are emerging in active new areas of research. Volumes will appear on an occasional basis as progress in the field dictates.

THOMAS G. SPIRO

Princeton, New Jersey

Preface

Zinc is second in abundance only to iron among the transition metals found in the body. Until recently, its biological importance has been undervalued because Zn^{2+} is invisible to most spectroscopies. However, this ion has been shown to be present as an essential constituent of a long and ever-growing list of enzymes. The chapters in this volume demonstrate the remarkable progress that has been made in understanding the structure and function of zinc enzymes, through careful chemical and physical studies, and the substitution of other metal ions, which offer more spectroscopic handles, for zinc. It is fair to say that zinc enzymes are understood as well as or better than any other class of metalloenzymes; yet many intriguing questions remain to be answered, as the ensuing chapters show.

Chapter 1 gives an overview of the field by the founder of modern zinc biochemistry, and one of the pioneers of bioinorganic chemistry, Bert L. Vallee. In Chapter 2, Vallee and his associates, Alphonse Galdes, David S. Auld, and James F. Riordan, give a thorough and critical account of carboxypeptidase, a prototypical zinc enzyme which they have long been engaged in studying. Carbonic anhydrase is perhaps the most ubiquitous zinc enzyme, its vital function being the interconversion of CO_2 and carbonic acid. The elementary nature of this reaction and the very high catalytic rate make carbonic anhydrase particularly fascinating. In Chapter 3, Sven Lindskog gives a masterful account of the current state of knowledge about this enzyme.

In Chapter 4, Hans Eklund and Carl-Ivar Brändén describe the role of zinc in alcohol dehydrogenase, focusing particularly on their elegant crystallographic studies. Alkaline phosphatase is the subject of Chapter 5, by Joseph E. Coleman and Peter Gettin, who provide a fascinating account, based to a large extent on the pioneering of Cd NMR studies of Coleman and his group, of a protein that has three different metal sites in close proximity, and strong anticooperative behavior between two, initially identical subunits. As dis-

cussed by Vallee in Chapter 1, the vital role of zinc in the enzymes of genetic replication, transcription, and translation is just beginning to be elucidated, and in Chapter 6 Coleman summarizes what is known about zinc in RNA and DNA polymerases. In Chapter 7, Nicholas E. Dixon and Alan M. Sargeson take up the question of the chemical mechanisms whereby zinc and other metal ions catalyze reactions of their coordinated ligands, a theme raised repeatedly in earlier chapters. In this realm, the penetrating studies of Sargeson and his colleagues, using substitution-inert Co(III) complexes to examine mechanistic details, come to the fore. Chapter 7 gives an informative discussion of these and related studies and their implications for enzymatic mechanisms.

These chapters give a picture of a mature but still evolving field, in which the complexities of the phenomena being uncovered are fully matched by the elegant experiments being designed to elucidate them.

THOMAS G. SPIRO

Princeton, New Jersey
August 1983

Contents

1 **Zinc in Biology and Biochemistry** 1

Bert L. Vallee

2 **Carboxypeptidase A** 25

Bert L. Vallee, Alphonse Galdes, David S. Auld, and James F. Riordan

3 **Carbonic Anhydrase** 77

Sven Lindskog

4 **The Role of Zinc in Alcohol Dehydrogenase** 123

Hans Eklund and Carl-Ivar Brändén

5 **Molecular Properties and Mechanism of Alkaline Phosphatase** 153

Joseph E. Coleman and Peter Gettins

6 **The Role of Zn(II) in RNA and DNA Polymerases** 219

Joseph E. Coleman

7 **Roles for the Metal Ion in Reactions of Coordinated Substrates and in Some Metalloenzymes** 253

Nicholas E. Dixon and Alan M. Sargeson

Index 353

CHAPTER 1

Zinc in Biology and Biochemistry

BERT L. VALLEE

Center for Biochemical and Biophysical Sciences and Medicine
Harvard Medical School
Boston, Massachusetts

CONTENTS

1 INTRODUCTION, 3

2 THE BIOLOGICAL ROLES OF ZINC, 7

3 PHYSIOLOGY OF ZINC, 9

4 PATHOLOGY OF ZINC, 10

5 ZINC ENZYMES AND BIOCHEMISTRY, 10

6 ZINC IN METALLOENZYMES, 13

7 POTENTIAL CATALYTIC MECHANISMS OF ZINC, 14

8 EXAMPLES OF ZINC ENZYMES, 15

9 ZINC AND CELL BIOLOGY, 17

REFERENCES, 22

1 INTRODUCTION

The presence of metals in biological matter has fascinated biologists for generations, although it has not proven easy to determine their function. This problem was accentuated by the fact that metals like zinc occur in such small amounts that even their very detection—and certainly their quantitative measurement—posed seemingly insurmountable problems only a generation ago. This focus of attention on the detection limits of analytical methods has become almost irrelevant now. Yet, the fact that the IIB metals and those of the first transition series all occur in biological matter in similarly low concentrations led to the tacit inference that all of them would be found to perform similar biological functions. Such an emphasis on concentrational similarities seemingly ignored the chemical properties of metals unique for each. Not too surprisingly, their biological roles have turned out to be similarly characteristic, enhanced by coordination of metals with a large variety of biological ligands. Such distinctive roles of metals are now well appreciated, methods for their detection are standardized, and approaches to the study of their functions have merged with those conventional in all other biochemical work. In short, metallobiochemistry is now established as an important subdivision of general biochemistry.

Among the transition and group II elements, the abundance of zinc in biology is second only to that of iron. It is present in and indispensible to all forms of life. Although the essentiality of zinc to normal life processes has been recognized widely since the 1930s, the underlying biochemistry has been elucidated only quite recently. Not counting the overriding analytical considerations mentioned above, the role of zinc in biological systems can be considered to have been investigated in four phases.

The first was nutritional and ultimately showed that zinc is essential for the development growth and differentiation of all species. Raulin (1) demonstrated that *Aspergillus niger* requires zinc for growth, as was shown subsequently for other microorganisms, plants, and animals. One hundred years later such studies culminated in the demonstration that zinc is indispensible for human growth and development. Table 1 summarizes the pertinent facts.

The second biochemical phase led to the realization that zinc is an integral component of many enzymes, indispensable to their catalytic function and structural stability. This phase was initiated by Keilin and Mann's (2) discovery of the first zinc enzyme, carbonic anhydrase. Thereafter, the rate of progress in this field accelerated remarkably as the result of conjoint progress in many disciplines. Technical advances in the isolation and characterization of enzymes in general and metalloenzymes in particular were basic to the rapid growth of metallobiochemistry and enzymology as is apparent from Table 2.

Table 1 Biological Roles of Zinc

	Growth	Development	Changes in Chemical Composition: Decrease	Changes in Chemical Composition: Increase	Decreased Enzymatic Activities
Microorganisms			Protein	DNA	Alkaline phosphatase
			RNA (ribosomal)	Amino acids	
	Retardation	Increase in cellular size	Pyridine nucleotides	Polyphosphates	Alcohol and D-lactate dehydrogenase
				Phospholipids	Tryptophan desmolase
				ATP	
				Organic acids	
Plants		Small abnormal leaves	Protein	Amino acids	Tryptophan desmolase
	Retardation	Chlorotic mottling	Auxin		Carbonic anhydrase
			Ethanolamine		Aldolase
		Decreased fruit production			Pyruvic carboxylase
Vertebrates		Testicular atrophy	Red blood cells	Uric acid	Alkaline phosphatase
	Retardation	Parakeratosis	Serum proteins		Pancreatic proteases
		Dermatitis			Malate, lactic alcohol dehydrogenase
		Coarse, sparse hair			NADH diaphorase

Table 2 Zinc Metalloenzymes 1982

Name	Number	Source	Role[a]
Class, I, Oxidoreductases			
Alcohol Dehydrogenase	9	Vertebrates, plants	A, D
Alcohol Dehydrogenase	1	Yeast	A
D-Lactate Dehydrogenase	1	Barnacle	?
D-Lactate Cytochrome Reductase	1	Yeast	?
Superoxide Dismutase	12	Vertebrates, plants, fungi, bacteria	(A)[b], D
Class II, Transferases			
Aspartate Transcarbamylase	1	*Escherichia coli*	B
Transcarboxylase	1	*Penicillium shermanii*	?
Phosphoglucomutase	1	Yeast	?
RNA Polymerase	10	Wheat germ, bacteria, viruses	A
DNA Polymerase	2	Sea urchin, T_4 phage	A
Reverse Transcriptase	3	Oncogenic viruses	A
Terminal dNT Transferase	1	Calf thymus	A
Nuclear poly (A) Polymerase	2	Rat liver, virus	A
Mercaptopyruvate Sulfur transferase	1	*E. coli*	?
Class III, Hydrolases			
Alkaline Phosphatase	8	Mammals, bacteria	A, (C)[b], D
Frutose-1,6-Biphosphatase	2	Mammals	C
Phosphodiesterase (exonuclease)	1	Snake venom	A
Phospholipase C	1	*Bacillus cereus*	A
Nuclease P_1	1	*Penicillium cirtrinum*	?
α-Amylase	1	*Bacillus subtilis*	B
α-D-Mannosidase	1	Jack bean	?
Aminopeptidase	10	Mammals, fungi, bacteria	A, C
Aminotripeptidase	1	Rabbit intestine	A
DD-Carboxypeptidase	1	*S. albus*	A
Procarboxypeptidase A	2	Pancreas	A
Procarboxypeptidase B	1	Pancreas	A
Carboxypeptidase A	4	Vertebrates, crustacea	A
Carboxypeptidase B	4	Mammals, crustacea	A
Carboxypeptidase (Other)	5	Mammals, crustacea, bacteria	A
Dipeptidase	3	Mammals, bacteria	A
Angiotensin-converting enzyme	3	Mammals	A
Neutral protease	16	Vertebrates, fungi, bacteria	A, (B)[b]

Table 2 (*Continued*)

Name	Number	Source	Role[a]
Collagenase	4	Mammals, bacteria	A
Elastase	1	*Pseudomonas aeruginosa*	?
Aminocylase	1	Pig kidney	?
β-Lactamase II	1	*B. cereus*	A
Creatininase	1	*P. putida*	?
Dihydropyrimidine aminohydrolase	1	Bovine liver	?
AMP deaminase	1	Rabbit muscle	?
Nucleotide pyrophosphatase	1	Yeast	A
Class IV, Lyases			
Fructose-1,6-biphosphate Aldolase	4	Yeast, bacteria	A
L-Rhammnulose-1-phosphate Aldolase	1	*E. coli*	A
Carbonic anhydrase	22	Animals, plants	A
δ-Aminolevulinic acid Dehydratase	2	Mammalian liver, erythrocytes	A
Glyoxalase I	4	Mammals, yeast	A
Class V, Isomerases			
Phosphomannose isomerase	1	Yeast	?
Class VI, Ligases			
t-RNA synthctase	3	*E. coli, Bacillus stearothermophilus*	A
Pyruvate carboxylase	2	Yeast, bacteria	?
Total	162*		

[a] A denotes a catalytic role, B a structural, C a regulatory, and D an undefined role. The ? indicates that available information is insufficient to make an assignment. See the text for further details.
[b] Letters in parentheses refers to roles fulfilled by metals other than zinc.
*Since this chapter was written, the total number of zinc enzymes has increased to beyond 200.

Zinc has now been found to be an essential component of more than 200 enzymes isolated from different species. They participate in biological reactions encompassing the synthesis and degradation of all major metabolites, that is, carbohydrates, lipids, proteins, and nucleic acids; at least one example of a zinc enzyme is found in each class of the six categories designated by the International Union of Biochemistry (Table 2). In addition, a number of

zinc proteins whose function is not as yet known, have also been described. Most recently, zinc enzymes have been shown to monitor the replication, transcription and translation of the genetic material of all species accounting for the absolute dependence of all forms of life on this element.

Recognition of the clinical manifestations associated with abnormal zinc metabolism in all vertebrates including man constitutes the third phase in the evolution of zinc biology and pathology. The symptoms of zinc deficiency are complex and vary markedly with the particular species studied (3) (Table 1). An increasing number of deficiency diseases occur spontaneously or can be induced experimentally. Finally, human disease can arise from conditioned zinc deficiencies (4), chronic dietary inadequacy (5), and congenital disease (6). In this respect, it is important to mention that when rats are rendered deficient during pregnancy, congenital malformations of the embryos occur in a very high percentage of the offspring with the primary effects on growing or proliferating tissues (see below).

Study of the fourth phase, the role of zinc in genetic expression, has just been initiated (7). A potential role of zinc in that sphere would seem to offer rich investigative opportunities.

Knowledge emerging from all four phases of investigation has given new incentives to experiments in all relevant disciplines. It has also sparked new interest in the possibility that recognition of the roles of metals in diseases might lead to their better management, a motivation over and above the scientific concern which is the focus of this overview.

2 THE BIOLOGICAL ROLES OF ZINC

The now very superior methods of analysis have markedly lowered the limits of zinc detection (8) with a corresponding gain in the understanding of its biological roles. However, once the zinc content of an organ is known quantitatively how can this knowledge be given functional significance?

The association of zinc (or lack of it) with specific biological functions can be detected by isolating cellular or subcellular components and individual molecules whose biological functions are known, measuring both activity and zinc.

The earliest indications that zinc plays a biological role date to 1869 when Raulin (1) found that *Aspergillus niger*, the common bread mold, does not grow in its absence. Shortly thereafter, zinc was found in plants and in animal tissues, and studies regarding its importance in plant growth began in the early part of the current century. In the subsequent two decades there was very little biological work that revealed any functional significance of zinc. Toxic effects of zinc, a result of inhalation of its fumes during smelting of its

ores, became of paramount interest to Public Health in the 1920s, however. Zinc was recognized to be an industrial hazard which was eliminated by suitable ventilation of work spaces and removal from processing plants. Thus, a toxic action of the element manifested as "metal fume fever" was established long before its essential biological functions were known. Yet, suitable credit should be given to an exceptionally alert team of investigators. While investigating the cause and basis of metal fume fever, Collier and Drinker (9) developed a chemical method—exceedingly imprecise and insensitive by present standards—for zinc analysis in organs. Nevertheless, they recognized that the presence of zinc in all organs of many species could not be accounted for by the then prevalent theories on toxicity. They realized and pointed out that this universal occurrence implied important biological functions rather than accidental contamination. This insight, epochal considering the state of the art and science of that time, has not lost any of its importance. It is relatively simple to implicate a metal in toxicity, but this does not allow the conclusion that it is not also biologically essential.

In this regard Gabriel Bertrand's "law" (10) should be recalled which states that the physiological and toxic effects of zinc and other metals should be considered a continuum. He pointed out that the physiological response to zinc would be expected to be triphasic as shown in Figure 1.

Potentially, the biological function of zinc and other metals can be as-

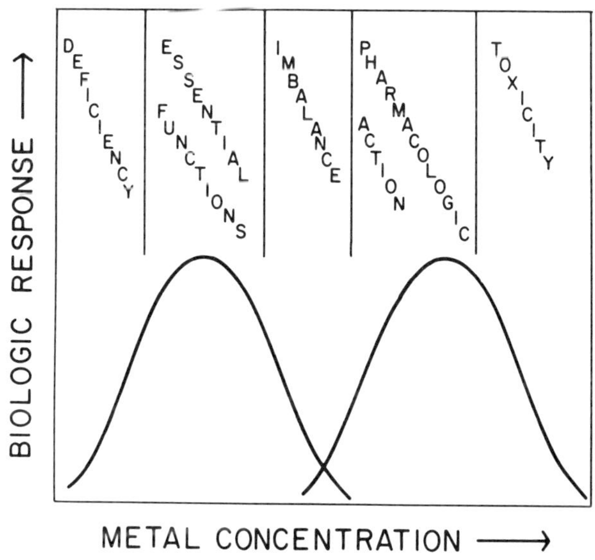

Figure 1 Potential dose response to a metal.

sessed properly only against the background of a deficiency state. Maximum functional response to zinc administration then constitutes phase I of this dose–response curve. Essential functions may be hampered, however, if the relationship of zinc to all other metals present is not balanced properly, resulting in imbalance, as indicated in the middle of the figure. Concentrations of zinc higher than those required to maintain essential functions may have secondary or pharmacological effects, phase II of the dose–response curve. At still higher concentrations, zinc—like every other element—is potentially toxic and eventually lethal, a pattern which pertains to virtually all elements for which adequate information now exists. Current knowledge has established the position of zinc in each portion of such a triphasic curve, and we are now left with the resolution of specific mechanisms leading to each of the responses.

3 PHYSIOLOGY OF ZINC

The tissues of a 70 kg human have been estimated to contain between 2 and 3 g of zinc (11) distributed in all tissues and varying from 10 to 200 μg zinc/g wet weight. Most organs contain 20–30 μg, while liver, voluntary muscle, and bone contain 60–180 μg zinc/g. The concentrations of zinc in the tissue of the eye, in particular of the iris, retina, and choroid, are somewhat higher. Zinc does not accumulate preferentially in any tissue. Although unexplained, its concentrations are remarkably high in the prostate, prostatic secretions, and spermatozoa, but they are unremarkable in the female reproductive tract. In most instances, the organ content of different species is quite similar.

Normal human whole blood contains about 900 μg zinc/100 ml and serum zinc concentrations average 96 ± 20 μg/100 ml (12). Normal erythrocytes contain 1.4 mg/100 ml of packed red blood cells. Three percent of all zinc in blood is found in leukocytes, 3.2×10^{-2} μg zinc/10^6 cells, about 25 times more than in a comparable number of erythrocytes. Variations of concentrations of zinc in blood are neither seasonal nor diurnal. In effect, all zinc in serum is protein bound and distributed in multiple—at least two and possibly more—fractions. Globulins bind zinc most firmly; an α_2-zinc macroglobulin, which does not seem to be a transport protein, accounts for 40–50% of all zinc in serum (12). Most of the remainder seems to be bound loosely to albumin or protein fractions electrophoretically close to the albumins. Whether or not albumin itself binds zinc specifically or nonspecifically is not clear. A specific transport protein remains the object of a persistent, yet so far frustrated, search. Carbonic anhydrase accounts for a large percentage of zinc in erythrocytes and alkaline phosphatase likely for some of that in leukocytes. Pregnancy and ACTH administration lower serum zinc concentrations (13).

Zinc is widely distributed in a variety of foods, particularly from animal sources. Shellfish is especially rich in this element. The average human dietary intake is about 10–15 mg zinc/day of which about 5 mg are retained.

The sites and mechanisms of absorption of zinc from the intestine are not known with certainty. A postulated role for metallothionein is unproven thus far. Phytic acid, a phosphorus storage compound of plant seeds which forms insoluble zinc·phytate complexes, decreases the availability of zinc for absorption, which calcium decreases further through the formation of a more insoluble mixed zinc·phytate·calcium complex. Once absorbed, zinc is excreted primarily in gastrointestinal and pancreatic secretions. Human urine contains about 0.5 mg Zn/day, an amount that is apparently independent of intake and urinary volume. Studies in the rat indicate that, in contrast to iron, the body stores of zinc are not mobilized readily and, hence, there is an unusual dependence upon a regular, exogenous supply of the element, particularly during periods of growth.

4 PATHOLOGY OF ZINC

Zinc-deficient female rats produce progeny with multiple congenital malformations including clubbed feet, adactyly, hydrocephalus, anencephaly, spina bifida, and cleft palate, among others. The decreased zinc content of the mothers is an index of the percentage of teratological defects observed (14).

In hogs a zinc deficiency called parakeratosis can either occur spontaneously (15) or can be induced experimentally (16).

In humans a number of zinc-deficiency diseases have been observed which are either nutritional, conditioned, or genetically determined. The conditioned zinc deficiency of post-alcoholic cirrhosis was the first to be observed (17). Dwarfism, first seen in Iranian and later on in Egyptian males, called attention to reversible growth failure in human adolescents (18, 19). Acrodermatitis enteropathica is the first human disease in which zinc reverses all pathological lesions and the accompanying clinical signs and symptoms (6, 20). The disease is inherited and transmitted as an autosomal recessive trait. It is characterized by alopecia, dermatitis, diarrhea, photophobia, psychological changes, and failure to thrive. Small daily supplements of zinc are curative. Similar signs and symptoms due to zinc deficiency have been observed in surgical postenterectomy patients (20a).

5 ZINC ENZYMES AND BIOCHEMISTRY

Zinc, a IIB element with a completed *d* subshell and two additional *s* electrons, chemically combines in the +2 oxidation state, but is neither oxidized

nor reduced in biological reactions. Its complex ions are generally tetrahedral although many octahedral ones are known and pentacoordinate ones are now being recognized with increasing frequency.

The characteristics of zinc in, for example, simple halo-, cyano-, and amino-complexes differ from those in metalloenzymes, the stereochemistry being determined largely by ligand size, electrostatic and covalent binding forces. The three-dimensional structure of proteins, heterogeneity of ligands, and the degree of vicinal polarity of the metal binding site jointly generate atypical coordination properties (21). An enzyme can generate zinc binding sites with unusual bond lengths, distorted geometries, and/or odd numbers of ligands, which when occupied by zinc or another metal, can be thermodynamically more energetic than a metal ion when free in solution where it is complexed to water or simple ligands. As a result, zinc is thought to be poised for its intended catalytic function in enzymes. In this context the term *entasis* indicates the existence of a condition of tension or stress in a zinc—or other metal—enzyme prior to combining with substrate (21).

In effect, the entatic state is thought to originate in the genetic heritage of the cell. The primary structure of the enzyme protein dictates the relative spatial positions of those amino acid side chains destined to serve as ligands when the apoprotein combines with the metal ion. Evidence suggests that the metal ion is not incorporated into the growing, ribosome-bound polypeptide chain (22), until the protein is fully formed. According to this view, the metal does not induce its own coordination site, but its interaction with the apoenzyme awaits the expression of the genetic message.

The lack of suitable physical-chemical probe properties of the diamagnetic zinc atom led to a search for a means to replace it with paramagnetic metals, for example, cobalt, which can signal information on the nature and environment of the active site (23). The spectra of such cobalt-substituted zinc enzymes (24–28) differ significantly from those of model Co^{2+} complexes and are thought to reflect the entatic state of the cobalt (and zinc) ions in these enzymes (21).

Identification of the metal ligands of metalloenzymes has been difficult by means other than X-ray crystallography. Cysteinyl, histidyl, and tyrosyl residues and the carboxyl groups of aspartic and glutamic acids have been implicated most commonly (29). Thus far, metal complex ions in which the mode of metal coordination is known quite precisely have not proven adequate to define metal binding in enzymes, invariably lacking the entatic environment, characteristic of metalloenzymes. Moreover, the water soluble complexes that are known most thoroughly and have been studied most extensively are bidentate, but all present evidence indicates that in metalloenzymes, metals are coordinated to at least three ligands (29). With few exceptions (30), multidentate complex ions, suitable for appropriate comparisons with metalloenzymes, have not been identified. Yet, absorption (31), circular dichroic and

magnetic circular dichroic (28, 32), and electron paramagnetic resonance spectra (33) combined with kinetic studies (26, 34, 35) of cobalt-substituted and chemically modified zinc metalloenzymes have enlarged the understanding both of the possible modes of interaction of zinc with the active sites of zinc metalloenzymes and their potential mechanisms of action.

The molecular details of metalloenzyme action have been elucidated greatly in the last few years (36). Crystal structures for bovine carboxypeptidase A (37), carbonic anhydrase (37a), thermolysin (38), horse liver alcohol dehydrogenase (39), superoxide dismutase (39a), alkaline phosphatase of *E. coli* (40) and aspartate transcarbomylase (40a) are now available, and chemical and kinetic studies have defined the role of zinc in substrate binding and catalysis (29, 36). In fact, many of the significant features elucidating the general modes of action of metals in enzymes have been defined at the hands of zinc metalloenzymes.

Work performed in the past 20 years has shown that zinc is essential to enzymatic reactions (29). The functional role of iron in oxidative processes recognized 100 years ago, suggested an essential role for that element and pointed the way to the discovery of metalloenzymes in general. However, the idea that other first-transition and IIB group metals might have significant roles remained unappreciated for a long time and biochemical knowledge in this area was almost nonexistent.

The search for an explanation of the physiologic role of zinc emphasized its interaction with enzymes. Such metalloenzymes are characterized by their stability constants, whose magnitudes have served as the operational basis for the two extremes of metal–protein interactions, that is, the very stable metalloenzymes and the more labile metal–enzyme complexes (41), the latter exemplified by sodium, potassium, calcium, and magnesium, the most abundant cations of mammalian species.

The transition metals and zinc, because of their electronic structures, tend to form stable complexes with proteins with characteristic functions. The lack of visible color of zinc proteins accounts for the long delay in their recognition, while the red iron (42) and blue copper (43) proteins called attention to themselves early. Carbonic anhydrase was the first zinc enzyme to be isolated and purified (2), bovine pancreatic carboxypeptidase A was the second to be recognized in 1954 (44). Additional zinc metalloenzymes were discovered quite rapidly (41, 45). Zinc is now known to be an integral component of a large variety of proteins and enzymes (Table 2) and, hence, the total zinc content of tissues cannot serve as a guide to the multiple functions of the metal. The number of functionally diverse types of zinc metalloenzymes that are now known document the importance of this metal in metabolism. The more than 200 zinc-containing enzymes and proteins now known have been isolated from multiple sources encompassing all phyla, and they participate in a wide

variety of metabolic processes including carbohydrate, lipid, protein, and nucleic acid synthesis or degradation. Each of the six major categories of enzymes, designated by the IUB Commission on Enzyme Nomenclature according to function, contains at least one example of a zinc enzyme, among them, several dehydrogenases, aldolases, peptidases, phosphatases, an isomerase, a transphosphorylase and aspartate transcarbamylase. Thus, there are zinc oxidoreductases, transferases, hydrolases, lyases, isomerases, and ligases (34, 35, 46) (Table 2). It may be conjectured that the biological abundance of zinc, its adaptability to interact with widely varied coordination geometries, and its resistance to oxidoreduction have generated selective evolutionary pressure for it to serve in biological catalysis.

In addition, there are a number of zinc proteins whose function are as yet unknown. Foremost among these are the α_2-macroglobulin of human serum and the metallothioneins. The metallothioneins are a remarkable set of small-molecular-weight (~6700) widely distributed proteins with exceptionally high sulfur content; one third of all amino acids are cysteines. Their zinc, cadmium (or copper) content is similarly high, 6–7 g atom/mol. Their function is unknown (47).

6 ZINC IN METALLOENZYMES

The roles of zinc in metalloenzymes can be divided into four categories: *catalytic, structural, regulatory*—or *modulatory*, and *noncatalytic*.

Zinc is said to have a *catalytic* role when it is essential for and *directly* involved in catalysis by the enzyme; carbonic anhydrase, carboxypeptidase, thermolysin, and aldolases are examples of this type. The removal of catalytic zinc results in an inactive apoenzyme which, however, often retains the native tertiary structure. The manner in which zinc participates in catalysis is detailed further below.

Zinc plays a *structural* role when it is required *solely* for the structural stability of the protein, being necessary for activity only to the extent that the overall conformation of the enzyme affects its action. Structural zinc often, but not exclusively, stabilizes the quaternary structure of oligomeric holoenzymes. Thus, a zinc atom serves to dimerize *B. subtilis* α-amylase without affecting its enzymatic activity. Similarly, zinc stabilizes the pentameric quaternary structure of aspartate transcarbamylase. The removal of zinc from this type of metalloenzyme (or from the dissociated subunits) prevents association.

A *regulatory* (or *modulatory*) role is indicated when zinc regulates, but is not essential for, enzymatic activity (which is present in the absence of metal) or for the stability of the protein. Regulatory zinc may act either as an activa-

tor, for example, bovine lens leucine aminopeptidase, or as an inhibitor, for example, porcine kidney leucine aminopeptidase or fructose-1,6-biphosphatase.

A fraction of zinc in certain metalloenzymes, for example, equine and human alcohol dehydrogenase and *E. coli* alkaline phosphatase, is neither involved directly in catalysis nor essential for the maintenance of the tertiary structure of the enzyme (although it may stabilize it), so that its function is as yet obscure. In the absence of specific knowledge on how the metal acts, it is noncommitally referred to as *noncatalytic*.

A given metalloenzyme may contain multiple numbers and types of zinc atoms. Thus, equine alcohol dehydrogenase contains both a catalytic and a noncatalytic zinc atom per subunit, and leucine aminopeptidase contains a catalytic and a regulatory zinc atom per subunit. In Table 2 zinc metalloenzymes known up to the present are categorized according to these roles of the metal, A denoting a catalytic, B a structural, C a regulatory, and D an undefined role. The appearance of multiple letters identifies different zinc atoms playing different roles in the same enzyme. In some instances the enzymes from unlike sources differ in this regard and are therefore listed separately on this basis, for example, alcohol dehydrogenase from yeast and vertebrates, respectively.

Metal exchange studies have shown that the replacement of catalytic and regulatory zinc with other metal ions can profoundly affect activity, while that of structural and noncatalytic zinc atoms seems to have only minor consequences. Moreover, the coordination geometry of catalytic zinc appears to differ significantly from that of structural and noncatalytic zinc atoms; the coordination properties of regulatory zinc atoms are as yet unknown. Thus, X-ray analysis shows catalytic zinc to be bound by three protein ligands and a water molecule, while structural and noncatalytic zinc are fully coordinated by four protein ligands. The presence of a water molecule bound to catalytic zinc, signifying an open coordination site, is considered essential for the function of zinc in catalysis. Furthermore, the coordination geometry of catalytic zinc is highly distorted and can fluctuate between tetracoordinate and pentacoordinate, for example, as in carbonic anhydrase, properties that reflect it to be *entatic* while the coordination geometry of noncatalytic zinc is more regular and, hence, not entatic. *Entasis* may be considered the keystone to catalysis by zinc metalloenzymes because it lowers the energy barrier for the transition state and hence accelerates the conversion of substrate to products.

7 POTENTIAL CATALYTIC MECHANISMS OF ZINC

In spite of intensive research, the precise mode of action of zinc is still not fully understood in any metalloenzyme. From the earliest days of the study of

zinc enzymes, two types of mechanisms have been thought to account for the manner in which the metal affects catalysis. One, termed the *zinc-carbonyl* mechanism, proposes that the substrate binds *directly* to zinc and displaces the metal-bound H_2O molecule in the process. The zinc is then envisioned to act as a Lewis acid and to polarize the bound substrate, thereby facilitating nucleophilic attack (e.g., aldolase, peptidases). Thus, according to this hypothesis, the function of zinc is thus to activate the *electrophile*. The second, known as the *zinc-hydroxide mechanism*, proposes that the substrate does *not* bind directly to the metal and that zinc mediates its function through the metal-bound-water molecule, which is *not* displaced on the binding substrate. Zinc is thought to lower the pK_a of the bound-water molecule from ~14 to ~7. The resultant metal-bound hydroxide ion can then attack the substrate, for example, carbonic anhydrase. Hence, in this hypothesis, the function of zinc is to activate the *nucleophile*.

These two hypotheses need not be mutually exclusive, and a mechanism that integrates the zinc-carbonyl and zinc-hydroxide mechanisms can be envisioned. According to this integrated hypothesis, the substrate binds *directly* to the metal but does *not displace* the metal-bound water, resulting in a *pentacoordinate-like* intermediate. The function of the zinc then would be *both* to polarize the substrate *and* to activate the water molecule that acts as a nucleophile. In addition, the metal, through its flexible coordination geometry, would act as a template to bring together the substrate and nucleophile; this suggestion is consistent with the entatic nature of the metal. This third mechanism, originally proposed for carbonic anhydrase, is thought to find support in recent crystallographic studies showing that CO_2 and H_2O are bound simultaneously to the catalytic zinc atom in this enzyme. However, at present the evidence for this mechanism is not decisive.

8 EXAMPLES OF ZINC ENZYMES

Alcohol dehydrogenases, alkaline phosphatases, carboxypeptidases, and carbonic anhydrases have been obtained from human tissues and erythrocytes, among other sources (Table 2). In these, as in other zinc metalloenzymes, the metal is indispensable for catalytic function and/or the structural integrity of the molecule.

Alcohol dehydrogenase of liver as well as other organs oxidizes ethanol, other primary and secondary alcohols, sterols, as well as vitamin A alcohol. Retinene reductase of the retina is apparently identical or very similar to the liver enzyme. Horse and human liver alcohol dehydrogenases contain four atoms of zinc per molecule of protein, and zinc is essential not only to the catalytic function of the enzyme, but also to maintain the subunit structure (48, 49). In addition to the above substrates, alcohol dehydrogenase from *hu-*

man liver also oxidizes methanol and ethylene glycol, serving as the primary mechanism of detoxification of these and other similar compounds (50). Human liver alcohol dehydrogenase has been purified (51, 52). It is composed of at least 20 isoenzymes, a remarkable polymorphism that remains to be explained.

Erythroycytes contain carbonic anhydrase in high concentrations. The enzyme catalyzes the reaction $CO_2 + H_2O \rightleftharpoons H_2CO_3$, and many tissues exhibit this enzymatic activity. In its absence the rate of carbon dioxide elimination is insufficient to sustain life; hence, this enzyme is as important to the elimination of carbon dioxide as is that of hemoglobin to the absorption and transport of oxygen. Carbonic anhydrases from ox, monkey, human, and other erythrocytes all contain 1 g atom zinc/mol enzyme. Under both normal and pathologic conditions there is a close correlation between the zinc content and this enzymatic activity of red blood cells. Substitution of cobalt for zinc results in an enzymatically active carbonic anhydrase (53).

Both carboxypeptidase A and B from pancreatic juice contain 1 g atom zinc/molecule of protein. The single zinc atom is indispensable for the catalytic activities of the enzymes which hydrolyze peptide bonds to liberate the amino acids from the carboxy-terminal ends of proteins and peptides. Zinc in carboxypeptidase A can be replaced *in vitro* by other metals, for example, Co, Mn, Ni, Cd, Hg, and Pb, with consequent dramatic alterations in catalytic activity and substrate specificity (54). Both carboxypeptidase A and B are excreted into the gastrointestinal tract and are essential to proteolysis and digestive processes.

Zinc is essential for the activity of alkaline phosphatase from *E. coli*. The enzyme consists of two subunits: in each of the two subunits one zinc atom is responsible for function and another for structure stabilization and—in addition—two Mg atoms control the activity of the holoenzyme (55). A decrease in alkaline phosphatase activity has long been noted in zinc-deficient experimental animals, a change which may now be attributed to a failure of synthesis of the active holoenzyme. The alkaline phosphatase of leukocytes may well be identical with the zinc-containing protein of human leukocytes which contain 0.3% zinc/g dry weight of protein, and is responsible for 80% of all zinc found in human leukocytes (56).

There are unusual opportunities to study the manner in which the specific function of zinc proteins and enzymes is achieved through interaction of the metal with the protein. By removal and restoration of zinc with a concomitant loss and restitution of enzymatic activity, it is possible to achieve the differential chemical labeling of the sites of the protein to which the metal binds. This permits physical-chemical approaches to discern what contribution zinc may make to overall protein structure and stability. Other metals can be substituted successfully in a number of enzymes for the native zinc with correspond-

ing characteristic alterations of physical properties and/or catalytic function. These, as well as other approaches, have assisted in the delineration of the role of zinc in the function and structure of enzymes. In addition to metal substitutions, these include inhibition by chelating agents, kinetics of metal exchange, absorption, circular dichroism, and magnetic circular dichroism, electron spin and nuclear magnetic resonance as well as resonance Raman spectroscopy, chemical modifications, and, last but not least, X-ray crystallographic analysis (28).

Substantial quantities of firmly bound zinc in RNA and DNA have revealed additional important avenues of its role in biology. Zinc appears to stabilize the secondary and tertiary structure of RNA and to play an important role in protein synthesis (see below). To date there is little or no information on the association of zinc with lipids or carbohydrates, glyco- and lipoproteins, but systematic analysis of enzymes involved in their metabolism might turn out to be important.

9 ZINC AND CELL BIOLOGY

The essentiality of zinc to the growth and development of *all* living matter was just recognized within the last decade in spite of abundant, long-standing evidence pointing in this direction (see above). A series of major, closely interrelated biochemical advances in nucleic acid chemistry changed the perspective rapidly.

Histochemical observation indicated early that zinc has a role in nucleic acid metabolism and is present in the nucleus, nucleolus, and chromosomes (57). Substantial quantities of firmly bound zinc stabilize the structures of RNA (58, 59), DNA (60), and ribosomes (61). EDTA inhibits DNA polymerase activity in cell cultures (62) and zinc deficiency lowers the activities of both DNA and RNA polymerases in various tissues and organisms, observations which intimated an extensive involvement of zinc in the metabolism of nucleic acids (63–65).

Chelating agents also inhibit the highly purified terminal deoxynucleotidyl transferase from calf thymus; on that basis Chang and Bollum (66) suggested that it is a zinc metalloenzyme. Metal analyses soon confirmed the presence of zinc in this enzyme and nucleotidyl polymerases isolated from viruses, prokaryotes, and eukaryotes. Thus, within a year of Bollum's observations, the RNA (67) and DNA polymerase (68) from *E. coli* were shown to contain 2 and 1 g atom zinc/mol, respectively. At present, approximately 20 nucleotidyl polymerases are known to contain zinc (Table 3). The role of zinc in these enzymes has been studied by standard approaches that include the concomitant reversible abolition and restoration of activity by chelating agents and of zinc.

Table 3 Zinc in Nucleic Acid and Protein Synthesis

	Zinc (g atom/mol)	Source
DNA Polymerases	1.0	*E. coli*[b]
	4.2	Sea urchin
	1.0	T4 phage
		Oncogenic viruses
	1.8–2.0	Avian
	1.4	Murine
	[a]	Feline
	1.0	Simian
	[a]	RD-144
RNA Polymerases	2.0	*E. coli*
	2.4	T7
	2.4	Yeast (I)
	1.0	Yeast (II)
	2.0	Yeast (III)
	2.1	*Euglena gracilis* (I)
	2.2	*E. gracilis* (II)
	2.2	*E. gracillis* (III)
	2.0	*B. subtilis*
	7.0	Wheat germ (II)
	5.5	N4 virion
tRNA Synthetases	2.0	*E. coli*
Deoxynucleotidyl transferases	1.0	Calf thymus
Elongation Factor I	1.3	Rat liver

[a]Molecular weight unknown.
[b]The original report (68) of the metal content of this polymerase has recently been retracted (78) subsequent to the finding of others that the enzyme does not contain zinc (79).

Zinc is thus indispensable to the function of enzymes critical to DNA, RNA, protein synthesis, and—as was long known—protein degradation. However, but a short time ago, little or nothing was known about any role of zinc in cellular metabolism.

The data of Chang and Bollum on the zinc dependence of deoxynucleotidyl transferase of calf thymus (66) pointed to transferases as a promising approach to study the metabolic consequences of zinc deficiency in *E. gracilis*. This unicellular protist has long served as a model system in cell biology.

Table 4 summarizes the characteristics of RNA polymerases I, II, and III of *E. gracilis*; they resemble those for the analogous enzymes obtained from other sources. All of them are zinc enzymes. In contrast to the *three* poly-

Table 4 Properties of DNA-Dependent RNA Polymerases from (+Zn) and (−Zn) *E. gracilis*

RNA polymerases	Molecular Weight (daltons)	$(V_i/V_c) \times 100$ (200 μg/ml α-amanitin)	OP[a] Inhibition	Zn (g atom/mol)
		+Zn		
I	630,000	100	+	2.1
II	700,000	0	+	2.2
III	650,000	25	+	(2.2)[c]
		−Zn		
"χ"[b]	650,000	40	+	(1.8)[c]

[a]Orthophenanthroline.
[b]The only polymerase found in −Zn cells.
[c]The bracket denotes a provisional value.

merases isolated from these (+Zn) eukaryotic cells, there is only *one* RNA polymerase in (−Zn) cells, dubbed "χ," reminiscent of prokaryotes. It contains a full complement of zinc and generates all three, that is, ribosomal, transfer, and messenger RNAs.

However, the properties of "χ" RNA polymerase from (−Zn) cells differ from those of all three RNA polymerases of the (+Zn) organisms, which zinc sufficiency restores completely. The DNA polymerases of *E. gracilis* also contain functionally essential zinc atoms (69). Thus, the entire enzymological machinery responsible for the synthesis of nucleic acids in *E. gracilis* is zinc dependent.

The mRNA of zinc-sufficient cells is composed of uracil (U), guanine (G), cytosine (C), and adenine (A). In addition, deficient cells contain three major and several minor additional bases which are as yet unidentified but not found in mRNA from normal cells. The (G + C)/(A + U) ratios of the mRNA samples, 1.6 for (+Zn) and 2.9 for (−Zn) cells, differ strikingly (70). Thus zinc deficiency also changes the composition of mRNA, but it does not alter its functional capacity as expressed by translation.

The capacity of purified *E. gracilis* mRNAs to initiate the synthesis of both large numbers and different types of peptides and proteins can be demonstrated in reticulocyte and wheat germ S_{30} lysate translation systems. The (+Zn) mRNA system produces a number of polypeptides of molecular weight greater than 10,000 daltons which are absent from that using mRNA of (−Zn) cells. Conversely, the (−Zn) system contains proteins either present in

much lesser amounts or not present at all in the (+Zn) system, as can also be shown with intact cells.

Fifteen years ago, it was shown that arginine-rich polypeptides, which were then not identified further, are characteristic of (−Zn) cells (71). In a fraction containing more than ten such arginine-rich peptides, four have now been sequenced. This material promptly vanishes when zinc sufficiency is restored; it contains nearly 50% arginine with asparagine, glutamine, and lysine accounting for the remainder.

Arginyl residues bind phosphate with particular avidity (72), and this peptide fraction markedly inhibits the activity of *E. gracilis* RNA polymerase II.

The simultaneous abolition of the synthesis of the three (−Zn) *E. gracilis* RNA polymerases and the formation of the "χ" form in zinc deficiency suggests a selective effect of zinc on *gene expression*, a process in which the RNA polymerases and histones play cardinal roles. Such a hypothesis calls for the study of both, particularly since the histones bind to DNA and facilitate or restrict its interaction with RNA polymerases to regulate transcription.

In an electric field the histones from (+Zn) *E. gracilis* (73) containing histones H_1, H_3, H_2B, H_2A, and H_4 (74) migrate in a manner that differs distinctly from that of material obtained from (−Zn) cells (Fig. 2). In addition, a fraction in the most cathodic region of the (−Zn) histones, absent from the comparable one of (+Zn) cells, comprises the arginine-rich peptide fraction.

The differences in the electrophoretic patterns of the histones and the accumulation of arginine-rich peptides in (−Zn) cells are entirely consistent

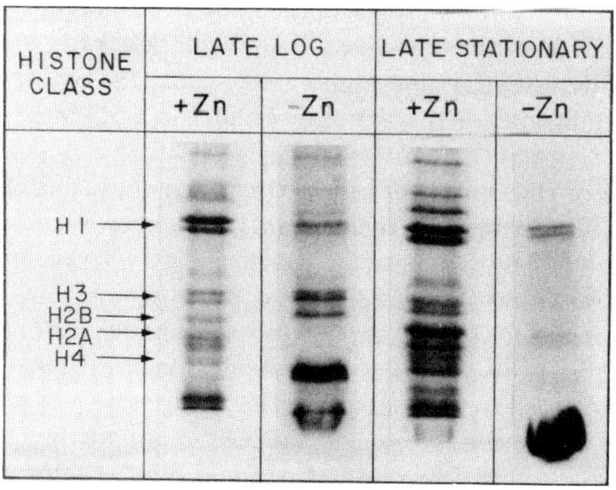

Figure 2 Equal amounts of histone fractions from (+Zn) and (−Zn) *E. gracilis* were electrophoresed on 15% polyacrylamide gels (0.1% SDS, pH 8.8). Identification of histones (left) by the criteria of Panyim and Chalkley (74) and Jardine and Leaver (73). The differences in histone patterns in the late log and late stationary growth phases are apparent.

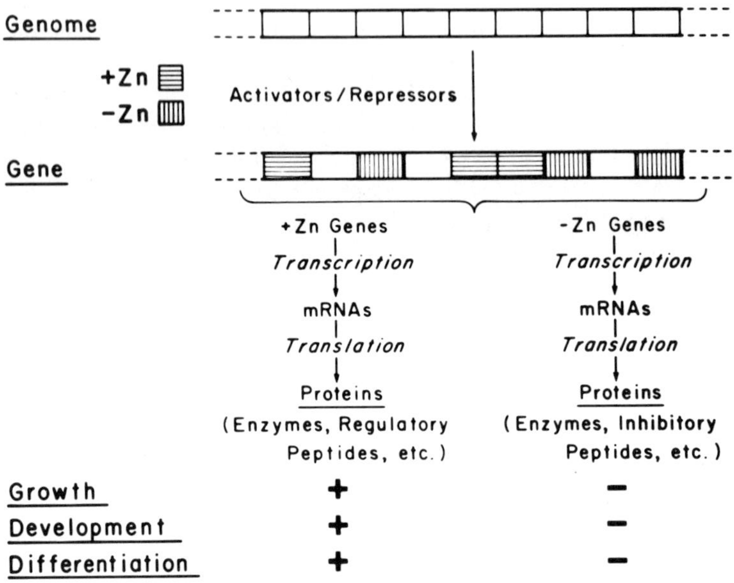

Figure 3 Expression of some genes is affected by binding of activators and/or repressors and this process is affected by zinc (±Zn). Activated (or repressed) genes are then transcribed (or not) into mRNA. This results in products or their failure to form. The consequences of (−Zn) manifest as alterations or cessation of growth, development, and differentiation.

with a crucial role for the element in gene expression. They further raise questions regarding its possible sites of action which might monitor the interaction of DNA with proteins known to regulate its function and structure.

Only a few of the large number of possible mechanisms that merit exploration can be mentioned here. When zinc becomes growth limiting, the biochemical responses will shift the cell from a state of active division to one of stationary growth. The consequent metabolic changes then will reduce protein synthesis, utilization of existent cellular energy sources, for example, β-1,3-glucan, and the stabilization, maintenance, and protection of existent inactive DNA from nucleases. Such metabolic transformations require selective gene activation and/or repression, perhaps as observed in the lac and tryp operons (75, 76) depending on molecules which bind to DNA and facilitate or hinder RNA polymerase binding to modulate transcription (77).

Zinc deficiency could selectively affect the transcription of genes through a number of such mechanisms. Thus, the histones of *E. gracilis* could be modified by, for example, metal-dependent methylation or phosphorylation or through interaction with arginine-rich peptides. The latter could also interact with DNA directly, of course, with arginyl residues binding to phosphate groups of the nucleic acid. In this manner, specific genes could be activated

ressed, selected proteins and enzymes synthesized or inhibited resulting arrest of cell division (Fig. 3).

These studies of the function of zinc in *E. gracilis* lay a foundation for the interpretation of relevant data yet to be acquired systematically for other eukaryotes, including vertebrates. They do clearly show the pervasive role of this metal in the overall growth, development, and differentiation which seem to be understood best by postulating a regulatory function through gene activation and/or repression. That hypothesis provides a rationale for the teratological, genetic, and medical abnormalities observed in zinc deficiency and predicts yet other, similar defects that have not as yet been recognized as being attributable to zinc and its metabolism. It may be predicted with confidence that the next decade will see major strides and achievements to consolidate these new insights.

REFERENCES

1. J. Raulin, *Ann. Sci. Natl. Botan. et Biol. Vegetale*, **11,** 93 (1869).
2. D. Keilin and T. Mann, *Biochem. J.*, **34,** 1163 (1940).
3. T.-K. Li and B. L. Vallee, in *Modern Nutrition in Health and Disease*, 6th ed., R. S. Goodhart and M. E. Shils, Eds., Lea and Febinger, Philadelphia, PA., 1980, p. 408.
4. B. L. Vallee, *Physiol. Rev.*, **39,** 443 (1959).
5. A. S. Prasad, Ed., *Zinc Metabolism*, C. C. Thomas, Springfield, Ill., 1966.
6. E. J. Moynahan, *Lancet*, **2,** 399 (1974).
7. B. L. Vallee and K. H. Falchuk, *Philos. Trans. R. Soc. Lond., B294*, 185 (1981).
8. C. Veillon and B. L. Vallee, *Methods Enzymol.*, **49,** 446 (1978).
9. E. S. Collier and K. P. Drinker, *J. Ind. Hyg.*, **8,** 257 (1926).
10. G. Bertrand, *8th Int. Congr. Appl. Chem.*, **28,** 30 (1912).
11. R. A. McCance and E. W. Widdowson, *Biochem. J.*, **36,** 392 (1942).
12. A. F. Parisi and B. L. Vallee, *Biochemistry*, **9,** 2421 (1970).
13. K. H. Falchuk, *N. Eng. J. Med.*, **296,** 1129 (1977).
14. L. S. Hurley, *Developmental Nutrition*, Prentice-Hall, Englewood Cliffs, N. J., 1980, p. 335.
15. H. C. H. Kernkamp and E. F. Ferrin, *J. Am. Vet. Med. Assoc.*, **123,** 217 (1953).
16. H. F. Tucker and W. D. Salmon, *Proc. Soc. Exp. Biol. Med.*, **88,** 613 (1955).
17. B. L. Vallee, W. E. C. Wacker, A. F. Bartholomay, and E. D. Robin, *New Engl. J. Med.*, **255,** 403 (1956).
18. A. S. Prasad, J. A. Halstead, and M. Nadimi, *Am. J. Med.*, **31,** 532 (1961).
19. A. S. Prasad, A. Miale, Z. Farid, and H. H. Sandstead, *J. Lab. Clin. Med.*, **61,** 537 (1963).
20. E. J. Moynahan and P. M. Barnes, *Lancet*, **2,** 676 (1973).

20a. R. G. Kay and C. Tasman-Jones, *Lancet 2*, 605 (1975).

21. B. L. Vallee and R. J. P. Williams, *Proc. Natl. Acad. Sci. USA*, **59,** 498 (1968).

22. M. I. Harris and J. E. Coleman, *J. Biol. Chem.*, **243,** 5063 (1968).
23. B. L. Vallee, in *Metal Ions in Biological Systems*, Vol. 40, S. K. Dhar, Ed., Plenum, New York, 1973, p. 1.
24. S. Lindskog and P. O. Nyman, *Biochim. Biophys. Acta*, **85,** 141 (1964).
25. R. T. Simpson and B. L. Vallee, *Biochemistry*, **7,** 4343 (1968).
26. B. L. Vallee and S. A. Latt, in Structure-Function Relationships of Proteolytic Enzymes, P. Desnuelle, H. Neurath and M. Ottesen, Eds., Academic, New York, 1970, p. 144.
27. B. Holmquist, T. A. Kaden, and B. L. Vallee, *Biochemistry*, **14,** 1454 (1975).
28. D. W. Darnell and R. G. Wilkins, Eds., Elsevier/North Holland, New York, N.Y., Amsterdam, Oxford, *Methods for Determining Metal Ion Environments in Proteins*, Las Cruces, N. M., 1980, pp. 27 and 265.
29. B. L. Vallee and W. E. C. Wacker in *The Proteins*, Vol. 5, 2nd ed., H. Neurath, Ed., Academic, New York, 1970.
30. T. Herskovitz, B. A. Averill, R. H. Holm, J. A. Ibers, W. D. Phillips, and J. F. Weiher, *Proc. Natl. Acad. Sci. USA*, **69,** 2437 (1972).
31. S. A. Latt and B. L. Vallee, *Biochemistry*, **10,** 4263 (1971).
32. T. A. Kaden, B. Holmquist, and B. L. Vallee, *Inorg. Chem.*, **13,** 2585 (1974).
33. F. S. Kennedy, H. A. O. Hill, T. A. Kaden, and B. L. Vallee, *Biochem. Biophys. Res. Commun.*, **48,** 1533 (1972).
34. B. L. Vallee, *Carlsberg Res. Commun.*, **45,** 423 (1980).
35. B. L. Vallee, in *Structural and Functional Aspects of Enzyme Catalysis*, H. Eggerer and R. Huber, Eds., Proceedings from 32nd Colloquium, Mosbach, Springer Verlag, Heidelberg, 1981, p. 75.
36. J. F. Riordan and B. L. Vallee, *Adv. Exp. Med. Biol.*, **48,** 33 (1974).
37. F. A. Quiocho and W. N. Lipscomb, *Adv. Prot. Chem.*, **25,** 1 (1971).
37a. K. K. Kannan, A. Liljas, I. Vaara, P. C. Bergsten, S. Lovgren, B. Standberg, U. Bengtsson, U. Carlbom, K. Fridborg, L. Jarup, and M. Petef, *Cold Spring Harbor Symp. Quant. Biol.* **36,** 221 (1971).
38. P. M. Colman, J. N. Jansonius, and B. W. Matthews, *J. Mol. Biol.*, **70,** 701 (1972).
39. C. I. Branden, H. Jörnvall, H. Eklund, and B. Furugren, in *The Enzymes*, P. D. Boyer, Ed., Academic, New York, 1975, p. 104.
39a. J. S. Richardson, K. A. Thomas, and D. C. Richardson, *Biochem. Biophys. Res. Commun.* **63,** 986 (1975).
40. J. M. Sowadski, B. A. Foster, and H. M. Wyckoff, *J. Mol. Biol.*, **150,** 245 (1981).
40a. H. L. Monaco, J. L. Crawford, and W. N. Lipscomb, *Proc. Natl. Acad. Sci. USA,* **75,** 5276 (1978).
41. B. L. Vallee, *Adv. Prot. Chem.*, **10,** 317 (1955).
42. W. T. Preyer, *De Haemoglobiono Observationes et Experimentia*, M. Cohen and Sons, Bonn, 1866, p. 27.
43. E. Harless, *Müller's Arch. F. Anat. Physiol.*, 148 (1847).
44. B. L. Vallee and H. Neurath, *J. Biol. Chem.*, **217,** 253 (1955).
45. B. L. Vallee, *Physiol. Rev.*, **39,** 443 (1959).
46. B. L. Vallee and W. E. C. Wacker, in G. D. Fasman, Ed., *Handbook of Biochemistry and Molecular Biology*, 3rd ed., Vol. 2, CRC Press, Cleveland, 1976, p. 276.

47. J. H. R. Kägi and M. Nordberg, Eds., *Metallothionein*, Birkhaüser Verlag, Basel, 1979, p. 378.
48. D. E. Drum, J. H. Harrison, T.-K. Li, J. L. Bethune, and B. L. Vallee, *Proc. Natl. Acad. Sci. USA*, **57,** 1434 (1967).
49. T.-K. Li, *Adv. Enzymol.*, **45,** 427 (1977).
50. A. H. Blair and B. L. Vallee, *Biochemistry*, **5,** 2026 (1966).
51. L. G. Lange and B. L. Vallee, *Biochemistry*, **15,** 4681 (1976).
52. L. G. Lange, A. J. Sytkowski, and B. L. Vallee, *Biochemistry*, **15,** 4687 (1976).
53. S. Lindskog and B. J. Malmstrom, *Biol. Chem.*, **237,** 1129 (1962).
54. B. L. Vallee, J. F. Riordan, and J. E. Coleman, *Proc. Natl. Acad. Sci. USA*, **49,** 109 (1963).
55. W. F. Bosron, F. S. Kennedy, and B. L. Vallee, *Biochemistry*, **14,** 2275 (1975).
56. F. L. Hoch and B. L. Vallee, *J. Biol. Chem.*, **195,** 531 (1952).
57. T. Fujii, *Nature Lond.*, **174,** 1108 (1954).
58. W. E. C. Wacker and B. L. Vallee, *J. Biol. Chem.*, **234,** 3257 (1959).
59. K. Fuwa, W. E. C. Wacker, R. Druyan, A. F. Bartholomay, and B. L. Vallee, *Proc. Natl. Acad. Sci. USA*, **46,** 1298 (1960).
60. Y. A. Shin and G. L. Eichhorn, *Biochemistry Wash.*, **7,** 1026 (1968).
61. J. A. Prask and D. J. Plocke, *Pl. Physiol.*, **48,** 150 (1971).
62. I. Lieberman and P. Ove, *J. Biol. Chem.*, **237,** 1634 (1962).
63. H. Swenerton, R. Shrader, and L. S. Hurley, *Science, N.Y.*, **166,** 1014 (1969).
64. H. H. Sandstead and R. A. Rinaldi, *J. Cell Physiol.*, **73,** 81 (1969).
65. K. H. Falchuk, A. Krishan, and B. L. Vallee, *Biochemistry, Wash.*, **14,** 3439 (1975).
66. L. M. S. Chang and F. J. Bollum, *Proc. Natl. Acad. Sci. USA*, **65,** 1041 (1970).
67. M. C. Scrutton, C. W. Wu, and D. A. Goldwait, *Proc. Natl. Acad. Sci. USA*, **68,** 2497 (1971).
68. J. P. Slater, A. S. Mildvan, and L. A. Loeb, *Biochem. Biophys. Res. Commun.*, **44,** 37 (1971).
69. A. G. McLennan and H. M. Keir, *Biochem. J.*, **151,** 227 (1975).
70. K. H. Falchuk, C. Hardy, L. Ulpino, and B. L. Vallee, *Proc. Natl. Acad. Sci. USA*, **75,** 4175 (1978).
71. W. E. C. Wacker, W. Kornicker, and L. Pothier, Abstr. Am. Chem. Soc. 150th Meetings 88C, (1965).
72. J. F. Riordan, *Molec. Cell Biochem.*, **26,** 71 (1979).
73. N. J. Jardine and J. L. Leaver, *Biochem. J.*, **169,** 103 (1978).
74. S. Panyim and R. Chalkley, *J. Biol. Chem.*, **246,** 7557 (1971).
75. R. C. Dickson, J. Abelson, W. M. Barnes, and W. S. Reznikoff, *Science*, **187,** 27 (1975).
76. K. Bertrand, L. Korn, F. Lee, T. Platt, C. L. Squires, C. Squires, and C. Yanofsky, *Science*, **189,** 22 (1975).
77. M. Ptashne, A. Jeffrey, A. D. Johnson, R. Mauver, B. J. Meyer, C. O. Pabo, T. M. Roberts, and R. T. Sauer, *Cell*, **19,** 1 (1980).
78. L. J. Ferrin, A. S. Mildvan and L. A. Loeb, *Biochem. Biophys. Res. Commun.*, *112*, 723, (1983).
79. K. E. Walton, P. C. FitzGerald, M. S. Herrmann, and W. D. Behnke, *Biochem. Biophys. Res. Commun. 108*, 1353, (1982).

CHAPTER 2
Carboxypeptidase A

BERT L. VALLEE, ALPHONSE GALDES, DAVID S. AULD, and JAMES F. RIORDAN

Center for Biochemical and Biophysical Sciences and Medicine
Harvard Medical School and Brigham and Women's Hospital
Boston, Massachusetts

CONTENTS

1 INTRODUCTION, 27

2 SOURCES AND CHARACTERISTICS, 27

2.1 Primary Structure, 28
2.2 Tertiary Structure, 29
2.3 Enzyme Complexes, 32
2.4 Metal Substitution, 34

3 SPECTROSCOPIC PROPERTIES, 35

3.1 Cobalt(II) Carboxypeptidase, 36
3.2 Cobalt(II) Procarboxypeptidase, 37
3.3 Other Metallocarboxypeptidases, 37
3.4 Spectral Perturbations of Cobalt(II) Carboxypeptidase, 38
3.5 EPR Spectra of Cobalt(II) Carboxypeptidase, 43
3.6 Cobalt(III) Carboxypeptidase, 00

4 CHEMICAL MODIFICATION, 44

4.1 Tyrosyl Residues, 44
4.2 Other Residues, 45

5 ARSANILAZO CARBOXYPEPTIDASE A, 45

6 CARBOXYPEPTIDASE A IN SOLUTION AND IN CRYSTALS, 51

6.1 Spectral Studies, 51
6.2 Kinetic Studies, 52

7 CATALYTIC PROPERTIES, 54

8 MECHANISTIC DIFFERENCES BETWEEN ESTERS AND PEPTIDES, 56

9 MECHANISM OF CARBOXYPEPTIDASE A, 59

9.1 The Role of Zinc, 60
9.2 The Role of Glu-270, 63
9.3 Kinetic and Spectroscopic Detection of Catalytic Intermediates, 63
9.4 Hydrolysis of Peptide Substrates, 67
9.5 Hydrolysis of Ester Substrates, 67

10 REMAINING ISSUES, 69

ACKNOWLEDGMENTS, 70

REFERENCES, 70

1 INTRODUCTION

This chapter describes the composition, structure, and function of carboxypeptidase A, one of the most intensively studied of the zinc metalloenzymes. The work to be described has been carried out over a great number of years and, accordingly, many investigators have participated in acquiring the knowledge that is available today. Not all of that work will be cited in this summary, which of necessity will concentrate largely on recent efforts, but those contributions not included were nevertheless tremendously important in laying the foundations for what has ensued and everyone now active in the field appreciates their value.

The activity of carboxypeptidase was first recognized more than half a century ago and the enzyme itself was isolated in pure form as early as 1937 (1). Considerable effort and attention were devoted to unravelling the complexities of its kinetic properties in subsequent years but it was not until two decades had elapsed that the enzyme was recognized to contain a stoichiometric amount of zinc essential to activity (2). Thus, carboxypeptidase became the second known zinc metalloenzyme, the prototype, carbonic anhydrase, having been discovered some 15 years earlier. The zinc atom has remained the enzyme's center of attraction ever since, despite the fact that in 1963 carboxypeptidase earned the distinction of being the first tyrosyl enzyme to be identified (3). Soon thereafter work began on the elucidation of its primary and three-dimensional structure to be correlated with its function and that culminated in reports presented at the Brookhaven Symposium in June, 1968 (4–6) and at the Royal Society meetings held in London later that same year (7–9). Numerous chapters and reviews on all this early work have appeared since that time (refs. 10–14, among others) and many additional details of the catalytic process have been uncovered. Considerable insight has been generated and interest in understanding the mechanism of action of the enzyme remains high. This present chapter is an attempt to summarize current views on the subject and offers our own perspectives and suggestions on the many questions and some answers that have emerged.

2 SOURCES AND CHARACTERISTICS

Mammalian and fish pancreas produce two zinc carboxypeptidases that catalyze the hydrolysis of C-terminal amino acid residues from peptides and proteins; that designated "A" (EC 3.4.17.1) is specific for aromatic residues, the other, called "B" (EC 3.4.17.2), prefers basic residues (15). Both are synthesized in the pancreas as zymogens, and their activation involves the limited

proteolytic action of trypsin. Enzymes with similar characteristics have been isolated from invertebrates (16–18), and *Streptomyces griseus* (19, 20).

Since the carboxypeptidase A isolated from bovine pancreas is characterized most thoroughly, the present discussion will focus on this enzyme. Procarboxypeptidase A, its precursor, is a trimeric protein with a molecular weight of 89,000 which is synthesized by the acinar cells of the pancreas (21). One of the subunits has a molecular weight of 41,000 and is the direct precursor of carboxypeptidase A. The other two subunits are identical and on treatment with trypsin give rise to a chymotrypsin-like endopeptidase activity, initially detected with acetyl tyrosine ethyl ester (ATEE),* and hence referred to as ATEEase. Both the proteolytic action of trypsin and the activation of ATEEase jointly generate carboxypeptidase A, which has a molecular weight of approximately 34,600. There is additional evidence that the activation process requires the presence of Ca(II) (22, 23). In contrast to the bovine zymogen, procarboxypeptidases of some other species are thought to be monomeric consisting solely of the immediate precursor of the enzyme (24, 25).

The amino acid sequence of rat preprocarboxypeptidase A has been determined from the corresponding mRNA nucleotide sequence and consists of a 15-residue signal peptide and a 94-residue propeptide (26). Activation of the zymogen involves the cleavage of an Arg—Ala peptide bond. The rat propeptide sequence is reported to be 70% homologous with that of the bovine zymogen (26), while the enzyme itself displays 78% homology with the sequence of the mature bovine enzyme. It is interesting that the highest degree of homology between the rat and bovine propeptide sequences is not immediately adjacent to the site of activation.

2.1 Primary Structure

Three different forms of the bovine enzyme have been identified, A_α, A_β, and A_γ, consisting of 307, 305, and 300 amino acid residues, respectively (7). These arise from the action of proteases during activation of the zymogen since the additional residues occur at the amino terminal region of the molecule. The form of the enzyme present in commercial preparations depends on the procedure used for isolation (27). Thus, enzyme prepared according to the method of Cox et al. (22) is predominantly, but not exclusively, A_α, while that prepared according to Allan et al. (28) is mainly A_β and that of Anson (1) is largely A_γ. The last of these preparations differs from the other two in the inability of its apoenzyme form to be reconstituted by zinc to a fully active species. Although the physicochemical basis for this difference has never been established, it appears to be related to the fact that the Anson enzyme is quite

*See Abbreviations section.

unstable in solution once its zinc ion has been removed. In addition, the crystal habit of the Anson form is said to differ from that of the Cox form (29) and it has been stressed that this distinction between the two forms must be recognized when comparisons of solution/crystal enzyme activities are made (30). As discussed below, the particular form of the enzyme used for kinetic or structural studies remains a critical point of contention.

The primary structure of the enzyme has been determined by Neurath and his collaborators (Table 1) (31). Four fragments are obtained by treatment with cyanogen bromide, in accordance with the presence of three methionyl residues in the enzyme (32). The fragments are referred to as F_N, F_{III}, F_I, and F_C respectively, with F_N being located at the N terminus and F_C at the C terminus of the protein. The F_C fragment consists of only six residues, but contains an allotypic replacement site, Val/Leu-305, linked to two additional sites in the molecule (Ile/Val-179 and Ala/Glu-228) (33). Hence, bovine carboxypeptidase A can occur in two distinct forms which differ at three positions in the sequence. These "Val" and "Leu" forms are equally distributed between the α, β, and γ forms, so that there can be six unique bovine carboxypeptidase A forms, A_α^{Val}, A_α^{Leu}, A_β^{Val}, A_β^{Leu}, A_γ^{Val}, and A_γ^{Leu}. While differing by only a few amino acid residues, they have remarkably different physical properties but their kinetic characteristics are virtually identical (34).

2.2 Tertiary Structure

X-ray crystallographic analysis of the α-carbon and side-chain positions in carboxypeptidase A (5) in conjunction with amino acid sequence analysis has led to the assignment of the active site residues. Thus, Arg-145 is postulated to be the site of interaction of the free α-carboxyl group of the substrate with the enzyme and Glu-270 the principle catalytic moiety. Two tyrosyl residues, Tyr-198 and Tyr-248, are also present in the active site. The binding site of the zinc atom consists of Glu-72, His-69, and His-196. A water molecule completes a distorted tetrahedral geometry about the metal atom.

Several reviews have described in much the same detail the structure of carboxypeptidase A_α, as discerned from X-ray crystallographic analysis at 2.0 Å resolution (5, 8, 10–12, 35). This structure has now been refined to a resolution of 1.75 Å (36).

The molecule is an ellipsoid with dimensions of about $50 \times 42 \times 38$ Å^3 (Fig. 1). Of the 307 residues 147 are in a random conformation, 115 in nine α-helical regions of various degrees of perfection (including some α_{II} helix), and 45 in a large β sheet—twisted by 120°—which occupies the center of the molecule. The β sheet contains four parallel and three antiparallel pairs of extended chains, comprising about 20% of the backbone atoms. Six of the nine helical regions are to one side of the pleated sheet, separated from it by hydro-

Table 1 Amino Acid Sequence of Bovine Carboxypeptidase A[a]

5 10 15 20 25
Ala-Arg-Ser-Thr-Asn-Thr-Phe-Asn-Tyr-Ala-Thr-Tyr-His-Thr-Leu-Asp-Glu-Ile-Tyr-Asp-Phe-Met-Asp-Leu-Leu-

30 35 40 45 50
Val-Ala-Glu-His-Pro-Glu-Leu-Val-Ser-Lys-Leu-Gln-Ile-Gly-Arg-Ser-Tyr-Glu-Gly-Arg-Pro-Ile-Tyr-Val-Leu-

55 60 65 60 75
Lys-Phe-Ser-Thr-Gly-Gly-Ser-Asn-Arg-Pro-Ala-Ile-Trp-Ile-Asp--u-Gly-Ile-His-Ser-Arg-Glu-Trp-Ile-Thr-

80 85 90 95 100
Gln-Ala-Thr-Gly-Val-Trp-Phe-Ala-Lys-Lys-Phe-Thr-Glu-Asn-Tyr-Gly-Gln-Asn-Pro-Ser-Phe-Thr-Ala-Ile-Leu-

105 110 115 120 125
Asp-Ser-Met-Asp-Ile-Phe-Leu-Glu-Ile-Val-Thr-Asn-Pro-Asn-Gly-Phe-Ala-Phe-Thr-His-Ser-Glu-Asn-Arg-Leu-

130 135 140 145 150
Trp-Arg-Lys-Thr-Arg-Ser-Val-Thr-Ser-Ser-Ser-Leu-Cys-Val-Gly-Val-Asp-Ala-Asn-Arg-Asn-Trp-Asp-Ala-Gly-

155 160 165 170 175
Phe-Gly-Lys-Ala-Gly-Ala-Ser-Ser-Ser-Pro-Cys-Ser-Glu-Thr-Tyr-His-Gly-Lys-Tyr-Ala-Asn-Ser-Glu-Val-Glu-

180 185 190 195 200
Val-Lys-Ser-Ile-Val-Asp-Phe-Val-Lys-Asn-His-Gly-Asn-Phe-Lys-Ala-Phe-Leu-Ser-Ile-His-Ser-Tyr-Ser-Gln-
Val

205 210 215 220 225
Leu-Leu-Leu-Tyr-Pro-Tyr-Gly-Tyr-Thr-Thr-Gln-Ser-Ile-Pro-Asp-Lys-Thr-Glu-Leu-Asn-Gln-Val-Ala-Lys-Ser-

230 235 240 245 250
Ala-Val-Ala-Ala-Leu-Lys-Ser-Leu-Tyr-Gly-Thr-Ser-Tyr-Lys-Tyr-Gly-Ser-Ile-Ile-Thr-Thr-Ile-Tyr-Gln-Ala-
Glu

255 260 265 270 275
Ser-Gly-Gly-Ser-Ile-Asp-Trp-Ser-Tyr-Asn-Gln-Gly-Ile-Lys-Tyr-Ser-Phe-Thr-Phe-Glu-Leu-Arg-Asp-Thr-Gly-

280 285 290 295 300
Arg-Tyr-Gly-Phe-Leu-Leu-Pro-Ala-Ser-Gln-Ile-Ile-Pro-Thr-Ala-Gln-Glu-Thr-Trp-Leu-Gly-Val-Leu-Thr-Ile-

305
Met-Glu-His-Thr-Val-Asn-Asn
Leu

[a]From ref. 31 with permission. The N-terminus of the α, β, and γ chains is at residues 1, 3, or 8, respectively. The upper sequence corresponds to the Val form and the lower one to the Leu form.

Figure 1 Diagram of the α-carbon backbone of carboxypeptidase A. Zinc atom is near the center, where the positions of the three protein ligands are shown by arrows. (From ref. 11 with permission.)

phobic side chains. The zinc atom is adjacent to the other side of the β sheet, associated with a groove and a pocket essential for substrate binding; this groove is lined on one side by the side chains of the pleated sheet. Below the active center is another bank of hydrophobic side chains that separate this side of the sheet from an area which is in a random coil, containing the disulfide bridge between Cys-138 and Cys-161 but with few hydrogen bonds.

Three *cis* peptide bonds not involving proline as the N donor, all located fairly near the active site, have been detected in the refined structure (36). These are located between Ser-197 and Tyr-198, Pro-205 and Tyr-206, and Arg-272 and Asp-273. This situation is unique among the protein X-ray structures solved to date and their functional significance, if any, is unknown. It might be noted, however, that Tyr-198 has been implicated as a critical component of the binding site for peptide substrates (37). In addition, His-196 is a

ligand for the active site zinc atom and it may well be that an unusual peptide bond in this region is related to the entatic state of that metal (see below).

2.3 Enzyme Complexes

The X-ray diffraction analyses have included high-resolution studies (2.0–2.8 Å) of the interaction of the crosslinked enzyme with the pseudosubstrates, Gly–Tyr (5, 8, 30, 36) and Phe–Gly–Phe–Gly (38); the ketone, 2-benzyl-3-*p*-methoxybenzoylpropionate (39); the inhibitor, indole-3-acetic acid; and the 39-amino acid inhibitor from potatoes (30, 40). In addition, low-resolution (6 Å) studies on the binding of the inhibitors *p*-iodo-β-phenylpropionate, L-Phe, and L-Lys–L-Tyr-NH_2 (35) have also been performed. The results of these high and low-resolution studies are stated to be in qualitative agreement (30).

The complex formed between Gly–Tyr and carboxypeptidase A has been used extensively as the model for the mechanism of the enzyme (5, 8, 11, 12, 30, 39, 41) even though such dipeptides have long been known to be atypical substrates (42). Gly–Tyr binds tightly to the enzyme in solution ($K_i = 1 \times 10^{-4} M$) (43), but its binding to the crystalline enzyme is much weaker resulting in only 33–43% of the enzyme molecules being associated with the pseudosubstrate (8). The difference diffraction maps between the free enzyme and the Gly–Tyr·carboxypeptidase complex have been interpreted to suggest the following interactions:

1. The phenolic side chain of Gly–Tyr occupies a pocket in the active center, displacing several water molecules.

2. The carboxylate group of the pseudosubstrate is aligned with the guanidinium group of Arg-145, such that two hydrogen bonds can be formed. The guanidium group of Arg-145 moves by about 2 Å through a rotation of the C_α—C_β bond to accommodate this interaction (Fig. 2).

3. The carbonyl oxygen atom of Gly–Tyr is coordinated to the zinc atom, displacing the metal-bound water molecule of the free enzyme. This involves the movement of the zinc atom by about 0.6 Å. The carbonyl carbon atom is situated about 3.5 Å from the carboxyl group of Glu-270 (Fig. 2).

4. The phenolic oxygen of Tyr-248 lies within hydrogen bonding distance of the peptide NH of the pseudosubstrate and may also interact with its carboxylate group. This involves a 12 Å movement of Tyr-248, as it rotates from its position on the surface of the free enzyme into the active site (Fig. 2). This motion has been cited (5) as the example *par excellence* of the induced fit theory of enzyme catalysis (44).

5. The amino group of Gly–Tyr is equally distributed through a rotation of 180° about the glycyl C_α-carbonyl bond alternately interacting with the

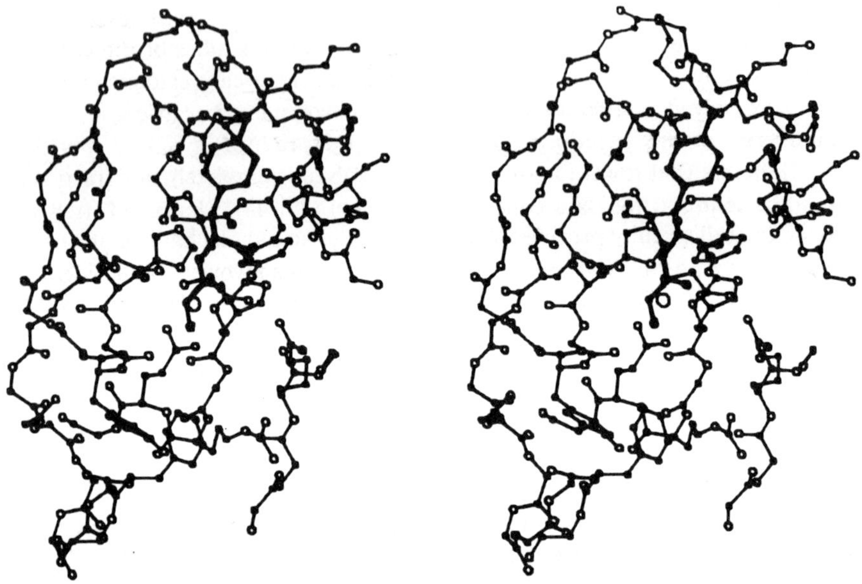

Figure 2 Stereodiagram of the active-site region in carboxypeptidase A·Gly–Tyr complex. Gly–Tyr molecule is indicated by the darker bonds. (From ref. 30 with permission.)

metal atom and with Glu-270. The latter interaction requires a displacement of 2 Å for the carboxylate group of Glu-270.

Interactions 1–4 have been considered critical to the binding and hydrolysis of substrate, whereas 5 has been described as "unproductive" (12, 36); the reason for this distinction is obscure.

The mechanism proposed for the catalysis of peptide hydrolysis has been based on the above model and incorporates a series of concerted structural changes that are thought to occur during substrate binding and ultimate hydrolysis of the susceptible peptide bond (5, 12, 30, 41). Analogous data for carboxypeptidase·ester complexes do not exist and hence mechanistic conclusions for ester hydrolysis lack a corresponding structural basis. The ketone purported to be a model for an ester substrate (39) (see below) is neither an ester nor a hydrolytic substrate.

The above conclusions are critically dependent on the assumptions that:

1. The carboxypeptidase·Gly–Tyr complex is very similar to the kinetically significant E·S complex(es).

2. The structural details of the enzyme as determined by X-ray crystallography in the crystalline phase are identical to those in solution.

Neither of these assumptions appear to be borne out, as is discussed in detail below. Indeed, the very fact that crystallographic studies of the carboxypeptidase Gly–Tyr complex are possible provide significant evidence that the structure solved is that of an *unproductive* complex, because the kinetic parameters of this pseudosubstrate ($K_m \approx 1 \times 10^{-4} M$, $k_{cat} \approx 1\ \text{min}^{-1}$ at 25°C) dictate that it would otherwise be hydrolyzed completely in a fraction of the time required for data collection (~60 hr) (35, 42). Hence, of itself the X-ray crystallography provides little guide to the actual mechanism of carboxypeptidase A, aside from identifying possible groups involved. Their identity, that is, the zinc atom and the side chains of Arg-145, Glu-270, and Tyr-248, has also been demonstrated by chemical modifications. It should be noted in addition that while the crystal structure, which has been known now for some 15 years, was supposed to have settled all questions regarding the catalytic mechanism, debate over this issue and the function subserved by the above residues still continues. Of course, this situation is not unique to carboxypeptidase and, in fact, appears to be the rule.

It is generally accepted that crystallography establishes the overall structural peptide backbone geometry of the protein molecule and specifies the relative positions of amino acid side chains. Such knowledge provides a major advantage when making mechanistic inferences based on structure and occasionally gives clues that might be impossible to acquire by other means, for example, distorted intermediates in lysozyme catalysis (46). It also seems clear that the crystal structure reflects many, although not necessarily all, features characteristic of the enzyme in solution, as evidenced by the fact that crystalline enzymes usually bind substrate and exhibit catalytic activity albeit often attenuated as in the case of carboxypeptidase A. Thus, the crystal structure can serve as a framework on which to assemble mechanistic hypotheses, but since it provides no unique insight to events beyond its spatial and temporal resolution, it clearly is neither a unique nor sufficient key to their verification. Furthermore, as long as one remains within the overall geometrical constraints set by this structure, there seems to be no valid obstacle to invoking what might be called "steric license" in repositioning certain side chains to accommodate physicochemical evidence acquired with the enzyme in solution. It is understood that this license is exercised to some extent by crystallographers as they interpret and later refine electron density maps, and all evidence indicates that it can also be employed by them and others who wish to make further extrapolations to specific mechanisms.

2.4 Metal Substitution

A zinc atom is essential to the catalytic activity of native carboxypeptidase A. As isolated, the enzyme contains 1 g atom zinc/mol (2). Its removal—either

by dialysis at low pH or with chelating agents at neutral pH—totally inactivates the enzyme, whereas its readdition, or the addition of a number of other divalent metal ions, restores the dual activities of carboxypeptidase toward peptides and esters (Table 2). The substrate specificity of the enzyme is very responsive to the particular metal ion which is substituted for zinc. Thus, the cobalt-substituted enzyme is more active than the native zinc enzyme toward peptides but its activity toward esters is nearly the same. The Mn(II)-substituted enzyme also hydrolyzes both esters and peptides while the Cd(II) enzyme is significantly active only against esters. In addition to those metals shown in Table 2, Ni(II), Hg(II), and Pb(II) restore some activity to apocarboxypeptidase, but the Cu(II) and Co(III) enzymes are completely inactive. The alkaline earth metal derivatives, and the enzymes substituted with most of the remaining transition-metal ions, are also totally inactive (47). Electron density difference maps have been inferred to suggest that in all the catalytically active metalloderivatives, save Hg(II) carboxypeptidase, the metal occupies the same site as zinc (10).

3 SPECTROSCOPIC PROPERTIES

The spectroscopic properties of the substituent metal ions, particularly those of Co(II), have been used extensively as probes for the active site of carboxypeptidase A to investigate the interaction of the enzyme with substrates and inhibitors (48–50; see also ref. 14, and references therein). The basis and ra-

Table 2 Metallocarboxypeptidase-Catalyzed Hydrolysis of Bz-(Gly)$_2$-L-Phe and Bz-(Gly)$_2$-L-OPhe[a]

	Bz-(Gly)$_2$-L-Phe		Bz-(Gly)$_2$-L-OPhe	
Metal	k_{cat} (min^{-1})	$10^{-3}K_m^{-1}$ (M^{-1})	$10^{-4}k_{cat}$ (min^{-1})	K_m^{-1} (M^{-1})
Cobalt	6000[b]	1.5[b]	3.9	3300
Zinc	1200	1.0	3.0	3000
Manganese	230[b]	2.8[b]	3.6	660
Cadmium	41	1.3	3.4	120

[a]From ref. 76 with permission. Assays performed at 25°C, pH 7.5, 1.0*M* NaCl, and a buffer concentration of 0.05*M* Tris for peptide hydrolysis and $10^{-4}M$ Tris for ester hydrolysis.

[b]Values are for carboxypeptidase A (Cox). All other values are for carboxypeptidase A (Anson).

tionale for this inorganic modification which is now practiced widely (51, 52) have been documented and discussed.

3.1 Cobalt(II) Carboxypeptidase

The absorption spectrum of cobalt carboxypeptidase (Fig. 3, bottom) is characterized by a shoulder at about 500 nm and maxima at 555 and 572 nm ($\epsilon \approx 150$). Two infrared bands are centered at 940 and 1570 nm ($\epsilon \approx 20$). Lowering the temperature increases the resolution of the visible bands but does not affect absorptivity (49). At 4°K two new shoulders are evident, and the maximum at 555 nm is shifted to 532 nm, while that at 572 nm is unaltered. The MCD spectrum (Fig. 3, top) is characterized by a pronounced negative, and a smaller, positive MCD band (50). These spectra closely resemble those of tetrahedrally coordinated Co(II) complex ions, suggesting an irregular (distorted) tetrahedral ligand geometry (49, 50). The EPR spectrum of the

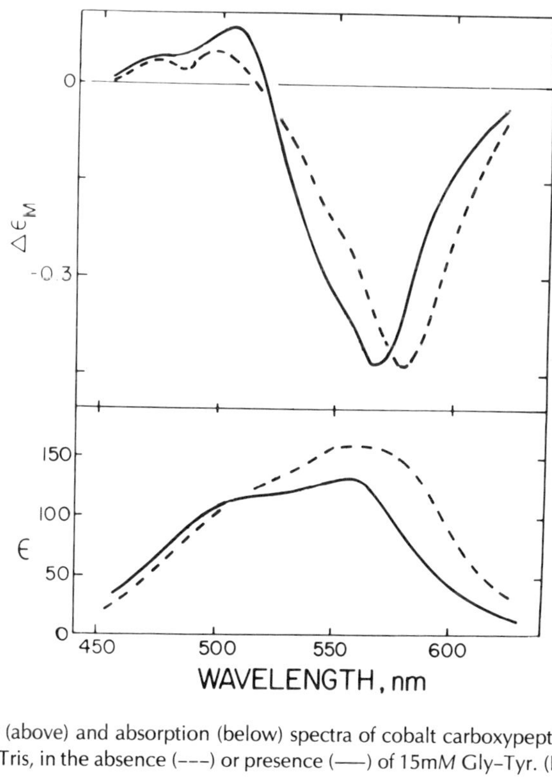

Figure 3 MCD (above) and absorption (below) spectra of cobalt carboxypeptidase A, pH 7.1, 1*M* NaCl, 5m*M* Tris, in the absence (---) or presence (—) of 15m*M* Gly–Tyr. (From ref. 50 with permission.)

cobalt enzyme is in agreement with this interpretation (see below) as were the early X-ray analyses of the native enzyme, which indicated that the zinc atom is coordinated by His-69, His-196, Glu-72, and H_2O in a distorted tetrahedral geometry (ref. 5; see above). On the other hand, magnetic susceptibility measurements on the cobalt enzyme have been thought to be more compatible with five coordination for the cobalt atom (53, 54) and recently the X-ray studies have been reinterpreted to indicate that Glu-72 may coordinate zinc *via both* its carboxylate O-atoms, making the metal pentacoordinate (36).

The spectral assignment of the metal geometry to rigid structures, whether tetrahedral or pentacoordinate, is probably an oversimplification and omits, moreover, the critically important contribution of the symmetry of the complex. The available evidence suggests that the metal coordination geometry and symmetry in metalloenzymes are highly unusual and rather flexible compared with model metal complex ions, so that geometry in metalloenzymes cannot be interpreted and/or classified readily as being any particular type based on models. This is the basic postulate of the entatic state hypothesis regarding active site metal geometry in metalloenzymes (55), which also emphasizes the functional significance of this arrangement. We therefore refer to metalloenzyme spectra as "tetrahedral-like" or "pentacoordinate-like" implying an empirical rather than theoretical assignment. Most inorganic chemists would agree that the theory for such macromolecular structures is not developed sufficiently to allow more definite conclusions at this time.

3.2 Cobalt(II) Procarboxypeptidase

The spectral properties of Co(II)-substituted procarboxypeptidase A and of cobalt carboxypeptidase are virtually identical (56, 57), that is, both are peculiar to those of enzymatically active proteins (55). This observation prompted an investigation of the catalytic properties of procarboxypeptidase A, which as a zymogen had been considered to be an inactive precursor (21). Remarkably, with certain substrates cobalt procarboxypeptidase was found to be as active, and in some cases even more active, than the native enzyme. The kinetic characteristics of the native zinc zymogen were examined subsequently in detail (58). These, as well as earlier observations (7, 59, 60), raised questions regarding the concept that zymogens are always inactive enzyme precursors, and confirmed the association between catalytic activity and entatic metal geometries.

3.3 Other Metallocarboxypeptidases

The coordination geometries of a number of other metalloderivatives, for example, Cu(II), Ni(II), Mn(II), VO(IV), and Cd(II), have also been probed by

a variety of techniques (e.g., electronic, PAC, EPR and NMR spectroscopy, and magnetic susceptibility measurements) and the results have been interpreted to suggest a coordination number ranging from 4 to 6 for these derivatives (54, 61-65), but as is the case for the zinc and cobalt enzymes, none of these assignments can be considered decisive. Hence, the exact coordination number of the metallocarboxypeptidases appears indeterminate and overall the data suggest considerable flexibility in the geometric arrangements at the metal ion binding site of carboxypeptidase.

3.4 Spectral Perturbations of Cobalt(II) Carboxypeptidase

In marked contrast to carbonic anhydrase, the addition of anions, inhibitors, and pseudosubstrates or altering pH does not induce major changes in the absorption, CD, and MCD spectra of cobalt carboxypeptidase (49, 50, 66, 67) indicating that binding of these agents does not affect the entatic coordination geometry. Thus, increasing pH from 7 to 9 progressively induces shoulders near 510 and 625 nm and the two maxima at 555 and 572 nm merge to become a single peak at 560 nm (66). A pH titration at 625 nm results in a sigmoidal curve with an inflection point at pH 8.8 (Fig. 4). The pH dependence of peptidase activity of the cobalt enzyme and the spectral changes superimpose, indicating that the ionization of a single group may be responsible for both effects (66). This group does not seem to be the coordinated water molecule, however (see below). Importantly, the spectra at low and high pH appear to reflect the same entatic geometry for the metal that exists at neutral pH. Similarly,

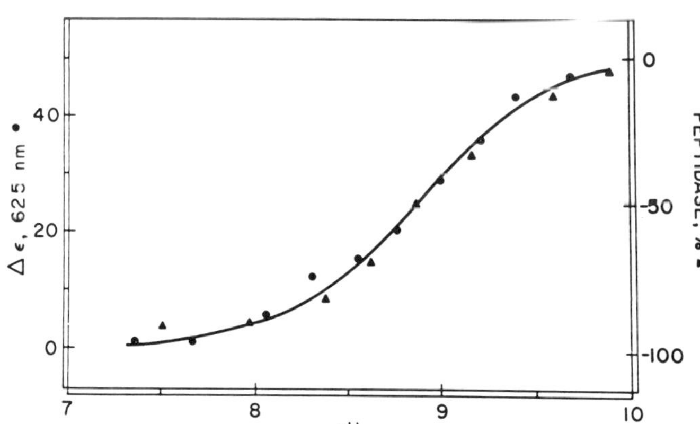

Figure 4 Dependence on pH of the change in absorptivity at 625 nm (●) and % peptidase activity (▲) of cobalt carboxypeptidase A. Peptidase activity was measured toward Dns–Gly–Trp, with the value at pH 7.1 taken as 100%. (From ref. 49 with permission.)

neither the replacement of 1*M* NaCl by NaBr or NaF, nor the addition of other anions (ranging from SO_4^{2-} to NO_2^-) significantly alters the spectrum of the cobalt enzyme (49, 66, 68), while inhibitors, for example, L-Phe, β-phenylpropionate, indole-3-acetate (Table 3), and the pseudosubstrate, Gly–Tyr, do cause some spectral shifts (Fig. 3) (49). Thus, upon addition of Gly–Tyr the two absorption maxima coalesce into a single broad peak at 550 nm, the shoulder at 505 nm remains, the 940 nm maximum splits into two bands, peaking at 850 and 1150 nm, and the maximum at 1570 nm shifts to 1420 nm (49). Gly–Tyr also induces CD bands at 505 and 550 nm (Fig. 5). Remarkably, the ellipticity at 550 nm is opposite in sign to that of the control and has a magnitude of +2100°, which is four times that of any band observed in the absence of substrate. These spectral perturbations potentially reflect participation of the metal in catalysis, but again do not indicate a drastic change in geometry or symmetry. Active-site directed inhibitors, for example, 2-mercaptoacetyl-D-Phe, cause large perturbations in the absorption and MCD spectra (Fig. 6), without apparently affecting the tetrahedral-like coordination geometry (67). In addition, the mercaptan inhibitors generate a distinct, intense ($\epsilon \approx 900$) transition at 338 nm which has been assigned to the formation of a sulfur → Co(II) charge transfer band. Its presence demonstrates direct binding of the thiol group to the metal atom of carboxypeptidase A.

The spectroscopic properties of the cobalt enzyme are altered drastically after modification of Glu-270 by the carbodiimide reagent, *N*-cyclohexyl-*N'*-morpholinoethylcarbodiimide metho-*p*-toluenesulfonate (CMC) (see below). In marked contrast to the unmodified enzyme, CMC-modified (and inactive) carboxypeptidase A exhibits spectrochemical behavior that parallels that of cobalt carbonic anhydrase, and its absorption and MCD spectra are very sensitive to pH and the presence of anions (68). Thus, the absorption spectrum of the CMC-modified enzyme in 1*M* NaCl has a maximum at 580 nm (~300) and prominent shoulders at 550 and 605 nm (Fig. 7). This spectrum and that previously reported for the complex of chloride with cobalt carbonic anhydrase are strikingly similar. Consistent with this, replacement of Cl^- with another anion (e.g., Br^-; Fig. 7) slightly decreases intensity and shifts the 605 nm shoulder to higher wavelength. Moreover, with increasing pH the spectra give way to one which is widely split and of lower intensity, with maxima at 535, 569, 608, and 628 nm (Fig. 8). This spectrum is again reminiscent of that seen for cobalt carbonic anhydrase at high pH. Also, as with that enzyme, MCD spectroscopy signals a change in the overall coordination geometry accompanying this pH-induced transition. Hence, these data as well as those obtained for the Cd(II) enzyme with PAC (65) demonstrate several interconvertible states of metal coordination in carboxypeptidase, in agreement with the entatic hypothesis.

Table 3 Inhibitor Effects on Cobalt Carboxypeptidase Absorption and Circular Dichroic Spectra[a]

Inhibitor	Spectral K_I [mM] or I (mM)	Absorption (ϵ, cm^{-1} M^{-1})		Circular Dichroism [θ] (deg cm^2 dmol^{-1})
β-Phenylpropionate	(0.5)	505(110), 555(160), 575(155), 610(80)	1040	490(−250), 543(−650)
	(2)	510(125), 555(150), 590(120)	1100	490(−450), 536(−900), 580(−800)
β-Iodopropionate	(0.5)	505(110), 558(178), 575(165)	1030	490(−200), 551(−650)
	(5)	520(140), 555(160), 590(130)	1090	495(−375), 555(−850), 590(−550)
Indole-3-acetate	(0.8)	510(120), 557(180), 573(180), 610(90)	1010	401(50), 555(−175)
	(25)	520(130), 565(160), 590(130)	1070	492(−300), 537(−500)
Phenylacetate	(2)	510(125), 555(155), 575(165), 605(95)	1030	495(−200), 539(−400)
	(20)	515(160), 575(180)	1090	510(100), 536(−275), 600(125)
Butyrate	(1)	505(105), 555(165), 570(170), 605(100)	1010	495(−150), 573(−450)
	(20)	515(130), 575(150)	1090	495(−150), 585(−700)
Acetate	(1000)	515(135), 575(155)	1050	510(100), 555(−300)
D-Phe	(0.5)	545(165), 580(205), 602(180)	1030	510(100), 575(100)
L-Phe	[3]	510(130), 555(200), 574(205), 610(140)	1000	510(200), 545(0), 580 (100)
Benzoate	[20]	505(105), 560(165), 575(160), 610(90)	1000	485(−50), 505(25), 560 (−400)
Thiobenzoate	[5]	505(115), 555(275), 574(300), 625(125)	1080	475(−100), 510(0), 570(−700), 370(−2000)
L-β-Phenyllactate	[<0.5]	510(135), 560(195), 572(190)	970	495(−300), 527(−325), 585(100)
L-Lys-L-Tyr-NH_2	[7]	515(175), 555(175)	1030	505(1500), 550(1050)

[a] From ref. 49 with permission. 1 M NaCl, 0.005M Tris-Cl, pH 7.1, 25°C. The effects of inhibitors on the cobalt spectra were examined in H_2O. Hence, in the infrared only effects on the band at 940 nm could be examined.

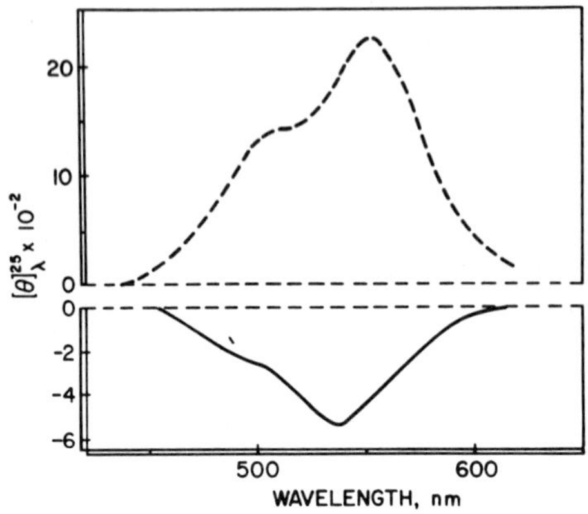

Figure 5 CD spectrum of 3mM cobalt carboxypeptidase A, pH 7.5, 1M NaCl, 5mM Tris, before (—) and after (---) the addition of 10mM Gly–Tyr. (From ref. 49 with permission.)

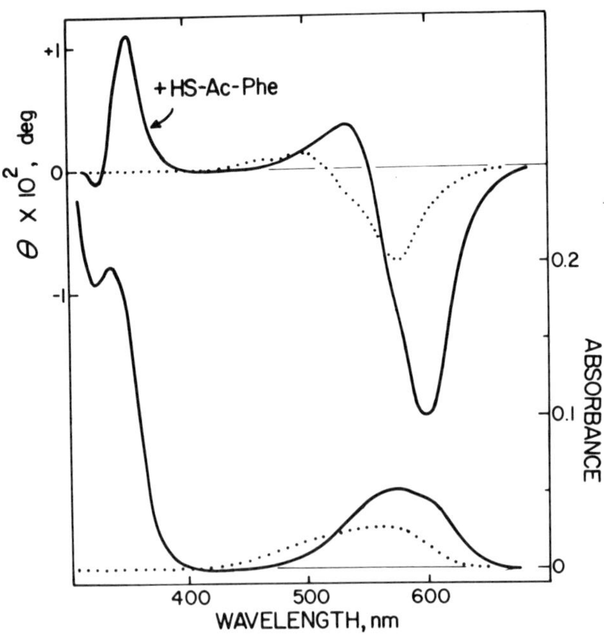

Figure 6 Absorption (lower) and MCD (upper) spectra of cobalt carboxypeptidase A (0.28mM), pH 7.5, 1M NaCl, 50mM Hepes, in the absence (. . .) and presence (—) of mercaptoacetyl-D-Phe (0.28mM). (From ref. 67 with permission.)

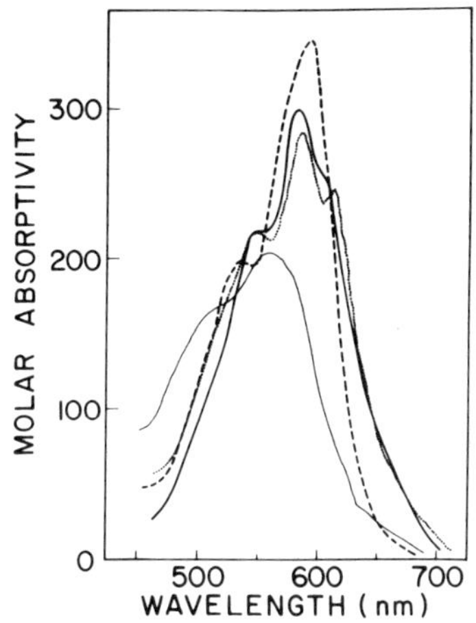

Figure 7 Effect of different anions on the absorption spectrum of CMC-modified cobalt carboxypeptidase A in 10m*M* Hepes, pH 7.0. Spectra of the modified enzyme in the presence of 1*M* NaCl (—), 1*M* $NaNO_3$ (—), 1*M* NaBr (. . .), and upon addition of 20m*M* NaN_3 to any of the foregoing (---).

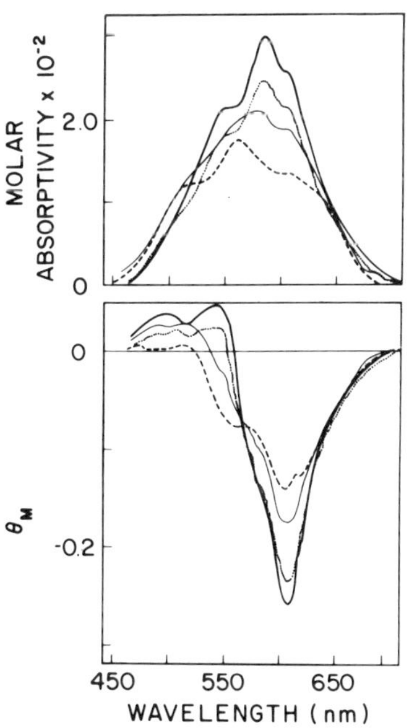

Figure 8 Effect of pH on the absorption (above) and MCD (below) spectra of CMC-modified cobalt carboxypeptidase A in 10m*M* Hepes, 1*M* NaCl, at pH 7.0 (—), 8.6 (. . .), 9.5 (—), and 10.4 (---), respectively, in order of decreasing absorptivity at 570 nm.

3.5 EPR Spectra of Cobalt(II) Carboxypeptidase

The EPR spectra of cobalt carboxypeptidase support the above conclusions. The spectrum of free, unmodified enzyme in solution (Fig. 9) is broad and characterized by three apparent g values ($g_1 = 5.53$; $g_2 = 2.94$; $g_3 = 2.01$) (69). It resembles that of tetrahedrally coordinated high-spin Co(II) complex ions, and, hence, suggests that the free enzyme is four coordinate-like (see above).

Addition of inhibitors or pseudosubstrate to the enzyme perturbs the EPR spectrum as reflected mainly in alteration of the g_1 and g_2 values, and a sharpening of the g_2 resonance (J. Thompson and B. L. Vallee, unpublished data). The spectra, however, suggest that these agents do not alter the *overall* metal coordination geometry. In contrast, the EPR spectra of Glu–270-modified cobalt enzyme suggest that its coordination geometry is altered (R. Martinelli, K. Geoghegan, and B. L. Vallee, unpublished data).

A combination of cryoenzymological and EPR techniques has been used to demonstrate that binding of Gly–Tyr to cobalt carboxypeptidase displaces the metal-bound water molecule (70). However, it is not clear if water displacement is due to coordination by the peptide carbonyl oxygen or by the amino group. In contrast, the spin-lattice relaxation of the active-site cobalt ion is perturbed in enzyme that has been acylated with *O*-(*trans*-*p*-chlorocinnamoyl)-L–β–phenyllactate when ^{17}O is incorporated either into the cinnamoyl carbonyl oxygen or into solvent water. This has been interpreted to indicate that the geometry of the metal ion in the acylated enzyme is five coordinate. In fact, it could also mean that the water molecule and the carbonyl oxygen could be coordinated alternately to the metal. Moreover, any inferences regarding overall metal coordination must be considered tenuous since they are based on the assumption that the remaining ligands are unchanged.

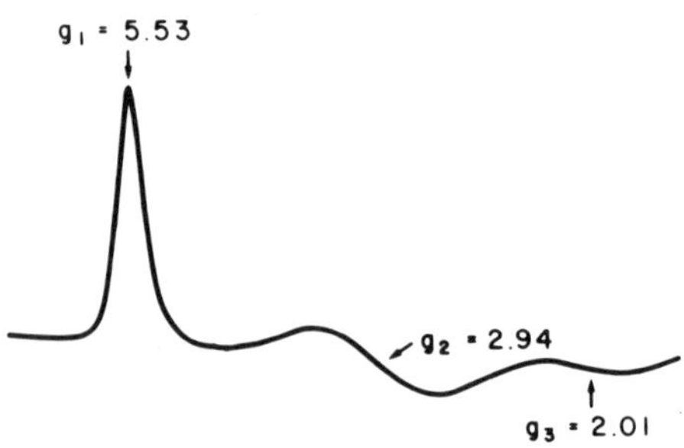

Figure 9 EPR spectrum of cobalt carboxypeptidase in solution in 2*M* NaCl, 50 m*M* Hepes, pH 7.5.

3.6 Cobalt(III) Carboxypeptidase

The active-site directed oxidizing agent, *m*-chloroperbenzoic acid, oxidizes Co(II) carboxypeptidase to the inactive, exchange inert Co(III) enzyme (71). The absorption, CD, MCD, and EPR spectra of this derivative are consistent with those characteristic of octahedral Co(III) complex ions. It is to be emphasized that contrary to earlier reports (72, 73), Co(III) carboxypeptidase is totally devoid of any catalytic activity toward both peptides and esters, nor does it interact with either type of substrate to yield a single turnover (71). Obviously, previous mechanistic conclusions based on the assumption that it is an active esterase (74, 75) cannot be valid. Co(III) carboxypeptidase binds peptides but not esters (71), a property it shares with the apoenzyme (76). Since the exchange-inert character of Co(III) precludes ligand (i.e., water) displacement, this finding demonstrates that inner-sphere coordination to the active-site metal is essential for the binding of esters but not of peptides, and highlights the mechanistic difference between these two classes of substrates (see below).

4 CHEMICAL MODIFICATION

4.1 Tyrosyl Residues

Through the use of a number of site-specific reagents, at least three, and perhaps four, different amino acid residues have been implicated in the catalytic function of carboxypeptidase A (Table 4). Acetylation with acetic anhydride or *N*-acetyl imidazole modifies the phenolic hydroxyl group of two tyrosyl residues that can be protected by the presence of the inhibitor, β-phenylpropionate (3, 77). Other monocarboxylic acid anhydrides also modify these tyrosyl residues as do dicarboxylic acid anhydrides. The phenolic hemiesters formed with the latter reagents undergo spontaneous hydrolysis at characteristic rates identifying the residues modified as tyrosines (78). The specific residues involved have not been identified, but since acetylation prevents subsequent modification of Tyr-248 by other reagents, this residue would seem to be one of them. Chemical modifications that are known to modify Tyr-248 include iodination (79, 80), nitration (81), and azo coupling (37, 82). These reactions are all *ortho*-position substitutions that do not block the phenolic OH group although they do alter its p*K* to varying extents. Urdea and Legg (83) were able to specifically block the phenolic OH of Tyr-248 by addition of Co(II) EDDA to arsanilazotyrosine-248 carboxypeptidase to form an azophenolate–cobalt complex. Oxidation of the metal to Co(III) rendered it exchange inert, blocking the OH group and abolishing all peptidase activity but retaining at

least some esterase activity. Most of the other modifications of Tyr-248 result in decreased peptidase but increased esterase activity. Evidence for functional significance of a second tyrosyl residue has also been obtained. Further modification of tetrazolylazotyrosine-248 carboxypeptidase A with tetranitromethane results in nitrationof Tyr-198, accompanied by a marked decrease in peptidase activity, suggesting that this residue may also be involved in catalytic activity (37).

4.2 Other Residues

There have been some reports of histidine modifications in carboxypeptidase but the specific residue involved and its possible role in catalysis have not been defined (84). Butanedione in borate buffer reversibly modifies a single arginyl residue with concomitant loss of peptidase and enhancement of esterase activity (85); the residue modified is thought to be Arg-145. If this modification is carried out in the absence of borate buffer, peptidase activity again decreases, although not as much as when borate is present, but esterase activity is not altered at all under these conditions. Subsequent addition of borate to the modified enzyme instantaneously changes activities to those observed on modification with butanedione/borate directly. Since borate merely stabilizes the butanedione–arginine adduct, it would seem that the same residue is modified under both conditions but the nature of the modified product determines the effect on activity.

In contrast to the above, the modification of Glu-270 by carbodiimide and alkylating reagents completely abolishes *both* peptidase and esterase activities (86–89). The carbodiimide reaction with Glu-270 occurs at pH 6.0 and in the absence of an added nucleophile, conditions that are unusual for carboxyl group modification. It has been thought that an intermediate O-acyl urea, which is the typical product of such a reaction, rearranges quickly to a stable N-acyl urea. However, it could be that the carboxylate–carbodiimide reaction proceeds directly to the N-acyl urea via a mechanism that mimics catalysis of substrate hydrolysis. The significance of these chemical modification studies to the mechanism of the enzyme will be commented upon below.

5 ARSANILAZO CARBOXYPEPTIDASE A

Another chemical modification that has proved exceptionally valuable in generating a spectral probe that helps elucidate the interaction of inhibitors and pseudosubstrates with carboxypeptidase A and also makes zinc visible is coupling with diazotized arsanilic acid. The properties of this derivative have established structural differences between the enzyme in solution and that in

Table 4 Chemical Modification of Residues in the Active Center of Carboxypeptidase A

			Activity of Product (% Native Enzyme)	
Residues or Groups Modified	Chemical Agent	Product	Peptidase Cbz-Gly-Phe	Esterase Bz-Gly-OPhe
Tyrosyl Residues				
2 Tyr (probably includes 248)	*N*-Acetylimidazole	O-Acetyl tyrosine	0	600
2 Tyr	Acetic anhydride	O-Acetyl tyrosine	0	600
Tyr	*i*-Butyric anhydride	O-Acyl tyrosine	83	134
Tyr	*n*-Butyric anhydride	O-Acyl tyrosine	80	162
Tyr	Propionic anhydride	O-Acyl tyrosine	49	342
Tyr-248 + other Tyr	Iodine	Iodotyrosine	0	500–600
Tyr-248	5-Diazo-1-H-tetrazole (eight fold molar excess)	Azotyrosine	100	200
Tyr-248	Tetranitromethane (TNM)	Nitrotyrosine	<15	180
Tyr-248	Diazotized arsanilic acid	Arsanilazotyrosine	50	100
Double modification Tyr residues	*N*-Acetylimidazole + TNM	O-Acetyl tyrosine	0	600
	N-Acetylimidazole + 5-diazo-1-H-tetrazole (eight fold molar excess)	O-Acetyl tyrosine	0	600

Tyr-248 and Tyr-198	5-Diazo-1-H-tetrazole + TNM	Azotyrosine-248 + nitrotyrosine-198	100 < 15	200 200
Tyr	Iodine + TNM	Nitrotyrosine (TNM may displace I)	0	200
Tyr-248	Diazotized arsanilic acid	Arsanilazotyrosine	50	100
	+ [Co(III)(EDDA)]	[Co(III)(EDDA)] (arsanilazotyrosine)	0	< 10
Histidyl Residues				
1 His—uncertain which residue (Tyr also modified)	5-Diazo-1-H-tetrazole (30-fold molar excess)	Azohistidine + Azotyrosine	0	100
His residues, perhaps Zn (II) ligands	Photooxidation with methylene blue	Oxidation products	Decrease (Protected by Zn)	Decrease
Arginyl Residues				
Arg—Possibly 145	Butanedione/borate		20	300–400
	Butanedione		50	100
Carboxyl Groups				
Glu-270	Cyclohexylmorpholinoethyl carbodiimide (CMC)	*N*-Acyl urea	0	0
Glu-270	*N*-Bromoacetyl-*N*-methyl-L-Phe	Esterifies the γ-carboxyl	0	0
Glu-270	Woodward's Reagent K	*N*-Acyl urea	0	0

the crystalline state. Modification occurs specifically at Tyr-248 (82), and therefore the enzyme contains a chromophoric probe at its active center. In solution, the azocarboxypeptidase is red with characteristic absorption and CD spectra (Fig. 10) and is almost fully active toward both peptides and esters. The absorption spectrum has a maximum at 510 nm and the CD spectrum has one positive and two negative extrema at 420, 335, and 510 nm, respectively (Fig. 10). Interaction of the azoenzyme with substrate, inhibitors, and denaturants, or removal of zinc, abolishes the red color (82, 90–92), suggesting that it is due to the formation of a zinc–azophenol coordination complex; this has now been confirmed directly by resonance Raman (RR) and NMR spectroscopy (45, 93–96). The red color is maximal at pH 8.3, above which the band at 510 nm shifts progressively to 485 nm, characteristic of the free azophenolate ion. In contrast, the metal-free azoenzyme forms only the band at 485 nm. These differences become more apparent on plotting absorbance versus pH for the zinc and apo azoenzymes. The titration of the zinc azoenzyme at 560 nm reveals two pK_{app} values at 7.7 and 9.5, respectively, whereas the titration of the apoenzyme at 485 nm results in a sigmoid curve with a midpoint at pH 9.4, characteristic of the ionization of arsanilazophenol (Fig. 11). Similarly, the CD/pH-titration curve at 510 nm for the zinc enzyme reveals the same two pK_{app} values, 7.7 and 9.5, while that of the apoenzyme at 485 nm again reveals a single pK_{app} at about 9.4.

The introduction of the azo group into Tyr-248 generates two discrete probes: azoTyr-248 itself and the azoTyr-248·zinc coordination complex. Each of these can monitor binding of ligands and supply direct information about their interaction (92). On interacting with the azoenzyme, a wide variety of ligands encompassing substrates, products, inhibitors, and activators disrupt the azoTyr-248·zinc complex and variously affect the spectrum of azoTyr-248. Stopped-flow pH- and temperature-jump studies demonstrate that the rate of this disruption (64×10^3 sec^{-1}) is orders of magnitude faster than catalysis (~ 100 sec^{-1}) and these techniques can quantitate the binding of ligands to the enzyme (90, 91, 97). It is to be emphasized that previous studies (29) which are still being cited (30, 41) clearly failed to detect this rapid process, and erroneously concluded that the disruption of the Tyr-248·zinc complex would be rate-limiting for catalysis.

Equilibrium CD titrations of azocarboxypeptidase have identified four separate ligand binding sites on the enzyme (92). The competitive inhibitors of peptide hydrolysis (43), Gly–Tyr and L-Phe, compete for the same site, designated site 1 (Fig. 12); the noncompetitive inhibitors (43), indole 3-acetate and β-phenylpropionate, bind at both sites 2 and 3; and the activator, cyclohexanol, binds at site 4 (Fig. 12). Ligand association with sites 1, 3, or 4 significantly perturbs the affinity of the two other sites, while association with site 2 has only a small effect on binding at site 1 or 4 and no effect on binding at site

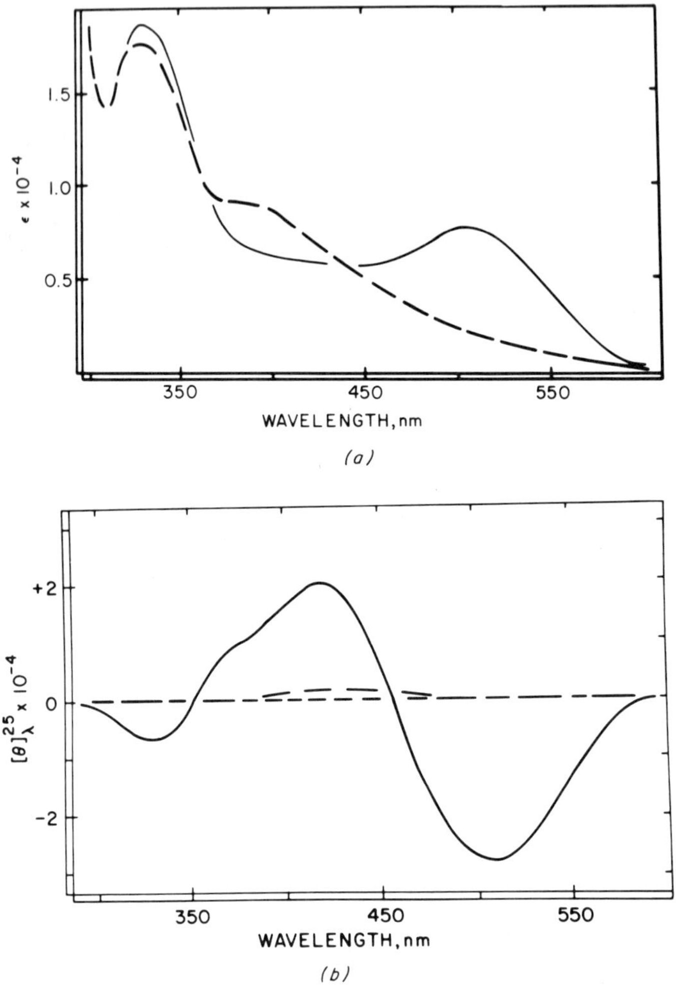

Figure 10(a) Absorption spectra of zinc arsanilazocarboxypeptidase in 50 m*M* Tris, pH 8.2: (——) dissolved in 1*M* NaCl; (---) crystal suspension. (*b*), CD spectra of zinc arsanilocarboxypeptidase (——) and of the corresponding apoenzyme (---) dissolved in 50m*M* Tris, 1*M* NaCl, pH 8.2. (From ref. 99 with permission.)

3 (Fig. 12) (92). These results are consistent with the presence of multiple, overlapping binding sites in carboxypeptidase, as previously proposed from kinetic evidence (98) (see below). This hypothesis is further supported by RR spectroscopy studies of azocarboxypeptidase that indicate that ligand association with sites 2 and 3 affects the conformational sites of Tyr-248, whereas that with site 1 does not (96).

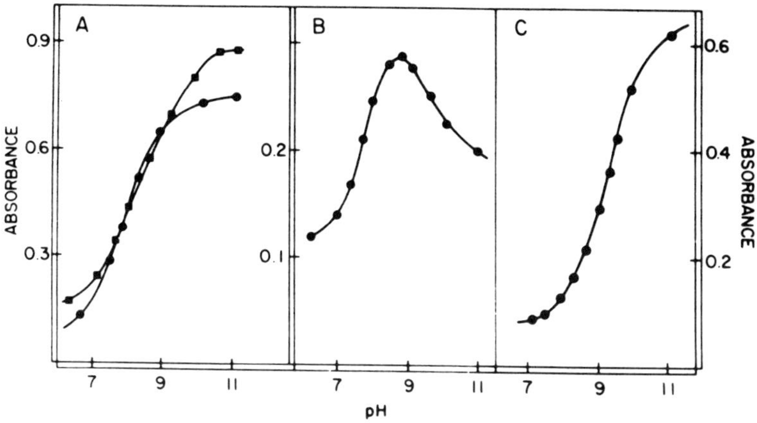

Figure 11 Absorbance–pH titration of (*A*), zinc arsanilazocarboxypeptidase $A_{\alpha,\beta,\gamma}$ at 485, 510, and (B) 560 nm, respectively, and of (C) apoarsanilazo carboxypeptidase A_{α} at 485 nm. (From ref. 99 with permission.)

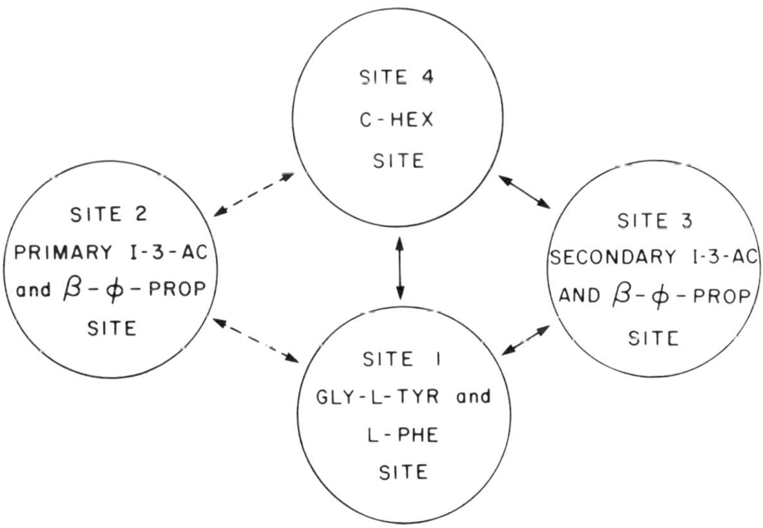

Figure 12 Schematic representation of ligand binding sites in the active center of carboxypeptidase A. Circles linked by solid arrows represent interacting sites while circles linked by broken arrows represent weakly interacting sites and circles which are not linked by arrows represent independent and noninteracting sites. I-3-AC(indole-3-acetate), β, ø-PROP(β-phenylpropionate), C-HEX(cyclohexanol). (From ref. 92 with permission.)

6 CARBOXYPEPTIDASE A IN SOLUTION AND IN CRYSTALS

6.1 Spectral Studies

The first indication that the structure of carboxypeptidase A in solution differs from that in the crystals came through studies of the arsanilazoTyr-248 modified enzyme. In solution, at pH 8.5 this derivative is red due to the formation of an azoTyr-248·zinc coordination complex but its crystals are yellow, indicating that crystallization prevents the formation of this complex and, hence, affects the conformation of Tyr-248 (99, 100). The spectral characteristics of all three forms of arsanilazo carboxypeptidase A—α, β, and γ—are identical; their solutions are red but their crystals are yellow (100). X-ray crystallographic analysis of the unmodified enzyme first placed Tyr-248 17 Å away from the zinc atom, then later placed it liganded to the metal in 15–25% of the molecules in the crystals (101), but most recently reversed this claim (30). As of now, based on the latest interpretation of the diffraction data, Tyr-248 is said to be distant from the metal in the crystals; there has been no change in the interpretation of the fact that the metal and azoTyr-248 are directly juxtaposed in solution since every study using different techniques has come to this same conclusion. Most importantly, this has been confirmed both by RR and NMR spectroscopy (45, 95). In solution azoTyr-248 is predominantly hydrogen bonded *intramolecularly* at pH 6.5 and chelated to the zinc atom at pH 8.5, whereas in the crystals it forms an *intermolecular* hydrogen bond to a group on the protein, which prevents its coordination to zinc (95).

Studies on nitroTyr-248 carboxypeptidase A, a chemical derivative with completely different characteristics (102; see above) confirm the principles of these findings. Nitrocarboxypeptidase A exhibits a visible absorption band with a maximum at 428 nm, which in solution titrates with a pK_a of about 6.4 (see above), but in the crystals with a pK_a of 8.3 (Fig. 13) (102). This dramatic shift in pK_a, solely due to a change in the physical state of the enzyme, confirms that in nitrocarboxypeptidase, as in the azoenzyme, the conformation of Tyr-248 differs in solution from that in the crystals.

Several spectroscopic techniques also demonstrate that the solution and crystal structure of the unmodified enzyme differ. Thus the EPR spectra of the Co(II) enzyme in the crosslinked crystal and solution states differ significantly (J. Thompson and B. L. Vallee, unpublished data), and the PAC spectra of ^{111}Cd(II) carboxypeptidase indicate that in the crystals one predominant state exists, whereas in solution several interconvertible forms are present (65).

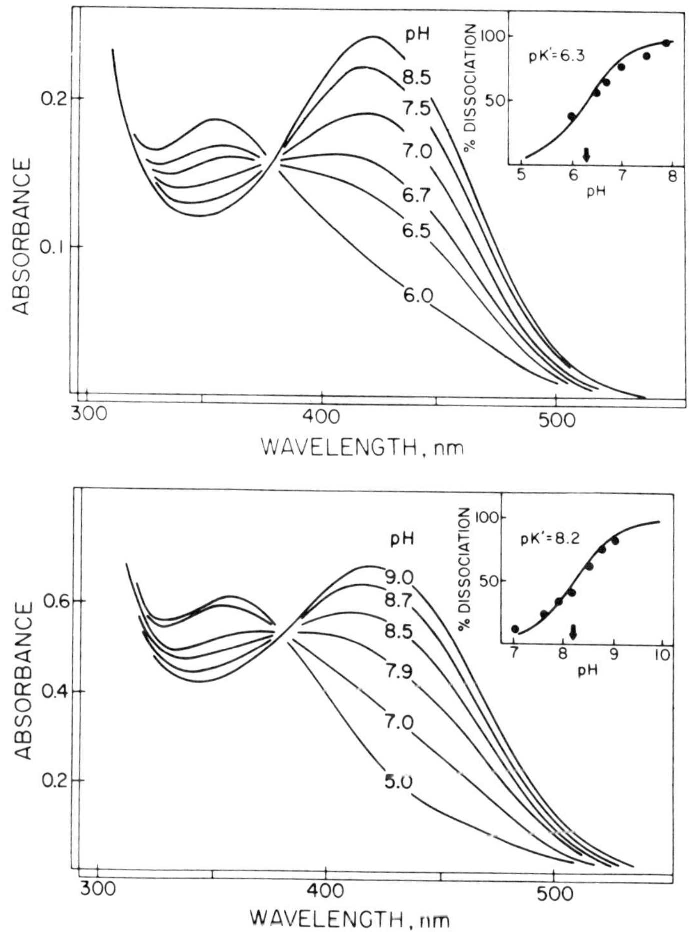

Figure 13 Absorption spectra of nitrocarboxypeptidase A (top) solution, in 2m*M* Tris-acetate, 0.5*M* NaCl, and (bottom) crystal suspension, in 2m*M* Tris-acetate, at the pH values indicated. Insets: degree of ionization as a function of pH. Solid lines describe normal ionization curves with midpoints (arrows, inset) at pH 6.3 and 8.2, respectively. (From ref. 102 with permission.)

6.2 Kinetic Studies

In addition, kinetic studies on carboxypeptidase A crystals and solutions, conducted under conditions which obviate artifacts due to diffusion limitation of substrate or products, have revealed critical differences (103, 104).

A series of peptides and esters varying in blocking groups and chain length were examined. In all instances crystallization of the enzyme markedly reduced catalytic efficiency, k_{cat}, from 20- to 1000-fold, depending on the substrate.

Substrate inhibition, apparent in solution for some di- and depsipeptides, is abolished in the crystals, and longer substrates which exhibit normal kinetics in solution may exhibit activation when hydrolyzed by crystals. The physical state of the enzyme also affects the mode of action of known modifiers of peptidase activity of the enzyme. In solution, addition of benzoylglycine or cinnamic acid, for example, markedly increases the rate of hydrolysis of Cbz-Gly-Phe but with the crystalline enzyme their addition hardly alters the activity. This is in accord with the weakening or absence of inhibitory enzyme-substrate binding modes.

The differences in the kinetic behavior of carboxypeptidase crystals on the one hand, and of their solutions on the other, bear importantly on efforts to interpret the function of the enzyme in structural terms. Hypothetical modes of substrate-enzyme interaction deduced from the superimposition of the substrate on the crystal structure of carboxypeptidase, hence purporting to simulate the kinetics of the enzyme in solution, have failed to detect all of the changes which affect inhibitory or activating binding modes (12). Since it is apparent that the kinetic properties of these two phases differ drastically, such simulation lacks an experimental basis.

It has been claimed (30) that the discrepancies between the X-ray crystal structure and the properties of the enzyme in solution are due to the use of different forms of the enzyme by the research groups involved—carboxypeptidase A (Cox) used by the crystallographers, and carboxypeptidase A (Anson) purportedly used by others. This is not so! Kinetic studies over the past 15 years have been performed on all three common forms of the enzyme, A (Cox), A (Allan), and A (Anson). All three forms display the same complex kinetic behavior toward dipeptides and their ester analogs and Michaelis-Menten kinetics toward longer peptide and ester analogs (43, 76, 105). The comparative studies between the physical and kinetic properties of carboxypeptidase A in crystals and solution described above were carried out on all these forms precisely to avoid such ambiguities. The results obtained with these enzyme forms were identical (90, 91, 95, 103, 104, 106). Indeed the enzyme now employed for crystallography (30) is from the same source, Sigma Chemical Co., that has been used by others for many years (90, 91, 100). The initial X-ray crystallography, however, was performed on an enzyme, apparently isolated from the pancreatic juice of a single cow, which was said to differ from all other preparations (29). The crystals of the arsanilazo derivative of this enzyme were claimed to be red, rather than yellow, and the activity of the native crystals greatly enhanced compared to other crystalline forms although apparently not so great as to preclude collection of diffraction data with Gly-Tyr as pointed out above. Moreover, the spectrum reported for the red crystals is not one that is characteristic of monoarsanilazotyrosine (29, 101). The properties of the special form of the enzyme used in these studies, which is now

unavailable, can only be speculated about; all the forms now available, however, behave differently from that crystalline enzyme.

In conclusion, the marked differences that exist in the physical and kinetic properties of carboxypeptidase A in crystals and solution preclude detailed mechanistic inferences based solely on the crystal structure of a peculiar form, and instead require structural studies performed synchronously with catalysis; these have been initiated. This difference between solution and crystals, once minimized, denied, or neglected by many, has meanwhile been emphasized for other systems by Blake (107) and Weber (108) among others and is now no longer thought to question the "dogma" of their identity.

7 CATALYTIC PROPERTIES

Early kinetic studies on the hydrolysis of acylamino acids, dipeptides, and their ester analogs by carboxypeptidase A were characterized by varying degrees of substrate activation and inhibition, making the determination of meaningful kinetic constants difficult (105, 109–111). In effect these phenomena constituted artifacts generated by the particular substrates then in general use. The synthesis and introduction of longer substrates, in particular tri- and tetrapeptides and depsipeptides, obviated the problem (43, 76, 112). The results obtained with shorter substrates led to the conclusion that they bind in multiple but overlapping modes, thereby causing the observed kinetic abnormalities (98). This hypothesis has now been verified directly (see below).

Longer substrates follow Michaelis–Menten kinetics over a wide concentration range, with k_{cat} values of 500–20 sec^{-1} and K_m values of 0.01–1mM (43, 76, 112). The catalytic efficiency of the enzyme toward esters is not much greater than that toward peptides, the k_{cat}/K_m values for esters being generally much less than 100 times that for peptides and in some cases, the same (76; D. S. Auld, unpublished results). The pH profiles for acylamino acid, di-, tri-, and tetrapeptide hydrolysis are bell-shaped (Fig. 14, top) and characterized by pK_a values of 6.1 and 9.0 (112). The individual profiles for k_{cat} and K_m reveal that the former is characterized by a pK_a of 6.1 and the latter of 9.0 as shown in Figure 14, bottom for Cbz-$(Gly)_2$-L-Phe. Hence, it appears that for peptides the protonation of a group with a pK_a of 6.0 abolishes catalysis but does not affect binding, whereas the ionization of a group with a pK_a of 9.0 abolishes binding but does not affect catalysis (112). These pH-rate profiles have also been interpreted in terms of a "reverse" protonation scheme (113). In this scheme ionization of the group with a pK_a of 6.0 abolishes peptide binding, while catalysis depends on ionization of the group with a pK_a of 9.0.

$$\begin{array}{ccccccc}
(EH_2) & & (EHS) & & & & \\
\updownarrow & & \updownarrow & & & & \\
EH + S & \rightleftharpoons & ES & \xrightarrow{k_2} & EY & \xrightarrow{k_n} & \text{Products} \\
\updownarrow & & & & & & \\
(E) & & & & & &
\end{array}$$

This scheme requires a pH-dependent formation of a second intermediate, EY, and a pH-independent breakdown of it into products. The rate constant for k_2 must be 1000 times that for k_n. The pH-rate profiles for ester hydrolysis are dependent on the substrate used; the k_{cat}/K_m vs. pH profiles of some esters are bell-shaped with pK_a's of 6.0 and 9.0 (114–116), while the pK_a's of others are in the 6.9–7.9 range (117, 118).

The groups responsible for the pK_a values observed for peptide hydrolysis have not been identified conclusively. Potentially the most likely candidates are the metal-bound H_2O, Glu-270, or a combination of both for the pK_a of 6.0 and either Tyr-248 or, less likely, the metal-bound H_2O for the pK_a of 9.0 (112, 119). The acid limb of the pH-rate profiles is perturbed by metal substitution and shifts from 6.1 for the zinc enzyme to 5.3 and 6.4 for cobalt and manganese carboxypeptidases, respectively (112). This suggests that this pK_a is due to the metal-bound H_2O molecule. Its kinetically determined heat of ionization, 3–7 kcal mol^{-1}, is not that expected for a carboxyl group (112, 119). However, values in the range of 6–8 kcal mol^{-1} have been observed for the ionization of the water ligand in model five-coordinate metal complexes (120), and decreasing the coordination number further would be expected to strengthen the metal–ligand bonding interactions and facilitate proton ionization (121). Alternatively, since according to the crystallographic evidence the metal-bound water and the carboxylate of Glu-270 are within hydrogen-bonding distance of one another (13), the pK_a at 6.0 could be regarded as a cooperative ionization of these two groups.

The pK_a of 9.0 is *not* metal dependent, and its heat of ionization, 5.4 kcal mol^{-1}, is very close to that of phenol (119). In addition, this pK_a is shifted to 6.6 when Tyr-248 is nitrated (113), and this latter value is close to that of nitrophenol (7.0). Taken together it could be inferred from these data that the alkaline pK_a is due to Tyr-248, although this conclusion must be regarded as tentative. In terms of the reverse protonation scheme, Tyr-248 would be acting as a general base catalyst (113), while in terms of the "normal" scheme it would assist peptide binding in its acidic form. In contrast, although the k_{cat}/K_m profiles for many esters show a pK_a of 9.0, nitration or azocoupling of Tyr-248 does *not* perturb this value (122), indicating not only that Tyr-248 is

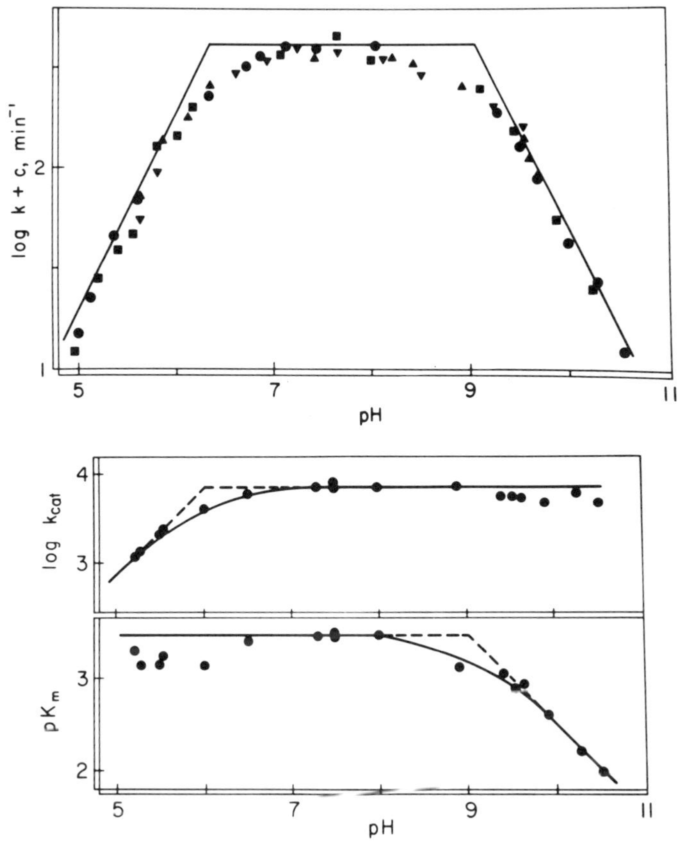

Figure 14 (Top): pH dependence of log (k_{cat}/K_m) for chloroacetyl—L-Phe (▼, C = 0.25), Bz–Gly–L-Phe (▲, C = 0.5), Bz–$(Gly)_2$–L-Leu (●, C = 1.0), and Cbz–$(Gly)_3$–L-Phe (■, C = 0) for carboxypeptidase A. (Bottom): pH dependence of log (k_{cat}) and pK_m for the hydrolysis of Cbz–$(Gly)_2$–L-Phe by carboxypeptidase A. Conditions of assay: 1*M* NaCl, 50m*M* Mes, Hepes, Tris, or carbonate buffer, 25°C. Dotted lines have slopes of 1, 0, and −1, and their intercepts yield the values of pK_a. (From ref. 112 with permission.)

not involved in ester hydrolysis, but that the protonated form of another group with a pK_a of 9 is required for esterase but not for peptidase activity, that is, that peptides and esters require different catalytic groups.

8 MECHANISTIC DIFFERENCES BETWEEN ESTERS AND PEPTIDES

There is overwhelming evidence that peptides and esters are hydrolyzed by different mechanisms (43, 76, 98, 112, 123). Mechanism is here meant to include all steps involved in substrate binding and catalysis.

1. Many chemical modifications of carboxypeptidase affect the two activities differently (see above). This applies not only to the modification of Tyr-248, where it has been argued (12, 41) that this group functions as a proton donor to peptides but not to esters (since RNH^-, compared to RO^-, is a much poorer leaving group), but also to the modification of Tyr-198 (37) and of arginine (probably Arg-145) (85) and metal substitution (76, 123).

2. The effects of metal substitution on the two activities differ significantly. For both peptides and esters the order of activity for different metallo derivatives is Co > Zn > Mn > Cd (Table 2) (43, 48, 76, 105). For peptides all these metallo derivatives have comparable K_m's, but their k_{cat}'s differ 100-fold (Table 2). In contrast, for esters the corresponding k_{cat}'s are identical but the K_m's differ 30-fold (Table 2). This strongly indicates that the metal is required for the hydrolysis, but not the binding of peptides, while it is critical for the binding as well as hydrolysis of esters as is further corroborated by (3).

3. Stopped-flow radiationless energy transfer (RET) studies, which visualize enzyme·substrate complexes directly (76) demonstrate that peptides bind tightly to the apoenzyme (Fig. 15) and thereby prevent restoration of activity by metal ions (124, 125), whereas esters do not (76).

4. Carboxylic acids that are *noncompetitive* inhibitors of peptide hydrolysis are *competitive* inhibitors with esters (Table 5) although the K_i values are identical (43, 76, 126). Short-chain peptides are noncompetitive inhibitors of ester hydrolysis (127, 128), demonstrating that peptides and esters bind differently to the enzyme.

5. The k_{cat} for ester hydrolysis shows a deuterium isotope effect of 1.9 (129), whereas no effect is observed for peptides (112).

These differences between peptides and esters have been thought to be due not so much to different binding modes, but to differences in rate-limiting steps (30, 41, 75). While such a view also considers their mechanisms to differ, of course, this hypothesis does not appear sufficient. It proposes that for both peptides and esters the initial binding mode does not involve interaction with either the metal or the specificity pocket. For these interactions to occur the substrate must rotate subsequent to initial binding and then undergo hydrolysis. The conversion of the binding modes is postulated to be rapid for esters, but rate limiting for peptides.

This hypothesis is inadequate on the following grounds:

1. It is proposed that peptides occupy the specificity pocket only *after* the rate-limiting step. The *specificity* of the enzyme toward peptides which have hydrophobic side chains (43, 130) cannot be accounted for under these circumstances.

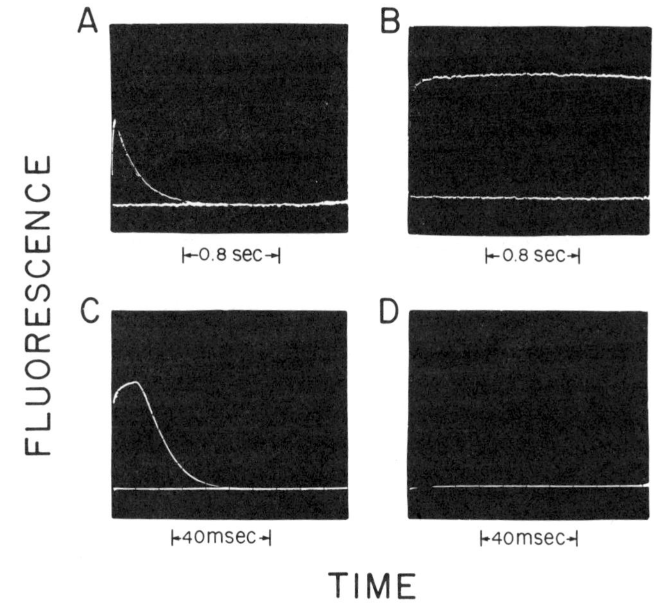

Figure 15 Stopped-flow RET measurements of $1 \times 10^{-4}M$ Dns-(Gly)$_3$-Phe, (*A, B*) and Dns-(Gly)$_3$-OPhe (*C, D*) binding to zinc (*A, C*) and apo (*B, C*) carboxypeptidase A_α. Enzyme tryptophans were excited at 285 nm, and the emission of the dansylated substrates was measured above 430 nm. (From ref. 76 with permission.)

Table 5 Carboxylic Acid Inhibitors of Ester and Peptide Hydrolysis[a]

Substrate	Inhibitor	10^4K_I (*M*)	Type of Inhibition
Bz-(Gly)$_2$-L-Phe	β-Phenylpropionate	1.2	Noncompetitive
Bz-(Gly)$_2$-L-OPhe	β-Phenylpropionate	1.2	Competitive
Cbz-(Gly)$_2$-L-Phe	Indole-3-acetate	1.7	Noncompetitive
Bz-(Gly)$_2$-L-OLeu	Indole-3-acetate	1.6	Competitive
Dns-(Gly)$_3$-L-Phe	Phenylacetate	3.3	Noncompetitive
Dns-(Gly)$_3$-L-OPhe	Phenylacetate	3.2	Competitive

[a]From ref. 76 with permission. Assays performed at 25°C, pH 7.5, 1.0*M* NaCl, 0.05*M* Tris except for the phenylacetate study where the conditions were pH 6.5, 1.0*M* NaCl, 0.03*M* Mes.

2. If both types of substrate utilize identical binding sites and catalytic steps, the modification of arginine should affect *both* peptidase *and* esterase activities; instead it specifically affects peptidase activity only (85).

3. The hypothesis does not explain the differential effects of carboxylic acid inhibitors. These inhibitors are said to occupy the specificity pocket and to interact with the metal (38) so as to interfere with the second binding mode proposed above, but not with the first. The noncompetitive inhibition of peptides predicted by such a scheme is indeed observed. However, the formation of an unproductive EIS complex with esters in the second step of the reaction (75) would manifest in *mixed inhibition* (131) rather than *competitive inhibition*, which is what actually is observed (76).

4. The hypothesis cannot account for the noncompetitive inhibition of ester hydrolysis by peptides (132).

X-ray diffraction analysis of enzyme crystals indicates that Gly–Tyr and indole-3-acetate (a noncompetitive inhibitor of peptides; Table 5) bind at the same site on the enzyme (30). This has been cited as evidence that peptide substrates and their noncompetitive inhibitors can bind at identical sites during catalysis, and hence to support the above hypothesis (30). In actuality, the mode of inhibition that would be observed when the inhibitor and substrate *both* bind to the enzyme at the *same time* has to be noncompetitive (131, 133); obviously, simultaneous binding cannot occur at the *same site*. Ligand binding studies of indole-3-acetate and Gly–Tyr to the arsanilazo enzyme confirm that they do not compete for the same site on the enzyme in solution (92). Hence, if it does bind to the same site as the substrate in the crystals, indole-3-acetate, a noncompetitive inhibitor of peptides *in solution* (Table 5), must be competitive *in the crystals*, adding another dramatic difference between these two phases to those already known.

9 MECHANISM OF CARBOXYPEPTIDASE A

A familiar scheme depicting carboxypeptidase catalysis of peptide bond cleavage appears in virtually all biological and enzyme chemistry textbooks. Much as it may be preferable to simplify complexity for didactic purposes, it is obvious from the literature that among specialists in the field there is as yet no unanimity of opinion regarding either the mode of action of the enzyme, in general, or the specific roles played by each of the critical residues, in particular.

9.1 The Role of Zinc

The earliest speculations regarding the mechanism of action of carboxypeptidase A, and indeed almost all succeeding proposals, envision the active-site zinc atom in the role of a Lewis acid polarizing the carbonyl group of the susceptible peptide bond of the substrate thereby making it more labile to hydrolysis. Prior to knowledge of the identity of essential amino side chains it could be speculated that this interaction might be sufficient to allow solvent water molecules alone to carry out the hydrolytic process. But it became clear from chemical and X-ray crystallographic evidence that the protein also contributes acidic and basic groups critical to catalysis. These three components of the enzyme—the metal, a proton donor, AH, and a base, B—constituted the essential features of a scheme suggested by this laboratory in 1963 (123). The latter two are now known, or at least considered by some, to be Tyr-248 and Glu-270.

The metal has also been envisioned as providing a coordinated OH^- that could serve as a nucleophile attacking the substrate carbonyl carbon in the catalytic process, a so-called metal-hydroxide mechanism (134). Such mechanisms have been demonstrated to occur in the hydrolysis of amide or ester bonds in model compounds (135–137) and, indeed, have been found to be far more efficacious than the zinc-carbonyl mechanism (135). The metal-hydroxide acyl intermediate that forms in these reactions has been termed a metal-carboxyl anhydride and would be kinetically indistinguishable from an anhydride involving Glu-270 (136). In addition, since this metal-carboxyl anhydride would undergo hydrolysis to form products via C—O bond cleavage rather than metal—O bond cleavage, it seemed to provide a reasonable explanation for the apparent anomalous activity of Co(III)-carboxypeptidase. However, as detailed above, the Co(III) enzyme is devoid of any catalytic activity when prepared unambiguously and, hence, such an explanation is not required. This lack of activity of the Co(III) enzyme, not even for a single turnover, suggests, in fact, that the metal-hydroxide mechanism, as generally defined, does not pertain in the case of carboxypeptidase. It might be argued that the changed to octahedral geometry on oxidation of Co(II) would reorient the metal-bound hydroxyl group relative to the substrate, thereby preventing hydrolysis. However, metal substitutions resulting in a range of coordination geometries all generate active carboxypeptidases. These include Zn(II)(distorted tetrahedral (11), Co(II)(distorted tetrahedral or five-coordinate-like) (50, 53), Ni(II)(octahedral) (53), and VO(IV)(square pyramidal) (63). On the other hand, the results of Co(II) oxidation do not rule out a metal-hydroxide nucleophilic reaction as part of a two-step catalytic mechanism involving initial inner-sphere coordination of the substrate to the metal. Importantly, this

coordination must be essential for catalysis and ester binding although it is not required for peptide binding (71).

If a metal-hydroxide is to participate in the mechanism of hydrolysis, the question of the pK_a of the coordinated water molecule becomes critical. As discussed above and previously (112, 119) evidence from the pH dependence of k_{cat} for Zn, Co, and Mn carboxypeptidase indicates that the group on the enzyme having a pK of 6.0 likely involves the metal-bound water. Enthalpies of ionization are consistent with this view as has been discussed in detail (115). Other investigators have assigned the metal-bound water a much higher pK_a since the hexaaquo-coordinated zinc(II) complex has a pK_a of about 10.2 (121). However, the pK_a of 8.9 that affects the upper limb of the pH-rate profile is not metal dependent (112) and the pK_a of metal-bound water in the enzyme is probably well below that in octahedral complexes since it is known to decrease as the coordination number is lowered (120). In addition, it has been demonstrated that a metal-hydroxide can serve as an effective nucleophile (120). Thus, a metal-hydroxide species becomes an important consideration for the carboxypeptidase mechanism.

Substitution of cobalt for zinc at the active site of the enzyme introduces a sensitive spectroscopic probe of the metal environment. Therefore, the visible absorption spectrum of the cobalt enzyme should be able to signal changes in the state of ionization of a coordinated water molecule. However, it is not clear how much of a change would occur since the spectroscopic effects of water and hydroxide are said to be almost identical (120). As discussed above, the absorption spectrum of Co(II) carboxypeptidase is pH dependent above pH 7 and the spectral changes titrate with a pK_a of 8.8. The identity of the group responsible for this effect is not known and requires further investigation. However, since the spectral changes correlate with loss of peptidase activity (Fig. 4) and since the group involved in peptidase activity is thought to be Tyr-248, these changes may reflect the state of ionization of this residue and its proxmity to the active-site metal.

Lowering pH also affects the absorption spectrum of Co(II) carboxypeptidase (68) but this is an indirect rather than a direct effect of metal–water ionization. At neutral pH the spectrum is anomalously insensitive to the addition of anions of the spectrochemical series (48, 49) but at pH 5.5 it is readily perturbed by them (68). Since carboxypeptidase requires high concentrations of salt (typically $> 1M$ NaCl) to achieve the solubility necessary for these spectral studies, the spectral changes at mildly acidic pH are due to an acquired anion sensitivity effect. Apparently, protonation of the metal hydroxide either facilitates anion displacement at the fourth coordination position or else opens up a fifth site. MCD spectra at neutral and acidic pH favor the displacement hypothesis (68).

Anion sensitivity can be conferred upon the cobalt absorption spectrum at neutral pH by chemically modifying Glu-270 with the carbodiimide reagent, CMC (Fig. 7) and similar effects are seen by ^{35}Cl NMR (138). This leads to the inference that Glu-270 contributes importantly to the anion insensitivity present in the native enzyme. It could do this merely by serving as an electrostatic barrier to approaching anions or it could interact directly with the metal-bound water through a hydrogen bond. The latter possibility is particularly attractive in terms of a plausible catalytic mechanism and has been offered as a rationalization for other observations. It is consistent with the 4.5 Å distance between the carboxylate of Glu-270 and the metal (101) and with calculations of coordination geometry from PAC spectra and angular overlap theory (65). Nuclear quadrupole interaction parameters for the various species detected by PAC spectroscopy indicate that two of these species differ by the equivalent of a half unit of charge on the solvent molecule that functions as a metal ligand. This has been accounted for in terms of a proton bridge shared between this metal-coordinated water molecule and an ionized carboxyl group (65). *Ab initio* molecular orbital calculations on model systems of carboxypeptidase reveal that proton transfer from metal-bound water to Glu-270 may indeed occur and that the minimum energy of the form $Zn^{2+}—OH^{-}—COOH$ is lower than that of the form $Zn^{2+}—H_2O—COO^{-}$ by 12 kcal/mol. Thus, the proximity of Glu-270 would promote the ionization of metal-bound water (139). Stephens et al. (138) concluded that chloride displacement of water from the active-site zinc atom measured by ^{35}Cl NMR occurs at low pH due to protonation of Glu-270. They suggested that water remains coordinated to zinc throughout the pH region from 5–10 and that its stability with respect to chloride displacement is related to the hydrogen-bonding opportunities generated by ionization of Glu-270. Another interpretation would be that a protonated Glu-270 is hydrogen bonded to metal-bound hydroxide, thus accounting for the stability.

Further evidence in favor of a Glu-270–water–metal interaction comes from a comparison of the active-center geometries of four different zinc metalloenzymes, including carboxypeptidase A (140). In each case a water molecule was proposed to occupy an identical position in the coordination sphere of the zinc atom. Further, there is an ionizable group in each of the active centers—Glu-270 in carboxypeptidase, Glu-43 in thermolysin, Ser-48 in alcohol dehydrogenase, and Thr-199 in carbonic anhydrase—that is suitably located to abstract a proton from the metal-bound water. The last two require additional assistance by associated residues. The function of this proton abstractor would be to generate an effective nucleophile in the form of a metal-hydroxide. It is remarkable that all four zinc enzymes display this feature despite their varying catalytic functions. Importantly, it is a feature that supports the hypothesis that metal-bound water is the nucleophile in these

enzymes, in general, and that Glu-270 does not act as a nucleophile in carboxypeptidase.

9.2 The Role of Glu-270

The role of Glu-270 and how it acts to promote hydrolysis is a mechanistic feature that has received much recent attention (11, 12, 74, 115, 136, 141). It is viewed as either a general base catalyzing attack of water, or, a nucleophile, directly forming a covalent anhydride with the N-terminal acyl segment of the substrate, analogous to some nonmetalloproteases (140, 142). However, despite many and persistent experimental efforts to demonstrate the existence of a covalent intermediate, no data to this effect have ever been observed or claimed for a peptide substrate and there is a large amount of evidence to the contrary (74, 89, 143–146; also see below). Only one group has claimed detection of a covalent intermediate (115, 141) based on the accumulation and ultraviolet spectrum of an intermediate at −60°C during the hydrolysis of the ester *O*-(*trans*-*p*-chlorocinnamoyl)-L-β-phenyllactate which could not be displaced by the potent inhibitor benzylsuccinate but which is not generated during hydrolysis of the corresponding peptide (74). However, of itself the accumulation of a compound neither identifies it as an intermediate nor as a *covalent* species, and the lack of reaction with benzylsuccinate could simply indicate a slow displacement reaction under the conditions chosen. In addition, there is always the possibility that this substrate represents a unique case, even among ester substrates, as it remains to be ascertained whether or not its acylation properties are due to the particular stereochemistry or electronic properties of the *trans*-cinnamoyl group.

9.3 Kinetic and Spectroscopic Detection of Catalytic Intermediates

Direct visualization and characterization of intermediates in enzyme catalysis has been a difficult task, since for good substrates the formation and breakdown of these intermediates generally occurs very rapidly. We have solved this problem by developing a computerized, low temperature stopped-flow apparatus that incorporates a high speed spectrophotometric detector. It has allowed us to study the mechanism of carboxypeptidase A by simultaneously providing the spectra and kinetic parameters of catalytic intermediates. Thus, the use of subzero temperatures extends the lifetime of these transient species and, hence, their spectra can be obtained readily by the rapid mixing and scanning techniques. The reaction dynamics of such intermediates can be monitored directly by radiationless energy transfer (RET) (126, 147–152), while electronic and EPR spectra of the cobalt-substituted enzyme can iden-

tify alterations of the metal coordination sphere that occur during the course of catalysis.

The mechanism of a series of *N*-dansylated peptide and depsipeptide substrates of carboxypeptidase A has been investigated in this manner (146). All exhibit Michaelis–Menten kinetics and are excellent substrates of the enzyme, with k_{cat}/K_m values of $\sim 10^9 M^{-1}$ min^{-1} at room temperature. The inclusion of 4.5*M* NaCl as a cryosolvent does not significantly affect the kinetic parameters, and linear Arrhenius plots are observed over wide temperature ranges (Fig. 16) (146). At −20°C two intermediates are observed during the hydrolysis of both peptides and depsipeptides (Fig. 17), which obey the kinetic scheme:

$$E + S \underset{}{\overset{K_s}{\rightleftharpoons}} ES_1 \underset{k_{-2}}{\overset{k_2}{\rightleftharpoons}} ES_2 \xrightarrow{k_3} E + P$$

whose individual rate and equilibrium constants have been determined (Table 6) (146). The formation of the second intermediate is not affected by the addition of products, and the intermediates cannot be trapped by nucleophiles. Hence, there is no evidence for a covalent acyl enzyme intermediate for any of the peptides or esters we have studied.* The results demonstrate that the C-terminal product is not released prior to the rate-limiting step as occurs in conventional acyl enzyme mechanisms. They are also inconsistent with the mechanism proposed by Cleland (75) which predicts only one pre-steady-state intermediate for peptides, whereas, in fact, two are observed (146).

Low-temperature stopped-flow rapid-scanning spectroscopy with Co(II) carboxypeptidase reveals that the formation of ES_2 is accompanied by perturbation of the Co(II) visible absorption spectrum (Fig. 18). Importantly, the Co(II) absorption spectrum obtained for the intermediate of peptide hydrolysis (Fig. 18 left panel) differs markedly from that obtained with its exact ester analog (Fig. 18 right panel). The EPR spectra of these intermediates display resolved hyperfine structure that is not observed for the free enzyme or enzyme plus products. The EPR spectrum of the peptide intermediate differs from that of the ester, again accentuating differences between the mechanisms of hydrolysis of peptides and esters. The absorption and EPR spectra of the intermediates identify catalysis-related, dynamic alterations in the active site metal coordination sphere that are clearly distinct from those observed under static conditions. The systematic changes in the cobalt spectra of carboxypeptidase accompanying formation of the peptide and ester intermedi-

*Trapping of intermediates was attempted with nucleophiles such as methanol or hydroxylamine and products were analyzed by high-performance liquid chromatography. The method was demonstrated to be successful with carboxypeptidase Y, a nonmetallo serine protease (146).

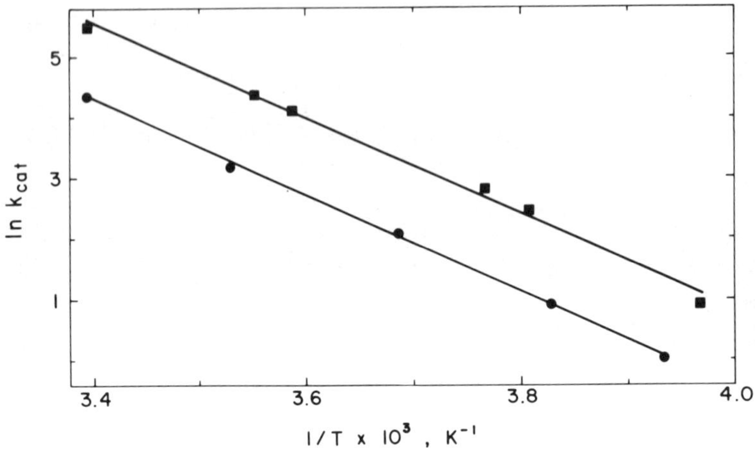

Figure 16 Arrhenius plots for the k_{cat} of (●) Dns–(Ala)$_2$–Phe in 10m*M* Hepes, 4.5*M* NaCl, pH 7.5 (E_a = 15.8 kcal/mol) and (■) Dns–(Gly)$_3$–OPhe in 10m*M* cacodylate, 4.5*M* NaCl, pH 7.0 (E_a = 15.6 kcal/mol). (From ref. 146 with permission.)

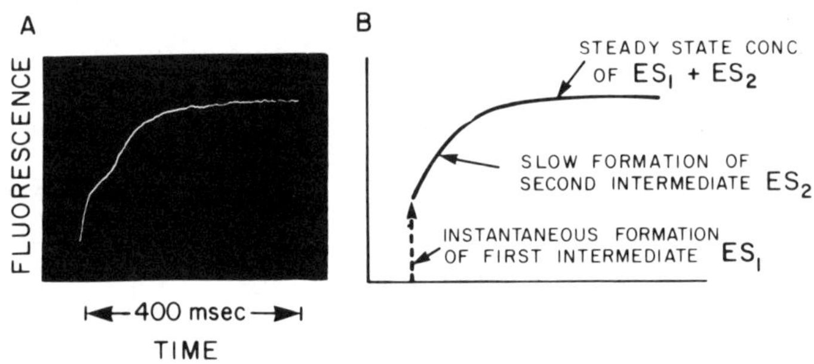

Figure 17 (A) Stopped-flow, pre-steady-state RET trace for hydrolysis of 100μ*M* Dns–(Gly)$_3$–OPhe by 0.5μ*M* carboxypeptidase A in 10m*M* Hepes-4.5*m* NaCl, pH 7.5, at −20°C. (B) Schematic representation of this trace showing the two intermediates observed. (From ref. 146 with permission.)

ates correlate with the formation of the ES$_2$ complex and thus underscore the mechanistic significance of the entatic-state hypothesis. Together the cryokinetic and cryospectroscopy studies of carboxypeptidase A show that peptides and esters generate structurally distinct metallointermediates during hydrolysis. This is in accord with previous kinetic and chemical modification studies that indicated that the metal plays a critical but different role in the mechanism of peptide and ester catalysis (see Section 8).

Table 6 Michaelis-Menten Parameters and Rate and Equilibrium Constants for the Hydrolysis of Dansylated Substrates by Carboxypeptidase A at −20°C[a]

Substrate	K_m (μM)	k_{cat} (sec^{-1})	k_2 (sec^{-1})	k_{-2} (sec^{-1})	k_3 (sec^{-1})	K_s (μM)
Peptides						
Dns-(Ala)$_2$-Phe	13.5	1.18	40.0	3.48	1.32	102
Dns-Gly-Ala-Phe	12.9	3.72	228.0	6.12	3.88	418
Esters						
Dns-(Ala)$_2$-OPhe[b]	1.6	0.06	53.3	0.50	0.06	129
Dns-Gly-Ala-OPhe	5.6	0.14	15.6	0.66	0.14	75
Dns-(Gly)$_3$-OPhe	22.3	3.13	32.6	1.41	3.61	164

[a]From ref. 146 with permission. Assays performed in 10m*M* Hepes, 4.5*M* NaCl, pH 7.5.
[b]At −10°C.

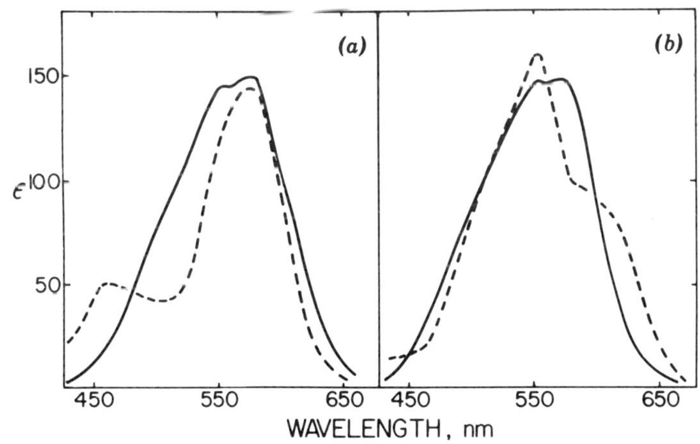

Figure 18. Rapid-scanning stopped-flow spectra of the intermediate ES_2 formed on the reaction of cobalt carboxypeptidase, $1 \times 10^{-4}M$, with peptides and esters: (left panel) Dns-Gly-Ala-Phe, $2.7 \times 10^{-4}M$, recorded at 250 ms (– – –) and 60 s (——); (right panel) Dns-Gly-Ala-OPhe, $1.2 \times 10^{-4}M$, recorded at 250 ms (---) and 10 s (——). Conditions for both are −20°C and 4.5*M* NaCl—10m*M* Hepes, pH 7.5. In both instances, the second spectrum (——) is that of enzyme plus products. (From ref. 68 with permission.)

9.4 Hydrolysis of Peptide Substrates

What can be offered as a mechanistic scheme that would be consistent with most of the pertinent experimental data? We assume that peptide substrates will bind to the active center of carboxypeptidase in much the same way as has been described for Gly–Tyr with Arg-145 and a hydrophobic pocket determining substrate specificity. The fourth coordination position of the metal is occupied by hydroxide, hydrogen bonded to protonated Glu-270. During catalysis a series of concerted steps ensues. As the carbonyl oxygen of the scissile peptide bond enters the metal coordination sphere, the metal-bound hydroxide would be displaced accepting the proton from Glu-270 and proceeding via a transient 5-coordinate-like metal species in the transition state. The displaced hydroxide, which would now be a trapped water molecule, is acted upon by the carboxylate of Glu-270 and attacks the carbonyl carbon of the peptide bond leading to a tetrahedral–like intermediate which, assisted by proton donation from Tyr-248, collapses into products. The attack of hydroxide may not require initial displacement but may actually be a concerted reaction promoted by the entrance of the substrate oxygen into the metal coordination sphere. Nakagawa et al. (139) have calculated that the geometrical arrangements necessary for such a mechanism would be allowed by the published coordinates (11) and Scheiner and Lipscomb (153) have shown that carbonyl oxygen coordination to the metal would facilitate catalysis. Desolvation would actually enhance this process since it would increase the net charge on the metal. The utilization of bound water could account for the absence of trapping by nucleophiles (74). An alternative scheme involving "reverse" protonation (113) requires further investigation. The ester substrate of Makinen and coworkers (115, 141) likely reflects the ability of the enzyme to utilize different mechanisms with substrates having unique characteristics much as it can catalyze proton exchange in a ketone (154).

This metal-hydroxide mechanism gains support from structural comparisons with human carbonic anhydrase B (140). The close similarities of the active sites of that enzyme and carboxypeptidase imply that both utilize transient pentacoordination to the metal with water acting as the lytic species. Similar comparisons with thermolysin (155) are also consistent with a Glu-270–catalyzed general base mechanism.

9.5 Hydrolysis of Ester Substrates

The role of the metal, Glu-270, and metal-bound water would likely be the same for ester and peptide hydrolysis. In contrast to initial views of the carboxypeptidase mechanism (123), we have considered for some time that the

major differences between esters and peptides relate to their mode of binding to the active site (98). These differences have been discussed above and are the subject of considerable documentation (76, 98). X-ray crystallography of a satisfactory enzyme–ester substrate complex has not been carried out and, hence, structural details of the mode of ester binding are not known. Much of the evidence suggests that esters require the metal for binding and that they interact through their C-terminal carboxyl group with it rather than with Arg-145 (156). Esters do not seem to require the proton-donating ability of Tyr-248 either for binding or for hydrolysis. It could be that by binding to the active-site metal atom, the C-terminal carboxyl group of the ester performs the same function as the peptide carbonyl oxygen in displacing water. Inhibitors that mimic ester binding (Fig. 12 and Table 5) and that interact with the metal through their carboxyl group, for example, β-phenylpropionate, are known to displace water from the first coordination sphere. Presumably the displaced water molecule would again be trapped and participate in a Glu-270-catalyzed general base mechanism. Whether Glu-72 remains coordinated to the metal under such circumstances cannot be ascertained (cf. ref. 139). Moreover, the carbonyl oxygen of the substrate might also coordinate with the metal with the resulting carbonyl bond polarization assisting catalysis. It is not yet clear why esters must bind in this manner in order to be hydrolyzed but as others have noted (74) for the reverse reaction, that is, peptide bond synthesis, one of the substrates must coordinate to the metal via its carboxyl group. It may well be that as substrates approach the active site they first interact with Arg-71 and Arg-127 and then "slide smoothly" to their final binding site. Molecular orbital calculations indicate this to be a reasonable scheme (157) although it is not clear how the results of arginine modification (85) relate to this hypothesis. In sliding toward Arg-145, the C-terminal carboxylate of the substrate may also encounter the metal since this sliding may be but the reverse of product dissociation. However, the stereochemistry of the peptide bond may preclude this as a productive interaction. In addition, when the carboxylate of a peptide substrate coordinates to the metal the scissile peptide NH may encounter unfavorable interactions, that is, the same ones that promote product dissociation.* On the other hand, esters may find carboxylate binding to the metal catalytically favorable compared to binding to Arg-145 (157) and, hence, this becomes the dominant mode of productive binding. Thus, ease of product dissociation may be mechanistically significant and account in part for the peptide versus ester differences.

*The influence of residues and peptide bonds in the penultimate, antepenultimate, and preantepenultimate positions has been documented (158) and likely accounts for the kinetic anomalies seen with short but not long peptide substrates (43, 76, 119).

10 REMAINING ISSUES

Clearly, the mechanistic proposals outlined above contain critical questions that remain to be verified. This will require structural and kinetic information obtained during the course of the catalytic process and in this regard the low-temperature stopped-flow RET and rapid-scanning techniques that we have described (146–152) will be especially useful. Indeed, cryokinetic and cryospectroscopic methods now available have identified for the first time, the absorption spectra of intermediates in an enzyme-catalyzed reaction (159). But beyond the present questions relating to mechanisms are such intriguing problems as the structural basis for zymogen activation. Procarboxypeptidase is resistant to CMC modification of Glu-270 and its pH-rate profile has an inflection well below that of the native enzyme (58). While this might suggest that activation affects the metal environment, the spectral properties of the cobalt enzyme and zymogen are remarkably similar (56).

What is the functional role of Tyr-248? Does it act as proton donor, or as a nucleophile, or does it merely exert torsional strain on the peptide bond to stabilize the nonplanar, tetrahedral intermediate? What, too, is the role of Tyr-198? It is close to the active center but resists chemical modification except in special circumstances (37) and yet, clearly, it cannot be dismissed from mechanistic considerations. The same is true for Arg-71 and Arg-127 which have been implicated (157) largely because they are there (at the active center) but for which direct physicochemical evidence for involvement is lacking.

Histidine modification inactivates the enzyme (84) but the specific residue affected has not been identified. The only histidyl residues within the active-center region are metal ligands and there is no evidence that either His-69 or His-196 participates directly in catalysis. If either or both were modified, however, metal binding and activity would almost certainly be altered. It has been noted (139) that His-69 is hydrogen bonded to Asp-142 and that this interaction may be important for the deprotonation of the metal-bound water. The role of hydrophobicity (130) in both substrate binding and catalysis is just one of the many features in need of additional examination.

It was stated at the beginning that this chapter would present a summary of recent studies on the mechanism of carboxypeptidase A and offer our own perspectives and current views. That done, we would also like to comment briefly on the questions remaining, such as whether or not Glu-270 is a nucleophile or shares a proton with the metal-bound water, which may appear to have come down to trivial details of limited academic interest.

For many perhaps the textbook mechanism is close enough, and thus acceptable. Carboxypeptidase, however, is but one of a large number of zinc metalloproteases that includes thermolysin and related bacterial neutral pro-

teases, the mammalian dipeptidyl carboxypeptidases such as angiotensin converting enzyme, Leu- and Met-enkephalinases, aminopeptidases from many species, collagenases, β-lactamase II, and perhaps other proteolytic enzymes. All of these likely share common mechanistic features and, at least in one instance, this has already led to the design of important therapeutic agents for the treatment of human disease. As we proceed toward a complete understanding of the carboxypeptidase mechanism, generalizations will undoubtedly emerge to enhance significantly the scope of such efforts. Perhaps the reluctance with which these details yield to elucidation is directly related to their eventual importance.

ABBREVIATIONS

ATEE acetyl tyrosine ethyl ester

CD circular dichroism

CMC *N*-cyclohexyl-*N*'-morpholinoethylcarbodiimide metho-*p*-toluene sulfonate

MCD magnetic circular dichroism

PAC perturbed angular correlation

RET radiationless energy transfer

RR resonance Raman

TNM tetranitromethane

ACKNOWLEDGMENTS

The authors are indebted to their colleagues, Kieran Geoghegan, Barton Holmquist, and Richard Martinelli, for many helpful discussions. This work has been supported by the National Institutes of Health (GM 15003, GM 24967, and GM 24968).

REFERENCES

1. M. L. Anson, *J. Gen. Physiol.*, **20**, 633 (1937).
2. B. L. Vallee and H. Neurath, *J. Am. Chem. Soc.*, **76**, 5006 (1954).
3. R. T. Simpson, J. F. Riordan, and B. L. Vallee, *Biochemistry*, **2**, 616 (1963).

4. H. Neurath, R. A. Bradshaw, L. H. Ericcson, D. R. Babin, P. H. Petra, and K. A. Walsh, *Brookhaven Symp. Biol.*, **21**, 1 (1968).
5. W. N. Lipscomb, J. A. Hartsuck, G. N. Reeke, F. A. Quiocho, P. H. Bethge, M. L. Ludwig, T. A. Steitz, H. Muirhead, and J. C. Coppola, *Brookhaven Symp. Biol.*, **21**, 24 (1968).
6. B. L. Vallee and J. F. Riordan, *Brookhaven Symp. Biol.*, **21**, 91 (1968).
7. H. Neurath, R. A. Bradshaw, P. H. Petra, and K. A. Walsh, *Philos. Trans. R. Soc. London Ser. B*, **257**, 159 (1970).
8. W. N. Lipscomb, G. N. Reeke, J. A. Hartsuck, F. A. Quiocho, and P. H. Bethge, *Philos. Trans. R. Soc. London Ser. B*, **257**, 177 (1970).
9. B. L. Vallee, J. F. Riordan, D. S. Auld, and S. Latt, *Philos. Trans. R. Soc. London Ser. B*, **257**, 215 (1970).
10. J. A. Hartsuck and W. N. Lipscomb in *The Enzymes*, Vol. 3, 3rd ed., P. D. Boyer, Ed., Academic, New York, 1971, p. 1.
11. F. A. Quiocho and W. N. Lipscomb, *Adv. Prot. Chem.*, **25**, 1 (1971).
12. M. L. Ludwig and W. N. Lipscomb in *Inorganic Biochemistry*, Vol. 1, G. Eichhorn, Ed., Elsevier, Amsterdam, 1971, p. 438.
13. W. N. Lipscomb, *Tetrahedron*, **30**, 1725 (1974).
14. J. F. Chlebowski and J. E. Coleman in *Metal Ions in Biological Systems*, Vol. 6, H. Sigel, Ed., Dekker, New York, 1976, p. 2.
15. H. Neurath in *The Enzymes*, Vol. 4, 2nd ed., P. D. Boyer, H. Lardy, and K. Myrback, Eds., Academic, New York, 1960, p. 11.
16. B. J. Gates and J. Travis, *Biochemistry*, **12**, 1867 (1973).
17. R. Zwilling, F. Jakob, H. Bauer, and H. Neurath, *Eur. J. Biochem.*, **91**, 223 (1979).
18. G. M. Hass, *Arch. Biochem. Biophys.*, **198**, 247 (1979).
19. K. Breddam, T. J. Bazzone, B. Holmquist, and B. L. Vallee, *Biochemistry*, **18**, 1563 (1979).
20. Y. Narahashi and K. Yoda, *J. Biochem. (Tokyo)*, **86**, 683 (1979).
21. H. Neurath, *Fed. Proc. Fed. Am. Soc. Exp. Biol.*, **23**, 1 (1964).
22. D. J. Cox, F. C. Bovard, J.-P. Bargetzi, K. A. Walsh, and H. Neurath, *Biochemistry*, **3**, 44 (1964).
23. J. R. Uren and H. Neurath, *Biochemistry*, **11**, 4483 (1972).
24. P. J. Keller and B. J. Allan, *J. Biol. Chem.*, **242**, 281 (1967).
25. G. R. Reeck, W. P. Winter, and H. Neurath, *Biochemistry*, **9**, 1398 (1970).
26. C. Quinto, M. Quiroca, W. F. Swain, W. C. Nikovits, Jr., D. N. Standring, R. L. Pictet, P. Valenzuela, and W. J. Rutter, *Proc. Natl. Acad. Sci. USA*, **79**, 31 (1982).
27. P. H. Petra in *Methods Enzymol.*, Vol. XIX, G. E. Perlmann and L. Lorand, Eds., Academic, New York, 1970, p. 460.
28. B. J. Allan, P. J. Keller, and H. Neurath, *Biochemistry*, **3**, 40 (1964).
29. F. A. Quiocho, C. H. McMurray, and W. N. Lipscomb, *Proc. Natl. Acad. Sci. USA*, **69**, 2850 (1972).
30. D. C. Rees and W. N. Lipscomb, *Proc. Natl. Acad. Sci. USA*, **78**, 5455 (1981).
31. R. A. Bradshaw, L. H. Ericsson, K. A. Walsh, and H. Neurath, *Proc. Natl. Acad. Sci. USA*, **63**, 1389 (1969).

32. J.-P. Bargetzi, K. S. V. Sampath Kumar, D. J. Cox, K. A. Walsh, and H. Neurath, *Biochemistry*, **2**, 1468 (1963).
33. P. H. Petra, R. A. Bradshaw, K. A. Walsh, and H. Neurath, *Biochemistry*, **8**, 2762 (1969).
34. P. H. Petra and H. Neurath, *Biochemistry*, **8**, 2466 (1969).
35. T. A. Steitz, M. L. Ludwig, F. A. Quiocho, and W. N. Lipscomb, *J. Biol. Chem.*, **242**, 4662 (1967).
36. D. C. Rees, M. Lewis, R. B. Honzatko, W. N. Lipscomb, and K. D. Hardman, *Proc. Natl. Acad. Sci. USA*, **78**, 3408 (1981).
37. L. Cueni and J. F. Riordan, *Biochemistry*, **17**, 1834 (1978).
38. F. A. Quiocho, P. H. Bethge, W. N. Lipscomb, J. Studebaker, R. D. Brown, and S. H. Koenig, *Cold Spring Harbor Symp. Quant. Biol.*, **36**, 561 (1971).
39. D. C. Rees, R. B. Honzatko, and W. N. Lipscomb, *Proc. Natl. Acad. Sci. USA*, **77**, 3288, (1980).
40. D. C. Rees and W. N. Lipscomb, *Proc. Natl. Acad. Sci. USA*, **77**, 4633 (1980).
41. W. N. Lipscomb, *Proc. Natl. Acad. Sci. USA*, **76**, 3875 (1973).
42. S. Yanari and M. A. Mitz, *J. Am. Chem. Soc.*, **79**, 1150, 1154 (1957).
43. D. S. Auld and B. L. Vallee, *Biochemistry*, **9**, 602 (1970).
44. D. E. Koshland, *Proc. Natl. Acad. Sci. USA*, **44**, 98 (1958).
45. W. W. Bachovchin, K. Kanamori, B. L. Vallee, and J. D. Roberts, *Biochemistry*, **21**, 2885 (1982).
46. C. C. F. Blake, L. N. Johnson, G. A. Mair, A. C. T. North, D. C. Phillips, and V. R. Sarma, *Proc. R. Soc. London Ser. B*, **167**, 378 (1967).
47. B. L. Vallee, T. L. Coombs, and F. L. Hoch, *J. Biol. Chem.*, **235**, PC45 (1960).
48. J. E. Coleman and B. L. Vallee, *J. Biol. Chem.*, **235**, 390 (1960).
49. S. A. Latt and B. L. Vallee, *Biochemistry*, **10**, 4263 (1971)
50. B. Holmquist, T. A. Kaden, and B. L. Vallee, *Biochemistry*, **14**, 1454 (1975).
51. B. L. Vallee and S. A. Latt, in *Proceedings of the International Symposium on Structure and Function of Proteolytic Enzymes*, P. Desnuelle, H. Neurath, and M. Ottesen, Eds., Munksgaard, Copenhagen, 1970, p. 144.
52. B. L. Vallee and B. Holmquist, in *Methods for Determining Metal Ion Environments in Proteins*, D. W. Darnall and R. Wilkins, Eds., Elsevier/North Holland, New York, 1980, p. 27.
53. R. C. Rosenberg, C. A. Root, R.-H. Wang, M. Cerdonio, and H. C. Gray, *Proc. Natl. Acad. Sci. USA*, **70**, 161 (1973).
54. R. C. Rosenberg, C. A. Root, and H. B. Gray, *J. Am. Chem. Soc.*, **97**, 21 (1975).
55. B. L. Vallee and R. J. P. Williams, *Proc. Natl. Acad. Sci. USA*, **59**, 498 (1968).
56. W. D. Behnke and B. L. Vallee, *Proc. Natl. Acad. Sci. USA*, **69**, 2442 (1972).
57. B. L. Vallee in *Trace Element Metabolism in Animals*, Vol. 2, W. G. Hoekstra, J. W. Suttie, H. E. Ganther, and W. Mertz, Eds., University Park Press, Baltimore, 1974, p. 5.
58. T. J. Bazzone and B. L. Vallee, *Biochemistry*, **15**, 868 (1976).
59. R. Piras and B. L. Vallee, *Biochemistry*, **6**, 348 (1967).
60. J. Kay and B. Kassell, *J. Biol. Chem.*, **246**, 6661, (1971).

61. G. Navon, R. G. Shulman, B. J. Wyluda, and T. Yamane, *Proc. Natl. Acad. Sci. USA*, **60**, 86 (1968).
62. G. Navon, R. G. Shulman, B. J. Wyluda, and T. Yamane, *J. Mol. Biol.*, **51**, 15 (1970).
63. R. J. DeKoch, D. J. West, J. C. Cannon, and N. D. Chasteen, *Biochemistry*, **13**, 4347 (1974).
64. R. C. Rosenberg, C. A. Root, P. K. Bernstein, and H. B. Gray, *J. Am. Chem. Soc.*, **97**, 2092 (1975).
65. R. Bauer, G. Christensen, J. T. Johansen, J. L. Bethune, and B. L. Vallee, *Biochem. Biophys. Res. Commun.*, **90**, 679 (1979).
66. S. A. Latt, Ph.D. Thesis, Harvard University, Cambridge, Mass., (1971).
67. B. Holmquist and B. L. Vallee, *Proc. Natl. Acad. Sci. USA*, **76**, 6216 (1979).
68. K. F. Geoghegan, B. Holmquist, C. A. Spilburg, and B. L. Vallee, *Biochemistry*, **22**, 1847 (1983).
69. F. S. Kennedy, H. A. O. Hill, T. A. Kaden, and B. L. Vallee, *Biochem. Biophys. Res. Commun.*, **48**, 1533 (1972).
70. L. C. Kuo and M. W. Makinen, *J. Biol. Chem.*, **257**, 24 (1982).
71. H. E. Van Wart and B. L. Vallee, *Biochemistry*, **17**, 3385 (1978).
72. E. P. Kang, C. B. Storm, and F. W. Carson, *J. Am. Chem. Soc.*, **97**, 6723 (1975).
73. M. M. Jones, J. B. Hunt, C. B. Storm, P. S. Evans, F. Carson, and W. J. Pauli, *Biochem. Biophys. Res. Commun.*, **75**, 253 (1977).
74. R. Breslow and D. L. Wernick, *Proc. Natl. Acad. Sci. USA*, **74**, 1303 (1977).
75. W. W. Cleland, *Advan. Enzymol.*, **45**, 273 (1977).
76. D. S. Auld and B. Holmquist, *Biochemistry*, **13**, 4355 (1974).
77. J. F. Riordan and B. L. Vallee, *Biochemistry*, **2**, 1460 (1963).
78. J. F. Riordan and B. L. Vallee, *Biochemistry*, **3**, 1768 (1964).
79. R. T. Simpson and B. L. Vallee, *Biochemistry*, **5**, 1760 (1966).
80. O. A. Roholt and D. Pressman, *Proc. Natl. Acad. Sci. USA*, **58**, 280 (1967).
81. G. Muszynska and J. F. Riordan, *Biochemistry*, **15**, 46 (1976).
82. J. T. Johansen, D. M. Livingston, and B. L. Vallee, *Biochemistry*, **11**, 2584 (1972).
83. M. S. Urdea and J. I. Legg, *Biochemistry*, **18**, 4984 (1979).
84. M. Sokolovsky and B. L. Vallee, *Biochemistry*, **6**, 700 (1967).
85. J. F. Riordan, *Biochemistry*, **12**, 3915 (1973).
86. J. F. Riordan and H. Hayashida, *Biochem. Biophys. Res. Commun.*, **41**, 122 (1970).
87. G. M. Hass and H. Neurath, *Biochemistry*, **10**, 3535 (1971).
88. P. H. Petra and H. Neurath, *Biochemistry*, **10**, 3171 (1971).
89. H. Nau and J. F. Riordan, *Biochemistry*, **14**, 5285 (1975).
90. L. W. Harrison, D. S. Auld, and B. L. Vallee, *Proc. Natl. Acad. Sci. USA*, **72**, 3930 (1975).
91. L. W. Harrison, D. S. Auld, and B. L. Vallee, *Proc. Natl. Acad. Sci. USA*, **72**, 4356 (1975).
92. G. M. Alter and B. L. Vallee, *Biochemistry*, **17**, 2212 (1978).
93. R. K. Scheule, H. E. Van Wart, B. L. Vallee, and H. A. Scheraga, *Proc. Natl. Acad. Sci. USA*, **74**, 3273 (1977).

94. R. K. Scheule, H. E. Van Wart, B. O. Zweifel, B. L. Vallee, and H. A. Scheraga, *J. Inorg. Biochem.*, **11**, 283 (1979).
95. R. K. Scheule, H. E. Van Wart, B. L. Vallee, and H. A. Scheraga, *Biochemistry*, **19**, 759 (1980).
96. R. K. Scheule, S. L. Han, H. E. Van Wart, B. L. Vallee, and H. A. Scheraga, *Biochemistry*, **20**, 1778 (1981).
97. L. W. Harrison and B. L. Vallee, *Biochemistry*, **17**, 4359 (1978).
98. B. L. Vallee, J. F. Riordan, J. L. Bethune, T. L. Coombs, D. S. Auld, and M. Sokolovsky, *Biochemistry*, **7**, 3457 (1968).
99. J. T. Johansen and B. L. Vallee, *Proc. Natl. Acad. Sci. USA*, **68**, 2532 (1971).
100. J. T. Johansen and B. L. Vallee, *Biochemistry*, **14**, 649 (1975).
101. W. N. Lipscomb, *Proc. Natl. Acad. Sci. USA*, **76**, 3797 (1973).
102. J. F. Riordan and G. Muszynska, *Biochem. Biophys. Res. Commun.*, **57**, 447 (1974).
103. C. A. Spilburg, J. L. Bethune, and B. L. Vallee, *Proc. Natl. Acad. Sci. USA*, **71**, 3922 (1974).
104. C. A. Spilburg, J. L. Bethune, and B. L. Vallee, *Biochemistry*, **16**, 1142 (1977).
105. R. C. Davies, J. F. Riordan, D. S. Auld, and B. L. Vallee, *Biochemistry*, **7**, 1090 (1968).
106. J. T. Johansen and B. L. Vallee, *Proc. Natl. Acad. Sci. USA*, **70**, 2006 (1973).
107. C. C. F. Blake, *FEBS Lett.*, **62**, E30 (1976).
108. G. Weber, *Adv. Prot. Chem.*, **29**, 2 (1975).
109. W. O. McClure, H. Neurath, and K. A. Walsh, *Biochemistry*, **3**, 1897 (1964).
110. E. T. Kaiser, S. Awazu, and F. W. Carson, *Biochem. Biophys. Res. Commun.*, **21**, 444 (1965).
111. R. C. Davies, D. S. Auld, and B. L. Vallee, *Biochem. Biophys. Res. Commun.*, **31**, 628 (1968).
112. D. S. Auld and B. L. Vallee, *Biochemistry*, **9**, 4352 (1970).
113. W. L. Mock and J.-T. Chen, *Arch. Biochem. Biophys.*, **203**, 542 (1980).
114. P. L. Hall, B. L. Kaiser, and E. T. Kaiser, *J. Am. Chem. Soc.*, **91**, 485 (1969).
115. M. W. Makinen, L. C. Kuo, J. J. Dymowski, and S. Jaffer, *J. Biol. Chem.*, **254**, 356 (1979).
116. J. W. Bunting and S. S.-T. Chu, *Biochemistry*, **15**, 3237 (1976).
117. F. W. Carson and E. T. Kaiser, *J. Am. Chem. Soc.*, **88**, 1212 (1966).
118. J. W. Bunting, J. Murphy, C. D. Myers, and G. G. Cross, *Can. J. Chem.*, **52**, 2648 (1974).
119. D. S. Auld and B. L. Vallee, *Biochemistry*, **10**, 2892 (1971).
120. P. Wooley, *Nature*, **258**, 677 (1975).
121. G. Zundel, *Hydration and Intermolecular Interaction*, Academic, New York, 1969, p. 69.
122. J. Suh and E. T. Kaiser, *J. Am. Chem. Soc.*, **98**, 1940 (1976).
123. B. L. Vallee, J. F. Riordan, and J. E. Coleman, *Proc. Natl. Acad. Sci. USA*, **49**, 109 (1963).
124. J.-P. Felber, T. L. Coombs, and B. L. Vallee, *Biochemistry*, **1**, 231 (1962).
125. J. E. Coleman and B. L. Vallee, *J. Biol. Chem.*, **237**, 3430 (1962).
126. D. S. Auld, S. A. Latt, and B. L. Vallee, *Biochemistry*, **11**, 4994 (1972).
127. J. W. Bunting and C. D. Myers, *Can. J. Chem.*, **53**, 1993 (1975).
128. M. S. Urdea and J. I. Legg, *J. Biol. Chem.*, **254**, 11868 (1979).

129. B. L. Kaiser and E. T. Kaiser, *Proc. Natl. Acad. Sci. USA*, **64**, 36 (1969).
130. A. Klyosov and B. L. Vallee, *Bioorganic Chemistry (Moscow)*, **3**, 806 (1977).
131. A. Cornish-Bowden, *Fundamentals of Enzyme Kinetics*, Butterworths, London, 1979, p. 73.
132. J. W. Bunting and S. H. Kabir, *J. Am. Chem. Soc.*, **99**, 2775 (1977).
133. I. R. Segel, *Enzyme Kinetics*, Wiley, New York, 1975, p. 100.
134. R. P. Davies, *Enzymes*, **5**, 595 (1961).
135. D. A. Buckingham, D. M. Foster, and A. M. Sargeson, *J. Am. Chem. Soc.*, **92**, 6151 (1970).
136. M. A. Wells and T. C. Bruice, *J. Am. Chem. Soc.*, **99**, 5341 (1977).
137. R. Breslow, D. E. McClure, R. S. Brown, and J. Eisenach, *J. Am. Chem. Soc.*, **97**, 194 (1975).
138. R. S. Stephens, J. E. Jentoft, and R. G. Bryant, *J. Am. Chem. Soc.*, **96**, 8041 (1974).
139. S. Nakagawa, H. Umeyama, K. Kitaura, and K. Morokuma, *Chem. Pharm. Bull.*, **29**, 1 (1981).
140. P. Argos, R. M. Garavito, W. Eventoff, M. G. Rossman, and C. I. Brändèn, *J. Mol. Biol.*, **126**, 141 (1978).
141. M. W. Makinen, K. Yamamura, and E. T. Kaiser, *Proc. Natl. Acad. Sci. USA*, **73**, 3882 (1976).
142. T. C. Bruice and S. J. Benkovic, *Bioorganic Mechanisms*, Vol. 1, Benjamin, New York, 1966, p. 212.
143. L. M. Ginodman, N. I. Mal'tsev, and V. N. Orekhovich, *Biokhimiya*, **31**, 1073 (1966).
144. P. L. Hall and E. T. Kaiser, *Biochem. Biophys. Res. Commun.*, **29,** 205 (1967).
145. G. Tomalin, B. L. Kaiser, and E. T. Kaiser, *J. Am. Chem. Soc.*, **92,** 6046 (1970).
146. A. Galdes, D. S. Auld, and B. L. Vallee, *Biochemistry*, **22,** 1888 (1983).
147. S. A. Latt, D. S. Auld, and B. L. Vallee, *Proc. Natl. Acad. Sci. USA*, **67**, 1383 (1970).
148. S. A. Latt, D. S. Auld, and B. L. Vallee, *Biochemistry*, **11**, 3015 (1972).
149. R. R. Lobb and D. S. Auld, *Proc. Natl. Acad. Sci. USA*, **76**, 2684 (1979).
150. R. R. Lobb and D. S. Auld, *Biochemistry*, **19**, 5297 (1980).
151. D. S. Auld in *Bioorganic Chemistry*, Vol. 1, E. E. Van Tamelin, Ed., Academic, New York, 1977, p. 1.
152. D. S. Auld in *Methods in Enzymology*, Vol. 61, C. M. W. Hirs and S. N. Timasheff, Eds. Academic, New York, 1979, Part H, p. 318.
153. S. Scheiner and W. N. Lipscomb, *J. Am. Chem. Soc.*, **99**, 3466 (1977).
154. T. Sugimoto and E. T. Kaiser, *J. Am. Chem. Soc.*, **101**, 3946 (1979).
155. W. R. Kester and B. W. Matthews, *J. Biol. Chem.*, **252**, 7704 (1977).
156. J. F. Riordan in *Food Related Enzymes*, J. R. Whitaker, Ed., Advances in Chemistry Series, American Chemical Society, Washington, D.C., 1974, p. 220.
157. S. Nakagawa and H. Umeyama, *J. Am. Chem. Soc.*, **100**, 7716 (1978).
158. N. Abramowitz, I. Schechter, and A. Berger, *Biochem. Biophys. Res. Commun.*, **29**, 862 (1967).
159. K. F. Geoghegan, A. Galdes, R. A. Martinelli, B. Holmquist, D. S. Auld, and B. L. Vallee, *Biochemistry*, **22,** 2255 (1983).

CHAPTER 3
Carbonic Anhydrase

SVEN LINDSKOG

Department of Biochemistry
Umeå University
Umeå, Sweden

CONTENTS

1 INTRODUCTION, 79

2 ISOENZYMES, 80

3 PRIMARY STRUCTURE AND EVOLUTION, 82

4 STRUCTURE OF ACTIVE SITE, 83

5 METALLOCARBONIC ANHYDRASES, 86

5.1 Spectroscopic Properties of Metallocarbonic Anhydrases and Their Inhibitor Complexes, 88

5.1.1 Cadmium(II), 88
5.1.2 Mercury(II), 87
5.1.3 Copper(II), 90
5.1.4 Nickel(II), 90
5.1.5 Cobalt(II), 91
5.1.6 Cobalt(III), 94
5.1.7 Manganese(II), 94
5.1.8 Vanadium(IV), 95

5.2 Metal Ion Binding: Rates and Equilibria

6 KINETICS AND INHIBITION, 97

6.1 Kinetics of CO_2—HCO_3^- Interconversion, 97
6.2 Anions, 101
6.3 Sulfonamides, 105
6.4 Other Inhibitors, 107

7 CHEMICAL MODIFICATIONS OF ACTIVE-SITE RESIDUES, 108

8 MOLECULAR MECHANISM, 109

ACKNOWLEDGMENTS, 113

REFERENCES, 113

1 INTRODUCTION

About 50 years ago, Meldrum and Roughton (1, 2) demonstrated that red blood cells contain an enzyme catalyzing the interconversion of CO_2 and HCO_3^-. They named the enzyme carbonic anhydrase. The enzyme is widespread in nature. It occurs in many vertebrate and invertebrate tissues (3, 4), in plants (5–7), and in a few bacteria (8, 9).

Carbonic anhydrase has an important position in the history of inorganic biochemistry. The discovery by Keilin and Mann (10, 11) in 1939 that the enzyme contains stoichiometric quantities of zinc was the first indication on the molecular level of a biocatalytical role for this metal ion. Although some early reports suggested that plant carbonic anhydrase may lack zinc, it is now well established that carbonic anhydrases from animal, plant, and bacterial sources are zinc metalloenzymes.

The purpose of this chapter is to summarize the current knowledge of the relations between structure and activity in carbonic anhydrase with special emphasis on the metal ion center. Since practically all work pertaining to such problems has been performed on enzyme from human or bovine red cells, the discussion will essentially be limited to these forms of carbonic anhydrase. There is a large number of excellent review articles available for those readers who want to extend their knowledge beyond the scope of this chapter or want a different point of view (3–7, 12–22). Several papers on the biochemistry and physiology of carbonic anhydrase are found in a recent symposium volume (23).

Before going into details it might be useful to summarize briefly some basic properties of the enzyme. In addition to the very efficient catalysis of the reversible hydration of CO_2, carbonic anhydrase catalyzes a variety of other reactions including the hydration of aldehydes and the hydrolysis of certain esters. Some examples of the types of reactions catalyzed by the enzyme are given in Table 1. A common feature of these reactions is the splitting of H_2O into OH^- and H^+ and the combination of OH^- with an electrophilic center, such as a carbonyl carbon, in the substrate molecule. The catalytic versatility of carbonic anhydrase has been investigated in detail by Pocker and collaborators in Seattle, and the subject has been competently reviewed by Pocker and Sarkanen (20).

It was shown by Roughton and Booth (40) in their pioneering work on the kinetic properties of carbonic anhydrase that the catalytic activity depends both on pH and on the ionic composition of the reaction medium. Extensive subsequent work has shown that the activity is controlled by an ionizing group having an apparent pK_a near 7 under usual experimental conditions (41). Most monovalent anions are inhibitors and bind to the enzyme when the activity-linked group is protonated. Thus, in the presence of anionic in-

hibitors, the apparent pK_a of this group is shifted to higher values. The most potent inhibitors are typical metal-binding agents, such as CN^- and SH^-, but many anions that do not normally bind to metal ions inhibit, for example NO_3^- and ClO_4^- (Sec. 6.2).

In addition to their discovery of the zinc metalloenzyme nature of carbonic anhydrase, Mann and Keilin (42) made the important finding that aromatic sulfonamides having an unsubstituted $R—SO_2NH_2$ function are powerful and specific inhibitors of the enzyme. Sulfonamide inhibitors are useful tools in studies of the molecular properties of the enzyme but also in studies of the role of carbonic anhydrase in a variety of physiological processes (4). Sulfonamides, covalently attached to an appropriate solid matrix, are used for the purification of the enzyme by affinity chromatography (43, 44). The pH dependence of sulfonamide binding complies with the complex being a combination of ionized inhibitor and enzyme having a protonated activity-linked group (Sec. 6.3).

The most straightforward model that has been used to explain these properties of carbonic anhydrase is the zinc–hydroxide mechanism summarized in Figure 1. However, this mechanism has not been accepted by all investigators working on the enzyme, and a variety of alternative models have been proposed. In the following sections the structure, ligand-binding properties, and kinetics of carbonic anhydrase will be discussed with some bias toward the zinc–hydroxide mechanism, while the relative merits of this model and alternative models are considered in Section 8.

2 ISOENZYMES

Mammals have at least three different forms of carbonic anhydrase that are genetically and immunologically distinct. One of these forms, designated C or II, is present in the blood of all investigated mammals. It also occurs in many other tissues, and it has been shown to be the only soluble form of the enzyme in some tissues, for example, kidney (45, 46). The II (or C) form is one of the most efficient of all known enzymes with a maximal turnover number for CO_2 hydration of 1×10^6 sec^{-1} at 25°C.

The blood of most mammals also contains another form, designated B or I, with a maximal turnover number for CO_2 hydration of about 1×10^5 sec^{-1} at 25°C. It also occurs in some parts of the alimentary canal, and large amounts are found in the rumen epithelium of ruminants (47) and the caecal mucosa (3, 48). All studied ruminants, cat, and dolphin lack this isoenzyme in the blood. Dog blood was reported to contain only the high-activity-type of carbonic anhydrase (49, 50) but later studies have shown that type I (or B) is also present (51, 52).

Table 1 Some Reaction Types Catalyzed by Carbonic Anhydrase

Reaction			Reference
$COS + 2\,H_2O$	$\leftrightarrow$	$HCO_3^- + HS^- + 2\,H^+$	24
$R{-}CHO + H_2O$	$\leftrightarrow$	$R{-}CH(OH)_2$	25–28
$CH_3COCOO^- + H_2O$	$\leftrightarrow$	$CH_3C(OH)_2COO^-$	29
$R{-}COOAr + H_2O$	$\leftrightarrow$	$R{-}COO^- + ArOH + H^+$	30–32
$R{-}COSAr + H_2O$	$\leftrightarrow$	$R{-}COO^- + ArS^- + 2\,H^+$	33
$R{-}OCOOAr + H_2O$	$\leftrightarrow$	$ROH + CO_2 + ArOH$	34, 35
$R{-}OCOCl + H_2O$	$\leftrightarrow$	$ROH + CO_2 + Cl^- + H^+$	36
$R{-}SO_2OAr + H_2O$	$\leftrightarrow$	$R{-}SO_3^- + ArOH + H^+$	37
$R{-}SO_2Cl + H_2O$	$\leftrightarrow$	$R{-}SO_3^- + Cl^- + 2\,H^+$	38
$Ar{-}F + H_2O$	$\leftrightarrow$	$ArOH + F^- + H^+$	39

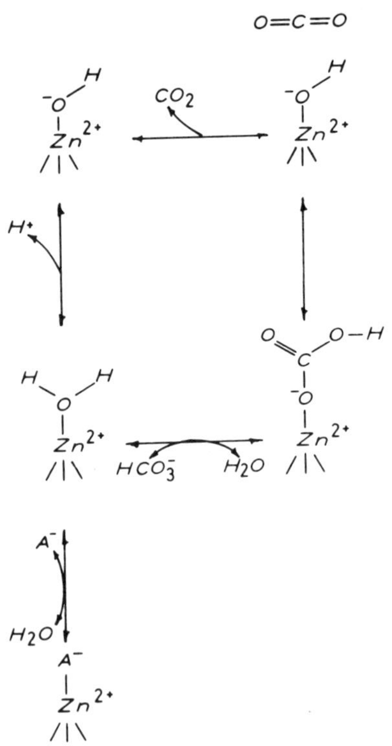

Figure 1 Zinc-hydroxide mechanism. A^- symbolizes an anionic inhibitor.

A third type of carbonic anhydrase is found in skeletal muscle (53, 54). It is characterized by a CO_2 turnover number of about 1×10^4 sec^{-1} at 25°C. This isoenzyme has hardly any esterase activity, and it binds sulfonamides much less strongly than the blood isoenzymes. A carbonic anhydrase with similar properties has been isolated from livers of male rats (55). The biosynthesis of the rat liver enzyme seems to be under hormonal control (56). Sequence data have been presented showing that the rat liver enzyme probably is identical to rat muscle enzyme (57).

Possibly, there are additional types of carbonic anhydrase in mammalian tissues. However, these have not been fully characterized. Sheep parotid gland is reported to contain a high-molecular-weight form of the enzyme (58). Membrane fractions from kidney cortex have carbonic anhydrase activity. The kinetic properties of this enzyme are similar, but probably not identical, to those of the high-activity blood isoenzyme (59–62), and it is not known whether it represents a new isoenzyme type or is a modified form of II (or C). Mitochondria, at least from certain tissues, contain carbonic anhydrase with unknown molecular properties (63).

The letter notation that is used for the blood isoenzymes in most biochemical literature was originally based on relative electrophoretic mobilities (41). This nomenclature has sometimes caused confusion. Thus, bovine blood usually contains two enzyme forms that were called A and B (64). They are both of the high-activity type, C or II. It is now known that the A form is a genetic variant of the B form with Gln for Arg-56 (65). The human blood form previously called A (41) is probably a deamidated variant (66) of B (or I). Recently, Holmes (53) suggested that the muscle isoenzyme should be called A. However, to avoid further confusion a change of nomenclature seems necessary. I suggest that the roman numeral notation used by Tashian (15–17) and others is generally adopted. Thus, in this review I shall call the B-type isoenzymes including the bovine rumen enzyme, CA I; the C-type isoenzymes including the bovine blood forms, CA II; the muscle isoenzyme, CA III. As new isoenzyme types become characterized they should be named CA IV, etc. Genetic variants or modified forms should be given an additional index, for example, CA I_1 for human A or CA $II^{56Arg \rightarrow Gln}$ for bovine A.

3 PRIMARY STRUCTURE AND EVOLUTION

The complete amino acid sequences of 11 mammalian carbonic anhydrases have been published. These include the CA I forms from humans (67–69), chimpanzee (70), rhesus macaque (71), ox (72), and horse (73), the CA II forms from humans (74, 75), ox (76), sheep (77), horse (78), and rabbit (79) and bovine CA III (72). The amino acid sequences of the bovine isoenzymes

I, II and III are shown in Figure 2. It is apparent that these sequences are homologous. If the sequences are aligned so that CA I has insertions at positions 1 and 126, the bovine CA I-CA II pair has 58% identical residue positions, the CA I-CA III pair 52%, and the CA II-CA III pair 55%. A comparison of all known complete sequences, partial sequences, and amino acid substitutions in genetic variants shows 75 invariant positions out of a total of 261 (29%). These are indicated in Figure 2.

It is probable that these isoenzymes have evolved from a common ancestor through gene duplication events and subsequent divergence. Holmes (53) has reported that isoenzymes of the CA I, CA II, and CA III types are present in chicken. Hall and Schraer (80) isolated isoenzymes similar to CA I and CA II from turtle red cells. The CA II type crossreacted with anti-chicken CA II antibodies. Tashian et al. (81) have shown that the sequence of turtle CA I is more closely related to mammalian CA I than to other mammalian isoenzyme types. These findings imply that the gene duplication events occurred before the divergence of the avian-reptilian and the mammalian lines. Maren et al. (82) found carbonic anhydrase isoenzymes with different specific activities in various tissues of an elasmobranch. However, structural information is required before it can be decided if these are related to the isoenzyme types found in mammals, birds, and reptiles.

Since all known physiological functions of carbonic anhydrase are associated with the catalysis of CO_2—HCO_3^- interconversion, it is puzzling that the less efficient CA I and CA III isoenzymes have retained their identity through at least 300 million years of evolution in the presence of the efficient CA II isoenzyme. Attempts have been made to find other enzyme activities of possible physiological significance associated with human CA I, but so far without success (83).

4 STRUCTURE OF ACTIVE SITE

The tertiary structures of human CA I and CA II have been determined by X-ray diffraction methods to a nominal resolution of 2 Å (84–86). The folding of the polypeptide chain is very similar in the two isoenzymes. A schematic drawing emphasizing secondary structure elements is shown in Figure 3. The structure is dominated by a twisted pleated sheet going through the center of the molecule and containing 10 β strands most of which are antiparallell. Most of the helical segments are short and irregular. The molecule appears compact with an ellipsoid overall shape of the dimensions 41 × 41 × 47 Å^3.

The active site is located in a cavity of roughly conical shape. The cavity is about 15 Å wide at the entrance and about 16 Å deep reaching almost to the

```
         1                          X             10                                20
BCA I    Ac-Ala-Ser-Pro-Asp-Trp-Gly-Tyr-Asp-Gly-Glu-Asn-Gly-Pro-Glu-His-Trp-Gly-Lys-Leu-Tyr-
BCA II      Ac-Ser-His-His-Trp-Gly-Tyr-Gly-Lys-His-Asn-Gly-Pro-Glu-His-Trp-His-Lys-Asp-Phe-
BCA III     Ac-Ala-Lys-Glu-Trp-Gly-Tyr-Ala-Asp-His-Asn-Gly-Pro-Asp-His-Trp-His-Glu-Leu-Phe-

                                         O   30                                       40
BCA I    -Pro-Ile-Ala-Asn-Gly-Asn-Asn-Gln-Ser-Pro-Ile-Asp-Ile-Lys-Thr-Ser-Glu-Thr-Lys-Tyr-
BCA II   -Pro-Ile-Ala-Asn-Gly-Glu-Arg-Gln-Ser-Pro-Val-Asn-Ile-Asp-Thr-Lys-Ala-Val-Val-Gln-
BCA III  -Pro-Asn-Ala-Lys-Gly-Glu-Asn-Gln-Ser-Pro-Ile-Glu-Leu-Asn-Thr-Lys-Glu-Ile-Ser-His-

                                             50                                       60
BCA I    -Asp-Pro-Ser-Leu-Lys-Pro-Arg-Ser-Val-Ser-Tyr-Asn-Pro-Ala-Thr-Ala-Lys-Glu-Ile-Val-
BCA II   -Asp-Pro-Ala-Leu-Lys-Pro-Leu-Ala-Leu-Val-Tyr-Gly-Glu-Ala-Thr-Ser-Arg-Arg-Met-Val-
BCA III  -Asp-Pro-Ser-Leu-Lys-Pro-Trp-Thr-Ala-Ser-Tyr-Asp-Pro-Gly-Ser-Ala-Lys-Thr-Ile-Leu-

                     X   X       X           70                                       80
BCA I    -Asn-Val-Gly-His-Ser-Phe-His-Val-Asn-Phe-Glu-Asp-Ser-Asp-Asn-Arg-Ser-Val-Leu-Lys-
BCA II   -Asn-Asn-Gly-His-Ser-Phe-Asn-Val-Glu-Tyr-Asp-Asp-Ser-Gln-Asp-Lys-Ala-Val-Leu-Lys-
BCA III  -Asn-Asn-Gly-Lys-Thr-Tyr-Arg-Val-Val-Phe-Asp-Asp-Thr-Tyr-Asp-Arg-Ala-Met-Leu-Arg-

                                             90  X   O       O       O               100
BCA I    -Gly-Gly-Pro-Leu-Ser-Glu-Ser-Tyr-Arg-Leu-Phe-Gln-Phe-His-Phe-His-Trp-Gly-Ile-Thr-
BCA II   -Asp-Gly-Pro-Leu-Thr-Gly-Thr-Tyr-Arg-Leu-Val-Gln-Phe-His-Phe-His-Trp-Gly-Ser-Ser-
BCA III  -Gly-Gly-Pro-Leu-Ala-Ala-Pro-Tyr-Arg-Leu-Arg-Gln-Phe-His-Leu-His-Trp-Gly-Ser-Ser-

                             O   O           110                         O       O   120
BCA I    -Asp-Glu-Asp-Gly-Ser-Glu-His-Leu-Val-Asp-Gly-Ala-Lys-Phe-Ser-Ala-Glu-Leu-His-Leu-
BCA II   -Asp-Asp-Gln-Gly-Ser-Glu-His-Thr-Val-Asp-Arg-Lys-Lys-Tyr-Ala-Ala-Glu-Leu-His-Leu-
BCA III  -Asp-Asp-His-Gly-Ser-Glu-His-Ser-Val-Asp-Gly-Val-Lys-Tyr-Ala-Ala-Glu-Leu-His-Leu-

          X                                  130 X                                   140
BCA I    -Val-His-Trp-Asn-Ser-Ala-Lys-Tyr-Pro-Ser-Phe-Ala-Asp-Ala-Ala-Ser-Lys-Ala-Asp-Gly-
BCA II   -Val-His-Trp-Asn-Thr-----Lys-Tyr-Gly-Asp-Phe-Gly-Thr-Ala-Ala-Gln-Gln-Pro-Asp-Gly-
BCA III  -Val-His-Trp-Asn-Ser-----Lys-Phe-Asn-Ser-Ile-Ala-Thr-Ala-Leu-Lys-His-Ala-Asp-Gly-

          X       X                          150                                     160
BCA I    -Leu-Ala-Leu-Ile-Gly-Leu-Leu-Val-Lys-Val-Gly-Gln-Ala-Asn-Pro-Asn-Leu-Gln-Lys-Val-
BCA II   -Leu-Ala-Val-Val-Gly-Val-Phe-Leu-Lys-Val-Gly-Asp-Ala-Asn-Pro-Ala-Leu-Gln-Lys-Val-
BCA III  -Ile-Ala-Val-Val-Gly-Val-Phe-Leu-Lys-Ile-Gly-Arg-Glu-Lys-Gly-Glu-Phe-Gln-Leu-Leu-

                                             170                                     180
BCA I    -Leu-Asp-Ala-Leu-Lys-Ala-Val-Lys-Asn-Lys-Asn-Lys-Lys-Ala-Pro-Phe-Thr-Asn-Phe-Asp-
BCA II   -Leu-Asp-Ala-Leu-Asp-Ser-Ile-Lys-Thr-Lys-Gly-Lys-Ser-Thr-Asp-Phe-Pro-Asn-Phe-Asp-
BCA III  -Leu-Asp-Ala-Leu-Asp-Lys-Ile-Lys-Thr-Lys-Gly-Lys-Glu-Ala-Pro-Phe-Asn-Asn-Phe-Asn-
                                                                                     200
                                             190                 O           X   O   X
BCA I    -Pro-Ser-Thr-Leu-Leu-Pro-Ser-Ser-Leu-Asp-Phe-Trp-Thr-Tyr-Ser-Gly-Ser-Leu-Thr-His-
BCA II   -Pro-Gly-Ser-Leu-Leu-Pro-Asn-Val-Leu-Asp-Tyr-Trp-Thr-Tyr-Pro-Gly-Ser-Leu-Thr-Thr-
BCA III  -Pro-Ser-Cys-Leu-Leu-Pro-Ala-Ser-Arg-Asp-Tyr-Trp-Thr-Tyr-His-Gly-Ser-Phe-Thr-Thr-

                                 X       O   210                                     220
BCA I    -Pro-Pro-Leu-Leu-Glu-Ser-Val-Thr-Trp-Ile-Ile-Phe-Lys-Glu-Thr-Ile-Ser-Ala-Ser-Ser-
BCA II   -Pro-Pro-Leu-Leu-Glu-Ser-Val-Thr-Trp-Ile-Val-Leu-Lys-Glu-Pro-Ile-Ser-Val-Ser-Ser-
BCA III  -Pro-Pro-Asp-Cys-Glu Cys-Ile-Val-Trp-Leu-Leu-Leu-Lys-Glu-Pro-Ile-Thr-Val-Ser-Ser-

                                             230                                     240
BCA I    -Glu-Gln-Leu-Ala-Gln-Phe-Arg-Cys-Leu-Leu-Ala-Asn-Ala-Glu-Gly-Asp-Gly-Glu-Leu-Leu-
BCA II   -Gln-Gln-Met-Leu-Lys-Phe-Arg-Thr-Leu-Asn-Phe-Asn-Ala-Glu-Gly-Glu-Pro-Glu-Leu-Leu-
BCA III  -Asp-Gln-Ile-Ala-Lys-Leu-Arg-Thr-Leu-Tyr-Ser-Ser-Ala-Glu-Asn-Glu-Pro-Pro-Val-Pro-

                      O                      250                                     260
BCA I    -Ile-Lys-Gln-Asn-His-Arg-Pro-Pro-Gln-Pro-Leu-Lys-Gly-Arg-Thr-Val-Lys-Ala-Ser-Phe
BCA II   -Met-Leu-Ala-Asn-Trp-Arg-Pro-Ala-Gln-Pro-Leu-Lys-Asn-Arg-Gln-Val-Arg-Gly-Phe-Pro-Lys
BCA III  -Leu-Val-Arg-Asn-Trp-Arg-Pro-Pro-Gln-Pro-Ile-Lys-Gly-Arg-Ile-Val-Lys-Ala-Ser-Phe-Lys
```

Figure 2 Amino acid sequences of bovine carbonic anhydrases I, II, and III. From refs. 72, 76. Residues 11, 14, 101, and 102, which are Asx or Glx in the published sequence of CA II, have been assumed to be identical to the corresponding residues in sheep CA II. Metal ion ligands and residues in sheep CA II. Metal ion ligands and residues involved in the hydrogen-bond networks shown in Figure 4 are marked ○. Active-site residues less than 10 Å from the zinc ion or in contact with bound sulfonamide inhibitors are marked X. Residues that are invariant in all published carbonic anhydrase sequences are boxed in.

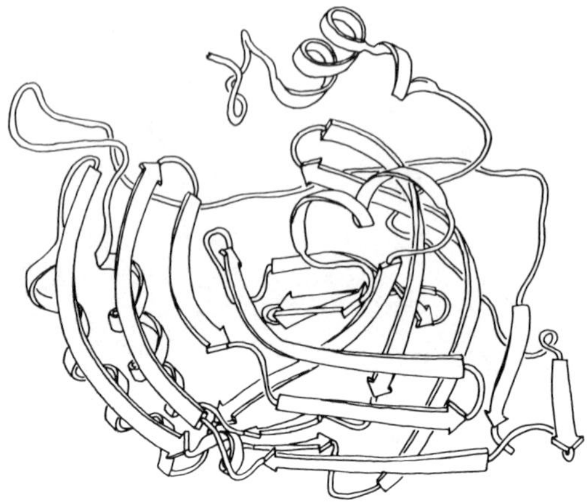

Figure 3 Schematic drawing of the tertiary structure of human CA II. By courtesy of Dr. Anders Liljas. Wallenberg Laboratory, Uppsala, Sweden.

center of the molecule. The zinc ion is near the apex of the cone and liganded to the imidazole groups of three histidines, His-94, His-96, and His-119. A water molecule appears to be bound to a fourth ligand position yielding an irregular, tetrahedral geometry. However, the coordination sphere of Zn(II) can accept more than one extraneous ligand. Clementi et al. (87) calculated by Monte Carlo methods that two water molecules may bind to the Zn(II) ion with bond distances of 1.9 and 2.2 Å, respectively, while a third H_2O position is 2.9 Å from Zn(II).

Two of the imidazole ligands, from His-94 and His-96, are bound via their $N\tau$ atoms, while His-119 is bound via its $N\pi$ atom. All the imidazole ligands appear to be hydrogen bonded to protein groups; His-94 to Gln-92 which is a solvent exposed residue in the active site cavity; His-96 to the peptide carbonyl oxygen of Asn-244; His-119 to the buried side chain of Glu-117 which takes part in a network of hydrogen bonds involving residues more or less shielded from the solvent, His-107, Tyr-194, Ser-29, and Trp-209. Glu-117 appears to accept three hydrogens in bonds to His-119, His-107, and the peptide NH of His-107. The metal-bound solvent molecule is linked to Thr-199 which is hydrogen bonded to another glutamic acid residue, Glu-106. The metal ion center and its hydrogen-bonded connections are shown schematically in Figure 4. The corresponding residues are labeled ○ in Figure 2. All of these 12 residues are invariant in the sequences of CA I, II, and III isoenzymes suggesting that the structure of the metal ion center is common to all three forms of carbonic anhydrase.

If a line through Thr-199, Zn, and Gln-92 in Figure 4 is imagined, the surface of the active-site cavity to the right of this line has an essentially hydro-

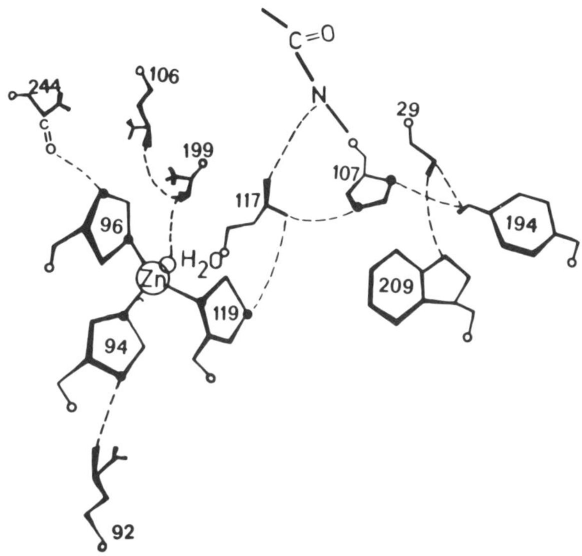

Figure 4 Schematic drawing of the metal ion center and the hydrogen-bond networks in the active site of carbonic anhydrase. From ref. 85 with permission.

phobic character, while the left-hand side is dominated by hydrophilic amino acid side chains. Residues within 10 Å of the zinc ion or within contact distance of bound sulfonamide inhibitors are marked X in Figure 2. Only one of these 12 residues, Tyr-7, is invariant in all studied sequences. The observed variations are listed in Table 2.

The only unique feature of the CA I isoenzymes that clearly distinguishes their active sites from those of the CA II and CA III isoenzymes is the presence of His at position 200. Bovine CA III has Lys-64 and Arg-67 in the hydrophilic part of the active site and Arg-91 in the outer portion of the hydrophobic part. These residues are also found in CA III from humans and gorillas (88), and may therefore be characteristic features of CA III isoenzymes and important modifiers of the functional properties of this isoenzyme type. The high specific activity of the CA II isoenzymes cannot be associated with the identity of a specific amino acid side chain in the active site. Perhaps the combination of His-64 and Thr or Asn at position 200 is important for this unique property of CA II.

5 METALLOCARBONIC ANHYDRASES

The zinc ion can be removed from CA I and CA II isoenzymes by dialysis against certain chelating agents. The original procedure utilized 1,10-phen-

Table 2 Variations of Active-Site Residues in Carbonic Anhydrase Isoenzymes.[a]

	CA I					CA II					CA III
Residue Number	Human/ Chimpanzee	Rhesus macaque	Ox	Horse	Turtle	Human	Ox	Sheep	Rabbit	Horse	Ox
64	His					His					Lys
65	Ser					Ala	Ser	Ser		Ser	Thr
67	His			Gln		Asn					Arg
91	Phe			Val	His	Ile	Val	Val			Arg
121	Ala	Val	Val	Val	Val	Val					Val
131	Leu		Phe	Phe	Phe	Phe					Ile
141	Leu					Leu					Ile
143	Val		Leu	Ile	—	Val					Val
198	Leu					Leu					Phe
200	His					Thr		Asn			Thr
207	Val					Val					Ile

[a]Blank space means identity with human CA I or CA II; dash means not determined.

anthroline at pH 5 (89), but it resulted in a substantial loss of material. An improved method employs pyridine-2,6-dicarboxylic acid (dipicoline) at neutral pH (90). Conditions for the preparation of apoCA III have not yet been found.

ApoCA I and II are catalytically inactive but can be completely reactivated by the addition of one equivalent of Zn(II) (89). Only small conformational differences between apoenzyme and holoenzyme have been detected (91–95). The apoenzymes are stable but are denatured at lower concentrations of denaturants than the native enzymes (94). Bovine apoCA II can refold spontaneously from random coil to a native-like conformation, but refolding is much slower than in the presence of zinc (94, 96). Thus, Zn(II) accelerates the folding process and contributes to the stability of the protein, but it is not necessary for the formation of the native conformation.

Various metal-ion-substituted derivatives are readily prepared by the addition of the appropriate metal salt to the apoenzyme. Most of these metallocarbonic anhydrases are catalytically inactive or have substantially altered catalytic properties. Only the Co(II) enzyme has functional properties closely similar to those of the Zn(II) enzyme. The Mn(II) enzyme and the Cd(II) enzyme have significant CO_2 hydration and esterase activities (97–100; L. Tibell and S. Lindskog, unpublished results), while early reports (99) that the Ni(II) and Fe(II) enzymes have weak catalytic activities have not been confirmed. The Cu(II) enzyme appears inactive in CO_2 hydration and 4-nitrophenyl acetate hydrolysis (91, 99, 101), but it has been reported that human Cu(II) CA I is an efficient catalyst of the hydrolysis of *S*-4-nitrophenyl thioacetate (33). Most of the metallocarbonic anhydrases can bind extraneous ligands, such as anions and sulfonamides.

5.1 Spectroscopic Properties of Metallocarbonic Anhydrases and Their Inhibitor Complexes

One important aim of the spectroscopic studies quoted in this section is to elucidate the detailed coordination geometry in the metallocarbonic anhydrases and their inhibitor adducts. Thus, structural aspects will be emphasized here whereas dynamic aspects are discussed in Sections 5.2 and 6. X-ray diffraction studies show that the three imidazole ligands from His-94, His-96, and His-119 are involved in anchoring various metal ions to the active site (86). However, the coordination sphere as a whole seems rather flexible, and four-, five-, or even six-coordination can occur depending on the metal ion and the identity of the extraneous ligand.

5.1.1 Cadmium(II). Bauer et al. (100, 102) studied ligand-induced perturbations of the angular correlation of γ rays emitted when a radioactive iso-

tope of ^{111}Cd decays to stable ^{111}Cd. Their results suggest that the metal ion in human ^{111}Cd(II) CA I is four-coordinated at low pH and probably 5-coordinated at high pH and in complexes with sulfonamides and imidazole.

NMR spectra of ^{113}Cd(II) carbonic anhydrases have been reported (103, 104). The results are partly conflicting, in some cases perhaps because of contaminating HCO_3^- coming from atmospheric CO_2. Human ^{113}Cd(II) CA I tends to give rather broad lines indicating that exchange processes contribute to the observed linewidths. Jonsson et al. (104) found evidence for a high pH form and a low pH form converting with pK_a values between 9.2 and 9.7 for different isoenzymes. The HCO_3^- complexes have a distinctly different chemical shift position. A large ^{113}Cd—^{13}C coupling is observed in the $^{13}CN^-$ complex. A two-bond ^{113}Cd—^{1}H coupling of 48 Hz is found in the complex with sulfide, showing that the ligand is the SH^- ion (257). In complexes with sulfonamides, labeled with ^{15}N in the R—SO_2NH_2 group, ^{113}Cd—^{15}N coupling constants of about 200 Hz have been observed (105). Since a two-bond coupling could not be detected in the complex with C $^{15}N^-$, it can be concluded that there is a direct Cd—N bond in the sulfonamide complex. Some ^{113}Cd NMR results are shown in Table 3.

5.1.2 Mercury (II). X-ray diffraction data show that the Hg(II) ion is centered about 0.8 Å from the Zn(II) position (86). The imidazole ligands are slightly shifted and the Hg(II) ion is in contact with the OH group of Thr-199. Two electron densities, assigned to Cl^- ions, complete a distorted, octahedral geometry. The NMR spectrum of the 1:1 complex of human ^{199}Hg(II) CA I with $^{13}CN^-$ shows a doublet with a 3700 Hz coupling constant (106). Addition of a large excess of CN^- does not change the spectrum, indicating that only one CN^- can bind.

Table 3 Some ^{113}Cd NMR Parameters for Human ^{113}Cd (II) CA II and Inhibitor Complexes[a]

Condition	δ (ppm)	J (Hz)
High-pH form	275	
Low-pH form	229	
HCO_3^- complex	216	
CN^- complex	354	1040 (^{113}Cd—^{13}C)
HS^- complex	374	48 (^{113}Cd—^{1}H)
Benzene sulfonamide complex	390	190 (^{113}Cd—^{15}N)

[a]From references 104, 105, and 257.

5.1.3 Copper(II). Cu(II) carbonic anhydrase and most of its inhibitor complexes have a broad absorption band at 700–800 nm with ϵ_{max} = 100–200M^{-1} cm^{-1} (91, 99, 107, 108). However, the N_3^-, Br^-, and I^- adducts have ϵ_{max} = 300–400M^{-1} cm^{-1} in this region and more intense, probably charge transfer, bands near 400 nm with ϵ_{max} about 3000M^{-1} cm^{-1} (107).

The EPR spectrum of the paramagnetic Cu(II) enzyme indicates a nearly axial symmetry (109). The anion complexes seem to have more or less pronounced rhombic distortions (107, 110). The Cu(II) enzyme enhances the relaxation rate (T_1^{-1}) of water protons indicating the presence of H_2O in the coordination sphere (108, 111). Most anions do not change the relaxation enhancement, but oxalate, which may act as a bidentate ligand, drastically reduces the effect. From these data one may assume that the metal ion in the Cu(II) enzyme is four-coordinated with one water ligand or perhaps five-coordinated with an additional water at a greater distance. The anion complexes are probably five-coordinated. A distorted trigonal bipyramidal structure has been suggested (112).

Sulfonamide complexes with human Cu(II) CA I have EPR spectra signifying axial symmetry (110). However, the complex of bovine Cu(II) CA II with *p*-toluene sulfonamide has an EPR spectrum that has been interpreted as indicative of a distorted tetrahedral geometry (108). In this case, the inhibitor appears to displace metal-bound H_2O.

The Cu(II) enzyme can form a dicyanide complex. The binding of the second CN^- is temperature dependent, and the complex is not fully formed at room temperature. Coleman et al. (109, 110, 113, 114) have studied this complex in detail by EPR in frozen solutions and propose the square pyramidal structure shown in Figure 5.

5.1.4 Nickel(II). Ni(II) carbonic anhydrase and its inhibitor complexes have absorption maxima at 640–730 nm (ϵ_{max} = 20–50M^{-1} cm^{-1}), at 390–440 nm (ϵ_{max} = 50–260M^{-1} cm^{-1}), and possibly near 1100 nm (115). Water proton relaxation measurements suggest the presence of metal-bound H_2O molecules. Inhibitors reduce the relaxation enhancement induced by the Ni(II) enzyme but do not abolish it. Bertini et al. (115) assume that the

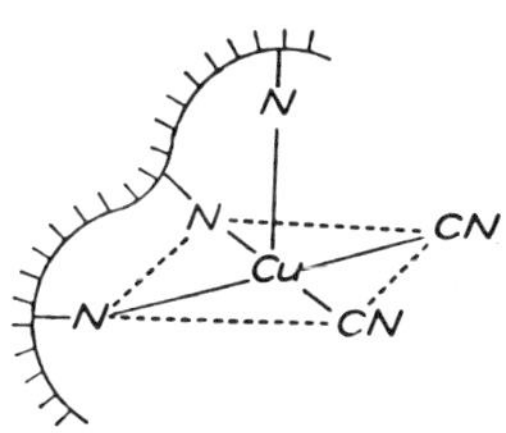

Figure 5 Dicyanide complex of Cu(II) carbonic anhydrase.

Ni(II) ion is six-coordinated and that inhibitors displace a water ligand (see, however, ref. 258).

5.1.5 Cobalt(II). The metal ion in Co(II) carbonic anhydrase and its 1:1 inhibitor complexes is in the high-spin state, $S = 3/2$ (116). The optical spectrum shows a remarkable variability depending on pH and extraneous ligands. This must reflect a variation in coordination geometry. Certain inhibitors, such as CN^-, NCO^-, SH^-, and thiadiazole sulfonamides yield intense spectra with $\epsilon_{max} > 500M^{-1}$ cm^{-1}, and those are clearly indicative of a pseudotetrahedral geometry (116, 117). Other inhibitors, such as SCN^-, NO_3^-, acetate, formate, and oxalate yield much less intense spectra with $\epsilon_{max} < 200M^{-1}$ cm^{-1}. A weak band near 1400 nm is present in these spectra, and this has been taken as evidence for a five-coordinate structure (117). Many ligands yield spectra of intermediate intensity. These are more difficult to interpret, but they may represent four- or five-coordinate structures. The Cl^- complex has the band at 1400 nm, but the spectrum is temperature dependent, suggesting a thermal equilibrium between four- and five-coordinated species (117).

The optical spectrum of the Co(II) enzyme depends on pH as shown in Figure 6. The titration curve suggests an equilibrium between a low pH form and a high pH form. However, the curve is complex (118) and may reflect the ionization behavior of more than one group (see Sec. 6.1). The spectral change parallels changes in the catalytic activity (101, 119). Since the low pH form, the high pH form, and the HCO_3^- complex probably are involved in the catalytic pathway of CO_2—HCO_3^- interconversion (cf. Fig. 1), their structures are of particular interest. However, their spectra are all of the interme-

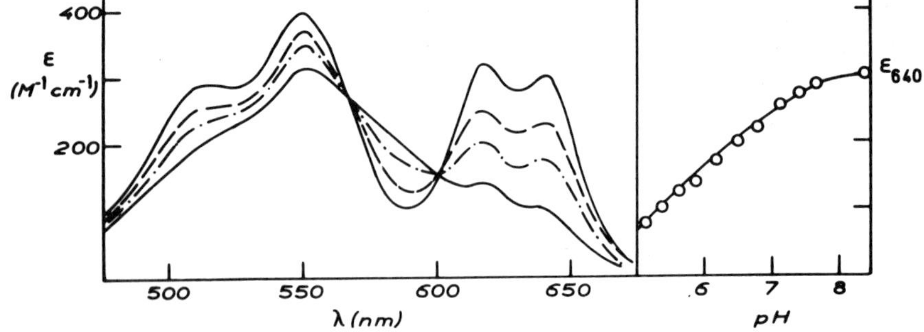

Figure 6 pH dependence of the visible spectrum of bovine Co(II) CA II. Conditions: 50mM Mes-NaOH or Hepes-NaOH buffers, room temperature. Left: spectra at pH 5.36, 6.18, 6.78, and 8.42. Right: ϵ_{640} as a function of pH.

diate intensity category. The spectra seem to lack the 1400 nm band and, therefore, they have been assumed to represent pseudotetrahedral species (117). While the spectra of the HCO_3^- complex and the low pH form are similar and rather featureless, the spectrum of the high pH form has a distinct structure with characteristic maxima at 617 and 640 nm. This signifies a rather special coordination geometry. Recently, it has been shown that the complexes with trichloroacetaldehyde (120) and phenol (121) yield spectra with similar features (Fig. 7). It would certainly be of great interest to determine the structures of these complexes by X-ray diffraction.

Bertini et al. (122) showed that the Co(II) complex of *tris*(3,5-dimethyl-1-pyrazolylmethyl)amine, which has a pseudotrigonal symmetry with 4 N ligands and H_2O at a fifth position, undergoes a pH-dependent spectral change reminescent of that of the Co(II) enzyme. The pH dependence is due to the ionization of H_2O to OH^-, but the structural change probably associated with the ionization has not been elucidated.

Magnetic circular dichroism (MCD) spectra of the low pH form of the Co(II) enzyme and the complexes with CN^-, NCO^-, and thiadiazole sulfonamides indicate pseudotetrahedral geometries. However, the MCD spectrum of the high pH form has been assigned to a distorted tetrahedral structure by Coleman and Coleman (123) and to a five-coordinated structure by Holmquist et al. (124).

The EPR spectra of Co(II) carbonic anhydrase and its inhibitor adducts are often not well resolved and indicate the presence of more than one complexed species. However, in some cases rather well-defined spectra have been obtained (125–128). These are usually highly anisotropic with g values ranging from 6.8 to 1.5 (Fig. 8). Cobalt hyperfine structure ($I = 7/2$) has been resolved in the low-field line for the I^- and NO_3^- complexes (25, 27, 28). The

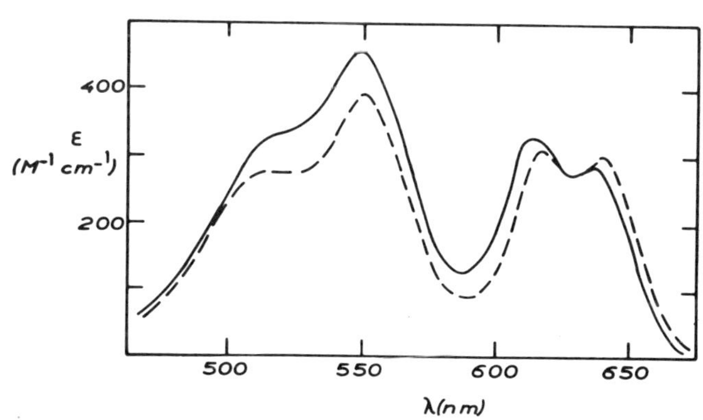

Figure 7 Visible spectra of phenol complex (—) and high pH form (---) of bovine Co(II) CA II.

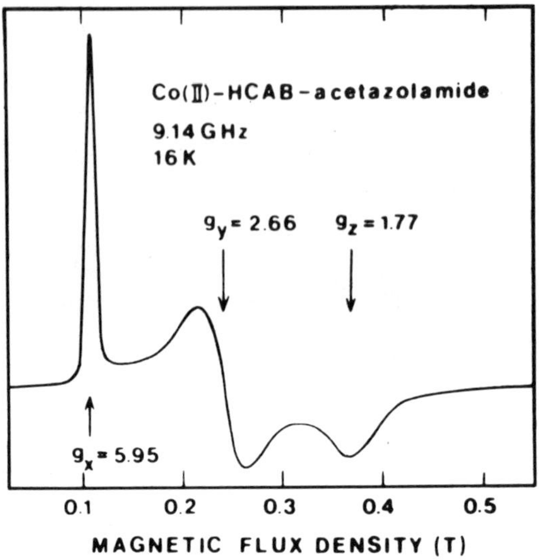

Figure 8 EPR spectrum of the complex of human Co(II) CA I with acetazolamide. From ref 256 with permission.

g values of high-spin Co(II) complexes are extremely sensitive to small variations in distortions from symmetrical geometry (29, 30). Therefore, the EPR spectra of the Co(II) enzyme and its derivatives have so far yielded very limited structural information.

In the presence of an excess of CN^-, a low spin, $S = 1/2$, Co(II) enzyme-dicyanide complex is formed. It has an EPR spectrum similar to that of Co(II)-cobalamin (Fig. 9). A square pyramidal structure analogous to that of the dicyanide complex of the Cu(II) enzyme (Fig. 5) has been postulated (110, 132).

The high pH form of the Co(II) enzyme enhances the relaxation rate of water protons (117, 133–135). From the magnitude of the effect and its temperature and frequency dependence it was calculated to correspond to one rapidly exchanging H_2O with a Co—H distance of 2.8 Å or, possibly, an OH^- with a Co—H distance of about 2.5 Å (133). However, the latter alternative would require a special mechanism by which metal-bound OH^- exchanges rapidly with bulk H_2O near neutral pH (see Sec. 8). In the presence of sulfate-containing buffers, the low pH form has only a small effect on water relaxation. Possibly, sulfate binds to the metal ion displacing H_2O or OH^- (see Sec. 6.2). However, Bertini et al. (117) and Fabry (134) have reported data which imply that in the absence of added buffers or salts, there is a low pH form affecting water proton relaxation. On the other hand, Jacob et al. (136) claim that the relaxation enhancement is entirely due to the high

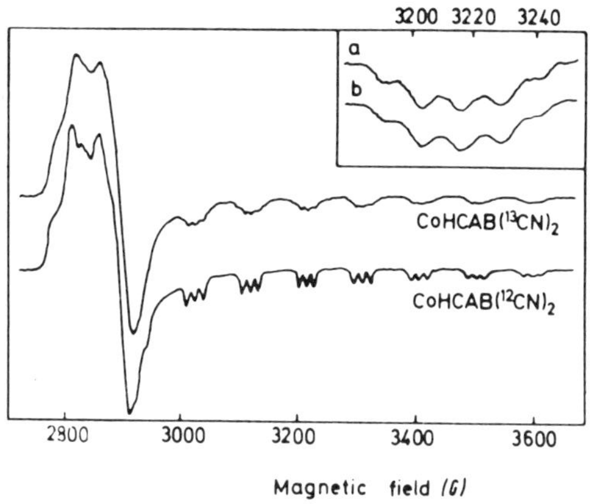

Figure 9 9.15 GHz EPR spectrum of the dicyanide complex of human Co(II) CA I at 77°K. Inset shows splittings from one ^{14}N and two equivalent ^{13}C in the $M_I = +1/2$ cobalt hyperfine line in the $g_{\parallel}$ region. (*a*) Experimental, (*b*) simulated. From ref. 131 with permission.

pH form of the enzyme, and that removal of salt simply results in a shift to a lower value of the pK_a for the equilibrium between high pH and low pH forms. Thus, there seems to be conflicting evidence concerning the presence of coordinated H_2O in the low pH form of the Co(II) enzyme. However, it should be pointed out that an undetectable or small relaxation effect does not rule out coordinated water. Anions generally decrease the Co(II) enzyme-induced relaxation enhancement to a low level. This also occurs with anions, such as I^- and N_3^-, which yield five-coordinated complexes according to the optical spectra. Since only one anion is bound, the putative fifth ligand is probably H_2O. Bertini et al. (117) have proposed that the small or absent relaxation effect depends on a smaller value of the electronic relaxation time in these complexes than in the high pH form (see also ref. 259).

5.1.6 Cobalt(III). The Co(II) enzyme can be oxidized by H_2O_2 to a diamagnetic Co(III) enzyme (137). The Co(III) enzyme is inactive, but active Co(II) enzyme can be reformed after reduction with dithionite, borohydride (138), or ascorbate (139). The Co(III) enzyme can bind anions. Ligand exchange rates are slow, but faster than for most simple Co(III) complexes (138). The electronic spectra of the Co(III) enzyme and its anion complexes have been interpreted in terms of octahedral coordination.

5.1.7 Manganese(II). Haffner et al. (140) found EPR signals near $g = 2$ and $g = 4$ in human Mn(II) CA I, whereas Meirovitch and Lanir (141) detected a signal in the $g = 2$ region in bovine Mn(II) CA II. An irregularly coordinated Mn(II) ion was assumed in both cases.

Like other paramagnetic metallocarbonic anhydrases, the Mn(II) enzyme enhances water proton relaxation. Lanir et al. (42), who used Tris-sulfate buffers, report that the effect depends on pH. In analogy with the Co(II) enzyme there seems to be a low pH form with a small effect and a high pH form with an effect corresponding to one rapidly exchanging H_2O. Thus, four-coordination seems likely. Bertini et al. (43), using unbuffered solutions, report a pH-independent effect. In both cases corrections for the incomplete binding of Mn(II) to the apoenzyme were applied. The reason for this apparent discrepancy is not clear; perhaps sulfate binds to Mn(II) displacing H_2O. However, anions such as N_3^- and NO_3^- do not change the relaxation effect, and this indicates that these ions do not displace H_2O. Sulfonamides and the bidentate oxalate dianion cause a considerable decrease of the effect. From these data it appears as if the anion complexes of the Mn(II) enzyme are five-coordinated.

5.1.8 Vanadium(IV). The EPR spectrum of bovine VO^{2+} CA II suggests an axially symmetric coordination (144). The spectrum is slightly pH dependent with a high pH form and a low pH form. Anions and sulfonamides do not affect the spectrum, but anions yield small changes of the optical spectrum (145). By the same criterion as for other paramagnetic metallocarbonic anhydrases, the VO^{2+} enzyme appears to have metal-bound H_2O. A square pyramidal structure with the oxygen atom of the VO^{2+} ion in the axial position has been proposed (145).

5.2 Metal Ion Binding: Rates and Equilibria

At neutral pH, the Zn(II) ion in carbonic anhydrase exchanges very slowly with extraneous, radioactive Zn(II). Romans et al. (146) measured such exchange rates with bovine CA II in the pH range 5 to 7. They estimated a half time of about 3 years for Zn(II) dissociation at 4°C and pH 7. However, the rate increases on lowering the pH and/or increasing the temperature. A half time of about 10 hr was estimated at 40°C and pH 5.2. Henkens and Sturtevant (93) measured Zn(II) association rate constants and found that these increase with increasing pH. The data yield a pH dependence of the apparent stability constant, K_{EZn} (M^{-1}), according to

$$\log K_{\mathrm{EZn}} = 3.7 + 1.1 \text{ pH} \tag{1}$$

This is in reasonable agreement with stability constants obtained by equilibrium dialysis using 1,10-phenanthroline as metal buffer (99). The observed pH dependence indicates that Zn(II) binding is associated with the dissociation of the H^+ ion.

$$EH^+ + Zn^{2+} \rightleftharpoons EZn^{2+} + H^+ \tag{2}$$

The source of this H^+ ion is not known. It may be derived from the imidazole ligands. However, it has not been possible to follow the titration of these ligands in the apoenzyme by NMR (147, 148).

Rate data have also been obtained for the reaction of apocarbonic anhydrase with Co(II), Ni(II), Mn(II), Cu(II), and Cd(II) (149–152). In general, these metal ions, like Zn(II), react relatively slowly with the apoenzyme in comparison with rates observed in low-molecular-weight complexes. However, the VO^{2+} ion appears to behave differently, and an association rate constant of $10^8 M^{-1}$ sec^{-1} was estimated at 25°C and pH 7 (144).

The binding of Zn(II) to bovine apoCA II at 25°C and pH 7 is associated with an unfavorable, positive enthalpy change and a large favorable, positive entropy change (153). The Zn(II) association and dissociation reactions have large activation enthalpies, but the association reaction has a positive $\Delta S‡$, while the dissociation reaction has a negative $\Delta S‡$ (93, 146). The association reaction probably involves the expulsion of ordered water molecules from the hydration sphere of the metal ion and perhaps also from the active site cavity.

Certain chelating agents promote a relatively rapid dissociation of Zn(II) from the enzyme. It was first suggested by Carpy (154) that these chelators, including 1,10-phenanthroline, form a ternary complex with the enzyme-bound metal ion, and that the rate-limiting step is the dissociation of the metal–chelator complex from the active site.

$$\text{E—Zn}^{2+} + \text{L} \underset{k_{-1}}{\overset{k_1}{\rightleftharpoons}} \text{E—Zn}^{2+} - \text{L} \underset{k_{-2}}{\overset{k_2}{\rightleftharpoons}} \text{E} + \text{Zn}^{2+} - \text{L} \tag{3}$$

Kidani and Hirose (155–157) have studied the Zn(II) dissociation mechanism in detail using a variety of chelating agents. They find that EDTA and nitrilotriacetate are inefficient and probably act by lowering the free metal ion concentration so that Zn(II) dissociates by the unassisted pathway of eq. (2). Various di- and tridentate chelators accelerate dissociation according to eq. (3). In particular, pyridine-2,6-dicarboxylate proved efficient in removing the metal ion. At pH 6.6 and 25°C this chelator is 5000 times more efficient than 1,10-phenanthroline (158). The large differences in the Zn(II) dissociation rates by the unassisted and the chelator-mediated pathways are largely associated with different activation entropies for the rate-limiting steps.

The formation of the ternary complexes is rapid. An exception is the pyridine-2,6-dicarboxylate complex with bovine Co(II) CA II. Harrington and Wilkins (159) could estimate values at pH 5.0 and 25°C of $1.6 \times 10^3 M^{-1}$ sec^{-1} and 2 sec^{-1}, respectively, for the rate constants k_1 and k_{-1} of eq. (3). These authors also report the interesting observation that a monodentate ligand, SCN^-, accelerates the reaction of apoenzyme with metal ion [k_{-2} of

eq. (3)] while metal ion dissociation (k_2) is not accelerated relative to the rate of the unassisted dissociation [eq. (2)]. Thus, the stabilization of the holoenzyme by the anionic inhibitor SCN^- is, perhaps unexpectedly, not expressed in the metal ion dissociation rate.

The optical spectra of ternary complexes with bovine Co(II) CA II suggest that 3- and 4-pyridine carboxylates, which are not capable of removing the metal ion, act as monodentate ligands (160). The relatively inefficient 2-pyridine carboxylate behaves probably as a bidentate ligand, while the efficient pyridine-2,6-dicarboxylate may act as a tridentate ligand.

At pH 5, Co(II) dissociation by the unassisted pathway has about the same rate as Zn(II) dissociation (152). The lower stability of the Co(II) complex is reflected in the association rate constant which is about 1000 times smaller for Co(II) than for Zn(II). On the other hand, the chelator-mediated dissociation is faster for Co(II) than for Zn(II). Co(II)-5-methyl-1,10-phenanthroline dissociates from the ternary complex with $k_2 = 7.7 \times 10^{-4}\ \text{sec}^{-1}$ while the corresponding value for the Zn(II) complex is $1.2 \times 10^{-5}\ \text{sec}^{-1}$ (157). This difference is essentially expressed in different activation entropies.

It has recently been found that Zn(II) can rapidly be transferred from Zn(II)-metallothionein to apocarbonic anhydrase (161, 162).

6 KINETICS AND INHIBITION

6.1 Kinetic Mechanism of CO_2—HCO_3^- Interconversion

Since the CA II isoenzymes catalyze the reversible hydration of CO_2 with extraordinary efficiency,

$$CO_2 + H_2O \rightleftharpoons HCO_3^- + H^+ \tag{4}$$

all elementary steps in the reaction mechanism must be very rapid. Consequently, it has so far not been possible to investigate the reaction in a transient state, but the kinetic information comes from studies of the reaction in the steady state or at chemical equilibrium.

Catalyzed rates of exchange between CO_2 and HCO_3^- at equilibrium can be estimated from the enzyme-induced broadening of the NMR resonances of $^{13}CO_2$ and $H\,^{13}CO_3^-$ (163, 164). These exchange rates follow a simple Michaelis-Menten-type behavior,

$$v_{\text{exch}} = \frac{k_{\text{cat}}^{\text{exch}} \cdot [S]}{K_{\text{eff}}^{S} + [S]} \tag{5}$$

Here K_{eff}^{S} is an apparent substrate dissociation constant. Simonsson et al. (165) showed that $k_{\text{cat}}^{\text{exch}}$ for human CA II is only slightly pH dependent in the

6.2–7.6 range whereas $K_{\mathrm{eff}}^{\mathrm{HCO}_3^-}$ behaves as if the acidic form of a group with pK_a near 7 is required for the binding of HCO_3^- (and, since these are equilibrium conditions, the basic form is required for the binding of CO_2). If the small pH dependence of $k_{\mathrm{cat}}^{\mathrm{exch}}$ is neglected, these results fit a very simple scheme:

$$E^- + CO_2 \rightleftharpoons E^- - CO_2 \rightleftharpoons EH{-}HCO_3^- \rightleftharpoons EH + HCO_3^- \quad (6)$$

An equivalent scheme involving the formation of $E^- + H_2CO_3$ can be written (165). However, the catalyzed rates are so rapid that this would require a second-order rate constant for the reaction of E^- with H_2CO_3 greater than $2 \times 10^{10} M^{-1}$ sec^{-1} (167–169). This value exceeds the calculated rate constant for a diffusion-limited reaction by one order of magnitude and, therefore, H_2CO_3 is not likely to be involved in the main catalytic pathway.

In the complete catalytic cycle involving HCO_3^-, the enzyme forms E^- and EH must be rapidly interconverted (cf. Fig. 1):

$$EH \rightleftharpoons E^- + H^+ \quad (7)$$

Although the rate of this reaction should not affect the rate of $CO_2{-}HCO_3^-$ exchange at equilibrium, the observed steady-state turnover rates require that the reaction has apparent first-order rate constants of at least 10^6 sec^{-1} in both directions. With H_2O, OH^-, and H_3O^+ as H^+ acceptors or donors at least one of these rate constants cannot be greater than about 10^3–10^4 sec^{-1} at neutral pH (170). Therefore, some alternative mechanism must operate. It was proposed independently by Khalifah (167), Prince and Woolley (168), and Lindskog and Coleman (169) that a sufficiently rapid H^+ transfer could occur via the buffer:

$$EH + B \rightleftharpoons E^- + BH^+ \quad (8)$$

Jonsson et al. (171) subsequently showed experimentally that the turnover rates for CO_2 hydration depend, indeed, on the buffer concentration in the predicted manner. Rate constants for the transfer of H^+ from EH to B were estimated to be about $10^8 M^{-1}$ sec^{-1}. However, at high buffer concentrations ($[B] \gg K_m^B$), when the buffer-dependent H^+ transfer step should not limit the turnover rate, the steady-state kinetic parameters do no behave as predicted from a combination of eqs. (6) and (8). Thus, while the maximal value of $k_{\mathrm{cat}}^{\mathrm{exch}}$ is about 2.5×10^6.sec^{-1} at 25°C, the maximal CO_2 turnover number, $k_{\mathrm{cat}}^{CO_2}$, is only 1×10^6 sec^{-1} (172–175). Furthermore, $k_{\mathrm{cat}}^{\mathrm{exch}}$ has the same value in H_2O and in D_2O, whereas the steady-state parameters have isotope effects (k_H/k_D) of 3–4 (174, 175). Therefore, it seems likely that the interconversion between EH and E^- involves at least two steps, a buffer-dependent step and a buffer-independent step, the latter of which limits the turnover rates at high buffer concentrations.

For example:

$$\mathrm{EH} \longleftrightarrow \mathrm{EH'} \underset{\mathrm{B}\quad\ \mathrm{BH^+}}{\longleftrightarrow} \mathrm{E^-} \tag{9}$$

Working at high buffer concentrations, Venkatasubban and Silverman (176) observed an exponential dependence of $k_{cat}^{CO_2}$ on the atom fraction of deuterium in mixtures of H_2O and D_2O. This is in accordance with a buffer-independent, rate-limiting step involving the transfer of two or more protons, but it is also possible that the isotope effect is caused by solvation changes in the active site.

From a reaction sequence involving eqs. (6) and (9) one predicts that $k_{cat}^{CO_2}$ should be independent of pH. However, it is found that $k_{cat}^{CO_2}$ varies with pH as if dependent on the basic form of a group with pK_a near 7, while $K_m^{CO_2}$ is independent of pH (172–175). Similarly, the pH dependence of the HCO_3^- reaction is essentially located in $k_{cat}^{HCO_3^-}$ (174, 175). To account for these observations as well as the deuterium isotope effects and product inhibition patterns (177), Steiner et al. (174) proposed that two ionizing groups with similar pK_a values are involved in catalysis as illustrated in Figure 10. In this scheme, one of the groups, the catalytic group, becomes protonated during a CO_2 hydration cycle. This proton is then transferred in a rate-limiting step to a proton-transfer group which can react rapidly with buffers. It was suggested that His-64 might act as the proton-transfer group in the CA II isoenzymes. Proton NMR studies have shown that the pK_a of His-64 in human and bovine CA II is near 7 (178, 179).

In the discussion above the second substrate, H_2O, has been neglected. Clearly, H_2O must exchange with the active site with an apparent first-order rate constant of at least 10^6 sec^{-1} to be kinetically competent. Using mixtures of $^{12}C—^{18}O$ and $^{13}C—^{16}O$ labeled substrates at chemical equilibrium, Silverman and Tu (180, 181) demonstrated the transient formation of $^{13}C—^{18}O$ labeled species before the ^{18}O isotope was ultimately delivered to bulk H_2O. This means that E^- in eq. (6) must contain the oxygen atom of the H_2O substrate either as a bound H_2O molecule or as a bound OH^- ion. From the rate of formation of $^{13}C—^{18}O$ species and the rate of exchange of ^{18}O labeled substrate with H_2O, Silverman et al. (182) calculated the rate of dissociation of the $H_2\,^{18}O$ product. At pH 8, where the enzyme form E^- dominates, a rate constant of about 2×10^5 sec^{-1} is estimated from their data. Obviously, this value is too small to account for the turnover rate of product H_2O dissociation. However, buffers accelerate the process, and at pH 7 and in the presence of 50mM buffers a rate constant of about 1×10^6 sec^{-1} is found. Furthermore, the H_2O dissociation rate shows a large deuterium isotope effect. Silverman et al. (182) interpreted these data in terms of the mechanism of Figure 10 and assumed that the HE $\rightleftharpoons$ EH step, which lim-

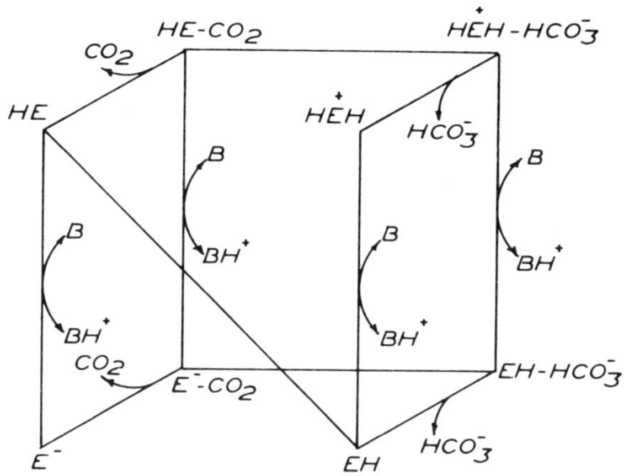

Figure 10 Kinetic mechanism scheme proposed by Steiner et al. (174). H to the right of E symbolizes protonated catalytic group. H to the left of E symbolizes protonated proton transfer group, probably His-64 in CA II. Diagonal line represents the intramolecular proton transfer step which limits the turnover rate at high substrate and buffer concentrations.

its the turnover rate at saturating buffer and substrate concentrations, also limits the rate of product H_2O dissociation. However, Koenig and Brown (183) have provided a completely different interpretation.

All the kinetic data discussed above were obtained in reaction media containing SO_4^{2-} ions to keep ionic strength under control. While SO_4^{2-} was previously thought to be completely "innocent," Bertini et al. (117) and Koenig et al. (184) have presented evidence to the contrary (cf. Sec. 5.1.5). Recent results from this laboratory (260) suggest that MES and HEPES buffers are practically noninhibitory. The pH-rate profile for the 4-nitrophenyl acetate hydrolase activity of bovine CA II in these buffers is complex as shown in Figure 11, but can be described in terms of two electrostatically interacting, ionizing groups (Fig. 12). One of these is the metal-linked catalytic group, the basic form of which is required for esterase activity. The other group is probably His-64 as evidenced by its titration behavior observed by ^{1}H NMR. The effect of SO_4^{2-} is discussed in Section 6.2. These results suggest that His-64 might not only function as a proton-transfer group in CO_2—HCO_3^- interconversion as indicated in Figure 10, but might also exert a modifying effect on the properties of the catalytic group.

The kinetic properties of CA I isoenzymes have not yet been fully characterized. However, available data suggest that CA I and CA II operate by analogous mechanisms, but that the relative rates of various reaction steps are different in the two isoenzyme types (261). The pH-rate profile for CO_2 hydration catalyzed by human CA I suggests that more than one ionizing

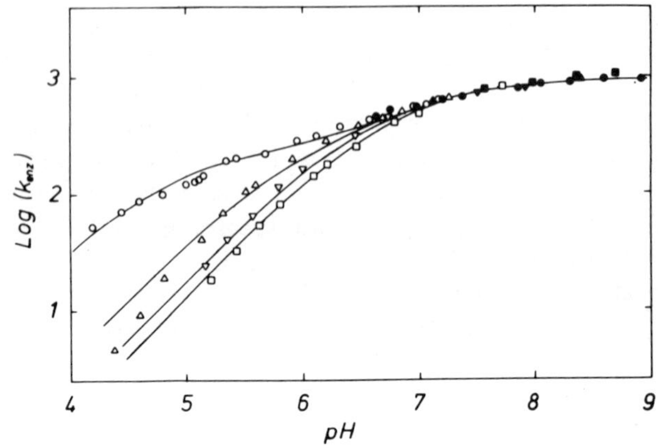

Figure 11 Logarithmic plot of the pH-rate profile for the 4-nitrophenol acetate hydrolase activity of bovine CA II. Conditions: 50m*M* Mes-NaOH buffers (open symbols), 50m*M* Hepes-NaOH buffers (filled symbols), 25°C. (○) No added sulfate; (▵) 5 m*M* Na_2SO_4; (▿) 25m*M* Na_2SO_4; (□) 50m*M* Na_2SO_4.

group affects the activity (173). Presumably, His-200 is an important modifier of the catalytic activity in CA I. The steady-state kinetic parameters of CO_2 hydration have isotope effects ≤ 2 in case of human CA I in contrast to 3–4 in case of CA II. NMR data of CO_2—HCO_3^- exchange indicate that this process is 50-fold slower in human CA I than in human CA II (262). Therefore, it seems probable that the H^+ transfer steps that limit the turnover rate for CA II are not strictly rate limiting for CA I.

6.2 Anions

There is overwhelming crystallographic and spectroscopic evidence that monovalent anions bind to the metal ion in carbonic anhydrase. A 1:1 anion–metalloenzyme complex is usually formed; exceptions are the dicyanide complexes of the Cu(II) and Co(II) enzymes and the dichloride complex of the

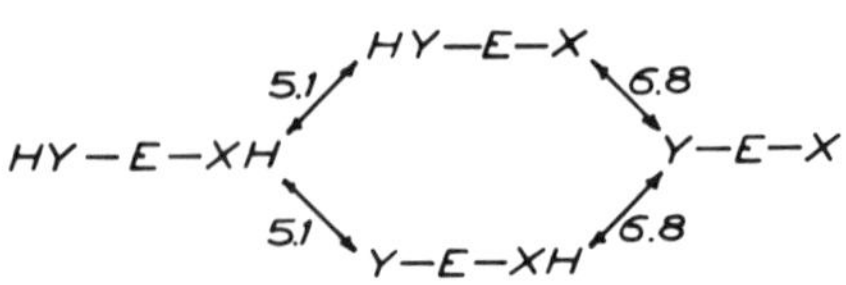

Figure 12 Microscopic ionization scheme describing the titration behavior of the catalytic group (X) and His-64 (Y) in bovine CA II in the absence of sulfate. Numbers are microscopic pK_a values obtained by fitting the pH-rate profile of Figure 11 and the titration curve of His-64 as observed by ^{1}H NMR. HY-E-XH and Y-E-XH are assumed to be inactive in the esterase reaction, whereas HY-E-X has 50% of the activity of Y-E-X.

Hg(II) enzyme (Sec. 5.1). The optical spectra of the Co(II) enzyme (Sec. 5.1.5) suggest that different anions may bind in different ways either displacing metal-bound H_2O and forming a four-coordinate complex or not displacing water and forming a five-coordinate complex. Furthermore, it is possible that the same anion may have different binding modes in the Zn(II) and the Co(II) enzymes. Thus, crystallographic data suggest that I^- displaces H_2O in the Zn(II) enzyme (86), while the optical spectra have been taken to indicate five-coordination in the Co(II) enzyme (117).

The anion affinity depends on pH. When SO_4^{2-} (>20mM) is used for control of ionic strength, this pH dependence is usually well described by the equation:

$$E^- \underset{H^+}{\rightleftarrows} EH \underset{A^-}{\rightleftarrows} EHA^- \tag{10}$$

However, a more complex pattern is observed for human CA I (185) and for bovine CA II in the absence of SO_4^{2-}. This is illustrated in Figure 13 showing the pH dependence of the inhibition of bovine CA II by I^-. Figure 14, which is an extension of Figure 12, can be applied to explain these and other observed deviations from the simple pattern generated by eq. (10). However, under most conditions the formation of the EI complex of Figure 14 can be neglected. Furthermore, if the effects of anions on the titration behavior of His-64 in human and bovine CA II (178, 179) and of His-67 and His-200 in human CA I (148) are taken into account, it seems as though HE has a much weaker anion affinity than EH. Thus, the protonated form of the catalytic group is required for strong anion binding. However, the data of Figure 13

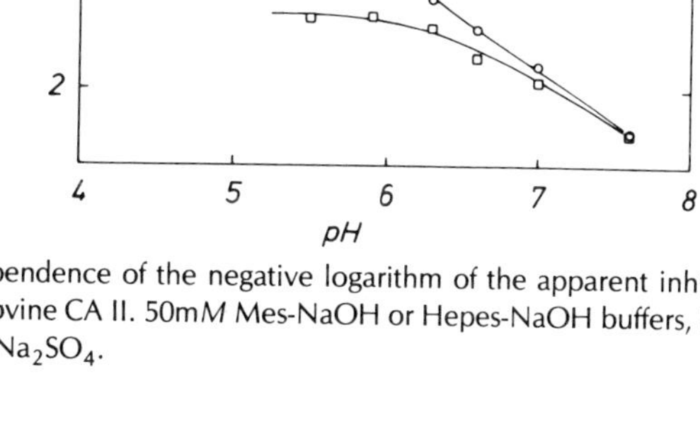

Figure 13 pH dependence of the negative logarithm of the apparent inhibition constant (M) for I^- binding to bovine CA II. 50mM Mes-NaOH or Hepes-NaOH buffers, 25°C. (○) No added sulfate; (□) 50mM Na_2SO_4.

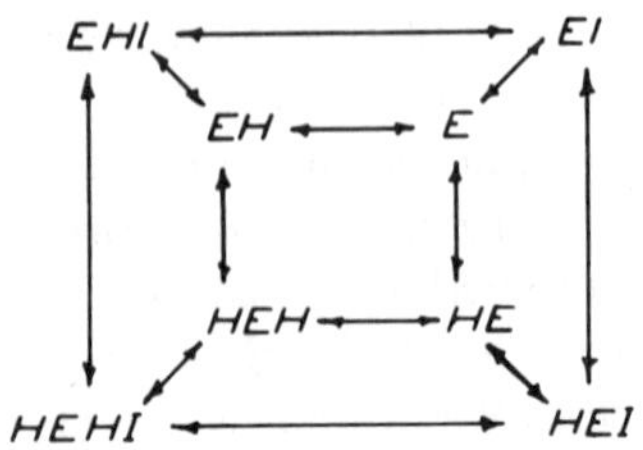

Figure 14 Generalized scheme for the binding of anionic inhibitors (I) and H^+ to the active site of carbonic anhydrase. H to the right of E symbolizes protonated catalytic group. H to the left of E symbolizes protonated His-64 in case of CA II and, presumably, His-200 in CA I.

are in accord with HEH having a 25-fold higher affinity for I^- than EH. Similarly, protonation of His-200 in human CA I appears to bring about an increase in the anion affinity (185, 186).

Koenig et al. (184) suggest that the apparent inhibition by sulfate (Fig. 11) is due to the binding of HSO_4^- to the metal ion. However, the effect of SO_4^{2-} can be adequately explained by the assumption that SO_4^{2-} is the inhibiting species which binds significantly only to the HEH form. The affinity of SO_4^{2-} appears strongly ionic strength dependent in contrast to the affinity of I^- as illustrated in Figure 15. Thus, if the SO_4^{2-} concentration is increased from 25 to 50m*M* (Fig. 11), the affinity decreases so that there is only a small change of the degree of inhibition. Sulfate and I^- behave as mutually competitive inhibitors, and in the presence of SO_4^{2-} the inhibition of bovine CA II by I^- follows the simple pattern of eq. (10) (Fig. 13).

Bertini et al. (187) have reported that the binding of the oxalate dianion to bovine Co(II) CA II depends more strongly on pH than predicted from

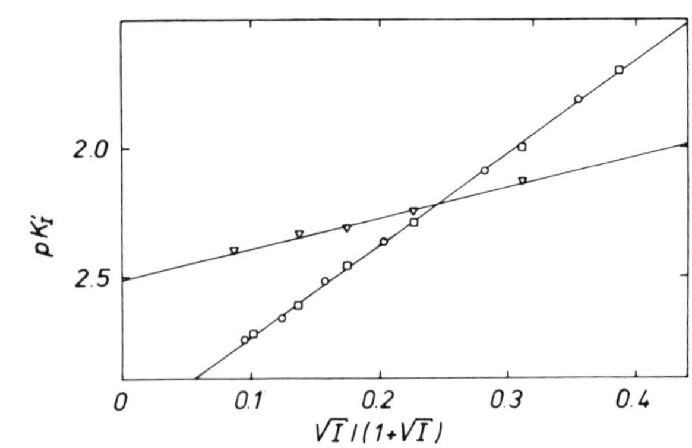

Figure 15 Ionic strength dependence of the apparent inhibition of the 4-nitrophenyl acetate hydrolase activity of bovine CA II by So_4^{2-} and I^- at pH 5.2. An apparent inhibition constant (K_I') was calculated for each salt concentration. Negative logarithms of K_I' are shown as functions of $\sqrt{I}/(1+\sqrt{I})$, where I is the ionic strength. 50m*M* Mes-NaOH buffer, pH 5.2, 25°C. (▿) NaCl, (○) Na_2SO_4, (□) $MgSO_4$.

eq. (10). Presumably, this divalent anion behaves like SO_4^{2-} and binds with a high affinity only when His-64 and the catalytic group are protonated.

Anions yield noncompetitive or mixed inhibition patterns with respect to CO_2 at neutral pH (119, 172, 175, 188, 189) and uncompetitive patterns at alkaline pH (263). Fewer studies have been made on the HCO_3^- reaction; the reported patterns vary from purely competitive (189) to mixed with a more or less prominent competitive component (175, 188). Wyeth and Prince (19) have extended the mechanism of Figure 10 to include anion inhibition in a general way, but the details remain to be worked out.

While the most potent anionic inhibitors, like CN^- and SH^-, are typical metal-binding agents, the relative affinities of other anions follow the Hofmeister lyotropic series which is related to the size, charge, and hydration of the anion. Thus, for such anions that usually do not form stable metal ion complexes, the binding can probably be regarded as resulting from the electrostatic interaction between the positively charged metal ion and an essentially unsolvated anion in a partially hydrophobic environment.

The Cd(II) and Cu(II) enzymes bind anions more strongly than the native enzyme (107, 190). The Cd(II) enzyme has a particularly high affinity for I^- in accordance with known properties of Cd(II). The stability constant for I^- binding to human Cd(II) CA II is about $10^6 M^{-1}$ at pH 8.6 and 25°C (191). In a study of the binding of 4-nitrothiophenolate anion to bovine metallocarbonic anhydrases II, Harrington and Wilkins (190) obtained the following order of affinities: Cd(II) > Cu(II) > Zn(II) ≈ Co(II) > Mn(II) ≈ Ni(II). The CA I isoenzymes bind many anions with a higher affinity than CA II isoenzymes (192).

Anions equilibrate rather rapidly with carbonic anhydrase. Typical values for the association rate constants are 10^8–$10^9 M^{-1}$ sec^{-1}, whereas the dissociation rate constants vary depending on the binding affinity. However, Taylor and Burgen (193) estimated $k_d > 100$ sec^{-1} for the CN^- complex of bovine CA II. This would yield $k_a > 2.8 \times 10^9 M^{-1}$ sec^{-1} for the reaction of CN^- with the low pH form of the enzyme. Since this value of k_a might exceed the diffusion-controlled limit, it was suggested that the complex can also be formed in a reaction between HCN and the high pH form of the enzyme. Nitrothiophenols react very rapidly with carbonic anhydrase, and k_a values around $10^9 M^{-1}$ sec^{-1} were calculated for the reaction of nitrothiophenolate anion and low pH enzyme form (190, 194). The rate constant calculated on the basis of the reaction of neutral thiophenol with high pH enzyme form would exceed the diffusion-controlled limit (192).

Taylor et al. (195) found by ^{1}H and ^{19}F NMR that formate and mono-, di-, and trifluoroacetates are in slow exchange with human Co(II) CA II on the NMR time scale. They estimated k_d values of 6×10^4 sec^{-1} and 1–2×10^5 sec^{-1} for formate and the fluoroacetates, respectively. From these values and

the affinity constants, k_a values in the range 2–4 × $10^8 M^{-1}$ sec^{-1} were calculated. From the temperature dependence of the $^{35}Cl^-$ NMR linewidths in carbonic anhydrase solutions, Ward and Cull (196) concluded that the relaxation rates are determined by Cl^- exchange. Their data yield $k_d \approx 1 \times 10^6$ sec^{-1} for the Cl^- adduct of human CA I. Taking a low pH limit stability constant of $300 M^{-1}$ from Whitney and Brandt (185), $k_a \approx 3 \times 10^8 M^{-1}$ sec^{-1} is obtained.

In ^1H NMR studies of acetate binding to native bovine CA II and its Mn(II) derivative, Lanir and Navon (98, 197) showed that, in addition to the metal ion site, acetate binds to another, noninhibitory site located about 5 Å from the metal ion. Acetate is displaced from this site by N_3^- but not by sulfonamides. Also this secondary acetate binding depends on pH, and an apparent pK_a of 7.4 was reported. Bertini et al. (198, 199) have demonstrated this acetate binding site also in the Co(II) enzyme. In addition, a secondary binding site for HCO_3^- has been postulated by Yeagle et al. (200) in ^{13}C NMR studies of $H^{13}CO_3^-$ binding to bovine Co(II) CA II. The location of the putative secondary anion binding site is unknown.

6.3 Sulfonamides

The interactions of a bound sulfonamide, acetazolamide, with the active sites of human CA I and CA II are shown schematically in Figure 16. According to the interpretation of the electron density maps presented by Kannan et al. (201), the sulfonamide group is a bidentate Zn(II) ligand. However, the optical spectrum of the acetazolamide complex with the Co(II) enzyme is indicative of a pseudotetrahedral geometry (Sec. 5.1.5). Possibly, acetazolamide is a monodentate ligand in the Co(II) enzyme, or a bidentate ligand with one bond longer than the other. Evidence that the N atom of the sulfonamide group is bonded to the metal ion comes from the observed ^{113}Cd—^{15}N coupling described in Sec. 5.1.1. In addition, there is evidence that the bound inhibitor has an ionized sulfonamide function, R—SO_2NH^-. This comes from ultraviolet difference spectra (202), fluorescence emission spectra (203), and resonance Raman spectra (204). However, Petersen et al. (205) who measured resonance Raman spectra of the azosulfonamide, neoprontosil, bound to bovine CA II, concluded that the un-ionized R—SO_2NH_2 group is involved. Carey and King (206) and Evelhoch et al. (105) have later presented data implying that the conclusion of Petersen et al. might be in error.

The sulfonamide group also interacts with the OH group and the amide NH group of Thr-199. The bond arrangement is slightly different in the two isoenzymes. The thiadiazole ring and the acetamido side chain of acetazolamide are located in the hydrophobic part of the active site. According to Kannan et al. (201), these groups do not form polar bonds with the protein

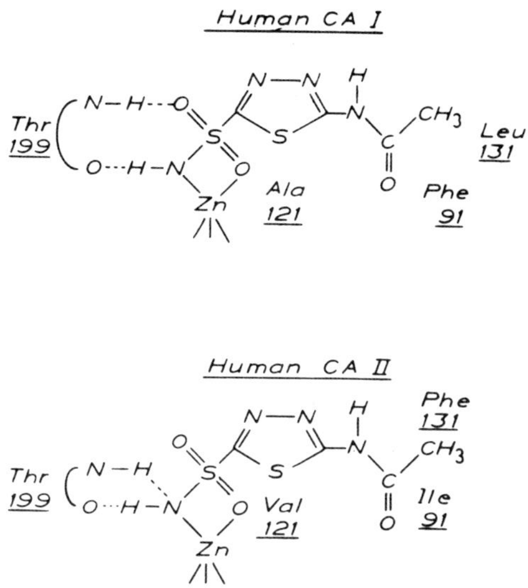

Figure 16 Schematic drawing of the interactions of acetazolamide with the active sites of human CA I and II as observed by X-ray crystallography. See ref. 201 for details.

but they are in contact with the hydrophobic side chains of residues, 91, 121, and 131.

The metal ion specificity of sulfonamide binding differs from that of anion binding. The following relative affinities have been found for acetazolamide in the bovine CA II system: Zn(II) ≈ Co(II) > Cu(II), Cd(II) > Mn(II), Ni(II) (190).

The pH dependence of sulfonamide binding is in accord with an equilibrium between the ionized inhibitor and the low pH form of the enzyme, or the equivalent equilibrium between the un-ionized inhibitor and the high pH form of the enzyme (207, 208).

$$\begin{array}{ccccc} EH & + & In^- & \nwarrow\!\!\searrow & \\ \updownarrow & & \updownarrow & & EHIn^- \\ E^- & + & HIn & \swarrow\!\!\nearrow & \end{array}$$

Sulfonamides equilibrate rather slowly with carbonic anhydrase. Typical values for the dissociation rate constants, k_d, are around 0.1 sec^{-1} at 25°C. Kinetic studies of sulfonamide binding have shown that the pH dependence is almost completely contained in the association rate constant, k_a, while k_d is practically independent of pH (207, 209, 210). King and Burgen (211) have proposed that sulfonamide binding is a two-step process. First, a rapid for-

mation of a weak complex independently of the ionization states of the sulfonamide group and the catalytic group in the enzyme. This step is followed by a rate-determining step in which the metal ion-sulfonamide bond is formed. King and Burgen could not detect the transient formation of an intermediate, but they showed that the apoenzyme binds sulfonamides with a pH-independent affinity which is 10^3–10^4 times lower than for the native enzyme. In addition, they showed that a substituted sulfonamide, 4-nitrobenzene sulfacetamide, binds to human CA II competitively with its unsubstituted analog but independently of pH and with a low affinity. The substituted inhibitor does not perturb the visible spectrum of the Co(II) enzyme, indicating that it is not bound to the metal ion.

Because of the slow equilibration rates, sulfonamides often behave as pseudoirreversible inhibitors of CO_2 hydration and HCO_3^- dehydration. Thus, noncompetitive patterns are usually observed with respect to these substrates (175, 212), but a competitive behavior with respect to HCO_3^- has been reported in some cases (213).

6.4 Other Inhibitors

Methanol is a weak inhibitor of the esterase activity of carbonic anhydrase (121, 214–216). This inhibition is only slightly dependent on pH (214, 215). Methanol does not seem to compete with anions or affect the optical spectrum of the Co(II) enzyme (215, 216). However, the Co(II) enzyme enhances the relaxation rates of nuclei in the methanol molecule (121, 216). Although these data suggest that methanol binds to the active site of carbonic anhydrase, the binding is very weak and it is probable that much of the observed inhibition is not a result of binding but comes from the effect of methanol on the thermodynamic activity coefficient of the ester substrate (121, 217).

The pH dependence of the interaction of aniline with carbonic anhydrase is simplest described as an equilibrium between the neutral inhibitor and the low pH form of the enzyme (121, 215, 218). The effect of aniline on the optical spectrum of the Co(II) enzyme suggests that this inhibitor is bound to the metal ion. The spectrum of the aniline–Co(II) enzyme complex is indicative of a pseudotetrahedral metal environment.

Since the phenol complex of bovine Co(II) CA II has an optical spectrum similar to that of the high pH form (Fig. 7), it appears as if phenol shifts the activity-linked pK_a to lower values (121). However, the binding of phenol can probably simplest be described as an equilibrium between neutral phenol and high pH enzyme form (264), or equivalently, between phenolate anion and low pH enzyme form.

Imidazole inhibits human CA I. The inhibition appears competitive with respect to CO_2 (173). Kannan et al. (219) showed by X-ray crystallography

that imidazole is located near the metal ion without displacing metal-bound water. The metal ion–imidazole distance is rather long, 2.7 Å, and the metal–imidazole bond is not in the plane of the imidazole ring. This peculiar orientation of the inhibitor molecule is expressed in the imidazole-induced shift of the C-2 ^{1}H resonance from the ligand His-94 (148). The pH dependence of imidazole binding and the effects of imidazole on the spectrum of the Co(II) enzyme are in accord with the neutral inhibitor molecule binding with similar affinity to the high pH form and the low pH form of the enzyme. Thus, there is an equilibrium between low pH and high pH forms of the inhibitor complex with pK_a near 7.5 (84, 220, 221). Imidazole and I^- are mutually competitive despite the different binding modes, however (220).

Alberti et al. (221) have recently found that tetrazole, 1,2,4-triazole, and 1,2,3-triazole bind to bovine Co(II) CA II as well as to human Co(II) CA I. Also in these cases the neutral compounds seem to bind independently of the ionization state of the catalytic group. However, the spectra of the complexes of these inhibitors with the Co(II) enzymes appear independent of pH in contrast to the complex of imidazole with human Co(II) CA I.

Magid (222) found that micromolar concentrations of Cu(II) ions inhibit human CA II. In ^{18}O exchange studies, Tu et al. (223) have made the interesting observation that Cu(II) ions at these concentrations do not inhibit CO_2—HCO_3^- exchange at equilibrium, but the exchange rate of product H_2O is specifically inhibited. The binding site for the inhibitory Cu(II) ion remains to be elucidated.

7 CHEMICAL MODIFICATIONS OF ACTIVE-SITE RESIDUES

The CA II isoenzymes have one titratable active-site histidine, His-64. Its pK_a is about 7.1 in human and bovine CA II in the presence of SO_4^{2-} (178, 179). This residue reacts with bromoacetazolamide forming an Nτ-alkylated derivative which is inactive at neutral pH (224–226). The modified enzyme has esterase activity at high pH, but Wells et al. (227) showed that this activity is not associated with the active-site region. The Nτ atom of His-64 in human CA II also reacts with bromopyruvate (228). This derivative has about 30% of the 4-nitrophenyl acetate hydrolase activity and 25% of the CO_2 hydration activity of the native enzyme (228, 229).

Human CA I has three active-site histidines, His-64, His-67, and His-200. Their pK_a values obtained in ^{1}H NMR studies are 4.7, 6.0, and 6.1, respectively (148). His-64 has not been modified in human CA I. His-67 has been modified with reactive sulfonamide inhibitors. *N*-(Chloroacetyl)chlorothiazide reacts with the Nτ atom giving an inactive product (230). *N*-(Bromo-

acetyl)acetazolamide reacts with the Nπ atom giving a product with 35% of the esterase activity of the native enzyme at pH 7.6 (226, 231).

His-200 reacts with a variety of reagents, 1-fluoro-2,4-dinitrobenzene (232), iodo- or bromoacetate (38, 233, 234), bromopyruvate (228), and iodoacetamide (38, 234). In all these cases, modification occurs at the Nτ atom of His-200. The products have low, but significant esterase and CO_2 hydration activities. Even after destruction of His-200 by photooxidation of the dinitrophenyl derivative, some enzymatic activity remains (235).

The properties of human Nτ-carboxamidomethyl-histidine-200 CA I and Nτ-carboxymethyl-histidine-200 CA I have been studied in detail. They bind sulfonamides and anionic inhibitors. The anion affinity depends on two groups, one with a pK_a of 6.1 in both derivatives and another with pK_a values of 8.2 and 9.0, respectively, in the carboxamidomethyl and carboxymethyl enzymes (185). The latter group has a major influence on the activity and anion binding and on the visible spectra of modified Co(II) enzymes. Protonation of the former group results in an increased affinity of anions for the modified Zn(II) enzymes and further spectral changes of the Co(II) derivatives. NMR studies of ^{13}C-labeled carboxymethyl enzyme have shown that the lower pK_a corresponds to the modified imidazole group (45, 186, 236, 237). The higher pK_a probably corresponds to the metal-linked catalytic group. In addition, the NMR data suggest that the carboxymethyl group is immobilized near pH 8, presumably through coordination of the carboxylate group to the metal ion (236, 237).

Modification of arginine residues in bovine CA III leads to an approximately two-fold increase of the activity with HCO_3^- as substrate (238).

8 MOLECULAR MECHANISM

The splitting of a water molecule, formally into H^+ and OH^-, is a crucial step during the catalyzed hydration of CO_2. In human CA II the observed value of $k_{cat}^{CO_2}$ (Sec. 6.1) shows that this step must have a rate constant exceeding $10^6 sec^{-1}$. If a protein group with a pK_a of 7 acts as the proton acceptor, a rate constant of only 10^3–10^4 is feasible unless the H_2O molecule is activated in some manner. Eigen and Hammes (170) pointed out that such an activation might be achieved through interaction of H_2O with the metal ion. In view of such considerations, it seems necessary to postulate that the H_2O substrate of carbonic anhydrase is liganded to the metal ion. Therefore, one expects that the other substrates, CO_2 and HCO_3^-, have binding positions at or near the metal ion. Indeed, most of the mechanism models proposed for carbonic anhydrase incorporate these features.

Yeagle et al. (200) and Stein et al. (239) measured the effect of the Co(II) enzyme on the ^{13}C relaxation of $^{13}CO_2$ and H $^{13}CO_3^-$. They estimated a weighted average ^{13}C—Co(II) distance of 3.6 Å at pH 6.3. Since HCO_3^- behaves like other anions in perturbing the spectrum of the Co(II) enzyme, it seems likely that it is bonded to the metal ion. While calculations show that CO_2 might coordinate (240), there is no experimental support for this. Infrared difference spectra reported by Riepe and Wang (241) were taken as evidence that CO_2 binds competitively with N_2O to a hydrophobic site. However, N_2O does not inhibit the enzyme (173) so the functional relevance of this CO_2 binding is doubtful. Bertini et al. (242) calculated a ^{13}C—Cu(II) distance greater than 6 Å from T_1 measurements on $^{13}CO_2$ interacting with bovine Cu(II) CA II. However, since the Cu(II) enzyme is inactive in CO_2 hydration, it is not certain if this evidence is relevant for the catalytic mechanism.

An unsolved problem of major concern is the identity of the catalytic group that is linked to the metal ion and titrates with a pK_a near 7 under usual experimental conditions. Four distinct alternatives have been considered. However, none of them can explain all pertinent data in a simple, straightforward manner: (a) A titratable histidine was proposed by Pocker and Storm (30). (b) A ligand imidazole ionizing to an imidazolate anion was postulated by Pesando (147, 243) and by Appleton and Sarkar (224). Pesando's proposal also includes the dissociation of the ligand from the metal ion at low pH. More recently, Pesando and Gupta (245) have modified their model so that it involves an unbound imidazolium cation at low pH and liganded neutral imidazole at high pH. (c) Glu-106 was proposed by Kannan et al. (219). (d) A zinc-bound H_2O ionizing to OH^- was proposed by Davis (246) and formulated approximately as shown in Figure 1 by Coleman (101).

Alternative (a) is essentially based on the thermodynamic properties of the ionization of the catalytic group. However, the ^1H NMR results of Campbell et al. (148, 178, 179) show that none of the titratable histidines in CA I or CA II have the properties of the catalytic group. Nevertheless, titratable active-site histidines seem to have important functions as modifiers of the properties of the catalytic group and, possibly, as proton transfer groups during catalysis (Sec. 6.1, 6.2, and 7).

Alternative (b) is based on studies of model complexes (244) and on the peculiar pH dependence of a ligand resonance in the ^1H NMR spectrum of human CA I (147). However, Martin (247) has shown that Zn(II)-induced pyrrole ionizations in imidazoles occur with pK_a about 13. Therefore, the imidazolate model for carbonic anhydrase is unlikely. Since the crystal-structure studies as well as NMR results (179) indicate that the pyrrole NH protons of the three imidazole ligands are hydrogen bonded to protein groups and not exchanging rapidly with solvent, one does not expect these ligands to be very

flexible (cf. Fig. 4). Therefore, any mechanism model involving the dissociation of an imidazole ligand from the metal ion seems unattractive.

Alternative (c) is based on crystal-structure studies of human CA I and CA II, which show that Glu-106 is linked to the metal-bound H_2O via Thr-199 (Sec. 4). However, one important property of the catalytic group that is difficult to reconcile with the Glu-106 model is the shift of the apparent pK_a from near 5 when His-64 is protonated (Sec. 6.1) to very high values (>11) in inhibitor complexes. Unfortunately, there are no data available pertaining to the ionization behavior of Glu-106.

In alternative (d), the zinc-hydroxide model, the inhibitor-induced pK_a shift is simply accounted for as a competition between inhibitor and OH^- for a coordination site on the metal ion. The major argument against this model is the difficulty of rationalizing the pH dependence of the water proton relaxation enhancement induced by the Co(II) enzyme (Sec. 5.1.5). In a recent review (248), I discussed a zinc-hydroxide model in which the Zn(II) ion remained five-coordinated during turnover. However, it is also possible to account for the relaxivity data in the four-coordinate model shown in Figure 17. Two assumptions must then be made. Firstly, the metal-bound OH^- must exchange with H_2O at a sufficient rate either by a dissociative mechanism or by an associative mechanism such as

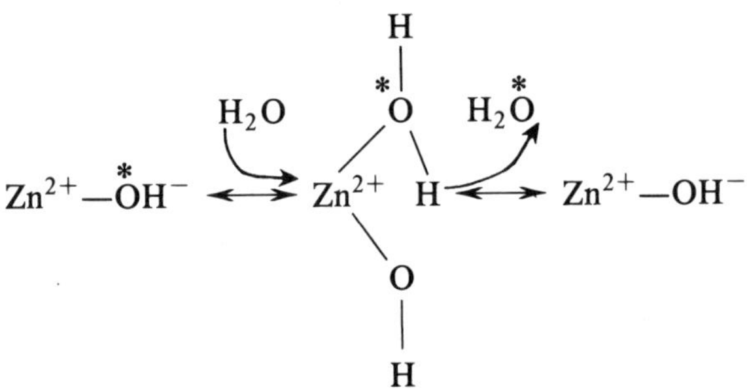

For native human CA II this water exchange appears to be characterized by a rate constant of $2 \times 10^5\ sec^{-1}$ (Sec. 6.1), and it is noteworthy that this rate is sufficient to yield fast exchange conditions in the relaxation experiment with the Co(II) enzyme (133, 134). Secondly, the model in Figure 17, like most mechanisms proposed for the enzyme, contains a low-pH form of the enzyme having metal-bound H_2O. From the data of Fabry (134) and Bertini et al. (117) it appears as if the low-pH form of the Co(II) enzyme has a relaxation effect on H_2O protons in the absence of sulfate, and this can be taken as support for the presence of metal-bound H_2O in the enzyme at low pH. However, this is a point that requires further, detailed measurements. Koenig et al. (184) have suggested that the low-pH form of the enzyme is always a com-

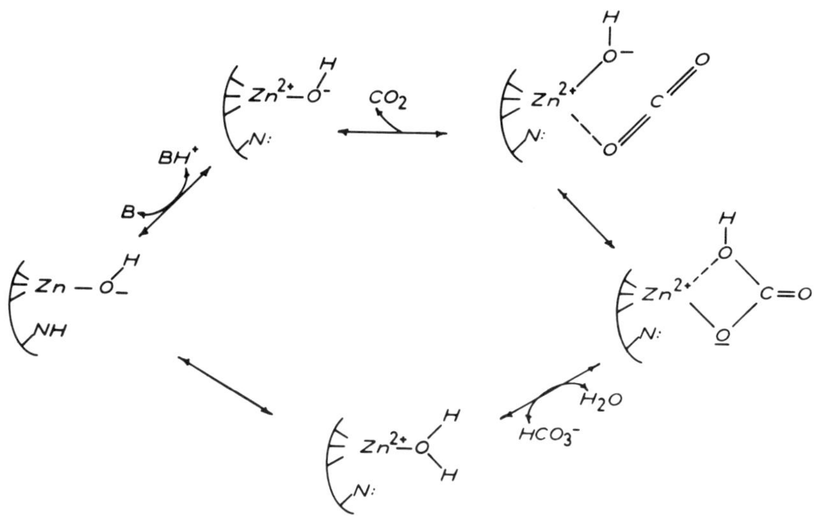

Figure 17 Extended zinc-hydroxide mechanism including a buffer-dependent and a buffer-independent step in the interconversion of high-pH and low-pH forms.

plex with some inhibiting anion, and that only the high-pH form partakes in the catalytic pathway. However, in my opinion their mechanism proposal has created more problems than it has solved (183).

Two features of the Zn(II)—OH^- mechanism have been considered to be particularly remarkable. Firstly, the enzymatic pK_a of 7 (or even 5 when His-64 is protonated) is low compared to the pK_a for the ionization of Zn(II)—H_2O in model complexes. Thus, the pK_a for H_2O in Zn(II)$(H_2O)_6$ is estimated to be 10 after application of a statistical correction factor (249). Secondly, the nucleophilicity of Zn(II)—OH^- in the enzyme must be very high despite the low basicity.

The following factors are thought to be important for promoting a low pK_a of Zn(II)—H_2O ionization: a low coordination number (249), neutral ligands transferring less electronic charge to the metal ion than anionic ligands (250), and a hydrophobic microenvironment (251, 252). Woolley (249) investigated metal complexes having tetradentate, macrocyclic ligands. In case of Zn(II) and Co(II) these complexes are five-coordinated with H_2O as an additional ligand at low pH. The pK_a for H_2O ionization was found to be 8.1 in some of the Zn(II) complexes. A pK_a of 7 for the corresponding equilibrium in the enzyme seems quite reasonable considering the special features of the active site.

Woolley's studies also showed that metal-bound OH^- in the macrocycle complexes is a much better catalyst of acetaldehyde hydration and CO_2 hy-

dration than metal-bound H_2O (253). In agreement with Martin (254), Woolley also found that the nucleophilicities of a series of metal ion-OH^- complexes varied relatively little despite large variations in the pK_a for H_2O ionization. While the Zn(II)-macrocycle-OH^- complex catalyzed acetaldehyde hydration with similar efficiency as carbonic anhydrase, the model complex was much less efficient than the enzyme in catalyzing CO_2 hydration. In this connection it is interesting to note than Jönsson et al. (255) calculated by *ab initio* methods that the gas-phase reaction between OH^- and CO_2 should proceed without any energy barrier. Thus, it seems probable that the barrier observed in solution is due to solvation effects. This suggests that a low degree of solvation of reactants in the active site of carbonic anhydrase could be an important catalytic factor.

ACKNOWLEDGMENTS

The author is grateful to Dr. I. Bertini, Dr. J. L. Evelhoch, Dr. D. N. Silverman, Dr. J. L. Sudmeier, and Dr. R. E. Tashian for communicating unpublished information. Many thanks also to Ms. Agneta Ehman for assistance in the preparation of the manuscript.

REFERENCES

1. N. U. Meldrum and F. J. W. Roughton, *J. Physiol.* (*London*), **75,** 15P (1932).
2. N. U. Meldrum and F. J. W. Roughton, *J. Physiol.* (*London*), **80,** 113 (1933).
3. M. J. Carter, *Biol. Rev.*, **47,** 465 (1972).
4. T. H. Maren, *Physiol. Rev.*, **47,** 503 (1967).
5. J. E. Lamb, *Life Sci.*, **20,** 393 (1977).
6. R. P. Poincelot in *Encyclopedia of Plant Physiology*, Vol. 6, M. Gibbs and E. Latzko, Eds., Springer, Berlin, 1979, p. 230.
7. M. L. Reed and D. Graham, *Progr. Phytochem.*, **7,** 47 (1981).
8. L. Adler, J. Brundell, S. O Falkbring, and P. O. Nyman, *Biochim. Biophys. Acta*, **284,** 298 (1972).
9. J. Brundell, S. O. Falkbring, and P. O. Nyman, *Biochim. Biophys. Acta*, **284,** 311 (1972).
10. D. Keilin and T. Mann, *Nature*, **144,** 442 (1939).
11. D. Keilin and T. Mann, *Biochem. J.*, **34,** 1163 (1940).
12. J. E. Coleman, *Ann. Rev. Pharmacol.*, **15,** 221 (1975).
13. M. F. Dunn, *Struct. Bonding* (*Berlin*), **23,** 61 (1975).
14. J. F. Chlebowski and J. E. Coleman in *Metal Ions in Biological Systems*, Vol. 6, H. Sigel, Ed., Dekker, New York, 1976, p. 1.
15. R. E. Tashian and N. D. Carter, *Adv. Hum. Genet.*, **7,** 1 (1976).

16. R. E. Tashian, *Isozymes: Curr. Top. Biol. Med. Res.*, **2,** 21 (1977).
17. R. E. Tashian, A. Kendall, and N. D. Carter, *Hemoglobin*, **4,** 635 (1980).
18. H. F. Bundy, *Comp. Biochem. Physiol. B*, **57,** 1 (1977).
19. P. Wyeth and R. H. Prince, *Inorg. Perspect. Biol. Med.*, **1,** 37 (1977).
20. Y. Pocker and S. Sarkanen, *Adv. Enzymol.*, **47,** 149 (1978).
21. R. H. Prince, *Adv. Inorg. Chem. Radiochem.*, **22,** 349 (1979).
22. I. Bertini, C. Luchinat, and A. Scozzafava, *Struct. Bonding (Berlin)*, **48,** 45 (1981).
23. C. Bauer, G. Gross, and H. Bartels, Eds., *Biophysics and Physiology of Carbon Dioxide*, Springer, Berlin, 1980.
24. C. P. Chengelis and R. A. Neal, *Biochem. Biophys. Res. Commun.*, **90,** 993 (1979).
25. Y. Pocker and J. E. Meany, *Biochemistry*, **4,** 2535 (1965).
26. Y. Pocker and J. E. Meany, *Biochemistry*, **6,** 239 (1967).
27. Y. Pocker and D. G. Dickerson, *Biochemistry*, **7,** 1995 (1968).
28. D. Chesnovsky and G. Navon, *Biochemistry*, **19,** 1866 (1980).
29. Y. Pocker and J. E. Meany, *J. Phys. Chem.*, **74,** 1486 (1970).
30. Y. Pocker and D. R. Storm, *Biochemistry*, **7,** 1202 (1968).
31. A. Thorslund and S. Lindskog, *Eur. J. Biochem.*, **3,** 117 (1967).
32. Y. Pocker and M. W. Beug, *Biochemistry*, **11,** 698 (1972).
33. P. C. Harrington and R. G. Wilkins, *J. Chem Soc. Chem. Commun.*, **1977,** 638.
34. Y. Pocker and L. J. Guilbert, *Biochemistry*, **11,** 180 (1972).
35. Y. Pocker and L. J. Guilbert, *Biochemistry*, **13,** 70 (1974).
36. B. G. Malmström, P. O. Nyman, B. Strandberg, and B. Tilander in *Structure and Activity of Enzymes*, T. W. Goodwin, J. I. Harris and B. S. Hartley, Eds., Academic, New York, 1964, p. 121.
37. E. T. Kaiser and K. -W. Lo, *J. Am. Chem. Soc.*, **91,** 4912 (1969).
38. P. L. Whitney, P. O. Nyman, and B. G. Malmström, *J. Biol. Chem.*, **242,** 4212 (1967).
39. P. Henkart, G. Guidotti, and J. T. Edsall, *J. Biol. Chem.*, **243,** 2447 (1968).
40. F. J. W. Roughton and V. H. Booth, *Biochem. J.*, **40,** 319 (1946).
41. S. Lindskog, L. E. Henderson, K. K. Kannan, A. Liljas, P. O. Nyman, and B. Strandberg, in *The Enzymes*, Vol. 5, 3rd ed., P. D. Boyer, Ed., Academic, New York, 1971, p. 587.
42. T. Mann and D. Keilin, *Nature*, 146 (1940).
43. W. R. A. Osborne and R. E. Tashian, *Anal. Biochem.*, **64,** 297 (1975).
44. R. G. Khalifah, D. J. Strader, S. H. Bryant, and S. M. Gibson, *Biochemistry*, **16,** 2241 (1977).
45. P. J. Wistrand, S. Lindahl, and T. Wahlstrand, *Eur. J. Biochem.*, **57,** 189 (1975).
46. T. Wåhlstrand and P. J. Wistrand, *Upsala J. Med. Sci.*, **85,** 7 (1980).
47. M. J. Carter, *Biochim Biophys. Acta*, **235,** 222 (1971).
48. N. D. Carter and J. Auton, *Biochim. Biophys. Acta*, **410,** 220 (1975).
49. P. Byvoet and A. Gotti, *Mol. Pharmacol.*, **3,** 142 (1967).
50. M. H. Soliman and I. Kluh, *Eur. J. Biochem.*, **44,** 611 (1974).
51. N. D. Carter and J. Auton, *Comp. Biochem. Physiol. B*, **53,** 461 (1976).
52. M. Sciaky and G. Laurent, *FEBS Lett.*, **63,** 141 (1976).

53. R. S. Holmes, *Eur. J. Biochem.*, **78,** 511 (1977).
54. A. M. Register, M. K. Koester, and E. A. Noltmann, *J. Biol. Chem.*, **253,** 4143 (1978).
55. R. W. King, L. C. Garg, J. Huckson, and T. H. Maren, *Mol. Pharmacol.*, **10,** 335 (1974).
56. L. C. Garg, *J. Pharmacol. Exp. Ther.*, **189,** 557 (1974).
57. N. D. Carter, D. Hewett-Emmett, S. Jeffrey, and R. E. Tashian, *FEBS Lett.*, **128,** 114 (1981).
58. R. T. Fernley, R. D. Wright, and J. P. Coghlan, *FEBS Lett.*, **105,** 299 (1979).
59. D. N. McKinley and P. L. Whitney, *Biochim. Biophys. Acta*, **445,** 780 (1976).
60. P. J. Wistrand and R. Kinne, *Pflügers Arch. Eur. J. Physiol.*, **370,** 121 (1977).
61. S. H. Vincent and D. N. Silverman, *Arch. Biochem. Biophys.*, **205,** 51 (1980).
62. G. Sanyal, N. I. Pessah, and T. H. Maren, *Biochim. Biophys. Acta*, **657,** 128 (1981).
63. S. J. Dodgson, R. E. Forster, B. T. Storey, and L. Mela, *Proc. Natl. Acad. Sci. U.S.A.*, **77,** 5562 (1980).
64. S. Lindskog, *Biochim. Biophys. Acta*, **39,** 218 (1960).
65. J. M. Gulian, N. Limozin, M. Charrel, G. Laurent, and Y. Derrien, *C. R. Acad. Sci., Ser. D*, **278,** 1123 (1974).
66. S. Funakoshi and H. F. Deutsch, *J. Biol. Chem.*, **244,** 3438 (1969).
67. B. Andersson, P. O. Nyman, and L. Strid, *Biochem. Biophys. Res. Commun.*, **48,** 670 (1972).
68. K.-T. D. Lin and H. F. Deutsch, *J. Biol. Chem.*, **248,** 1885 (1973).
69. N. Giraud, C. Marriq, and G. Laurent-Tabusse, *Biochimie*, **56,** 1031 (1974).
70. E. P. B. Contel, D. Hewett-Emmett, S. K. Stroup, and R. E. Tashian, *Isozyme Bull.*, **14,** 44 (1981).
71. D. Henriksson, R. J. Tanis, and R. E. Tashian, *Biochem. Biophys. Res. Commun.*, **96,** 135 (1980).
72. R. E. Tashian, D. Hewett-Emmett, S. K. Stroup, M. Goodman, and Y. -S. L. Yu in *Biophysics and Physiology of Carbon Dioxide*, C. Bauer, G. Gros, and H. Bartels, Eds., Springer, Berlin, 1980, p. 165.
73. J. R. Jabusch, R. P. Bray, and H. F. Deutsch, *J. Biol. Chem.*, **255,** 9196 (1980).
74. L. E. Henderson, D. Henriksson, and P. O. Nyman, *J. Biol. Chem.*, **251,** 5457 (1976).
75. K.-T. D. Lin and H. F. Deutsch, *J. Biol. Chem.*, **249,** 2329 (1974).
76. M. Sciaky, N. Limozin, D. Filippi-Foveau, J. M. Gulian, and G. Laurent-Tabusse, *Biochimie*, **58,** 1071 (1976).
77. R. J. Tanis, R. E. Ferrell, and R. E. Tashian, *Biochim. Biophys. Acta*, **371,** 534 (1974).
78. J. R. Jabusch and H. F. Deutsch, *Fed. Proc.*, **40,** 1677 (1981).
79. R. E. Ferrell, S. K. Stroup, R. J. Tanis, and R. E. Tashian, *Biochim. Biophys. Acta*, **533,** 1 (1978).
80. G. E. Hall and R. Schraer, *Comp. Biochem. Physiol. B*, **63,** 561 (1979).
81. R. E. Tashian, S. K. Stroup, and G. E. Hall, *Fed. Proc.*, **40,** 1677 (1981).
82. T. H. Maren, B. R. Friedland, and R. S. Rittmaster, *Comp. Biochem. Physiol. B*, **67,** 69 (1980).
83. S. K. Chapman and T. H. Maren, *Biochim. Biophys. Acta*, **527,** 272 (1978).
84. A. Liljas, K. K. Kannan, P. C. Bergstén, I. Waara, K. Fridborg, B. Strandberg, U. Carlbom, L. Järup, S. Lövgren, and M. Petef, *Nature*, **235,** 131(1972).

85. K. K. Kannan, B. Notstrand, K. Fridborg, S. Lövgren, A. Ohlsson, and M. Petef, *Proc. Natl. Acad. Sci. U.S.A.*, **72,** 51 (1975).

86. K. K. Kannan in *Biophysics and Physiology of Carbon Dioxide*, C. Bauer, G. Gros and H. Bartels, Eds., Springer, Berlin, 1980, p. 184.

87. E. Clementi, G. Corongiu, B. Jönsson, and S. Romano, *FEBS Lett.*, **100,** 313 (1979).

88. D. Hewett-Emmett, *Fed. Proc.*, **39,** 1915 (1980).

89. S. Lindskog and B. G. Malmström, *J. Biol. Chem.*, **237,** 1129 (1962).

90. J. B. Hunt, M.-J. Rhee, and C. B. Storm, *Anal. Biochem.*, **79,** 614 (1977).

91. J. E. Coleman, *Biochemistry*, **4,** 2644 (1965).

92. S. Lindskog, *Biochim. Biophys. Acta*, **122,** 534 (1966).

93. R. W. Henkens and J. M. Sturtevant, *J. Am. Chem. Soc.*, **90,** 2669 (1968).

94. K.-P. Wong and L. M. Hamlin, *Arch. Biochem. Biophys.*, **170,** 12 (1975).

95. P. Zavodszky, J. T. Johansen, and A. Hvidt, *Eur. J. Biochem.*, **56,** 67 (1975).

96. A. Yazgan and R. W. Henkens, *Biochemistry*, **11,** 1314 (1972).

97. A. Lanir and G. Navon, *Biochemistry*, **11,** 3536 (1972).

98. A. Lanir and G. Navon, *Biochim. Biophys. Acta*, **341,** 75 (1974).

99. S. Lindskog and P. O. Nyman, *Biochim. Biophys. Acta*, **85,** 462 (1964).

100. R. Bauer, P. Limkilde, and J. T. Johansen, *Biochemistry*, **15,** 334 (1976).

101. J. E. Coleman, *J. Biol. Chem.*, **242,** 5212 (1967).

102. R. Bauer, P. Limkilde, and J. T. Johansen, *Carlsberg Res. Commun.*, **42,** 325 (1977).

103. A. J. M. Schoot Uiterkamp, I. M. Armitage, and J. E. Coleman, *J. Biol. Chem.*, **255,** 3911 (1980).

104. N. B.-H. Jonsson, L. A. E. Tibell, J. L. Evelhoch, S. J. Bell, and J. L. Sudmeier, *Proc. Natl. Acad. Sci. U.S.A.*, **77,** 3269 (1980).

105. J. L. Evelhoch, D. F. Bocian, and J. L. Sudmeier, *Biochemistry*, **20,** 4951 (1981).

106. J. L. Sudmeier and T. G. Perkins, *J. Am. Chem. Soc.*, **99,** 7732 (1977).

107. L. Morpurgo, G. Rotilio, A. Finazzi Agrò, and B. Mondovi, *Arch. Biochem. Biophys.*, **170,** 360 (1975).

108. I. Bertini, G. Canti, C. Luchinat, and A. Scozzafava, *J. Chem. Soc. Dalton*, **1978,** 1269.

109. J. S. Taylor and J. E. Coleman, *J. Biol. Chem.*, **246,** 7058 (1971).

110. P. H. Haffner and J. E. Coleman, *J. Biol. Chem.*, **250,** 996 (1975).

111. S. H. Koenig and R. D. Brown, *Ann. N.Y. Acad. Sci.*, **222,** 752 (1973).

112. L. Morpurgo, R. Falcioni, G. Rotilio, A. Desideri, and B. Mondovi, *Inorg. Chim. Acta*, **28,** L141 (1978).

113. J. S. Taylor and J. E. Coleman, *J. Biol. Chem.*, **248,** 749 (1973).

114. P. H. Haffner and J. E. Coleman, *J. Biol. Chem.*, **248,** 6626 (1973).

115. I. Bertini, E. Borghi, and C. Luchinat, *Bioinorg. Chem.*, **9,** 495 (1978).

116. S. Lindskog and A. Ehrenberg, *J. Mol. Biol.*, **24,** 133 (1967).

117. I. Bertini, G. Canti, C. Luchinat, and A. Scozzafava, *J. Am. Chem. Soc.*, **100,** 4873 (1978).

118. I. Bertini, C. Luchinat, and A. Scozzafava, *Inorg. Chim. Acta*, **46,** 85 (1980).

119. S. Lindskog, *Biochemistry*, **5,** 2641 (1966).

120. I. Bertini, E. Borghi, G. Canti, and C. Luchinat, *J. Inorg. Biochem.*, **11,** 49 (1979).

121. G. S. Jacob, R. D. Brown, and S. H. Koenig, *Biochemistry*, **19**, 3754 (1980).
122. I. Bertini, G. Canti, C. Luchinat, and F. Mani, *Inorg. Chim. Acta*, **46**, L91 (1980).
123. J. E. Coleman and R. V. Coleman, *J. Biol. Chem.*, **247**, 4718 (1972).
124. B. Holmquist, T. A. Kaden, and B. L. Vallee, *Biochemistry*, **14**, 1454 (1975).
125. E. Grell and R. C. Bray, *Biochim. Biophys. Acta*, **236**, 503 (1971).
126. S. A. Cockle, S. Lindskog, and E. Grell, *Biochem. J.*, **143**, 703 (1974).
127. A. Desideri, L. Morpurgo, J. B. Raynor, and G. Rotilio, *Biophys. Chem.*, **8**, 267 (1978).
128. A. Bencini, G. Canti, D. Gatteschi, C. Luchinat, and I. Bertini, *J. Inorg. Biochem.*, **14**, 81 (1981).
129. W. DeW. Horrocks, J. N. Ishley, B. Holmquist, and J. S. Thompson, *J. Inorg. Biochem.*, **12**, 131 (1980).
130. A. Desideri, A. A. G. Thomlinson, and J. B. Raynor, *J. Inorg. Biochem.*, **12**, 221 (1980).
131. S. A. Cockle, *Chem. Commun.*, **1973**, 905.
132. S. A. Cockle, *Biochem. J.*, **137**, 587 (1974).
133. M. E. Fabry, S. H. Koenig, and W. E. Schillinger, *J. Biol. Chem.*, **245**, 4256 (1970).
134. M. E. Fabry, *J. Biol. Chem.*, **253**, 3568 (1978).
135. J. W. Wells, S. I. Kandel, and S. H. Koenig, *Biochemistry*, **18**, 1989 (1979).
136. G. S. Jacob, R. D. Brown, and S. H. Koenig, *Biochem. Biophys. Res. Commun.*, **82**, 203 (1978).
137. H. Shinar and G. Navon, *Biochim. Biophys. Acta*, **334**, 471 (1974).
138. H. Shinar and G. Navon, *Eur. J. Biochem.*, **93**, 313 (1979).
139. J. Hirose and Y. Kidani, *Chem. Pharm. Bull.* (*Tokyo*), **28**, 3189 (1980).
140. P. H. Haffner, F. Goodsaid-Zalduondo, and J. E. Coleman, *J. Biol. Chem.*, **249**, 6693 (1974).
141. E. Meirovitch and A. Lanir, *Chem. Phys. Lett.*, **53**, 530 (1978).
142. A. Lanir, S. Gradstajn, and G. Navon, *Biochemistry*, **14**, 242 (1975).
143. I. Bertini, C. Luchinat, and A. Scozzafava, *FEBS Lett.*, **87**, 92 (1978).
144. J. J. Fitzgerald and N. D. Chasteen, *Biochemistry*, **13**, 4338 (1974).
145. I. Bertini, G. Canti, C. Luchinat, and A. Scozzafava, *Inorg. Chim. Acta*, **36**, 9 (1979).
146. A. Y. Romans, M. E. Graichen, C. H. Lochmüller, and R. W. Henkens, *Bioinorg. Chem.*, **9**, 217 (1978).
147. J. M. Pesando, *Biochemistry*, **14**, 681 (1975).
148. I. D. Campbell, S. Lindskog, and A. I. White, *J. Mol. Biol.*, **90**, 469 (1974).
149. K. Gerber, F. T. T. Ng, and R. G. Wilkins, *Bioinorg. Chem.*, **4**, 153 (1975).
150. R. S. McQuate and J. E. Reardon, *J. Chem. Educ.*, **55**, 607 (1978).
151. R. G. Wilkins and K. R. Williams, *J. Am. Chem. Soc.*, **96**, 2241 (1974).
152. Y. Kidani and J. Hirose, *Chem. Lett.* (*Tokyo*), **5**, 475 (1977).
153. R. W. Henkens, G. D. Watt, and J. M. Sturtevant, *Biochemistry*, **5**, 1874 (1969).
154. S. Carpy, *Biochim. Biophys. Acta*, **151**, 245 (1968).
155. Y. Kidani, J. Hirose, and H. Koike, *J. Biochem.* (*Tokyo*), **79**, 43 (1976).
156. Y. Kidani and J. Hirose, *J. Biochem.* (*Tokyo*), **81**, 1383 (1977).
157. J. Hirose and Y. Kidani, *Chem. Pharm. Bull.* (*Tokyo*), **26**, 1768 (1978).
158. Y. Pocker and C. T. O. Fong, *Biochemistry*, **19**, 2045 (1980).

159. P. C. Harrington and R. G. Wilkins, *J. Inorg. Biochem.*, **12,** 107 (1980).
160. J. Hirose and Y. Kidani, *Biochim. Biophys. Acta*, **622,** 71 (1980).
161. T.-Y. Li, A. J. Kraker, C. F. Shaw, and D. H. Petering, *Proc. Natl. Acad. Sci, U.S.A.*, **77,** 6334 (1980).
162. A. O. Udom and F. O. Brady, *Biochem. J.*, **187,** 329 (1980).
163. S. H. Koenig, R. D. Brown, T. E. Needham, and N. A. Matwiyoff, *Biochem. Biophys. Res. Commun.*, **53,** 624 (1973).
164. S. H. Koenig, R. D. Brown, R. E. London, T. E. Needham, and N. A. Matwiyoff, *Pure Appl. Chem.*, **40,** 103 (1974).
165. I. Simonsson, B.-H. Jonsson, and S. Lindskog, *Eur. J. Biochem.*, **93,** 409 (1979).
166. S. H. Koenig and R. D. Brown, *Proc. Natl. Acad. Sci. U.S.A.*, **69,** 2422 (1972).
167. R. G. Khalifah, *Proc. Natl. Acad. Sci. U.S.A.*, **70,** 1986 (1973).
168. R. H. Prince and P. R. Woolley, *Bioorg. Chem.*, **2,** 337 (1973).
169. S. Lindskog and J. E. Coleman, *Proc. Natl. Acad. Sci. U.S.A.*, **70,** 2505 (1973).
170. M. Eigen and G. G. Hammes, *Adv. Enzymol.*, **25,** 1 (1963).
171. B.-H. Jonsson, H. Steiner, and S. Lindskog, *FEBS Lett.*, **64,** 310 (1976).
172. J. C. Kernohan, *Biochim. Biophys. Acta*, **96,** 304 (1965).
173. R. G. Khalifah, *J. Biol. Chem.*, **246,** 2561 (1971).
174. H. Steiner, B.-H. Jonsson, and S. Lindskog, *Eur. J. Biochem.*, **59,** 253 (1975).
175. Y. Pocker and D. W. Bjorkquist, *Biochemistry*, **16,** 5698 (1978).
176. K. S. Venkatasubban and D. N. Silverman, *Biochemistry*, **19,** 4984 (1980).
177. H. Steiner, B.-H. Jonsson, and S. Lindskog, *FEBS Lett.*, **62,** 16 (1976).
178. I. D. Campbell, S. Lindskog, and A. I. White, *J. Mol. Biol.*, **98,** 597 (1975).
179. I. D. Campbell, S. Lindskog, and A. I. White, *Biochim. Biophys. Acta*, **484,** 443 (1977).
180. D. N. Silverman and C. K. Tu, *J. Am. Chem. Soc.*, **98,** 978 (1976).
181. C. K. Tu and D. N. Silverman, *J. Biol. Chem.*, **252,** 3332 (1977).
182. D. N. Silverman, C. K. Tu, S. Lindskog, and G. C. Wynns, *J. Am. Chem. Soc.*, **101,** 6734 (1979).
183. S. H. Koenig and R. D. Brown, *Biophys. J.*, **35,** 59 (1981).
184. S. H. Koenig, R. D. Brown, and G. S. Jacob in *Biophysics and Physiology of Carbon Dioxide*, C. Bauer, G. Gros, and H. Bartels, Eds., Springer, Berlin, 1980, p. 238.
185. P. L. Whitney and H. Brandt, *J. Biol. Chem.*, **251,** 3862 (1976).
186. R. G. Khalifah, *Biochemistry*, **16,** 2236 (1977).
187. I. Bertini, C. Luchinat, and A. Scozzafava, *Bioinorg. Chem.*, **9,** 93 (1978).
188. Y. Pocker and N. Tanaka, *Science*, **199,** 907 (1978).
189. T. H. Maren and E. O. Couto, *Arch. Biochem. Biophys.*, **196,** 501 (1979).
190. P. C. Harrington and R. G. Wilkins, *Biochemistry*, **16,** 448 (1977).
191. L. Tibell and S. Lindskog, *Abstr. Commun. 12th Meet. Fed. Europ. Biochem. Soc.*, 1978, Abs 2262.
192. T. H. Maren, C. S. Rayburn, and N. E. Lidell, *Science*, **191,** 469 (1976).
193. P. W. Taylor and A. S. V. Burgen, *Biochemistry*, **10,** 3859 (1971).
194. J. Olander and E. T. Kaiser, *Biochem. Biophys. Res. Commun.*, **45,** 1083 (1971).

195. P. W. Taylor, J. Feeney, and A. S. V. Burgen, *Biochemistry*, **10,** 3866 (1971).
196. R. L. Ward and M. D. Cull, *Arch. Biochem. Biophys.*, **150,** 436 (1972).
197. A. Lanir and G. Navon, *Biochim. Biophys. Acta*, **341,** 65 (1974).
198. I. Bertini, C. Luchinat, and A. Scozzafava, *Biochim. Biophys. Acta*, **452,** 239 (1976).
199. I. Bertini, C. Luchinat, and A. Scozzafava, *J. Chem. Soc. Dalton*, **1977,** 1962.
200. P. L. Yeagle, C. H. Lochmüller, and R. W. Henkens, *Proc. Natl. Acad. Sci. USA*, **72,** 454 (1975).
201. K. K. Kannan, I. Vaara, B. Notstrand, S. Lövgren, A. Borell, K. Fridborg, and M. Petef in *Drug Action at the Molecular Level*, G. C. K. Roberts, Ed., University Park Press, Baltimore, 1977, p. 73.
202. R. W. King and A. S. V. Burgen, *Biochim. Biophys. Acta*, **207,** 278 (1970).
203. R. F. Chen and J. C. Kernohan, *J. Biol. Chem.*, **242,** 5813 (1967).
204. K. Kumar, R. W. King, and P. R. Carey, *Biochemistry*, **15,** 2195 (1976).
205. R. L. Petersen, T.-Y. Li, J. T. McFarland, and K. L. Watters, *Biochemistry*, **16,** 726 (1977).
206. P. R. Carey and R. W. King, *Biochemistry*, **18,** 2834 (1979).
207. J. C. Kernohan, *Biochim. Biophys. Acta*, **118,** 405 (1966).
208. S. Lindskog and A. Thorslund, *Eur. J. Biochem.*, **3,** 453 (1968).
209. P. W. Taylor, R. W. King, and A. S. V. Burgen, *Biochemistry*, **9,** 3894 (1970).
210. R. W. King and T. H. Maren, *Mol. Pharmacol.*, **10,** 344 (1974).
211. R. W. King and A. S. V. Burgen, *Proc. R. Soc. London Ser. B*, **193,** 107 (1976).
212. T. H. Maren and C. E. Wiley, *J. Med. Chem.*, **11,** 228 (1968).
213. K. C. Leibman and F. E. Greene, *Proc. Soc. Exptl. Biol. Med.*, **125,** 106 (1967).
214. J. A. Verpoorte, S. Mehta, and J. T. Edsall, *J. Biol. Chem.*, **242,** 4221 (1967).
215. D. W. Appleton and B. Sarkar, *Bioinorg. Chem.*, **4,** 309 (1975).
216. J. O'Connor Westerik, B. Sarkar, R. Deslauriers, and I. C. P. Smith, *J. Biol. Chem.*, **253,** 8499 (1978).
217. H. Steiner and S. Lindskog, *FEBS Lett.*, **24,** 85 (1972).
218. I. Bertini, C. Luchinat, and A. Scozzafava, *J. Am. Chem. Soc.*, **99,** 581 (1977).
219. K. K. Kannan, M. Petef, K. Fridborg, H. Cid-Dresdner, and S. Lövgren, *FEBS Lett.*, **73,** 115 (1977).
220. H. R. Wolpert, C. D. Strader, and R. G. Khalifah, *Biochemistry*, **16,** 5717 (1977).
221. G. Alberti, I. Bertini, C. Luchinat, and A. Scozzafava, *Biochim. Biophys. Acta*, **668,** 16 (1981).
222. E. Magid, *Scand. J. Haemat.*, **4,** 257 (1967).
223. C. K. Tu, G. C. Wynns, and D. N. Silverman, *J. Biol. Chem.*, **256,** 9466 (1981).
224. S. I. Kandel, S.-C. C. Wong, M. Kandel, and A. G. Gornall, *J. Biol. Chem.*, **243,** 2437 (1968).
225. M. Kandel, A. G. Gornall, S.-C. C. Wong, and S. I. Kandel, *J. Biol. Chem.*, **245,** 2444 (1970).
226. D. L. Cybulsky, S. I. Kandel, M. Kandel, and A. G. Gornall, *J. Biol. Chem.*, **248,** 3411 (1973).

227. J. W. Wells, S. I. Kandel, M. Kandel, and A. G. Gornall, *J. Biol. Chem.*, **250,** 3522 (1975).

228. P. O. Göthe and P. O. Nyman, *FEBS Lett.*, **21,** 159 (1972).

229. R. G. Khalifah and J. T. Edsall, *Proc. Natl. Acad. Sci. USA*, **69,** 172 (1972).

230. P. L. Whitney, G. Fölsch, P. O. Nyman, and B. G. Malmström, *J. Biol. Chem.*, **242,** 4206 (1967).

231. S.-C. C. Wong, S. I. Kandel, M. Kandell, and A. G. Gornall, *J. Biol. Chem.*, **247,** 3810 (1972).

232. P. Henkart and F. Dorner, *J. Biol. Chem.*, **246,** 2714 (1971).

233. S. L. Bradbury, *J. Biol. Chem.*, **244,** 2002 (1969).

234. P. L. Whitney, *Eur. J. Biochem.*, **16,** 126 (1970).

235. M. Kandel, A. G. Gornall, L. K. Lam, and S. I. Kandel, *Can. J. Biochem.*, **53,** 599 (1975).

236. P. K. Jeffers, W. M. Sutherland, and R. G. Khalifah, *Biochemistry*, **17,** 1305 (1978).

237. A. J. M. Schoot Uiterkamp, I. M. Armitage, J. H. Prestegard, J. Slomski, and J. E. Coleman, *Biochemistry*, **17,** 3730 (1978).

238. R. E. Tashian, J. T. Johansen, and E. Christiansen, *Fed. Proc.*, **39,** 1855 (1980).

239. P. J. Stein, S. P. Merrill, and R. W. Henkens, *J. Am. Chem. Soc.*, **99,** 3194 (1977).

240. D. Demoulin, A. Pullman, and B. Sarkar, *J. Am. Chem. Soc.*, **99,** 8498 (1977).

241. M. E. Riepe and J. H. Wang, *J. Biol. Chem.*, **243,** 2779 (1968).

242. I. Bertini, E. Borghi, and C. Luchinat, *J. Am. Chem. Soc.*, **101,** 7069 (1979).

243. R. K. Gupta and J. M. Pesando, *J. Biol. Chem.*, **250,** 2630 (1975).

244. D. W. Appleton and B. Sarkar, *Proc. Natl. Acad. Sci. USA*, **71,** 1686 (1974).

245. J. M. Pesando and R. K. Gupta in *Biophysics and Physiology of Carbon Dioxide*, C. Bauer, G. Gros and H. Bartels, Eds., Springer, Berlin, 1980, p. 273.

246. R. P. Davis, *J. Am. Chem. Soc.*, **81,** 5674 (1959).

247. R. B. Martin, *Proc. Natl. Acad. Sci. USA*, **66,** 874 (1974).

248. S. Lindskog in *Advances in Inorganic Biochemistry*, Vol. 4, G. L. Eichhorn and L. Marzilli, Eds., Elsevier/North Holland, New York, 1982, p.115.

249. P. Woolley, *Nature*, **258**, 677 (1975).

250. C. I. H. Ashby, C. P. Cheng, and T. L. Brown, *J. Am. Chem. Soc.*, **100,** 6057 (1978).

251. J. H. Coates, G. J. Gentle, and S. F. Lincoln, *Nature*, **249,** 773 (1974).

252. M. M. Werber and Y. Shalitin, *Bioorg. Chem.*, **4,** 149 (1975).

253. P. Woolley, *J. Chem. Soc. Perkin Trans. 2*, **1977,** 318.

254. R. B. Martin, *J. Inorg. Nucl. Chem.*, **38,** 511 (1976).

255. B. Jönsson, G. Karlström, and H. Wennerström, *J. Am. Chem. Soc.*, **100,** 1658 (1978).

256. R. Aasa, M. Hanson, and S. Lindskog, *Biochim. Biophys. Acta*, **453,** 211 (1976).

257. J. L. Sudmeier and D. B. Green in *The Coordination Chemistry of Metalloenzymes*, I. Bertini, R. S. Drago, and C. Luchinat, Eds., D. Reidel, Dordrecht, 1983, p. 35.

258. I. Bertini, E. Borghi, C. Luchinat, and R. Monnanni, *Inorg. Chim. Acta*, **67,** 99 (1982).

259. I. Bertini, G. Canti, and C. Luchinat, *Inorg. Chim. Acta*, **56,** 99 (1981).

260. I. Simonsson and S. Lindskog, *Eur. J. Biochem.*, **123,** 29 (1982).

261. S. Lindskog, S. A. Ibrahim, B.-H. Jonsson, and I. Simonsson in *The Coordination Chemistry of Metalloenzymes*, I. Bertini, R. S. Drago, and C. Luchinat, Eds., D. Reidel, Dordrecht, 1983, p. 49.

262. I. Simonsson, B.-H. Jonsson, and S. Lindskog, *Eur. J. Biochem.*, **129,** 165 (1982).

263. Y. Pocker and T. L. Deits, *J. Am. Chem. Soc.*, **104,** 2424 (1982).

264. I. Simonsson, B.-H. Jonsson, and S. Lindskog, *Biochem. Biophys. Res. Commun.*, **108,** 1406 (1982).

CHAPTER 4

The Role of Zinc in Alcohol Dehydrogenase

HANS EKLUND and CARL-IVAR BRÄNDÉN

Department of Chemistry and Molecular Biology
Swedish University of Agricultural Sciences
Uppsala, Sweden

CONTENTS

1 INTRODUCTION, 125

2 THE STRUCTURE OF THE MOLECULE, 125

2.1 General Fold, 125
2.2 The Zinc Atoms, 127
2.3 The Substrate-Binding Pocket, 129
2.4 Conformational Differences between Open and Closed Forms, 131
2.5 Factors which Influence the Conformational Change, 131
2.6 Dynamic Aspects of the Structure, 132

3 CRYSTALLOGRAPHIC STUDIES OF LIGAND BINDING, 134

3.1 Coenzyme Binding, 134
3.2 Substrate Binding, 136
3.3 Inhibitor Binding, 137

4 SPECTROSCOPIC INVESTIGATIONS OF THE ROLE OF THE CATALYTIC ZINC ATOM, 141

4.1 Metal Substitutions, 141
4.2 NMR Studies, 142
4.3 Resonance Raman Spectroscopy, 143
4.4 Electron Paramagnetic Resonance Studies, 144

5 ASPECTS OF THE CATALYTIC MECHANISM, 144

5.1 Proton Abstraction, 144
5.2 Hydride Transfer, 147

6 CONCLUSIONS ON THE ROLE OF ZINC IN LADH, 148

ACKNOWLEDGMENTS, 149

REFERENCES, 149

1 INTRODUCTION

Horse liver alcohol dehydrogenase (LADH) (EC 1.1.1.1.) is an NAD^+-dependent enzyme that catalyzes the oxidation of various primary and secondary alcohols to the corresponding aldehydes. The active enzyme has a molecular weight of 80,000 and is a dimer of two identical subunits. Each subunit has one coenzyme-binding site and firmly binds two zinc atoms, one of which is essential for catalysis. The sequence of the 374 amino acids in the single polypeptide chain of one subunit of the E isozyme has been determined by Jörnvall (1). Coenzyme binding induces a conformational change of the enzyme from an open to a closed form (2). The X-ray structure of the native enzyme (EE isozyme), which is in the open form, has been determined to 2.4 Å resolution (3) and refined to an *R* value of 21% (T. A. Jones, unpublished work). The structures of the complexes with NADH and dimethylsulfoxide (4), NAD^+ and pyrazole (5), NAD^+ and bromobenzyl alcohol (6) and tetrahydro-NAD and dimethylaminocinnamaldehyde (7), which are in the closed form, have been determined to 2.9 Å resolution with an *R* value for the NADH·dimethylsulfoxide complex of 25% (T. A. Jones and H. Eklund, unpublished work).

Different aspects of this enzyme have been reviewed during the last decade (8–16). In this chapter we will concentrate on the geometrical constraints that the protein part of this enzyme imposes on the catalytic role of its active-site zinc atom. Granted that the structure of the enzyme in the crystalline state is essentially the same as that in solution, the crystallographic studies reviewed here severely limit the number of possible mechanisms for the participation of the zinc atom to the catalytic action. A number of physicochemical solution studies are in accord with the crystallographic results, others have been interpreted differently. Some of these studies will be mentioned here but not extensively discussed.

It is our firm opinion that X-ray crystallographic studies of ligand binding to enzyme molecules provide a solid framework for the interpretation of spectroscopic signals from solution studies. When an apparent conflict arises it has usually been found that the conflict can be resolved by reinterpreting the spectroscopic signals rather than assuming radically different structures for the active site in solution versus crystals.

2 THE STRUCTURE OF THE MOLECULE

2.1 General Fold

The general structural features of the open and closed forms of the LADH molecule are similar. Each subunit is divided into two domains that are

separated by a cleft containing a deep pocket. This pocket accommodates the substrate and the nicotinamide moiety of the coenzyme. One of the domains is responsible for binding the coenzyme and the other provides ligands to the catalytic zinc atom as well as most of the groups that control substrate specificity. The two domains are unequal in size; the catalytic domain is larger and comprises 231 residues whereas the coenzyme-binding domain contains 143 residues. The coenzyme-binding domains of the two subunits are joined together into a central compact core of the dimeric molecule. A schematic diagram of the polypeptide fold and the bound zinc atoms is shown in Figure 1 and a stereodigram of one subunit in Figure 2.

The amount of secondary structure in the molecule is high. The two domains are, however, built up from quite different arrangements of secondary structural elements. The coenzyme-binding domain is a typical α/β domain (17) with a central pleated sheet of six parallel strands surrounded by helices on both sides. The arrangement of strands observed in such domains provides an excellent framework for ligand binding (18). In particular it was found (9, 19, 20) that the similar topological fold of the coenzyme-binding domains causes the coenzyme molecule to bind in a very similar way in all NAD^+-dependent dehydrogenases. This point is emphasized here and will be described in more detail later because these specific interactions provide

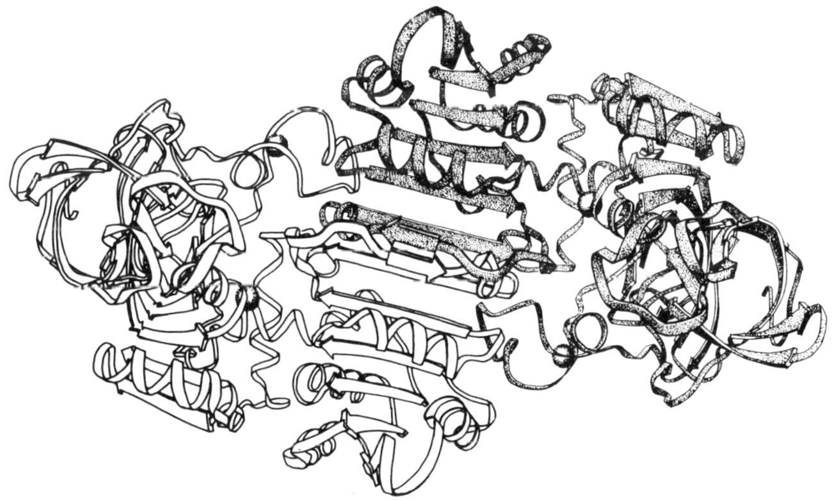

Figure 1 Schematic drawing of the dimeric LADH molecule. One subunit is shaded. Central part of the molecule is the two coenzyme-binding domains bound together across a twofold axis perpendicular to the plane of the paper. Two catalytic domains are separated from this central part by two clefts. Shaded circles are the four zinc atoms. (We are indepted to B. Furugren from the drawing.)

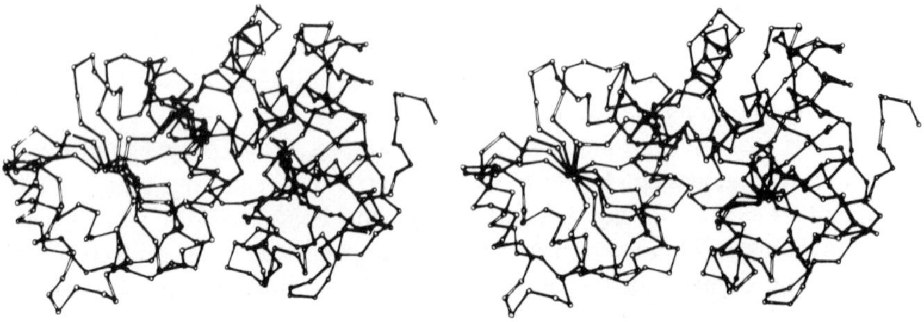

Figure 2 Stereodiagram of the C_α chain of one subunit of LADH.

rather stringent geometric conditions for proper positioning of the nicotinamide in relation to the substrate and the active-site zinc atom.

2.2 The Zinc Atoms

The catalytic domain is built up from an intricate network of antiparallel pleated sheet regions and a few short helices. This domain comprises residues 1–174 and 318–374. There are thus two covalent connections between the domains that are close together in space at the back of the molecule as seen in Figure 1.

This domain provides the ligands to the catalytic zinc atom; Cys-46, His-67, and Cys-174 (Fig. 3). The peptide units of all three zinc ligands are firmly positioned in secondary structural elements and none is part of a flexible loop

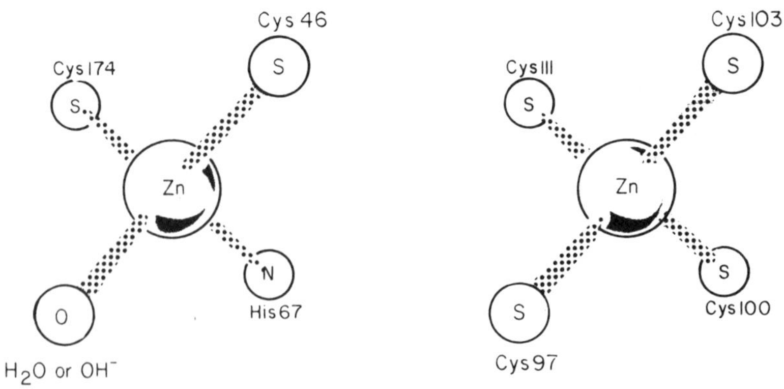

Figure 3 Schematic diagram showing the zinc ligands in LADH. (*a*) Ligands to the catalytic zinc atom; (*b*) Ligands to the second zinc atom.

region. Cys-46 is the first residue of a short helix, comprising residues 46–55 and at the same time the last residue of a β strand. His-67 is the first residue of a β strand in the middle of a pleated sheet region and Cys-174 is in the middle of the helix that joins the two domains.

There is also a second zinc atom in each subunit. This zinc is liganded in a tetrahedral arrangement by four sulfur atoms from cystein residues 97, 100, 103, and 111 (Fig. 3). Bond lengths and bond angles around this zinc are given in Table 1. These cystein residues are part of a lobe that projects out of the catalytic domain having only a few side-chain interactions with the re-

Table 1 Bond Lengths and Bond Angles around the Zinc Atoms in Both Forms of LADH[a]

(A) Bond Lengths		Open Form (Å)	Closed Form (Å)
(1) Catalytic zinc	Zn—S-46	2.3	2.3
	Zn—S-174	2.1	2.4
	Zn—N-67	2.2	2.1
	Zn—O	2.1	2.0
(2) Second zinc	Zn—S-97	2.4	2.3
	Zn—S-100	2.3	2.4
	Zn—S-103	2.3	2.2
	Zn—S-111	2.5	2.4
(B) Bond Angles		(degrees)	(degrees)
(1) Catalytic zinc	S-46—Zn—S-174	127	134
	S-46—Zn—N-67	113	101
	S-46—Zn—O	108	93
	N-67—Zn—S-174	105	108
	N-67—Zn—O	101	94
	S-174—Zn—O	100	115
(2) Second zinc	S-97—Zn—S-100	105	97
	S-97—Zn—S-103	119	103
	S-97—Zn—S-111	105	104
	S-100—Zn—S-103	107	117
	S-100—Zn—S-111	115	119
	S-103—Zn—S-111	107	114

[a]The nonprotein ligands to the catalytic zinc atoms are H_2O and bromobenzyl alcohol in the open and closed forms, respectively.

maining part of the subunit. Accessibility calculations show that this zinc is completely buried by its ligands and not accessible to the solution.

The function of this second zinc atom and the protein lobe that surrounds it is unknown, although a structural role has been suggested (21). The fact that all ligands to this zinc are close together along the polypeptide chain supports such a role. It is usually found that metal atoms that have a catalytic role in proteins have their ligands from different parts of the polypeptide chain, whereas ligands to atoms with a structural role are close together. Examples of this last type are calcium binding in thermolysin (22) and zinc binding in aspartate transcarbamylase (23).

2.3 The Substrate-Binding Pocket

The catalytic zinc atom is positioned deep inside the subunit close to the covalent junction between the two domains. It is not accessible to solution from the backside nor from the topside of the molecule as seen in Figure 1. In the open conformation it is accessible from the front side through the crevice between the domains. In the closed conformation this crevice is closed. From the bottom side there is a deep pocket, 5–10 Å wide and about 20 Å long from the protein surface to the zinc atom. This is the substrate-binding pocket. The zinc atom is accessible from solution through this pocket both in the open and closed forms. The bottom of this pocket is formed by the zinc atom and its ligands.

The three protein ligands are arranged in three corners of a distorted tetrahedral arrangement. The fourth corner of this distorted tetrahedron points into the substrate-binding pocket. Both in the open and closed forms of the enzyme this coordination position is occupied by a water molecule or, if present, bound inhibitor or substrate. Observed bond lengths and bond angles around the zinc atom in both forms are listed in Table 1.

The only polar atoms in the pocket near zinc are the zinc ligands, the nicotinamide moiety of the coenzyme, and the side chain of Ser-48. The remaining part of the inside wall of the pocket is lined with hydrophobic side chains from residues Phe-93, Phe-140, Leu-141, Leu-57, Leu-116, Ile-318, Val-294, Phe-110, Pro-295, and Pro-296, which are from the same subunit as the ligands to zinc and Met-306 as well as Leu-309 from the other subunit. In the open form there are several firmly fixed water molecules in the inner part of the pocket outside the coordination sphere of zinc. In ternary complexes in the closed form there are no such water molecules since the coenzyme and substrate occupy the space. Stereodiagrams of the substrate-binding pocket are given in Figure 4 and a schematic diagram of the hydrophobic barrel in Figure 5.

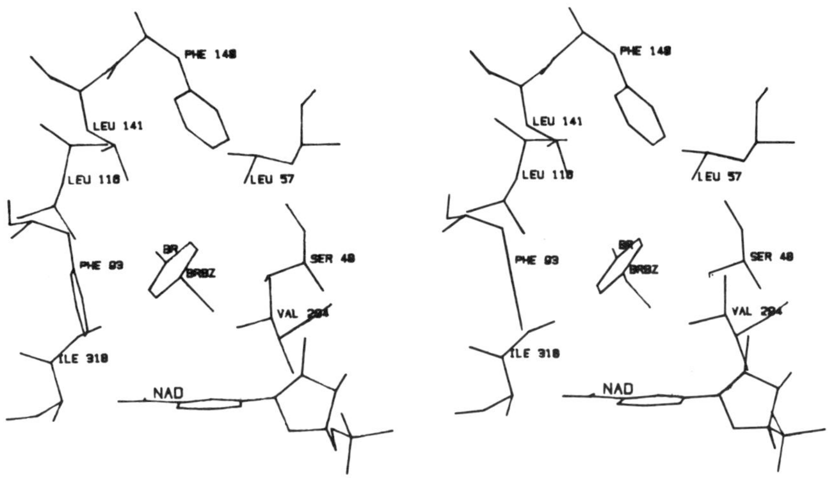

Figure 4 Stereodiagram illustrating the substrate-binding pocket in LADH viewed from the direction of the zinc atom. Bromobenzyl alcohol molecule is bound in the pocket.

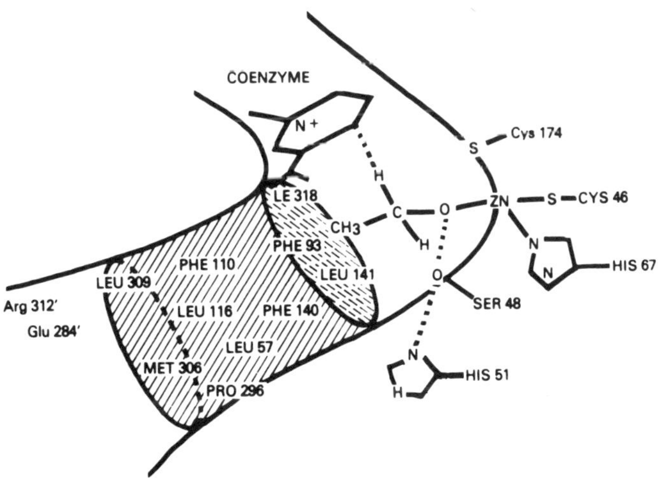

Figure 5 Schematic representation of the substrate-binding pocket in LADH.

2.4 Conformational Differences between Open and Closed Forms

The main conformational differences between the open and the closed forms of the enzyme consists of a rigid-body rotation of each catalytic domain by about 10° with respect to the central core of the dimer (4). In addition, there is a rotation of about 1° of the two coenzyme-binding domains with respect to each other around an axis approximately perpendicular to the molecular twofold axis. The internal structures of the domains are almost completely preserved with the exception of a loop region comprising residues 293 to 298 in the coenzyme-binding domain. This loop changes its structure drastically in order to allow the rotation of the catalytic domains.

The main effect of this conformational change from the open to the closed form is to narrow the crevice between the domains on the front side of the molecule. In addition, the catalytic zinc atoms that follow the rotation of the catalytic domains move approximately 1 Å closer to the coenzyme-binding domain and thus bring the substrate-binding site closer to the nicotinamide binding position. Furthermore the substrate-binding pocket becomes somewhat smaller and some of the side chains that line the pocket change their orientation, in particular Val-294.

2.5 Factors which Influence the Conformational Change

The conformational change from the open to the closed form is induced by coenzyme binding. The presence of one or the other form has so far only been studied by crystallographic methods. The open form gives rise to orthorhombic crystals whereas the closed form gives monoclinic or triclinic crystals (4, 14). The molecular structure is the same in these last two forms (24). Unfortunately there is no reliable spectroscopic signal by which these two forms can be differentiated in solution. Furthermore the low angle scattering curves are identical within the limits of error (Ptitsyn and Finkelstein, private communication) since both the radius of gyration and the shape of the molecule change only slightly by the domain rotation, in contrast to hexokinase (25) or phosphoglycerate kinase (26).

The nicotinamide moiety of the coenzyme is essential for the conformational change to occur since the complex with ADP-ribose is in the open form even though ADP-ribose binding to the coenzyme-binding domain is very similar to the binding of the corresponding part of NADH to the closed form (27). Small modifications of the nicotinamide moiety shift the equilibrium between the two forms. Thus the complex with NADH in the presence of the precipitating agent methylpentanediol produces crystals of the closed form whereas the similar complex with tetrahydro-NAD gives crystals of the open form (7). If, on the other hand, a ternary complex is formed with this inactive

coenzyme analogue and a large substrate such as dimethylaminocinnamaldehyde, the equilibrium shifts to the closed form. Ternary complexes with NAD^+ and alcohol or pyrazole as well as with NADH and inhibitors like dimethylsulfoxide or isobutyramide also give crystals of the closed form (14). In contrast, the ternary complex with NADH and imidazole crystallizes in the open form (E. Cedergren-Zeppezauer, unpublished results).

From these studies it emerges that the nature of the substrate or inhibitor bound to the catalytic zinc atom is an essential factor for proper positioning of the nicotinamide moiety of the coenzyme. This position seems to be important in determining the equilibrium between the open and the closed forms of the enzyme. In order to examine the role of the catalytic zinc atom in this process we studied complexes with the enzyme where the catalytic zinc was specifically (28) removed. Both the binary complex with NADH (85) and a ternary complex with NAD^+ and pyrazole were studied (G. Schneider, H. Eklund, and M. Zeppezauer, unpublished results). Both these complexes crystallized in the closed form. Thus the catalytic zinc atom plays only a minor role, if any, in the process of the conformational change. It would thus appear that the role of bound substrate or inhibitor in this process is essentially a steric role. Inhibitors such as imidazole that bind better to the open form due to the somewhat larger substrate-binding pocket shift the equilibrium to that form whereas large substrate molecules shift the equilibrium to the closed form. The nature of the interaction between the zinc atom and the inhibitor or substrate, for example, whether the ligand is a nitrogen or oxygen atom, is on the other hand of no importance.

2.6 Dynamic Aspects of the Structure

A crystallographic study of an enzyme molecule at a given temperature reflects the average structure of a distribution of a large number of different dynamic states of the molecule. In a few cases the distribution of such states have recently been studied theoretically (29, 30) and concerted as well as individual motions of atoms or groups of atoms have been deduced from experimental data (31, 32). These studies have recently been reviewed (33). It is thus relevant to ask the following question: To what extent does the observed averaged protein structure differ from the structure of a catalytically active transition state which might be at one extreme edge of the distribution of different possible dynamic states of the molecule? A quantitative answer to this question requires extensive studies that so far have not been done on a large molecule like LADH. We will try to address this question in a qualitative way with respect to the possible motions of the zinc ligands, substrate, and the nicotinamide part of the coenzyme molecule.

It should first perhaps be pointed out that all observed large conforma-

tional changes in proteins involve either changes in particular side-chain conformations, changes in the conformation of flexible loop regions, or rotation of complete domains with the internal domain structure preserved. In LADH there are examples of all three types of changes. In no case has a change in regular secondary structure of the protein been observed. Large (>1 Å) individual movements of main chain atoms which are parts of helices or pleated sheet structures are thus of low probability. All three residues that provide ligands to the catalytic zinc atom participate in such secondary structure. We would thus expect possible changes in the coordination sphere of the zinc atom to be restricted to changes in the conformation of the side chains with the position of Cα fixed. For cysteine such changes are restricted to rotation about the Cα—Cβ bonds and for histidine in addition rotation about the Cβ—Cγ bond. In order to examine experimentally such changes in the most drastic case we have studied the structure of the zinc-free enzyme, both in the open (34) and in the closed form (85). In the open form no changes of main chain atoms were found. The side chain of Cys-174 is in the same position as in the zinc enzyme. The imidazole ring of His 67 is rotated around the Cβ—Cγ bond about 20°. The largest change occurs for the sulfur atom of Cys-46 which has moved 2 Å in a direction away from the zinc site by a rotation around the Cα—Cβ bond.

In the closed form the largest changes occur for Cys-174 which is in the hinge region for the conformational change. Here the whole side chain including Cα has moved less than 1 Å in a direction such that the sulfur atom is closer to the zinc site. Cys-46 stays constant and the imidazole group of His-67 rotates in the same way as in the open form.

It is thus clear from these studies that the protein structure provides a rigid framework for the zinc binding site with a rather fixed orientation of the side chains. Only small adjustments of this geometry occur when the zinc atom is bound to this site. It is thus rather improbable that the coordination geometry of zinc to the protein is drastically different in any of the catalytic states from that of the complexes observed in the crystal. This argument severely restricts the number of possible mechanisms.

Since the main conformational change involves a rotation of domains, there is a definite possibility that there could be concerted vibrational motions of the complete domains as rigid bodies. This would affect the motions of substrate and coenzyme differently since they are bound to different domains. Since these motions start appearing in the picosecond range and catalysis occur in the millisecond range, such states could be very important for the actual process of hydride transfer.

Finally the substrate and the nicotinamide moiety could be subject to individual vibrational motions. There are few firm interactions between the nicotinamide moiety and the protein. Thus there seems to be steric possibil-

ities for small individual motions of this group, as well as for the substrate molecule.

3 CRYSTALLOGRAPHIC STUDIES OF LIGAND BINDING

3.1 Coenzyme Binding

The coenzyme NAD (nicotinamide adenine dinucleotide) binds to a well-defined binding site of the enzyme. The coenzyme-binding domain is built up in such a way that a cleft is created across the carboxyl edge of the parallel pleated sheet. The adenosine half of the coenzyme binds to one side of this edge while the nicotinamide half is bound to the other side.

The coenzyme molecule has a number of specific interactions that give hydrogen bonds from oxygen atoms of the phosphate and the two ribose moieties to the side chains of Asp-223, Arg-47, Ser-48, His-51, Arg-369, and main-chain atoms of the enzyme. These interactions are shown schematically in Figure 6. This coenzyme-binding site locks the coenzyme in a fixed conformation which is identical in all those ternary complexes of alcohol dehydrogenase which have the closed protein structure conformation (35).

The nicotinamide ring is positioned in a slot deep inside the protein. Atoms from the coenzyme-binding domain form one side of this slot and the zinc atom and its ligands form the other side. The nicotinamide is anchored to the protein by hydrogen bonds from the carboxyamide group to main chain protein atoms at the bottom of the slot. At the other end it is fixed by the nicotinamide ribose which is tightly bound between the two domains of the subunit.

When the nicotinamide is properly positioned in this site and a substrate or inhibitor molecule is bound in the substrate site, the zinc atom, its ligands, and the nicotinamide ring are no longer in contact with the solvent that surrounds the enzyme. Furthermore, all water molecules which are located in the active site in the open form are displaced in these ternary complexes. The whole active site of the enzyme is thus in a completely water-free environment in the closed form (36).

Binding of coenzyme is absolutely necessary for the conformational change to occur since the closed conformation of the enzyme has only been observed in the presence of coenzyme. In cases where the conformational change has not been induced by coenzyme binding, the coenzyme is not locked into this normal tight interaction with the enzyme. Furthermore, in these cases the conformations of bound coenzyme are quite different (7, 37).

The active-site zinc atom is not directly involved in coenzyme binding. There is no ligation of any part of the coenzyme to the zinc atom. However,

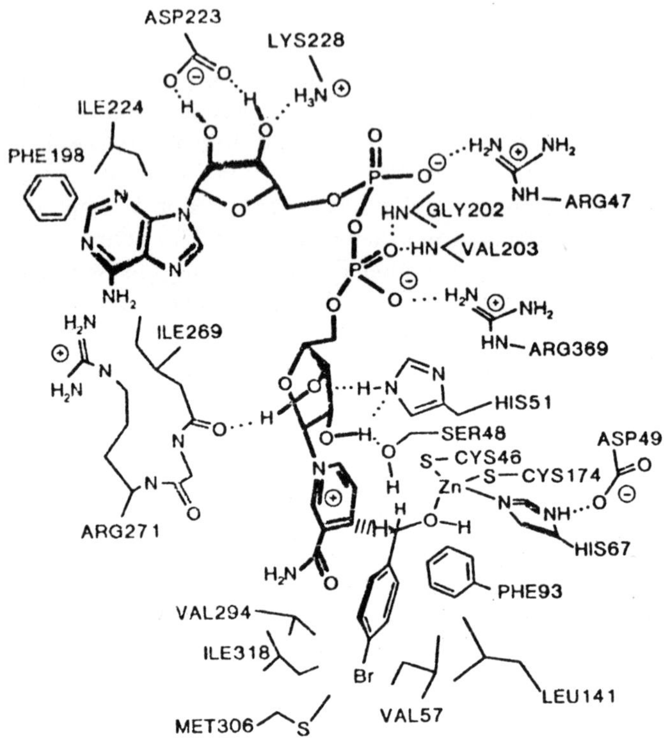

Figure 6 Schematic representation of interactions from coenzyme and substrate to the enzyme molecule LADH.

when the coenzyme is properly positioned, the C-4, C-5 and C-6 atoms of the nicotinamide are about 4 Å from the zinc atom and in van der Waals contact to the two cysteine sulfur ligands. Thus, the zinc atom, although it does not directly participate in coenzyme binding, might still influence the binding since the active-site zinc atom and its ligands, including the fourth free ligand, form one side of the slot into which the nicotinamide ring binds. However, the zinc-free enzyme binds NADH and undergoes the conformational change to the closed form. Furthermore, the conformation of bound NADH is the same as in the complex with native enzyme.

Since the nicotinamide ring is quite close to the active-site zinc atom, it influences both the accessibility of ligands to zinc and the electronic properties of the metal. In particular, the zinc atom should be influenced by the positive charge on the nicotinamide ring in complexes with the oxidized form of the coenzyme.

3.2 Substrate Binding

Crystallographic studies of substrate binding have concentrated on two complexes. Alcohol binding has been studied in crystals obtained from an equilibrium mixture of enzyme, coenzyme, and bromobenzyl alcohol/aldehyde (6, 38) where solution studies show that the ternary complex with alcohol is greatly favored. Aldehyde binding has been studied in a complex with tetrahydro-NAD and dimethylaminocinnamaldehyde (7). Ternary complex formation in solution with this chromophoric aldehyde and NADH or tetrahydro-NAD gives a characteristic spectroscopic signal from the conjugated double-bond system of the aldehyde (39), which is also present in the crystals (7). Both bromobenzyl alcohol and dimethylaminocinnamaldehyde bind directly to the active-site zinc atom giving zinc a roughly tetrahedral coordination. Since these studies are at 2.9 Å resolution, direct binding of the substrate oxygen is firmly established in these cases. Indirect binding via a water molecule as has been suggested by Sloan et al. (40) for substrate binding is completely incompatible with the crystallographic results. A low resolution study has also been done with trifluoroethanol (38) which also shows direct binding of alcohol to the active-site zinc atom.

The position and conformation of the bound bromobenzyl alcohol can be very accurately and distinctly deduced from its electron density maps (6) due to the heavy bromine atom and the planarity of the benzyl group. A benzyl alcohol molecule placed in the observed density has its oxygen atom bound to the zinc atom and its C-1 atom placed close to Ser-48, His-67, Phe-93, Cys-174, and the nicotinamide ring. The bromophenyl part of the molecule fits nicely in the hydrophobic substrate-binding pocket with many interactions with the side chains of the residues that line this pocket. Stereodiagrams of bound bromobenzyl alcohol is shown in Figures 4 and 7.

The size of the substrate-binding cleft is partly accommodated to the size of the substrate by changes in side-chain conformations. The side chain of Leu-116 in particular changes its conformation drastically in this substrate complex as well as in the complex with dimethylaminocinnamaldehyde.

The binding mode of bromobenzyl alcohol found in the crystallographic study represents the most stable form of the complex. This form differs slightly from the complex that should be present at the hydride transfer step, since the hydrogen to be transferred does not point toward C-4 of the nicotinamide ring. Furthermore, the distance between these atoms is longer than expected for an active complex. Model-building experiments show, however, that the position of the bromobenzyl alcohol molecule could easily be rotated around an axis through the oxygen and bromine atoms of the molecule such that it is in an ideal position for hydride transfer (6). This position is illustrated in Figure 7.

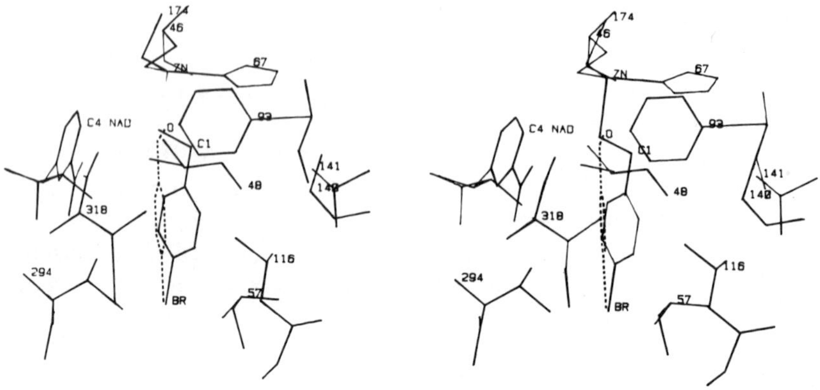

Figure 7 Stereodiagram of bound bromobenzyl alcohol to LADH. Solid lines represent observed positions of substrate, coenzyme, and enzyme atoms, dotted lines represent a model-built position of the substrate, poised for hydride transfer.

3.3 Inhibitor Binding

The crystallographic studies of inhibitor binding to LADH have shown that there are two main classes of inhibitors. One type binds in the coenzyme-binding crevice, preferably the region where the AMP part of the coenzyme binds. Inhibitors belonging to this class are the dye cibachron blue (41), which is used for affinity chromatography of nucleotide binding proteins; anilinonaphthalenesulfoxide (42), the spectral changes of which are frequently used as a measure of the hydrophobicity of binding regions in proteins, and salicylate derivatives that have been shown (43) to simulate binding of the adenosine moiety of nucleotides.

The second type of inhibitors binds in the substrate-binding pocket and competes with the substrate for ligation to the active-site zinc atom. The mode of binding of dimethylsulfoxide (4), pyrazole, and some of its derivatives (5) as well as imidazole and *ortho*-phenanthroline (44) have been studied in detail. They all bind to the free surface of the active-site zinc atom. They act as inhibitors either by competing with substrate binding to the closed form of the enzyme or by binding to zinc in the open form of the enzyme, thus preventing the conformational change and proper coenzyme binding. Pyrazole and *ortho*-phenanthroline are typical examples of these two different binding modes which give rise to quite different results in kinetic studies of inhibitor action.

The binding site for these inhibitors as well as for substrates in the closed form of the enzyme is narrow in the immediate vicinity of the zinc atom. The site is limited in space by the side chains of Ser-48, His-67, and Phe-93 as

well as the nicotinamide ring and allows only one molecule to bind at the same time. Figure 8, which shows imidazole binding, illustrates this point. Simultaneous binding of water and inhibitor to zinc is also excluded for steric reasons (36) as will be discussed later.

Pyrazole derivatives are very strong inhibitors to LADH with inhibition constants of $10^{-10}M$ for some derivatives (45). These inhibitors have been extensively used for metabolic studies (46, 47), kinetic experiments (13), purification (48), and different spectroscopic studies (49–51). Crystallographic studies of the complex between enzyme, NAD^+, and pyrazole or 4-iodopyrazole show that the pyrazole ring forms a bridge between the active-site zinc atom and the coenzyme (5). One nitrogen atom, N-2, forms a normal coordinate bond to zinc. The other nitrogen atom, N-1, is about 2 Å from atom C4 of the nicotinamide ring, thus somewhat longer than a normal covalent bond. This complex is shown in Figure 9. The C-4—N-1 bond is approximately normal to the plane of the nicotinamide ring.

On the basis of spectroscopic and kinetic studies it has previously been suggested (86) that this complex simulates a transition-state complex in the catalytic mechanism of LADH. The crystallographic results support this suggestion with the oxygen of the substrate replacing N-2 as ligand to zinc and N1 substituted by C-1 of the substrate from which a hydride ion is transferred to C-4 of the nicotinamide. Deviations from typical covalent bond lengths have earlier been reported both in organic compounds (52) and in enzyme complexes (53) and are presumably common in intermediate complexes in enzyme catalysis.

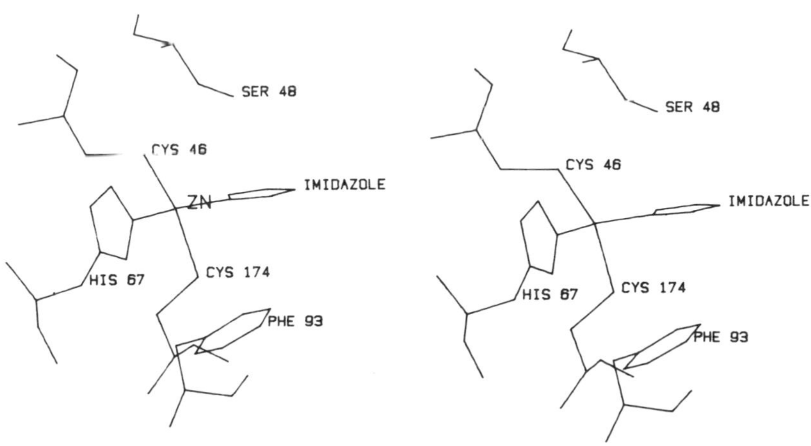

Figure 8 Stereodiagram of bound imidazole to LADH, illustrating the slot between the side chains of Ser-48 and Phe-96 where inhibitors and substrates bind.

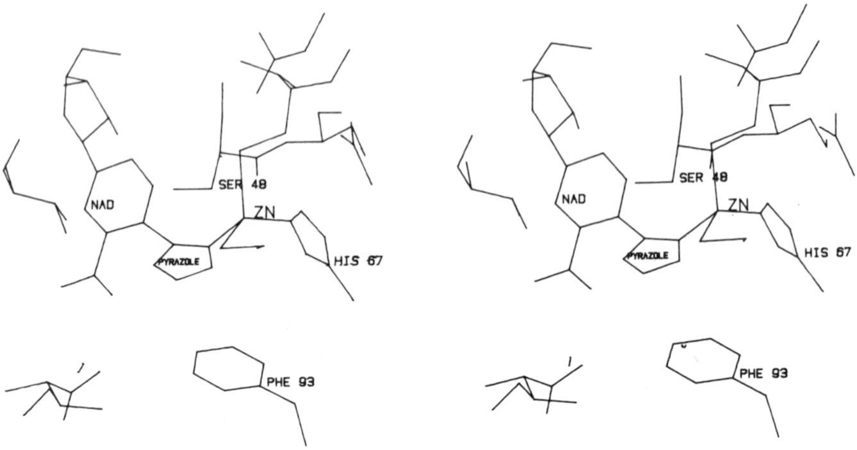

Figure 9 Stereodiagram of bound pyrazole to LADH and NAD^+, illustrating the bridge that is formed by the two nitrogen atoms of pyrazole between zinc and atom C-4 of NAD^+.

Substitutions in the 4 position of the pyrazole ring enhances the inhibitory power, whereas substitutions in the 3 and 5 positions reduces the binding strength (54). Substitutions in the 5 position are too close to the coenzyme and in the 3 position they interfere with His-67 and Ser-48 (55). Substituents in the 4 position on the other hand are positioned in the hydrophobic barrel of the substrate-binding pocket. Figure 10 shows the results of a crystallographic study of the complex with 4-iodopyrazole (5). The iodine atom is in van der Waals contact with side-chain atoms of Leu-57, Phe-93, Leu-116, Leu-141, and Val-294. Using the position of bound 4-iodopyrazole, Eklund et al. (5) model built in a computer display a possible binding mode of a large linear alkyl 4-substituent to illustrate the correlation of structural details of the substrate-binding pocket with the kinetic studies of a large number of different 4-substituted pyrazole derivatives (45). This model is shown in Figure 11. As the size of the hydrophobic alkyl chain increases, the inhibitory power of the derivative increases due to a larger number of hydrophobic interactions with side chains of the barrel. Charged groups on the substituent, on the other hand, decrease the affinity unless they are remote from the pyrazole ring. In that case they are outside the hydrophobic barrel and reach polar residues at the bottom of the substrate-binding site, close to the outer solution. Branched and bulky substituents are bad inhibitors since they sterically interfere with protein residues along the rather narrow hydrophobic channel.

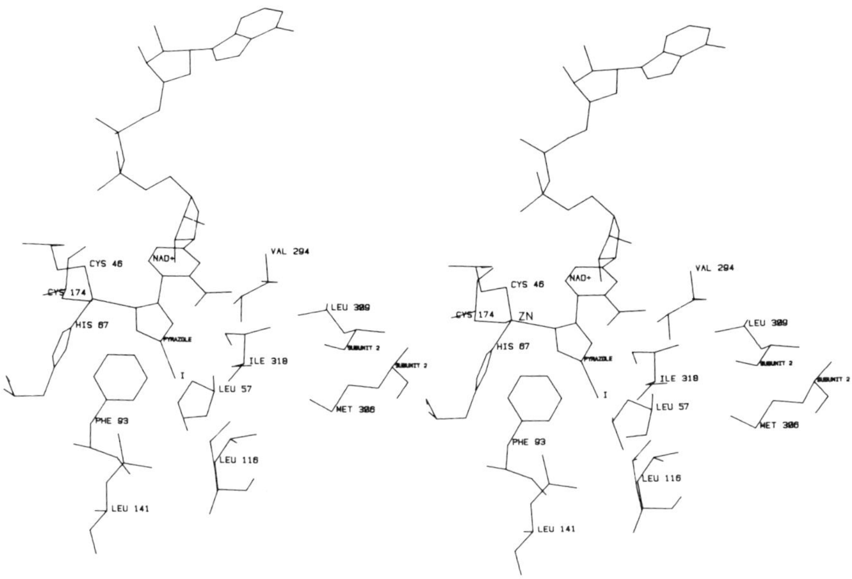

Figure 10 Stereodiagram of bound iodopyrazole illustrating the relation of bound coenzyme and inhibitor to zinc and the substrate-binding pocket.

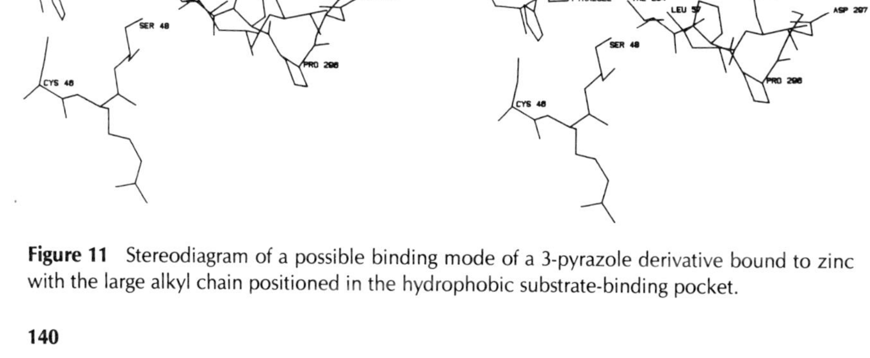

Figure 11 Stereodiagram of a possible binding mode of a 3-pyrazole derivative bound to zinc with the large alkyl chain positioned in the hydrophobic substrate-binding pocket.

4 SPECTROSCOPIC INVESTIGATIONS OF THE ROLE OF THE CATALYTIC ZINC ATOM

4.1 Metal Substitutions

A very useful tool for evaluating the role of catalytic zinc atoms in zinc metalloenzymes is to substitute the zinc atom with different metal atoms (56). In alcohol dehydrogenase such studies have been complicated by the fact that the enzyme has two different metal sites. Most studies have therefore previously been performed on enzyme where both sites were substituted. This situation has recently changed with the successful selective replacement of the active-site zinc atom performed by M. Zeppezauer and co-workers (28). Such enzyme species are now being used for a large number of different physicochemical studies.

Selective removal of the active-site zinc atom can be accomplished by treating crystals of native enzyme with specific chelating agents (28). The resulting catalytically inactive enzyme has no zinc atom at the active site, while the second zinc atom is present to a full extent. This was proven by X-ray diffraction studies of such crystals (34), which also show that small movements of the side chains of Cys-46 and His-67 are the only structural differences between this enzyme and the native protein. The activity of the enzyme can be fully reconstituted to that of the native enzyme by addition of zinc (28, 87). Crystals lacking the active-site zinc atoms can be specifically reconstituted at the active site with Co(II), Cd(II), Cu(II), Ni(II), Pb(II), and Hg(II) (28, 57–60, 87).

The UV-visible spectra and CD spectra of the Co(II) and Ni(II) enzymes are consistent with distorted tetrahedral coordination geometries (28, 59, 61). X-ray studies of the Co(II)-substituted enzyme (34) have proven that the Co atom has a coordination geometry very similar to that of the zinc enzyme. In a comparative study of the decay of the enzyme-bound dimethylaminocinnamaldehyde intermediate, Dunn et al. (62) found that the rate constants for the hydride transfer process obtained for the Zn(II), Co(II), and Ni(II) enzymes are very similar, while the value for the Cd(II) enzyme is much smaller. The Zn(II), Co(II), and Ni(II) enzymes yield the same kinetic isotope effect with deuterated NADH, while that of the Cd(II) enzyme appears to be slightly smaller. Thus, they conclude that for the Zn(II), Co(II), and Ni(II) enzymes the nature of the transition state for hydride transfer is almost unchanged, while substitution to Cd(II) changes the transition state. Zn(II), Co(II), and Ni(II) have similar distances to sulfur ligands in tetrahedral small molecule complexes, whereas Cd(II) has about 0.2 Å longer distances. The large ionic radius and consequent longer bond lengths makes Cd(II) a weaker Lewis acid than Zn(II), Co(II), and Ni(II). The spectral shifts

upon binding of dimethylaminocinnamaldehyde to the active-site metal of alcohol dehydrogenase roughly parallel the expected relative Lewis acid strengths of these metal ions in a tetrahedral ligand field. The longer bonds to Cd(II) may also result in a slight misalignment of the reacting atoms and perhaps a different coordination number and/or geometry (62).

Binding experiments with dimethylaminocinnamaldehyde show that both the metal and the coenzyme are absolutely necessary for substrate binding (61). Active-site depleted enzyme does not bind this aldehyde. For the enzyme with cobalt at the active site, the spectra of both the chromophore aldehyde and the metal ion are changed significantly upon complex formation, providing evidence for a direct coordination of the carbonyl oxygen atom of the substrate to the catalytic cobalt ion. Studies of metal directed affinity labeling of Zn(II), Co(II), and Cd(II) with bromoimidazolyl propionic acid have also been interpreted in terms of direct binding of the label to the active-site metal ion in competition with imidazole (63).

Substitution with Cu(II) ions at the active site creates an artificial blue copper center similar to that of plastocyanin and related copper proteins (64). By an intramolecular redox reaction a cuprous enzyme is formed from the cupric enzyme. Whereas the activity of the cupric enzyme is low (about 1% of the native enzyme), the cuprous enzyme is distinctly active (8% of the native enzyme). The low activity of the Cu(II) enzyme has been attributed to weak binding of alcohol (65).

4.2 NMR Studies

A number of NMR investigations have previously been made with fully substituted cobalt enzyme. From such studies, distances have been calculated using the paramagnetic effects of Co(II) at the catalytic site on the relaxation rates of methyl protons of the inhibitor isobutyramide (40). The distances obtained in this way for both the binary complex with enzyme and the ternary complex with enzyme and NADH are not compatible with direct binding of the inhibitor to the metal. These results were thus interpreted in terms of indirect binding of the ligand to the metal, with a hydroxyl ion or a water ligand between the metal and the inhibitor. Similar results were obtained for the binary enzyme–ethanol complex and the abortive ternary enzyme NADH–ethanol complex (40) as well as for complexes with imidazole (66), DMSO, and trifluoroethanol (67). However, Andersson et al. (51) could detect no paramagnetic contribution from the cobalt ions to the magnetic relaxation rate of solvent water and methanol protons, despite attempts to enhance the detection of these effects. Instead they found small differences in the diamagnetic contribution to the relaxation rates of the native, demetalized, and Co(II)-substituted enzyme that change sign with magnetic field. They point out that these

small variations readily can be mistaken for true paramagnetic effects when analysis of the relaxation data is limited to the high value of magnetic field strength usually used for measurements of relaxation enhancement. They conclude that previous high field data require reinterpretation since any paramagnetic effects that may be present are small and not easily separable from a variety of small diamagnetic effects that depend on solvent composition. Furthermore, it is now known that in the work of Sloan et al. (40) the substituted Co(II) ions were at the structural and not at the catalytic site (28, 58). These factors are obvious reasons for the discrepancy between these NMR studies and studies made with other techniques including X-ray diffraction, which show direct binding of the ligands to the active-site metal.

^{113}Cd NMR studies of the Cd(II)-substituted enzyme showed a shift of the spectrum when imidazole was present (68). This shift was large enough to be consistent with first-sphere coordination. However, pyrazole and trifluoroethanol ternary complexes with enzyme and NAD^+ gave large positive chemical shifts but also the same spectrum (50). Since ^{113}Cd NMR is known to be sensitive to oxygen versus nitrogen ligation, this result was interpreted as an outer-sphere coordination. Both these ligands are known from crystallographic studies to bind directly to the catalytic zinc atom.

Perturbed angular correlation spectra have been measured on ^{111}Cd inserted specifically in the catalytic site (69). Both coenzyme, pyrazole, and trifluoroethanol affect the spectra of the incorporated Cd ion. The data for the native enzyme and for binary complexes with coenzyme were interpreted to be consistent with a four-coordinated nearly tetrahedral geometry. The authors suggest that trifluoroethanol and pyrazole enter as a fifth ligand not displacing the solvent ligand to the metal.

4.3 Resonance Raman Spectroscopy

Resonance-enhanced Raman spectra show that the same stable intermediate is formed from *p*-dimethylaminobenzyl alcohol with NAD^+ as from the corresponding aldehyde with NADH, when the concentration of substrate is low (70, 71). This intermediate showed an essentially identical resonance-enhanced Raman spectrum to that of a complex between the aldehyde and Zn^{2+} in diethyl ether. Most striking in this spectrum is the total absence of the carbonyl vibration, which indicates that the C=O bond no longer exists, most probably due to the presence of a zinc–oxygen bond.

The resonance Raman spectrum has been recorded for a binary complex between 2-carboxy-2′-hydroxy-5′-sulfonazyl benzene (zincon) and liver alcohol dehydrogenase (72). The spectra for this complex and the zinc–zincon complex are different. The zincon molecule is completely displaced from its

binary complex with the enzyme when the NAD^+-pyrazole complex is formed. This shows that the zincon molecule binds only to the active-site zinc atom.

4.4 Electron Paramagnetic Resonance Studies

The coordination environment of the catalytically active metal ion has been investigated by electron paramagnetic resonance (EPR) methods using the active-site specific Co^{2+}-reconstituted enzyme (49). It was found that the spectra are compatible with a tetracoordinate ion in the active site of the free enzyme, but a pentacoordinated metal in the binary enzyme-NAD^+ complex and in the ternary enzyme-NAD^+-pyrazole complex. These interpretations are based on comparison with structurally defined small molecule complexes.

5 ASPECTS OF THE CATALYTIC MECHANISM

5.1 Proton Abstraction

The overall oxidation of alcohol to aldehyde involves the net release of one proton. The ultimate source of this proton is alcohol. The mechanism by which the proton is transferred from alcohol to the solvent around the enzyme has attracted more interest than any other feature in the mechanism of action of the enzyme (8, 13). An alcohol like ethanol has an apparent pK of 15.7 and thus no tendency to dissociate a proton under normal conditions.

One aspect of enzymatic alcohol oxidation is that the enzyme facilitates the removal of the proton from the alcohol molecule. In lactate dehydrogenase hydride transfer and proton abstraction from the alcohol group of the substrate occur simultaneously according to the elegant "oil-water-histidine" mechanism proposed by Parker and Holbrook (73). In alcohol dehydrogenase the picture is more confused and several different mechanisms have been proposed, some of which will be discussed here. In most of these mechanisms the zinc atom at the active site has been assigned an important role. Kvassmann and Pettersson (74) have pointed out that when an alcohol molecule is bound to the zinc atom in the preformed binary complex of enzyme and oxidized coenzyme, the positive charges of the zinc atom and the nicotinamide ring in a water-free environment give the alcohol a much stronger tendency to dissociate a proton. The pK value associated with ethanol in this situation has been measured to 6.4 and to 5.4 and 4.3 for 2-chloroethanol and triflouroethanol, respectively (75). This proton release from bound alcohol takes place in the center of the enzyme molecule in a region that is closed off from solution by the conformational change of the enzyme. The proton can be picked up by a group at the active site where it is kept to the end of the reaction and

released in the open form during NADH dissociation. Alternatively the proton could be transferred by groups in the closed form of the enzyme to the surrounding solution. Since there is no proton release associated with NADH dissociation, the last mechanism seems more likely. There is in fact a possible system for such a transfer in the closed form of alcohol dehydrogenase (6). A network of hydrogen bonds which involves the oxygen atom of the alcohol, the side chain of Ser-48, the O-2 oxygen atom of the nicotinamide ribose, and the side chain of His-51 (Fig. 12) could be used for proton transfer from the alcohol bound to the active-site zinc atom to the solution outside the enzyme. The negatively charged alcoholate ion that is then produced is favorable for hydride transfer. This mechanism also allows simultaneous proton release and hydride transfer analogous to the mechanism in lactate dehydrogenase.

A number of other mechanisms for the proton transfer step have been suggested. Here we will discuss two of these. The first one was originally described by Dworschack and Plapp (76) and involves a pentacoordinated zinc atom. NAD^+ binding is associated with a pK shift to 7.6 of a group that has been assumed to be the zinc bound-water/hydroxyl (8). The substrate is proposed to bind to the zinc atom in addition to the hydroxyl ion. The proton of the alcohol is then transferred to the hydroxyl ion and the hydride ion to the nicotinamide ring of the coenzyme. As has been described previously, the space around the zinc atom is very limited with the existing geometry of the zinc ligands. Careful model building has shown (36) that it is not possible to model build a pentacoordinate zinc atom in the apoenzyme unless the ligand is a chelator. Pentacoordination is even less likely to occur in the ternary complexes, where the nicotinamide ring covers a large part of the remaining

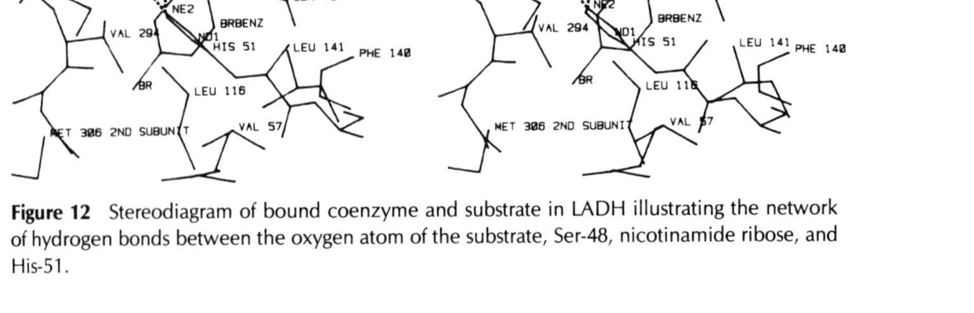

Figure 12 Stereodiagram of bound coenzyme and substrate in LADH illustrating the network of hydrogen bonds between the oxygen atom of the substrate, Ser-48, nicotinamide ribose, and His-51.

free space around the zinc atom (Table 2). Furthermore, it has been pointed out (75) that a pentacoordinated water molecule would be catalytically disadvantageous since it would tend to increase the pK of a zinc bound-alcohol substrate.

A second model for the proton transfer step was originally suggested by Theorell (77), modified by Shore et al. (78), and combined with the existing three-dimensional apoenzyme structural data by the authors (8, 79). In this model the zinc bound-water molecule is deprotonated upon binding of oxidized coenzyme. The zinc bound-hydroxyl ion then acts as a base that abstracts a proton from the alcohol. The resulting water molecule is then substituted by the alcoholate ion as ligand to zinc.

The difference between this proton release scheme and the one described above, first suggested by Kvassmann and Pettersson (74), is that water dissociation precedes alcohol binding in the latter case. In the former mechanism, which has been mainly adopted for a long time, an internal proton transfer occurs from the alcohol to the hydroxyl ion before binding of alcoholate to the zinc atom. The structure of the active site in alcohol dehydrogenase is so

Table 2 Distances Less than 5.5 Å from the Catalytic Zinc Atom in Both Forms of LADH to Nonliganded Neighbors[a]

Atoms	Open Form (Å)	Closed Form (Å)	Atoms	Open Form (Å)	Closed Form (Å)
Zn—BrBz C-1	—	3.0	Zn—Glu-68 OE	5.2	4.5
Zn—His-67 CD	3.3	3.1	Zn—Arg-369 NEH	5.2	4.9
Zn—His-67 CE	3.1	3.1			
Zn—Cys-174 CB	3.3	3.3	Zn—Phe-93 CE	5.1	4.6
Zn—Cys-46 CB	3.4	3.5	Zn—Phe-93 CZ	5.2	5.0
			Zn—Phe-93 CD-1	5.5	5.0
Zn—Ser-48 OG	3.9	4.0			
Zn—NAD C-5	—	3.9	Zn—BrBz C-3	—	5.2
			Zn—Cys-46 CA	4.8	4.9
Zn—His-67 ND	4.3	4.2	Zn—Ser-48 CB	4.5	4.7
Zn—His-67 CG	4.3	4.2	Zn—Glu-68 CB	5.4	5.0
Zn—BrBz C-2	—	4.3	Zn—Cys-174 CA	4.6	4.7
			Zn—Cys-174 C	5.2	5.0
Zn—NAD C-4	—	4.3	Zn—Gly-175 N	5.2	4.9
Zn—NAD C-6	—	4.5			

[a]The closed form is the complex with coenzyme NAD and substrate bromobenzyl alcohol, BrBz.

narrow that a prior dissociation of water seems more plausible from a structural viewpoint. Both schemes involve the formation of an alcoholate ion bound to the zinc atom.

The structural basis for the transfer of protons from the active site to the solvent was originally suggested to be composed of the system of hydrogen bonds between the oxygen of the substrate and the side chains of Ser-48 and His-51 (8). Refinement of the structures of the two different enzyme conformations has shown that this system is not present in the open form (36). The distance between the serine oxygen atom and the histidine nitrogen atom is too long for a hydrogen bond. However, in the ternary complex with bromobenzyl alcohol a system of hydrogen bonds is present (Fig. 12) which involves the same residues originally suggested, but also the O-2 atom of the nicotinamide ribose.

5.2 Hydride Transfer

No crystallographic study of alcohol dehydrogenase has demonstrated a complex where substrate and nicotinamide are positioned such that hydride ions are transferred back and forth between the two counterparts. However, model building has shown that only slight rearrangement of the observed position of bromobenzyl alcohol in the closed form of the enzyme is needed to position the alcohol such that direct hydride transfer can occur (6). A schematic diagram of this hypothetical transition state for ethanol oxidation is shown in Figure 13. The model can be built without any change of the protein structure, which shows that there is no large energy barrier to overcome. This repositioning of the substrate was done without changing the position of the substrate oxygen atom.

It is important to realize that hydride transfer occurs in a completely water-free environment. The inner part of the active site where the nicotinamide ring as well as the oxygen and C-1 atoms of the substrate are situated is inaccessible to water. The effect of the positive charges of zinc and NAD^+ to polarize the alcohol and thus facilitate hydride transfer is enhanced in a water-free environment. Furthermore, if water was present, there would be a high probability for side reactions between water and the hydride ion.

For an alcohol substrate the creation of an alcoholate ion greatly facilitates hydride transfer. One important role of the zinc atom in alcohol oxidation is therefore to stabilize the alcoholate ion for the hydride transfer step (75). Alcoholate ion formation has also been an important component in most mechanisms described for the enzyme (8, 74, 77–79). In the reverse reaction zinc plays the role of electron attractor, which gives rise to an increased electrophilic character of the aldehyde, thus facilitating the transfer of a hydride ion to the aldehyde.

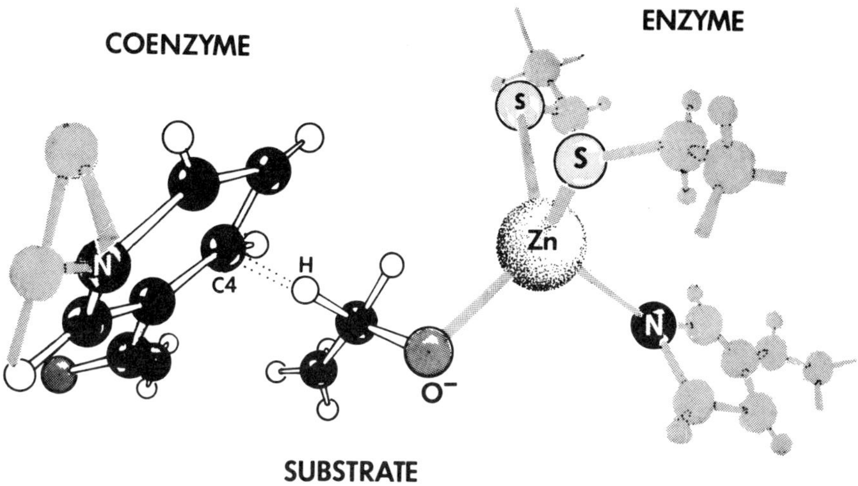

Figure 13 Schematic diagram illustrating a model of zinc, substrate, and coenzyme in LADH during hydride transfer. (We are indepted to B. Furugren for this drawing.)

Although transfer of hydrogen from the alcohol to the nicotinamide is generally described as a hydride transfer, it may occur stepwise with the transfer of a hydrogen atom and an electron or a proton plus two electrons. Such intermediate stages have been discussed generally (80) and in connection with alcohol dehydrogenase (13, 76, 81–83).

6 CONCLUSIONS ON THE ROLE OF ZINC IN LADH

The chemical reactions catalyzed by NAD^+-dependent dehydrogenases are in principle very simple reactions. An alcohol group of a metabolite is oxidized by the removal of a proton from the hydroxyl group and the transfer of a hydride ion from the reacting carbon to the oxidized form of the coenzyme. Most dehydrogenases catalyze this reaction without involving a zinc atom. In fact, even alcohol dehydrogenase from Drosophila (84), which is unrelated to the liver and yeast enzymes, requires no zinc atom to catalyze ethanol oxidation efficiently.

One important role for the active-site zinc atom in liver alcohol dehydrogenase is to bind the substrate in a suitable position relative to the coenzyme for direct hydride transfer to occur. The substrate binds by direct coordination of its oxygen atom to the zinc atom in the fourth corner of a distorted tetrahedral coordination around zinc. Water is excluded from the reactive

region of the active site and there is in all probability no water bound to zinc in addition to the substrate.

Furthermore, zinc plays an important role in the catalytic mechanism both in proton abstraction and hydride transfer. By binding to the zinc atom in its water-free environment close to the positively charged nicotinamide ring, the pK value of an alcohol group is drastically lowered to a value where proton abstraction readily occurs. In addition, zinc facilitates hydride transfer by acting as a Lewis acid. The protein ligands around zinc, two cysteine sulfur atoms, and one histidine nitrogen atom are preserved in all known zinc-containing alcohol dehydrogenases. This particular ligand arrangement must thus be important either for substrate binding or to provide a proper electronic structure for zinc in electrophilic catalysis. This ligand arrangement is quite different from that in carboxypeptidase or carbonic anhydrase.

Coenzyme binding and the associated conformational transition from the open to the closed form of the enzyme are important aspects of the catalytic mechanism. Zinc does not seem to play any role in these processes since they occur in the same way in enzyme species where the catalytically active zinc atom has been removed.

The function of the second zinc atom in LADH is still not known with certainty. It seems most likely that this zinc atom is required for structural stability of the enzyme.

ACKNOWLEDGMENTS

This work was supported by a grant from the Swedish Natural Science Research Council. The authors are grateful to Mrs. A. Rogelius for her assistance in the preparation of the manuscript.

REFERENCES

1. H. Jörnvall, *Eur. J. Biochem.*, **16,** 25 (1970)
2. C.-I. Brändén and H. Eklund in *Molecular Interactions and Activity in Proteins, Ciba Foundation Symposium 60 (new series)*, Excerpta Medica, Amsterdam, 1978, p. 63.
3. H. Eklund, B. Nordström, E. Zeppezauer, G. Söderlund, I. Ohlsson, T. Boiwe, B.-O. Söderberg, O. Tapia, C.-I. Brändén, and Å. Åkeson, *J. Mol. Biol.*, **102,** 27 (1976).
4. H. Eklund, J.-P. Samama, L. Wallén, C.-I. Brändén, Å. Åkeson, and T. A. Jones, *J. Mol. Biol.*, **146,** 561 (1981).
5. H. Eklund, J.-P. Samama, and L. Wallén, *Biochemistry*, **21,** 4858 (1982).
6. H. Eklund, J.-P. Samama, B. Plapp, and C.-I. Brändén, *J. Biol. Chem.*, **257,** 14349 (1982).

7. E. Cedergren-Zeppezauer, J.-P. Samama, and H. Eklund, *Biochemistry*, **21**, 4895 (1982).
8. C.-I. Brändén, H. Jörnvall, H. Eklund, and B. Furugren, in *The Enzymes*, Vol. 11, 3rd ed., P. D. Boyer, Ed., Academic, New York, 1975, p. 103.
9. M. G. Rossmann, A. Liljas, C.-I. Brändén, and L. J. Banaszak in *The Enzymes*, Vol. 11, 3rd ed., P. D. Boyer, Ed., Academic, New York, 1975, p. 61.
10. M. F. Dunn *Structure and Bonding*, Vol. 23, Springer, Berlin, New York, 1975, p. 61.
11. M. Hughes and R. H. Prince, *Bioorganic Chemistry*, **6**, 137 (1977).
12. W. Eventoff and M. G. Rossmann, *CRC Critical Reviews in Biochemistry*, **4**, 111 (1975).
13. J. P. Klinman, *CRC Critical Reviews in Biochemistry*, **10**, 39 (1981).
14. C.-I. Brändén and H. Eklund, in *Dehydrogenases Requiring Nicotinamide Coenzymes*, J. Jeffrey, Ed., Birkhäuser Verlag, Basel, 1980, p. 41.
15. U. Grau, in *Pyridine Nucleotide Coenzymes*, J. Everse, K. S. Andersson, and K. S. You, Eds., Academic, New York, 1981, p. 56.
16. C.-I. Brändén, in *Metalloproteins; Structure, Molecular Function and Clinical Aspects*, U. Weser, Ed., Georg Thieme Verlag Stuttgart, 1979, p. 220.
17. M. Levitt and C. Chothia, *Nature (London)*, **261**, 552 (1976).
18. C.-I. Brändén, *Quart. Rev. Biophys.*, **13**, 317 (1980).
19. I. Ohlsson, B. Nordström, and C.-I. Brändén, *J. Mol. Biol.*, **89**, 339 (1974).
20. M. G. Rossmann, D. Moras, and K. W. Olsen, *Nature (London)*, **250**, 194 (1974).
21. D. E. Drum, T.-K. Li, and B. L. Vallee, *Biochemistry*, **8**, 3792 (1969).
22. B. W. Matthews, L. H. Weaver, and W. R. Kester, *J. Biol. Chem.*, **249**, 8030 (1974).
23. H. L. Monaco, J. L. Crawford, and W. N. Lipscomb, *Proc. Natl. Acad. Sci. USA*, **75**, 5276 (1978).
24. J.-P. Samama, Dissertation, L'Université Louis Pasteur de Strasbourg, France, 1979.
25. R. C. McDonald, T. A. Steitz, and D. M. Engelman, *Biochemistry*, **18**, 338 (1979).
26. C. A. Pickover, D. M. McKay, D. M. Engelman, and T. A. Steitz, *J. Biol. Chem.*, **254**, 11323 (1979).
27. M. A. Abdallah, J.-F. Biellmann, B. Nordström, and C.-I. Brändén, *Eur. J. Biochem.*, **50**, 475 (1975).
28. W. Maret, I. Andersson, H. Dietrich, H. Schneider-Bernlöhr, R. Einarsson, and M. Zeppezauer, *Eur. J. Biochem.*, **98**, 501 (1979).
29. J. A. McCammon, B. R. Gelin, and M. Karplus, *Nature (London)*, **267**, 585 (1979).
30. S. H. Northrup, M. R. Pear, J. A. McCammon, M. Karplus, and T. Takamo, *Nature (London)*, **287**, 659 (1980).
31. H. Frauenfelder, G. A. Petsko, and D. Tsernglou, *Nature (London)*, **280**, 558 (1979).
32. P. J. Artymiuk, C. C. F. Blake, D. E. P. Grace, S. J. Oatley, D. C. Phillips, and M. J. E. Sternberg, *Nature (London)*, **280**, 563 (1979).
33. M. J. Karplus and J. A. McCammon, *CRC Critical Reviews in Biochemistry*, **9**, 293 (1981).
34. G. Schneider, H. Eklund, E. Cedergren-Zeppezauer, and M. Zeppezauer, *Proc. Natl. Acad. Sci. USA*, in press.
35. H. Eklund, J.-P. Samama, C.-I. Brändén, and T. A. Jones, unpublished results.
36. T. A. Jones and H. Eklund, unpublished results.

37. J.-P. Samama, E. Zeppezauer, J.-F. Biellmann, and C.-I. Brändén, *Eur. J. Biochem.*, **81,** 403 (1977).

38. B. V. Plapp, H. Eklund, and C.-I. Brändén, *J. Mol. Biol.*, **122,** 23 (1978).

39. M. F. Dunn, J.-F. Biellmann, and G. Branlant, *Biochemistry*, **14,** 3176 (1975).

40. D. L. Sloan, J. M., Young and A. S. Mildvan, *Biochemistry*, **14,** 1998 (1975).

41. J.-F. Biellmann, J.-P. Samama, C.-I. Brändén, and H. Eklund, *Eur. J. Biochem.*, **102,** 107 (1979).

42. C.-I. Brändén, H. Eklund, E. Zeppezauer, J.-P. Samama, and B. V. Plapp in Proceedings of FEBS 12th meeting Dresden, Vol. 52, 1978, p. 15.

43. R. Einarsson, H. Eklund, E. Zeppezauer, T. Boiwe, and C.-I. Brändén, *Eur. J. Biochem.*, **49,** 41 (1974).

44. T. Bowie and C.-I. Brändén, *Eur. J. Biochem.*, **77,** 173 (1977).

45. B.-R. Tolf, J. Piechaczek, R. Dahlbom, H. Theorell, Å. Åkeson, and G. Lundquist, *Acta Chem. Scand.*, **B33,** 483 (1979).

46. H. L. Deis and D. Lester, in *Biochemistry and Pharmacology of Ethanol*, Vol. 1, E. Mejchrowicz and P. Noble, Eds., Plenum, 1979, p. 303.

47. C. S. Lieber, *Metabolic Aspects of Alcoholism*, University Park Press, Baltimore, 1977, p. 1.

48. G. L. Lange and B. L. Vallee, *Biochemistry*, **15,** 4681 (1976).

49. M. W. Makinen and M. B. Yim, *Proc. Natl. Acad. Sci. USA*, **78,** 6221 (1981).

50. R. B. Bobsein and J. R. Myers, *J. Biol. Chem.*, **256,** 5313 (1981).

51. I. Andersson, W. Maret, M. Zeppezauer, R. D. Brown III, and S. H. Koenig, *Biochemistry*, **20,** 3424 (1981).

52. B. H. Bürgi, J. D. Dunitz, and E. Shefter, *J. Am. Chem. Soc.*, **95,** 5065 (1973).

53. M. N. G. James, A. R. Sielecki, G. D. Brayer, L. T. J. Delbaere, and C.-A. Bauer, *J. Mol. Biol.*, **144,** 43 (1980).

54. H. Theorell, T. Yonetani, and B. Sjöberg, *Acta Chem. Scand.*, **23,** 255 (1969).

55. C.-I. Brändén, in *Pyridine Nucleotide-dependent Dehydrogenases*, H. Sund, Ed., Walter de Gruyter, Berlin, New York, 1977, p. 325.

56. J. F. Chlebowski and J. E. Coleman, in *Metal Ions in Biological Systems*, Vol. 6, H. Sigel, Ed., Dekker, New York, 1976, p. A1.

57. L. Skjeldal, K. H. Dahl, and J. S. McKinley-McKee, *FEBS Letters*, **137,** 257 (1982).

58. I. Andersson, W. Maret, M. Zeppezauer, R. D. Brown III, and S. H. Koenig, *Biochemistry*, **20,** 3433 (1981).

59. H. Dietrich, W. Maret, H. Kozlowski, and M. Zeppezauer, *J. Inorg. Biochemistry*, **14,** 297 (1981).

60. W. Maret, Ph.D. Thesis, Saarbrücken University, Saarland, FRG, 1980.

61. H. Dietrich, W. Maret, L. Wallén, and M. Zeppezauer, *Eur J. Biochem.*, **100,** 267 (1979).

62. M. F. Dunn, H. Dietrich, A. K. H. MacGibbon, S. C. Koerber, and M. Zeppezauer, *Biochemistry*, **21,** 354 (1982).

63. K. H. Dahl and J. S. McKinley-McKey, *J. Inorg. Biochem.*, **15,** 79 (1981).

64. W. Maret, H. Dietrich, H.-H. Ruf, and M. Zeppezauer, *J. Inorg. Biochem.*, **12,** 241 (1980).

65. I. Andersson, D. R. Burton, H. Dietrich, W. Maret, and M. Zeppezauer, in *Metalloproteins* U. Weser, Ed., Georg Thieme Verlag, Stuttgart, 1979, p. 246.

66. J. M. Young and A. S. Mildvan in *Alcohol and Aldehyde Metabolizing Systems*, Vol. 2, R. G. Thurman, J. R. Williamson, H. R. Drott and B. Chance, Eds., Academic, New York, 1977, p. 109.

67. B.-E. Drysdale and D. P. Hollis, *Arch. Biochem. Biophys.*, **205,** 267 (1980).

68. B. R. Bobsein and R. J. Myers, *J. Am. Chem. Soc.*, **102,** 2454 (1980).

69. I. Andersson, R. Bauer, and I. Demeter, *Inorganica Chimica Acta.*, **67,** 53 (1982).

70. P. W. Jagodzinky and W. L. Peticolas, *J. Am. Chem. Soc.*, **103,** 234 (1981).

71. P. W. Jagodzinky, G. F. Funk, and W. L. Peticolas, *Biochemistry*, **21,** 2193 (1982).

72. J. T. McFarland, K. L. Watters, and R. L. Petersen, *Biochemistry*, **14,** 624 (1975).

73. D. M. Parker and J. J. Holbrook, in *Pyridine Nucleotide Dependent Dehydrogenases*, H. Sund, Ed., Walter de Gruyter, Berlin, 1977, p. 485.

74. J. Kvassmann and G. Pettersson, *Eur. J. Biochem.*, **103,** 557 (1980).

75. J. Kvassmann, A. Larsson, and G. Pettersson, *Eur. J. Biochem.*, **114,** 555 (1981).

76. R. T. Dworschack and B. V. Plapp, *Biochemistry*, **16,** 2716 (1977).

77. H. Theorell, *Harvey Lect.*, Series 61, Academic, New York, 1967, p. 17.

78. J. D. Shore, H. Gutfreund, R. L. Brooks, D. Santiago, and P. Santiago, *Biochemistry*, **13,** 4185 (1974).

79. H. Eklund, B. Nordström, E. Zeppezauer, G. Söderlund, I. Ohlsson, T. Boiwe, and C.-I. Brändén, *FEBS Letters*, **44,** 200 (1974).

80. G. A. Hamilton, in *Progress in Bioorganic Chemistry*, Vol. 1, E. T. Kaiser and F. J. Kzedy, Eds., Wiley-Interscience, New York, 1971, p. 83.

81. J. P. Klinman, Biochemistry, **15,** 2018 (1976).

82. K. M. Welsh, D. J. Creighton, and J. P. Klinman, *Biochemistry*, **18,** 2005 (1980).

83. P. F. Cook, N. J. Oppenheimer, and W. W. Cleland, *Biochemistry*, **20,** 1817 (1981).

84. D. R. Thatcher, *Biochem. J.*, **187,** 875 (1980).

85. G. Schneider, H. Eklund, E. Cedergren-Zeppezauer, and M. Zeppezauer, *EMBO Journal*, in press (1983).

86. H. Theorell, and T. Yonetani, *Biochem. Z.*, **338,** 537 (1963).

87. G. Schneider, and M. Zeppezauer, *J. Inorg. Biochem.*, **18,** 59 (1983).

CHAPTER 5

Molecular Properties and Mechanism of Alkaline Phosphatase*

JOSEPH E. COLEMAN and PETER GETTINS

Department of Molecular Biophysics and Biochemistry
Yale University
New Haven, Connecticut

*Original work from the authors' laboratories was supported by grant AM09070 from the National Institutes of Health and grant PCM 76-82231 from the National Science Foundation. The 200 MHz NMR facility is partially supported by grants PCM 77-18941 and CHE 79-16210 from the National Science Foundation.

CONTENTS

1 INTRODUCTION, 156

2 SYNTHESIS OF THE ALKALINE PHOSPHATASE MONOMER AS A PREENZYME, 156

3 PRIMARY STRUCTURE OF E. COLI ALKALINE PHOSPHATASE, 157

4 ALKALINE PHOSPHATASES FROM VARIOUS SPECIES, 157

5 ENZYMATIC REACTION AND KINETICS, 160

- 5.1 Phosphoenzyme Intermediates, 161
- 5.2 Transphosphorylation, 163
- 5.3 Enzymatic Activity of Mn(II), Co(II), Cu(II), and Cd(II) Alkaline Phosphatases, 164

6 METAL ION BINDING: Mn(II), Co(II), AND Cd(II) METALLOALKALINE PHOSPHATASES, 166

- 6.1 ^{115}Cd Binding to Apoalkaline Phosphatase as Functions of Phosphate, Magnesium, pH, and Cd(II) Concentration, 167
- 6.2 ^{65}Zn Binding to Apoalkaline Phosphatase as Functions of Phosphate, Mg, pH, and Zn(II) Concentration, 169
- 6.3 Summary of Metal Ion Binding Properties, 173
- 6.4 Binding and Functional Effects of Mg(II), 174

7 CRYSTAL STRUCTURE OF ALKALINE PHOSPHATASE, 175

- 7.1 Molecular Boundaries, 176
- 7.2 Metal-Binding Sites, 177
- 7.3 Three Potential Metal-Binding Sites at the Active Centers of Alkaline Phosphatase, 178

8 OPTICAL SPECTROSCOPY OF ACTIVE-SITE METAL IONS, 180

9 ELECTRON SPIN RESONANCE OF Mn(II), Co(II), and Cu(II) ALKALINE PHOSPHATASES, 182

10 ^{113}Cd AND ^{111}Cd NMR OF Cd(II) ALKALINE PHOSPHATASE, 185

- 10.1 Phosphorylation of ^{113}Cd(II)$_2$ Alkaline Phosphatase, 188
- 10.2 Slow Equilibria Associated with Phosphorylation of Cd(II)$_2$ Alkaline Phosphatase, 192

11 ^{13}C, ^{19}F, AND ^{35}Cl NMR OF ALKALINE PHOSPHATASE, 194

- 11.1 ^{13}C NMR Studies of $\beta,\beta-[\gamma-{}^{13}$C]Dideuterohistidine-labeled Alkaline Phosphatase., 194
- 11.2 ^{19}F NMR of Alkaline Phosphatase Containing Fluorinated Aromatic Amino Acids, 196
- 11.3 ^{35}Cl NMR of Alkaline Phosphatase, 198

12 ^{31}P NMR OF PHOSPHOENZYME INTERMEDIATES, 199

12.1 ^{31}P NMR Spectra of Homogeneous Zn(II)-Containing Alkaline Phosphatase Species, 200
12.2 Homogeneous Cd(II)-Substituted Alkaline Phosphatase, 203
12.3 E-P ⇌ E·P Equilibrium, 203
12.4 Factors Affecting the Chemical Shift of E·P and E-P, 205
12.5 ^{113}Cd—O—^{31}P Coupling in E·P, 208
12.6 Stoichiometry of Phosphate Binding, 208
12.7 The E-P ⇌ E·P ⇌ P_i Equilibrium, 209
12.8 Dissociation of E·P, 210
12.9 Subunit–Subunit Interactions, 211
12.10 Mechanistic Implications, 212

REFERENCES, 213

1 INTRODUCTION

The dimeric Zn(II) metalloenzyme alkaline phosphatase from *Escherichia coli* was first described in 1933 (1). Following the demonstration by Horiuchi et al. (2) in 1959 and Torriani (3) in 1960 that the *E. coli* enzyme was derepressed by phosphate starvation, investigations of the enzyme have continually expanded. The enzyme was demonstrated to be a Zn(II) metalloenzyme by Plocke and coworkers (4, 5) in 1962. While the first "alkaline phosphatase" to be studied in any detail was the enzyme from calf intestine (see below), the enzyme from *E. coli* has become the prototype for enzymes of this class and will be the object of most of the discussion in this chapter. All alkaline phosphatases isolated thus far share the basic properties of the *E. coli* enzyme. The purpose of this chapter is to cover the newest information on the structure and mechanism of this enzyme much of which clarifies some of the controversial features of this enzyme such as the metal ion stoichiometry required for activity, the stoichiometry of phosphointermediates, the molecular weight, and the presence or absence of the phenomenon of negative cooperativity between subunits, that is, half-the-sites reactivity. The literature prior to 1979 has been covered in previous reviews (6, 7) and an exhaustive compilation of the data on alkaline phosphatase prior to 1979 has been published in book form (8). Hence the literature prior to 1979 will not be exhaustively reviewed except as it specifically relates to current topics.

2 SYNTHESIS OF THE ALKALINE PHOSPHATASE MONOMER AS A PREENZYME

Alkaline phosphatase has been known since the late 1960s to be among the enzymes located in the periplasmic space of *E. coli*, since the enzyme is released on spheroplast formation or on osmotic shock of *E. coli* cells (9, 10). It now appears that the *E. coli* enzyme is synthesized on polysomes bound to the inner membrane and initially carries an N-terminal extension or signal peptide of hydrophobic character that enables the protein to traverse the inner membrane to the periplasmic space (11, 12). The molecular weight of the preenzyme on SDS gels is ~15% greater than that of the mature enzyme and it was initially believed that the processing peptidase was located primarily in the outer membrane (11). More recently, work has shown that the true processing peptidase that gives rise to the normal N terminal (probably exclusively the Arg isozyme) of alkaline phosphatase appears to be associated primarily with the inner membrane fraction (13).

3 PRIMARY STRUCTURE OF *E. COLI* ALKALINE PHOSPHATASE

The complete amino acid sequence of the monomer of alkaline phosphatase as determined by Bradshaw et al. (14) is given in Figure 1. The monomer is a single unglycosylated polypeptide chain of 449 amino acid residues. This sequence results in a calculated molecular weight of 47,029 or a native dimer molecular weight of 94,058. Isozyme 1 has an additional arginine residue at the amino terminus which presumably reflects some variation in the processing of the preenzyme, although the Thr N terminus of isozyme 3 may result from a further proteolytic cleavage after the primary processing event (12). There are no free SH groups, since all four cysteine residues are involved in the two intrachain disulfide bridges. The positions of the ten histidyl residues of the monomer are indicated in Figure 1, since several of these are ligands to the metal ions at the active center (see below). A secondary structure prediction by the method of Chou and Fasman (15) applied by Bradshaw et al. (14) predicts that this sequence will form approximately 45% α-helix and 13 and 16%, respectively, of β-sheet and β-turn conformations. The highest estimate of α-helical content determined from an analysis of circular dichroism data has been $\sim$40% α-helix (16). The crystal structure of the enzyme shows each monomer to consist of a large core of β-pleated sheet consisting of six strands sandwiched between stretches of α-helix. The helical segments of the α/β core represents 27% of the 449 residues, while the β strands represent $\sim$7% of the total residues (17). Much of the rest of the molecule is of irregular structure and precise evaluation of the number of additional helical residues outside the core will require further refinement.

4. ALKALINE PHOSPHATASES FROM VARIOUS SPECIES

Data on the molecular properties of the best characterized preparations of alkaline phosphatase from a variety of mammalian sources and several microorganisms are presented in Table 1. All of the highly purified enzymes have proved to be Zn(II) metalloenzymes, dimers of identical monomers. The mammalian enzymes are all significantly larger than their bacterial counterparts and in all cases studied appear to be rather closely associated with the membranes of the cell. In the case of bone, the enzyme is found in the membrane of the osteoblast as well as the membranes of the matrix vesicles formed from these cells. The latter vesicles are the sites of nucleation of the hydroxyapatite crystals (29–32). In the case of the matrix vesicles alkaline phosphatase appears to account for the ATPase and pyrophosphatase as well as the classical

10 20

Thr - Pro - Glu - Met - Pro - Val - Leu - Glu - Asn - Arg - Ala - Ala - Gln - Gly - Asn - Ile - Thr - Ala - Pro - Gly-

30 40

Gly - Ala - Arg - Arg - Leu - Thr - Gly - Asp - Gln - Thr - Ala - Ala - Leu - Arg - Asn - Ser - Leu - Ser - Asp - Lys-

50 60

Pro - Ala - Lys - Asn - Ile - Ile - Leu - Leu - Ile - Gly - Asp - Gly - Met - Gly - Asp - Ser - Glu - Ile - Thr - Ala-

70 80

Ala - Arg - Asn - Tyr - Ala - Glu - Gly - Ala - Gly - Gly - Phe - Phe - Lys - Gly - Ile - Asp - Ala - Leu - Pro - Leu-

90 100

Thr - Gly - Gln - Tyr - Thr - His - Tyr - Ala - Leu - Asn - Lys - Lys - Thr - Gly - Lys - Pro - Asp - Tyr - Val - Thr-

110 120

Asp - Ser - Ala - Ala - Ser - Ala - Thr - Ala - Trp - Ser - Thr - Gly - Val - Lys - Thr - Tyr - Asn - Gly - Ala - Leu-

130 140

Gly - Val - Asp - Ile - His - Glu - Lys - Asp - His - Pro - Thr - Ile - Leu - Glu - Met - Ala - Lys - Ala - Ala - Gly-

150 160

Leu - Ala - Thr - Gly - Asn - Val - Ser - Thr - Ala - Glu - Leu - Gln - Asp - Ala - Thr - Pro - Ala - Ala - Leu - Val-

170 180

Ala - His - Val - Thr - Ser - Arg - Lys - Cys - Tyr - Gly - Pro - Ser - Ala - Thr - Ser - Gln - Lys - Cys - Pro - Gly-

190 200

Asn - Ala - Leu - Glu - Lys - Gly - Gly - Lys - Gly - Ser - Ile - Thr - Glu - Gln - Leu - Leu - Asn - Ala - Arg - Ala-

210 220

Asp - Val - Thr - Leu - Gly - Gly - Gly - Ala - Lys - Thr - Phe - Ala - Glu - Thr - Ala - Thr - Ala - Gly - Glu - Trp-

230 240

Gln - Gly - Lys - Thr - Leu - Arg - Glu - Gln - Ala - Gln - Ala - Arg - Gly - Tyr - Gln - Leu - Val - Ser - Asp - Ala-

250 260

Ala - Ser - Leu - Asn - Ser - Val - Thr - Glu - Ala - Asn - Gln - Gln - Lys - Pro - Leu - Leu - Gly - Leu - Phe - Ala-

270 280

Asp - Gly - Asn - Met - Pro - Val - Arg - Trp - Leu - Gly - Pro - Lys - Ala - Thr - Tyr - His - Gly - Asn - Ile - Asp-

290 300

Lys - Pro - Ala - Val - Thr - Cys - Thr - Pro - Asn - Pro - Gln - Arg - Asn - Asp - Ser - Val - Pro - Thr - Leu - Ala-

310 320

Gln - Met - Thr - Asp - Lys - Ala - Ile - Glu - Leu - Leu - Ser - Lys - Asn - Glu - Lys - Gly - Phe - Phe - Leu - Gln-

330 340

Val - Glu - Gly - Ala - Ser - Ile - Asp - Lys - Gln - Asp - His - Ala - Ala - Asn - Pro - Cys - Gly - Gln - Ile - Gly-

350 360

Glu - Thr - Val - Asp - Leu - Asp - Glu - Ala - Val - Gln - Arg - Ala - Leu - Glu - Phe - Ala - Lys - Lys - Glu - Gly-

370 380

Asn - Thr - Leu - Val - Ile - Val - Thr - Ala - Asp - His - Ala - His - Ala - Ser - Gln - Ile - Val - Ala - Pro - Asp-

390 400

Thr - Lys - Ala - Pro - Gly - Leu - Thr - Gln - Ala - Leu - Asn - Thr - Lys - Asp - Gly - Ala - Val - Met - Val - Met-

410 420

Ser - Tyr - Gly - Asn - Ser - Glu - Glu - Asp - Ser - Gln - Glu - His - Thr - Gly - Ser - Gln - Leu - Arg - Ile - Ala-

430 440

Ala - Tyr - Gly - Pro - His - Ala - Ala - Asn - Val - Val - Gly - Leu - Thr - Asp - Gln - Thr - Asp - Leu - Phe - Tyr-

449

Thr-Met-Lys-Ala-Ala-Leu-Gly-Leu-Lys

Figure 1. Amino acid sequence of *E. coli* alkaline phosphatase (isozyme 3). Data from ref. 14.

Table 1 Alkaline Phosphatases from Eukaryotes and Prokaryotes

Source	Molecular Weight	Zn Content	Reference
Eukaryotes			
Human placenta	125,000	2–3 Zn/dimer	18,19
Human liver	136,000	*n* Zn	20
Bovine bone	216,000	*n* Zn	21
Human bone	~200,000	*n* Zn	22
Bovine intentine	138,000	4 Zn/dimer	23
Bovine kidney	172,000	4 Zn/dimer	24
Rat liver	154,000	*n* Zn/dimer	25
Prokaryotes			
E. coli	94,000	2 or 4 Zn/dimer	7,14
Bacillus subtilis	100,000	2 Zn/dimer	26
B. lichenformis	121,000	*n* Zn	27
Micrococcus sodonensis	80,000	*n* Zn, 8 Ca	28

alkaline phosphatase activity associated with these membranous structures (21, 31). In fact, this membrane-attached alkaline phosphatase may be the major enzyme associated with the vesicles (30). Prior to disintegrating and depositing their contained hydroxyapatite crystals, the matrix vesicles accumulate high concentrations of Ca^{2+} (98m*M*) and inorganic phosphate (50m*M*) (30). A number of phosphate monoesters will serve as the source of the inorganic phosphate when incubated with the isolated vesicles *in vitro* (30).

The alkaline phosphatase of mammalian intestine is associated with the cells of the brush border and is isolated along with membrane fragments (33). The enzyme is a glycoprotein of molecular weight 138,000 (69,000 subunits) (23, 36). This was the first alkaline phosphatase to be extensively investigated and Engström first demonstrated the formation of the phosphoseryl intermediate (E-P) with the calf intestinal enzyme (34, 35). The amino acid sequence around the phosphoseryl residue was shown to be Asp—Ser P—Ala (36). Later the *E coli* alkaline phosphatase was shown to contain the same sequence around the phosphorylatable seryl residue (37), thus supporting the conservation of the alkaline phosphatase structure and mechanism over a long period of evolution. It has been suggested that the intestinal enzyme may participate in Ca^{2+} transport by the intestinal cells (37–39). If the membrane-associated alkaline phosphatase is responsible for the Ca^{2+}-ATPase activity of the brush border (an activity not inhibited by the usual inhibitors of Na^+-K^+-ATPase), then the above hypothesis may be possible. Recent evidence suggests that Ca^{2+} transport in the intestine may be primarily by facilitated diffusion without a major contribution from active transport (38, 40).

5 ENZYMATIC REACTION AND KINETICS

A minimal reaction path for this enzyme is pictured in Scheme 1.

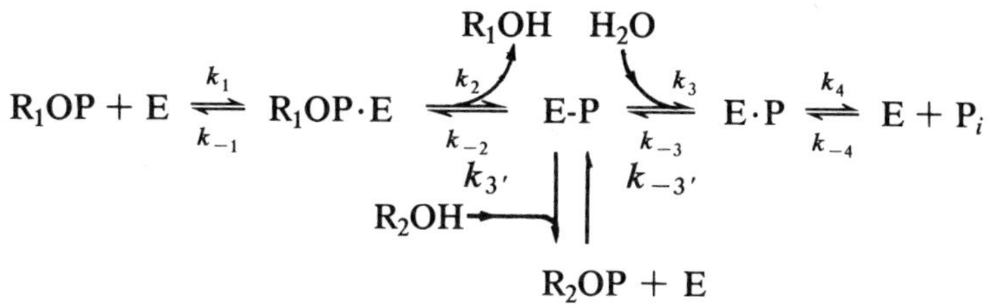

Scheme 1

Alkaline phosphatase is a nonspecific phosphomonoesterase in the sense that it will hydrolyze phosphate monoesters (ROP) at similar rates regardless of the size or chemical nature of R (6). This suggests that there is no specific binding pocket for the alcoholic or phenolic portion of the substrate. Likewise there is little change in the enzymatic rate as the pK_a of the leaving group is varied over six orders of magnitude, for example, from 2,4-dinitrophenylphosphate to methyl phosphate. This is readily explained, since it is now known that at acid pH dephosphorylation of the serylphosphate is rate limiting (6), while at alkaline pH, dissociation of the product phosphate from the noncovalent complex, E·P, is rate limiting (6, 41, 42) (see below). Hence the slow step always follows ejection of alcohol or phenol. There is indication from stopped-flow data at acid pH that if phosphorylation rates, which are on the millisecond time scale, could be conveniently measured for a wide range of pK_a values of the leaving group, a large dependency of the phosphorylation rate on pK_a would be observed. Significant differences in phosphorylation rate, k_2, appear to occur for even moderate changes in pK_a of the leaving group; for example, *p*-nitrophenylphosphate compared to 2,4-dinitrophenylphosphate (43). Since it is difficult or impossible to secure substrates absolutely free of product inorganic phosphate, P_i, it is also evident that true phosphorylation rates calculated from the pre-steady-state burst require measurements at several concentrations of added phosphate followed by extrapolation to zero phosphate concentration (43).

The enzyme hydrolyzes P—S bonds just as rapidly as P—O (44), but the O-phosphorothioates (ROPS) are hydrolyzed $\sim 10^3$ times more slowly, supporting a nucleophilic mechanism rather than a metaphosphate dissociative-type reaction (45–47). P—N bonds are cleaved only slightly less rapidly than P—O bonds (48). P—C bonds are not cleaved and the phosphonate dianions are competitive inhibitors, although they bind considerably less tightly than the phosphate dianion, HPO_4^{2-} (6, 8).

5.1 Phosphoenzyme Intermediates

An idealized pH-rate profile for the hydrolysis of *p*-nitrophenylphosphate or *p*-phenylazophenylphosphate (45) is shown in Figure 2. The reaction rate profile is sigmoid to alkaline pH with an apparent pK_a of ~7.5. The curve drawn in Figure 2 is that for a single proton dissociation. This adequately describes presently available experimental profiles, but the precision of activity data on the enzyme is not sufficient to detect moderate deviations from this idealized curve. The maximum velocity at pH 9.0 is ~70 sec^{-1}, although considerable variation is observed between various preparations that may reflect Zn(II) and Mg(II) concentrations as well as enzyme purity and phosphate concentrations (see below).

Only at acid pH are large equilibrium concentrations of E-P observed, and for the $Zn(II)_4AP$ or $Zn(II)_2Mg(II)_2AP$ [E-P] = [E·P] at pH 5.0, where k_3 (dephosphorylation) = k_{-3} (phosphorylation from E·P). In the hydrolysis of ROP at acid pH dephosphorylation of E-P is rate limiting, $k_3 = \sim 0.7\ sec^{-1}$, and burst kinetics are observed (45, 49). Stopped-flow studies can be pursued below pH 5.0, but equilibrium studies are difficult since the metal ions begin to dissociate (50). While phosphate is often considered an inhibitor of the enzyme with a kinetically determined $K_i \cong 10^{-6}M$ (very similar to K_m for ROP), the enzyme readily forms E-P from P_i in a slow pH-independent phosphorylation, $k_{-3} = \sim 0.7\ sec^{-1}$, where the phosphorylation step is rate limiting and at least 10^3 times slower than phosphorylation by ROP, k_2 (45).

As the pH is raised, equilibrium concentrations of E-P decline and by pH 7 no significant E-P is observed and the species present at equilibrium is E·P, the noncovalent Michaelis complex with P_i. The curve for this equilibrium is

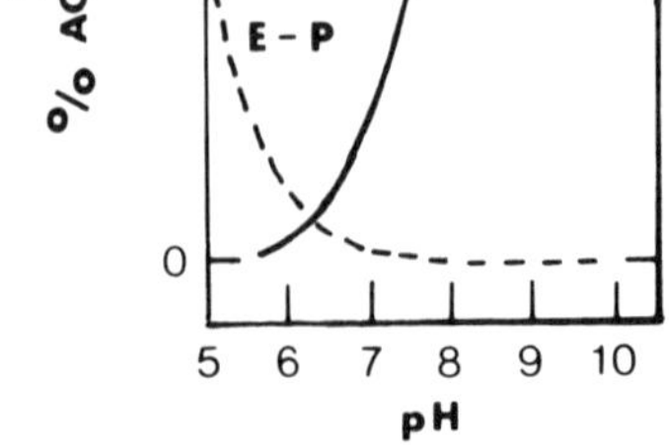

Figure 2 pH profile for the hydrolysis of *p*-nitrophenyl phosphate by *E. coli* alkaline phosphatase. Data from ref. 45.

also given in Figure 2. The latter function simply reflects the pH dependency of the ratio of two rate constants, k_3/k_{-3} (i.e., dephosphorylation of E-P/phosphorylation from E·P). This function does not in general represent the actual pK_a of the proton dissociation controlling this ratio which may be several pH units removed from the pH at which the ratio of rate constants happens to equal one. The pK_a governing the k_3/k_{-3} ratio in alkaline phosphatase may be the same one that determines the pH-rate profile with an apparent pK_a of 7.5–8.0 (Fig. 2). The E-P and E·P concentrations shown are taken from ^{31}P NMR spectra of enzyme solutions phosphorylated with P_i and the concentrations of the two intermediates determined from the amplitudes of their characteristic ^{31}P NMR signals (see below) (51). These functions are identical within the error of measurements to the equilibrium data determined by ^{32}P labeling or by kinetic means (52, 53).

While no significant E-P is observed at alkaline pH where the enzyme is maximally active, the enzyme does catalyze the exchange of ^{18}O from solvent $H_2{}^{18}O$ into P_i (52–54, 55). Knowles (56) has also prepared [^{16}O, ^{17}O, ^{18}O] ROP chiral phosphate and shown that hydrolysis by alkaline phosphatase proceeds with retention of configuration around the phosphorus, which very strongly implies a mechanism consisting of two sequential in line nucleophilic displacements involving a covalent intermediate, clearly the phosphoseryl residue.

While the burst kinetics at acid pH suggest rate-limiting dephosphorylation, the classic rapid kinetic studies of alkaline phosphatase have shown this burst to disappear at alkaline pH (45, 49). Hence it was thought that phosphorylation must become at least partially rate limiting at alkaline pH (45, 57). For S-phosphorothioates, however, the observation of a burst at alkaline pH, despite the absence of significant covalent intermediate, suggested that dissociation of thiophosphate was rate limiting (45). With the realization from ^{31}P NMR studies (see below) as well as ^{32}P-labeling studies (58), that dissociation of the product, phosphate, from E·P is rate limiting at alkaline pH for oxyphosphates as well, the disappearance of burst kinetics receives another explanation. As can easily be documented by checking the ^{31}P NMR spectrum, most standard preparations of alkaline phosphatase, unless the apoenzyme has been an intermediate, are contaminated with at least 1 mol/mol of phosphate. Since most stock enzyme solutions are above pH 7, this is present as E·P. Phosphorylation of the enzyme from E·P is so slow compared to that from E·ROP that rapid mixing of this species with substrate at pH 5.0 does not affect burst magnitude. E-P is formed from ROP and the effect of E·P is simply to inhibit the burst rate as has been documented by this laboratory (43). On the other hand, at alkaline pH, where dissociation of E·P is rate limiting, the prior presence of E·P can and does abolish the burst expected from rate-limiting dissociation of E·P. If phosphate-free enzyme is used, typical burst kinetics are observed at alkaline pH (59). The magnitude of these bursts and

the stoichiometry of the phosphoenzyme intermediates will not be discussed here but in a special section on negative cooperativity at the end of the chapter.

Two additional reactions catalyzed by the enzyme should be mentioned, thiophosphate hydrolysis and the phosphate transfer reactions. The enzyme catalyzes the formation of H_2S from the thiophosphate as shown in Scheme 2, where E-P is the normal phosphoseryl group (45).

$$HPSO_3^{2-} + E \rightleftharpoons HPSO_3^{2-}\cdot E \xrightleftharpoons{{}^{-}SH} E\text{-}P \xrightleftharpoons{H_2O} E\cdot P \rightleftharpoons E + P_i$$

Scheme 2

Not unexpectedly, since ^{-}SH can leave more readily than ^{-}OH, the preferred binding mode for thiophosphate places sulfur in the position of the leaving group and the seryl oxyphosphate is formed. This reaction is more rapid than phosphorylation of Ser-102 by E·P and the k_{cat} for H_2S release is two-fold higher, although still pH independent, suggesting that phosphorylation remains rate limiting (45).

On the other hand, in the case of the O-phosphorothioates, the first preferred leaving group is RO^- and the seryl O-phosphorothioate is the first covalent intermediate formed as shown by the incorporation of ^{35}S into the enzyme (45) (Scheme 3):

$$ROPSO_2^{2-}\cdot E \xrightleftharpoons{RO^-} E\text{-}OPSO_2^{2-} \xrightleftharpoons{H_2O} E\cdot HPSO_3^{2-} \rightleftharpoons E$$

$$+ HPSO_3^{2-} \rightleftharpoons E\cdot HSPO_3^{2-} \xrightleftharpoons{{}^{-}SH} E\text{-}P \xrightleftharpoons{H_2O} E\cdot P \rightleftharpoons E + P_i$$

Scheme 3

The first product, thiophosphate, is then hydrolyzed to H_2S and P_i.

5.2 Transphosphorylation

If an alcohol, R_2OH, is added to the reaction mixture in relatively high concentration (1*M*), the enzyme will transfer the phosphate from the substrate, R_1OP, to the acceptor to form the product R_2OP (48). Amino alcohols carrying the charged amino group on the carbon adjacent to that carrying the OH acceptor are the most effective; for example, ethanolamine and Tris buffer tris(hydroxymethyl)aminomethane (6). If a chromophoric assay of R_1O^- release is employed, 1*M* Tris will stimulate the apparent reaction rate of oxyphosphate monoesters by three to fourfold at pH 8.0. If the products are determined, the increased release of RO^- is found to be almost entirely due to the additional phosphotransferase reaction with product ratios, $R_1OH{:}P_i$, of 3:1

(48). A particularly striking result is the finding that the apparent rate of reaction of O-phosphorothioates is stimulated 15-20-fold by 1*M* Tris associated with a product ratio, R_1OH: (phosphate + thiophosphate) of 60:1 (47). This implies that the transphosphorylation ($TrisOPSO_3^{2-}$ product) is favored almost 60:1 over hydrolysis when the covalent intermediate is the seryl-O-phosphorothioate. This finding needs more investigation. Since the rate-limiting step for the normal hydrolysis reaction is now known to be dissociation of E·P, it appears likely that dissociation of E·R_2OP must be more rapid. No saturation of the reaction with the competing nucleophile has been observed; hence it is not clear whether a specific binding site is involved. It may be that the acceptor alcohol competes for the solvent (possibly HO^-) position in the hydrolysis of E-P.

5.3 Enzymatic Activity of Mn(II), Co(II), Cu(II), and Cd(II) Alkaline Phosphatases

While the above transition and IIB metal ion substitutions at the Zn(II) binding sites of alkaline phosphatase have all been characterized spectroscopically by absorption (60, 62), circular dichroism (CD) (63), EPR (64–66), or NMR spectroscopy (67), only the Co(II) derivative shows significant enzymatic activity [~30% that of the Zn(II) enzyme] as characterized by the complete turnover of *p*-nitrophenylphosphate (7, 62). The others, however, bind phosphate and will phosphorylate seryl residue 102 from the E·P complex (52). The pH dependency of the equilibrium formation of E-P by three of these [Mn(II), Co(II), and Cd(II)] has been determined and is pictured in Figure 3 and compared to the same function for the Zn(II) enzyme. Data on the apoenzyme are also included which illustrate the absolute dependence of the binding of phosphate and the phosphorylation of the seryl residue on the presence of the metal ion at the active center. One of the reasons for the relative inactivity of several of the metal-substituted enzymes may be either the very slow dephosphorylation of E-P or the slow dissociation of E·P. In the case of the Cd(II) enzyme, dephosphorylation clearly remains very slow until ~pH 9, since E-P remains the predominant species at equilibrium until well above pH 8 (Fig. 3) (see ^{113}Cd NMR below). This appears to imply that the pK_a of the group controlling the k_3/k_{-3} ratio and probably activity as well has shifted 2–3 pH units to the alkaline. It could well be that the Cd(II) enzyme is active above pH 9.0 where E·P is the predominant species. This has been observed for carbonic anhydrase where the Cd(II) derivative does become an active esterase at pH 9.5 and above (68, 69). This is difficult to test in the case of alkaline phosphatase because of the rapid hydroxide-catalyzed hydrolysis of *p*-nitrophenylphosphate and the dissociation of Cd(II) from the enzyme. The pK_a of a cooordinated solvent molecule at the open site on a mixed H_2O–metal–protein

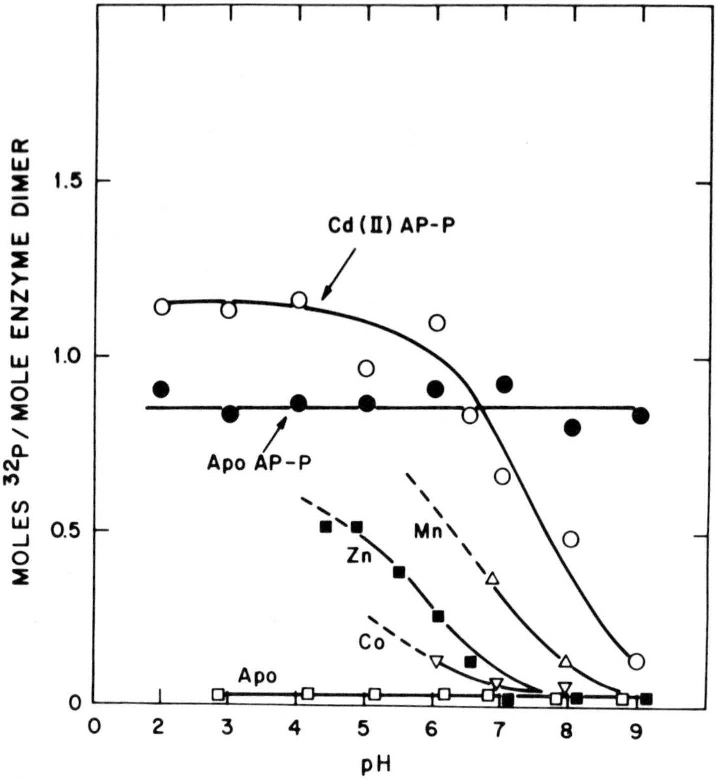

Figure 3 Equilibrium concentrations of E-P vs. pH for Mn(II) (△), Co(II) (▽), Zn(II) (■), Cd(ii) (○), and Apo (□) alkaline phosphatases. Apo AP-P (●) refers to phosphorylated apoenzyme prepared from Cd(II) AP-P by removal of Cd(II) by chelating agents. From ref. 52.

complex can be very dependent on the metal ion species and suggests that the metal-dependent pK_a may be that of a metal-coordinated solvent molecule which is shifted to the alkaline by the Cd(II) substitution (see Sec. 12). It is more difficult to see how the substitution of Cd(II) for Zn(II) could shift the pK_a of an amino acid side chain by ~3 pH units.

That the metal ion species present at the active center has a marked effect on the rate constants characterizing the E-P $\rightleftharpoons$ E·P $\rightleftharpoons$ E + P_i interconversions has been strikingly demonstrated by Bock and Cohn (70) who measured ^{18}O exchange out of inorganic phosphate into solvent as catalyzed by the Zn(II) and Co(II) enzymes. Their method uses the isotope chemical shift induced in the ^{31}P NMR signal by one, two, three, or four adjacent ^{18}O atoms which results in an NMR multiplet for ^{18}O-enriched phosphate. As the enzyme catalyzes the incorporation of ^{16}O from solvent via the formation and subsequent

hydrolysis of E-P, the relative amplitudes of the lines of the multiplet as a function of time will indicate whether phosphate binds to the enzyme and undergoes one turnover and then is lost to bulk phosphate or whether once bound it will undergo more than one phosphorylation–dephosphorylation (i.e., ^{16}O exchange) before dissociating. The Zn(II) enzyme catalyzes a single turnover for each binding step, thus producing a statistical loss of ^{18}O. On the other hand, Co(II) at the active center results in several oxygen exchanges every time E·P is formed. Thus Co(II) has significantly altered the probability of phosphorylation of the serine from E·P relative to the probability of dissociation to E and P_i. This may account for a previous observation that while overall turnover is less rapid, the transphosphorylation reaction of the Co(II) enzyme is more efficient relative to hydrolysis than it is for the Zn(II) enzyme (61).

6 METAL ION BINDING: Mn(II), Co(II), Cu(II), AND Cd(II) METALLOALKALINE PHOSPHATASES

The stoichiometry of Zn(II) binding to *E. coli* apoalkaline phosphatase has been the object of a number of studies. The results show the stoichiometry of the rebinding of Zn(II) to the metal-free apoenzyme to vary from two to six or more gram atoms of Zn(II) per mole of dimer depending on pH and buffer conditions (50, 62, 71). The Zn(II) content of the native enzyme reported in the literature has also varied between 2 and 4 g atom/mol of enzyme dimer, depending on the mode of preparation (4, 71, 72). The lower values appear to correlate with the use of ammonium sulfate precipitation in the absence of Zn(II)-containing buffers. The average Zn(II) content of enzyme released from *E. coli* by osmotic shock after treatment of the cells with EDTA and the cold water wash loaded directly onto a DEAE cellulose column, is 2.7 to 3.5 g atom/mol (71).

Alkaline phosphatase has an absolute requirement for Zn(II), but in common with many phosphoryl transferases the Zn(II) enzyme can be further activated by Mg(II) (6). Studies of the reactivation of the apoenzyme by metal ions have also yielded variable results concerning the minimum stoichiometry required for restoration of full activity with values from 2 to 4 g atom Zn(II)/mol of enzyme dimer required to restore maximal activity (50, 61, 73–76). Part of this variability appears to derive from the fact that restoration of the "native" state from metal ions and apoenzyme is not an instantaneous process and is subject to other variables in addition to the Zn(II) stoichiometry *per se*. The presence of Mg(II) and phosphate are additional factors that affect the restoration of activity. Mg(II) potentiates the restoration of activity to the apoenzyme by Zn(II). The addition of Mg(II) to enzymes containing two Zn(II) or two Co(II) ions per mol of dimer enhances the activity of such sam-

ples to that observed with excess Zn(II) or Co(II) (72, 77, 78). The native enzyme isolated by standard methods, employing osmotic shock, contains slightly more than 1 g atom Mg(II)/mol of dimer and has been observed to bind ~1.8 g atoms Mg(II)/mol at free Mg(II) concentrations of 0.45m*M* (62). The latter binding is cooperative with that of Zn(II) at pH values below 9, that is, maximum Mg(II) binding requires the presence of Zn(II).

Recent studies employing radioactive metal ions (^{65}Zn, ^{115m}Cd, and ^{60}Co) have given a more detailed picture of metal ion binding as functions of metal ion concentration, pH, phosphate, and magnesium (78). Cd(II) binds less tightly to alkaline phosphatase than the native zinc ion and in the presence of excess metal ion has fewer total binding sites at comparable free metal ion concentrations. Thus the differential binding affinities of the several pairs of potential metal-binding sites on alkaline phosphatase are most easily demonstrated with ^{115}Cd(II).

6.1 ^{115}Cd Binding to Apoalkaline Phosphatase as Functions of Phosphate, Magnesium, pH, and Cd(II) Concentration

Many of the studies of Cd(II) alkaline phosphatase, particularly NMR studies (see below), have been performed at pH 6.5. At pH 6.5, Cd(II) is relatively weakly bound; slightly less than 2 Cd(II) ions are bound per enzyme dimer at enzyme and Cd(II) concentrations of $5 \times 10^{-5}M$ or below (Fig. 4*A*). Equilibrium binding as the Cd(II) concentration is lowered to $10^{-6}M$ shows the apparent dissociation constant for this pair to be $\sim 5 \times 10^{-5}M$. The presence of 1m*M* phosphate at pH 6.5 induces the binding of a second pair of Cd(II) ions (Fig. 4*A*). The bound phosphate at this pH is present entirely as the phosphoseryl intermediate (E-P) (see ^{31}P NMR below). A total of 4 Cd(II) ions per enzyme dimer are bound and the dissociation constants for both pairs are $2 \times 10^{6}M$ or less (Fig. 4*A*).

Both pairs of metal binding sites on the alkaline phosphatase dimer bind Cd(II) more tightly at pH 8 than at pH 6.5. At enzyme and Cd(II) concentrations of $5 \times 10^{-5}M$, four Cd(II) ions are bound per dimer at pH 8, even in the absence of the phosphate ligand (Fig. 4*B*). Two regions of Cd(II) binding can be detected one pair with a dissociation constant, $K_d \cong 5 \times 10^{-7}M$, and one pair bound about ten times less tightly, $K_d \cong 5 \times 10^{-6}M$ (Fig. 4*B*). The presence of 10m*M* Mg(II) at pH 8 prevents the binding of the pair with $K_d \cong 5 \times 10^{-6}M$ (Fig. 4*C*), while the pair with a K_d of $\cong 5 \times 10^{-7}M$ is undisturbed by Mg(II).

The pH dependency of Cd(II) binding from pH 6.5 to 9 is shown in Figure 4*D*. In the presence of the phosphate ligand, the enzyme binds four Cd(II) ions per dimer at all pH values tested, while in the absence of the ligand, bind-

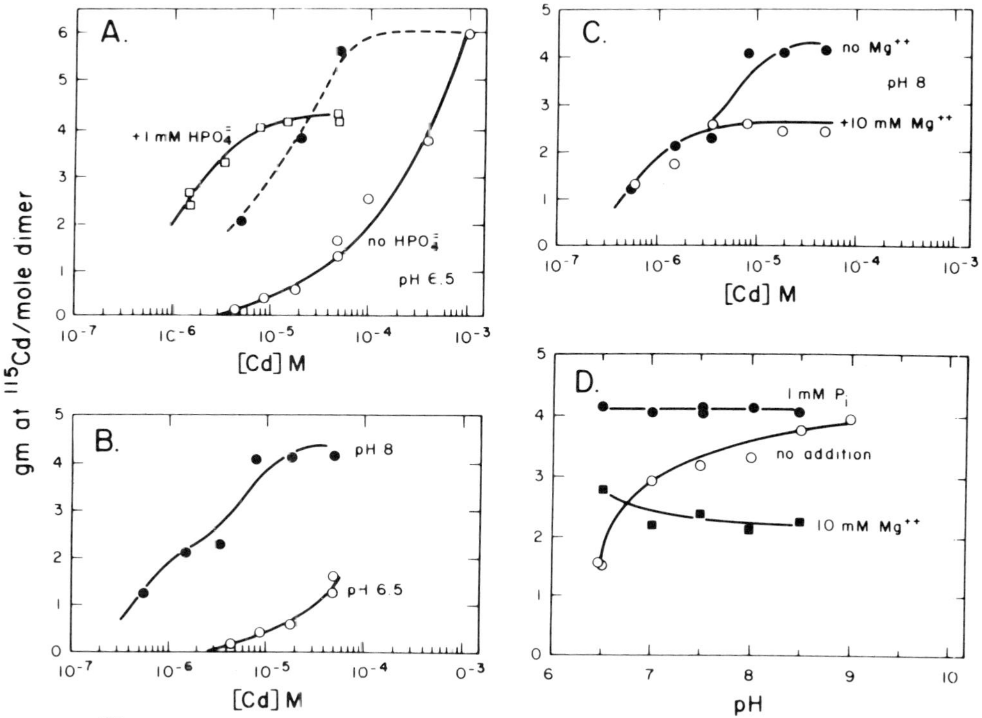

Figure 4 ^{115}Cd binding to apoalkaline phosphatase as a function of [Cd(II)], [HPO_4^{2-}], [Mg(II)], and pH. (*A*) Enzyme-bound ^{115}Cd/dimer versus [Cd(II)] at pH 6.5 (○); plus m*M* HPO_4^{2} (□); and following dialysis of samples labeled at $10^{-3}M$ Cd(II) (●). (*B*) ^{115}Cd/dimer versus [Cd(II)] at pH 6.5 (○) and pH 8.0 (●). (*C*) ^{115}Cd/dimer versus [Cd(II)] at pH 8.0 (●); plus 10m*M* Mg(II) (○). (*D*) ^{115}Cd/dimer versus pH with no addition (○) or in the presence of 1 m*M* HPO_4^{2-} (●) or 10 m*M* Mg(II) (■); $5 \times 10^{-5}M$ [^{115}Cd(II)], $5 \times 10^{-5}M$ enzyme. Points are missing at pH 9 because of the accumulation with time of a dialyzable ^{115}Cd-containing species in the dialysis bag at the very alkaline pH when P_i and Mg(II) and excess Cd(II) are present. In parts (*A*), (*B*), and (*C*), enzyme concentration was $1 \times 10^{-5}M$ in the lower concentration range, $5 \times 10^{-5}M$ in the mid-concentration range, and $2 \times 10^{-4}M$ in the high concentration range. All buffers were 0.01*M*

ing rises from ~1.5 Cd(II) ions bound at pH 6.5 to ~4 Cd(II) ions bound per dimer at pH 8.5 to 9.0. One of the two pairs of Cd ions bound to the dimer is displaced at all pH values by 10m*M* Mg(II).

If higher enzyme concentrations ($1.5 \times 10^{-4}M$) are used in order to measure Cd(II) binding at higher free-metal ion concentrations, binding in the absence of phosphate ligand can be increased to 6 g atom/mol dimer at $10^{-3}M$ Cd(II) (Fig. 4*A*). These two additional pairs, however, have binding constants between 10^{-4} and $10^{-3}M$ which suggest kinetic dissociation constants, k_{diss}, on the order of 10^3 to 10^5 sec^{-1}, the latter value applying if binding is assumed to be diffusion controlled, that is, ~10^8M^{-1} sec^{-1}. Such exchange rates between bound and free Cd(II) can easily be in the intermediate exchange range for ^{113}Cd(II) NMR signals of exchanging species and can have dramatic effects on ^{113}Cd(II) NMR signals (see Sec. 10.2).

The Cd(II) binding sites are shared by other transition and IIB metal ions binding to alkaline phosphatase. One pair of binding sites is also shared by Mg(II) as illustrated by the set of order of additions experiments listed in Table 2. The competing metal ions were added first and then the enzyme dialyzed against $5 \times 10^{-5}M$ ^{115}Cd(II) to label the remaining binding sites and the stoichiometry of the ^{115}Cd(II) label determined at 24 h and after 30 days. The latter stoichiometry reflects the slow exchange of ^{115}Cd(II) with the occupied metal binding sites. Addition of two Zn(II) ions and phosphate at either pH 6.5 or 8 blocks one pair of Cd(II) sites. Four Zn(II) ions and phosphate block both pairs of Cd(II) sites at both pH values, somewhat more effectively at pH 8 (Table 2). The fractional binding observed reflects a minimal exchange in 24 h. Without phosphate this exchange is more rapid. Mg(II) (10m*M*) added to the two Zn(II) or two Mn(II) enzyme blocks at least 50% of the second pair of Cd(II) binding sites. The combination of 2Mn(II) 2Mg(II) and phosphate is particularly effective. This is in contrast to 2Mn(II) and phosphate which does not appear particularly stable, since over 3 g atom of ^{115}Cd(II) are bound within 24 h. This must reflect the relative stabilities of the mixed metal-enzyme-phosphate complexes.

6.2 ^{65}Zn Binding to Apoalkaline Phosphatase as Functions of Phosphate, Magnesium, pH, and Zn(II) Concentration

^{65}Zn(II) binding to the apoenzyme as a function of pH is shown in Figure 5*A*. In contrast to Cd(II), 4 g atom Zn(II)/mol of dimer are bound at pH 6.5 in the absence of the phosphate ligand. This stoichiometry remains constant to pH 7.5. At higher pH values there is a slow increase in binding to ~5 g atom/mol dimer at pH 9. While the Zn(II) stoichiometry at pH 6.5 is not altered by m*M* P_i, the binding of extra Zn(II) at alkaline pH begins at lower pH, 7.0, and reaches ~6.5 g atom/mol dimer at pH 9. The bound ligand is almost ex-

Table 2 Competition between Cd(II), Zn(II), Mn(II), and Mg(II) for Metal-Binding Sites of Alkaline Phosphatase[a]

Additions to Apoenzyme	Additions to Medium	g atom Cd/mol Dimer 24 hr	30 days
(A) ^{115}Cd Binding, pH 6.5			
Apo	1mM P_i	4.1	4.1
Apo+2Zn+1mM P_i	1mM P_i	2.14	2.0
Apo+2Zn+10mM Mg^{2+} + 1mM P_i	1mM P_i, 10mM Mg^{2+}	1.47	2.2
Apo+2Mn+1mM P_i	1mM P_i	3.10	3.6
Apo+2Mn+10mM Mg^{2+} + 1mM P_i	1mM P_i, 10mM Mg^{2+}	0.69	0.44
Apo+4Zn+1mM P_i	1mM P_i	0.64	1.4
(B) ^{115}Cd Binding to Alkaline Phosphatase, pH 8.0			
Apo	1mM P_i	4.2	4.2
Apo+2Zn+1mM P_i	1mM P_i	1.75	2.4
Apo+2Zn+10mM Mg^{2+} + 1mM P_i	1mM P_i, 10mM Mg^{2+}	1.05	1.4
Apo+4Zn	none	0.88	1.6
Apo+4Zn+1mM P_i	1mM P_i	0.12	1.7
		g atom Zn/mol Dimer 24 hr	120 hr
(C) ^{65}Zn Binding to Alkaline Phosphatase, pH 6.5			
Apo	None	4.05	
Apo	1mM P_i	3.92	
Apo + 10mM Mg^{2+}	1mM P_i, 10mM Mg^{2+}	2.48	
Apo + 2Cd	None	1.60	2.30
Apo + 2Cd + 1mM P_i + 10mM Mg^{2+}	1mM P_i, 10mM Mg^{2+}	2.25	4.20
Apo + 4Cd	None	0.84	2.30
Apo + 4Cd +	1mM P_i	0.91	1.20

[a]Stoichiometric additions were made to the apoenzyme as indicated in the left hand column. One ml aliquots were then placed in 1 ml dialysis bags and the samples labeled with ^{115}Cd(II) or ^{65}Zn(II) as described in Materials and Methods. Conditions: $5 \times 10^{-5}M$ Me(II), $5 \times 10^{-5}M$ enzyme, 0.01M , 4°C.

clusively in the form of the noncovalent phosphate complex over this pH range in the case of the Zn(II) enzyme (see Sec. 12.1). Addition of 10mM Mg(II) alone to the labeling solution has little effect on the ^{65}Zn(II) binding stoichiometry at any pH (Fig. 5B). On the other hand, if the phosphate ligand is added in addition to 10mM Mg(II), the Zn(II) stoichiometry is depressed by ~2 g atom/mol dimer at all pH values compared to ^{65}Zn(II) binding in the

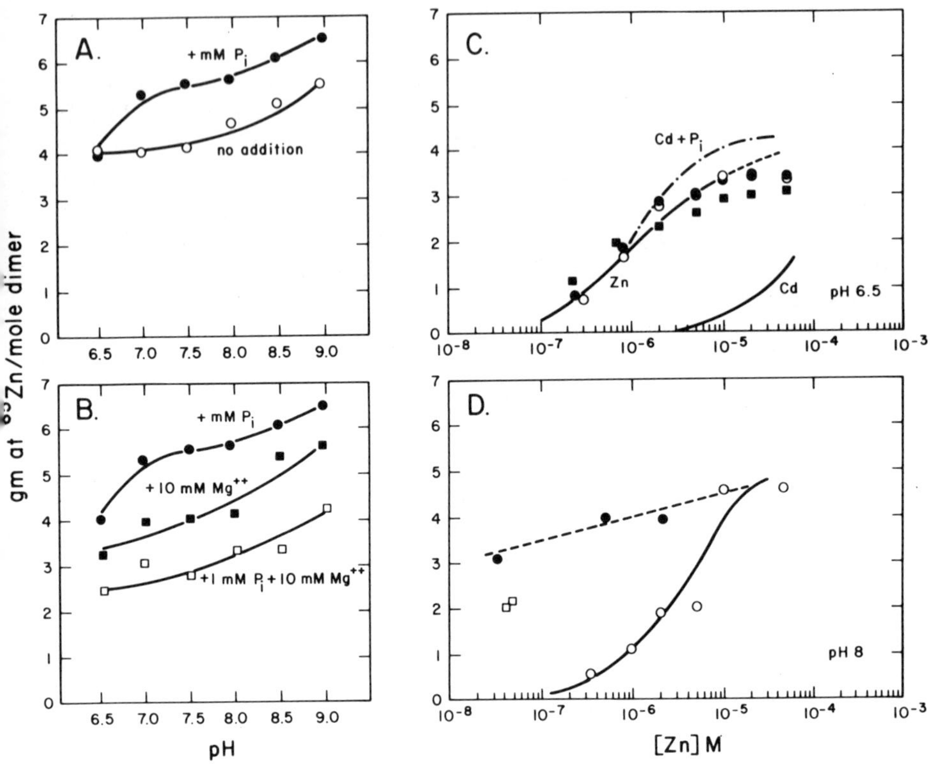

igure 5 $^{65}Zn(II)$ binding to apoalkaline phosphatase as a function of [Zn(II)], $[HPO_4^{2-}]$, [Mg(II)], and pH. A) Enzyme-bound $^{65}Zn(II)$/dimer versus pH, no addition (○); plus 1m*M* HPO_4^{2-} (●); 5 × $10^{-5}M$ Zn(II), 5 × 0^{-5}M enzyme. (*B*) Enzyme-bound $^{65}Zn(II)$/dimer versus pH; plus 10m*M* Mg(II) (■), identical data were bserved without Mg(II); (●) plus 1m*M* P_i; (□) plus 1m*M* P_i and 10m*M* Mg(II); 5 × $10^{-5}M$ Zn(II), 5 × 0^{-5}M enzyme. (*C*) Enzyme-bound $^{65}Zn(II)$/dimer versus [Zn(II)] pH 6.5, (○) no addition; (●) plus 1m*M* P_i; ■) plus 1m*M* P_i and 10m*M* Mg(II). Corresponding Cd(II) data in the absence of P_i (——); plus P_i (—·—). (*D*) nzyme-bound $^{65}Zn(II)$/dimer versus [Zn(II)], pH 8.0. (○) Initial labeling; (●), following dialysis of samples abeled at 5 × $10^{-5}M$ Zn(II) against metal-free buffer; points are plotted at final free Zn(II) concentration. □) Two samples were further dialyzed against 10m*M* Mg(II). Conditions as in Figure 4. From ref. 78.

presence of phosphate alone (Fig. 5*B*). The Zn(II), above 4 g atom/mol dimer, bound at the higher pH values dialyzes very quickly.

At pH 6.5 one pair of Zn(II) ions binds to apoalkaline phosphatase with a binding constant of $\sim 5 \times 10^{-7}M$, estimated from the Zn(II) concentration required for binding of an average of one Zn(II) ion per dimer (Fig. 5*C*). The second pair binding at pH 6.5 has a binding constant of $\sim 5 \times 10^{-6}M$. Phosphate has relatively little effect on the Zn(II) equilibrium stoichiometry or the apparent binding constants as determined by this direct equilibrium dialysis of the apoenzyme (Fig. 5*C*).

In the same manner as Zn(II) blocks ^{115}Cd(II) binding, Cd(II) can be shown to block ^{65}Zn(II) binding (Table 2). At 5 × $10^{-5}M$ enzyme and Zn(II), approximately 4 g atoms ^{65}Zn(II) are bound to the apoenzyme/mol of dimer after 24 hr dialysis in the presence or absence of P_i. The prior addition of 2 Cd(II) ions per mol dimer blocks a pair of these sites. Mg(II), 10mM, in the presence of phosphate will also block a pair for 24 hr if the Mg(II) and P_i are added to the enzyme prior to dialysis. After prolonged dialysis ^{65}Zn(II) exchanges with the Mg(II). A second pair of Cd(II) ions also blocks the second pair of potential Zn-binding sites, but after 24 hr of dialysis against ^{65}Zn(II) significant exchange has occurred (Table 2).

It is not possible to interpret metal ion binding to apoalkaline phosphatase as a simple reversible equilibrium process analogous to that occurring in simple ligand systems. The first indication of this is illustrated in Figure 5*D* where the concentration dependence of ^{65}Zn(II) binding at pH 8 is presented. While binding reaches 4 g atom/mol dimer at lower concentrations of free metal than at pH 6.5, the apparent stability of the first pair of Zn(II) ions bound is not appreciably increased by alkaline pH, judging by the direct binding study; a somewhat surprising result. Once four Zn(II) ions are bound, if the samples are resuspended in metal-free buffer and allowed to reequilibrate at 10^{-5}–$10^{-7}M$ free Zn(II) (as determined by the radioactivity of the final dialysate), equilibrium is reached at a much higher stoichiometry of ^{65}Zn(II), ~4 g atom/mol dimer. The stoichiometry does not fall even to 3 g atom/mol dimer until ~3 × $10^{-8}M$ free Zn(II) (Fig. 5*D*). This suggests that the binding of a metal ion to a second site on the monomer must induce conformational changes that stabilize both the first and second metal ion complexes, conformational changes that are not readily reversible by equilibrium dialysis. This "hysteresis" effect on metal-enzyme stability is also observed for $Cd(II)_4AP$ where a similar but not as dramatic increase in stability of the fully metalated enzyme is observed (Fig. 4*A*). At pH 6.5, 2 g atom Cd(II) remain bound at 5 × $10^{-6}M$ Cd(II), a condition where very little would bind in the forward arm of this hysteretic binding curve.

Cu(II) and Mn(II) binding to the enzyme have been most thoroughly investigated by ESR spectroscopy (64, 66, 79). The spectroscopic results will be discussed in Sec. 9, but the stoichiometric features of Mn(II) and Cu(II) binding are similar to Zn(II) and Cd(II). Two pairs of potential metal-binding sites exist. One pair occupied by two Mn(II) or two Cu(II) in the absence of Zn(II), gives rise to unusual ESR spectra (64, 66). In the case of Mn(II) this pair of sites has an unusual distortion from regular symmetry. Cu(II) ESR of this pair of sites shows ligand hyperfine structure arising from three equivalent nitrogen ligands (64). Both ions will also occupy at least one additional pair of binding sites if a second pair is added (see Sec. 9 for further details).

6.3 Summary of Metal Ion Binding Properties

Alkaline phosphatase can potentially bind three pairs of metal ions over a free-metal concentration range from 10^{-8}–$10^{-3}M$ (Figs. 4 and 5). Additional metal ions bound under some conditions can be readily dialyzed. Even the third pair of Cd(II) ions undergoes fairly rapid dissociation on dialysis, but less rapidly than the excess above 6 g atom/mol dimer. The magnitude of the binding constants of the successively more weakly bound pairs are often not sufficiently different to resolve the binding curves for the separate binding of the first three pairs of metal ions. Because of this, binding to the apoenzyme does not always produce a strictly homogeneous species (each type of site fully occupied) until all six metal ions are bound. While this is clearly a potential problem, several conditions of metal ion binding can be found which enable one to distinguish the properties of the different pairs during less than maximal metal ion binding. These distinguishing features will be used to designate the pairs of sites as A, B, and C in decreasing order of binding affinity. From the direct binding curves for ^{115}Cd(II), apoalkaline phosphatase binds approximately two Cd(II) ions at $5 \times 10^{-5}M$ Cd(II) and pH 6.5 (Fig. 4*A*). Since the binding of the first two Cd(II) ions in the absence of phosphate is part of a smooth binding curve achieving a stoichiometry of six Cd(II) ions per dimer at $10^{-3}M$ Cd (Fig. 4*A*), these data alone cannot identify the first two as a unique pair. The following additional data have to be used to argue that these represent homogeneous A-site Cd(II). A second pair of Cd(II) ions is induced to bind by phosphate, hence one of the pairs is not binding in the $< 10^{-5}M$ Cd(II) range. This latter pair involves binding to sites whose affinity is markedly increased by formation of the phosphoseryl intermediate (Fig. 4) and will be designated the B sites. The enzyme at pH 6.5 with only two Cd(II) ions bound per dimer shows a single ^{113}Cd(II) NMR signal corresponding to both ^{113}Cd(II) ions (see Sec. 10). Thus a unique pair of sites is occupied, which will be defined as the A sites.

Not unexpectedly both A- and B-site Cd(II) complexes are made more stable by raising the pH (Fig. 4), but only the less stable pair is successfully competed for by Mg(II) (Fig. 4*C*). Competition with Mg(II) is thus another operational definition of the B-site Cd(II) ions. In the presence of phosphate at pH 6.5, however, both sites are significantly stabilized and from the direct binding curves alone it would not be possible to distinguish the two types of site (Fig. 4*A*).

X-ray diffraction studies of crystals of the Cd(II) enzyme soaked in 50m*M* Cd(II) have recently identified two Cd(II) binding sites on each monomer separated by ~3.9Å (see Sec. 7) (17). These appear to correspond to the A and B sites defined here. One of these sites contains three histidyl ligands and ap-

pears to correspond to the tightest or A site which has been identified by Cu(II) ESR to contain at least three equivalent nitrogens as ligands (64). This A site is occupied by Zn(II) in crystals of the native enzyme and the two A sites, one in each monomer, are related by the twofold symmetry of the dimer and separated by 32 Å (17). The second pair of sites, 3.9 Å away from the A sites, would appear to be the sites to which Cd(II) binding is induced by phosphate at low pH. The differential binding of Cd(II) to these two sites and the direct interaction of one of them with phosphate will be detailed in Sec. 9 on ^{113}Cd NMR.

6.4 Binding and Functional Effects of Mg(II)

Activation of the "native" zinc alkaline phosphatase by Mg(II) ions has been more frequently demonstrated with the mammalian enzymes (6). The *E. coli* enzyme as isolated does contain at least 1.2 g atom Mg(II)/mol dimer and will, in the presence of $10^{-4}M$ Mg (II), bind 2 g atoms Mg(II)/dimer in the presence of 2 g atoms Zn(II)/mol dimer (62). Mg(II) binds weakly to the apoenzyme and does not activate it. It binds much more strongly, however, to the Zn(II) enzyme and will very substantially activate the apoenzyme to which two Zn(II) or two Co(II) ions are added (72, 77, 78). These results suggest that there is a site, separate from the primary of "catalytic" Zn(II) site, that binds Mg(II) as well as first-transition and IIB metal ions. Competition between Mg(II), Cd(II), Zn(II), and Co(II) for a phosphate-stabilized site is easily demonstrated by the binding studies with radioactive metal ions illustrated in Figures 4 and 5 (78).

The topographic relationship of this "Mg(II)" site to the primary Zn(II) or A site is not absolutely certain as yet. The B site identified in the crystal structure ~3.9 Å from the primary Zn(II) site by its high occupancy with Co(II) or Cd(II) when the crystals are soaked in 20–50 mM Cd(II) or Co(II) may be this "Mg(II)" site. A Mg(II) site very close to the primary Zn(II) site and adjacent to the phosphate-binding site would explain the cooperativity between Zn(II) and Mg(II) binding, the stabilization of both active-center metal ions by phosphate (be the second pair Cd(II), Co(II), or Mg(II) [Table 2 (51)]), as well as the marked effect of Mg(II) on the phosphorylation rate of the Cd(II) enzyme (78). (See below.)

The effects of Mg(II) can be graphically demonstrated in two ways. Restoration of complete activity to the apoenzyme by Zn(II) or Co(II) frequently requires from 4–6 g atom Me(II)/mol dimer in the absence of Mg(II) (Fig. 6). The curvature of the plot suggests that more than one metal-binding equilibrium may be involved. On the other hand, if the apoenzyme is in the presence of 10mM Mg(II), the restoration of activity by Zn(II) is a sharp titration curve reaching a maximum at 2 g atom Me(II)/mol dimer (Fig. 6).

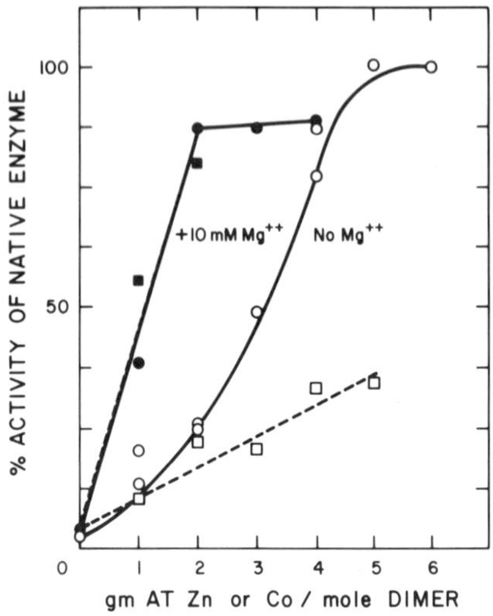

Figure 6 Reactivation of apoAP ($5 \times 10^{-5}M$) by Zn(II) and Co(II) as a function of mol Me(II) added/mol enzyme dimer in 0.01*M* TrisHCl-0.1*M* NaCl [○, Zn(II); □ Co(II)], and in the additional presence of 10m*M* Mg(II) [●, Zn(II); ■ Co(II)]. Following 1 hr of incubation activity was determined at pH 8.0. Activity is expressed as the percent recovered relative to the activity of the native enzymes from which the apophosphatase samples were prepared. Activity of individual preparations of the native enzyme varied from 2000 to 2500 μmol substrate hydrolyzed/hr/mg enzyme. From ref. 78.

Phosphorylation of $Cd(II)_2AP$ in the absence of Mg(II) is a relatively slow process with a phosphorylation time of hours (Fig. 7). $^{113}Cd(II)$ and ^{31}P NMR (Secs. 10 and 12) show that this slow formation of E-P takes place with the Cd(II) ions entirely in the A site, the B site remaining unoccupied. Addition of 5–10m*M* Mg(II) accelerates the phosphorylation rate such that phosphorylation is complete in 10 min (Fig. 7). The stoichiometric addition of a second pair of Cd(II) ions (shown by ^{113}Cd NMR to occupy the B site) also produces the same increase in the rate constant for phosphorylation of the $Cd(II)_2AP$ (Fig. 7). Thus is appears that occupancy of a second pair of metal-binding sites by either Mg(II) or Cd(II) accelerates the phosphorylation of Ser-102 by ~100-fold. The full effect of Mg(II) requires concentrations of 5–10m*M* in contrast to the stoichiometric addition of two Cd(II) ions, implying that the unnatural metal Cd(II) has considerably higher affinity for this site. The additional effect of Cd(II) on the final stoichiometry of E-P will be discussed below under negative cooperativity. Acceleration of phosphorylation by the binding of a metal ion at a second site in each active center suggests that it functions by further neutralizing the negative charge on the phosphate dianion, thus making the nucleophilic attack of the serine hydroxyl on the phosphorus a much more facile reaction.

7 CRYSTAL STRUCTURE OF ALKALINE PHOSPHATASE

Electron density maps of alkaline phosphatase have been obtained on two different crystal forms of the enzyme, one a trigonal form (space group $P3_121$)

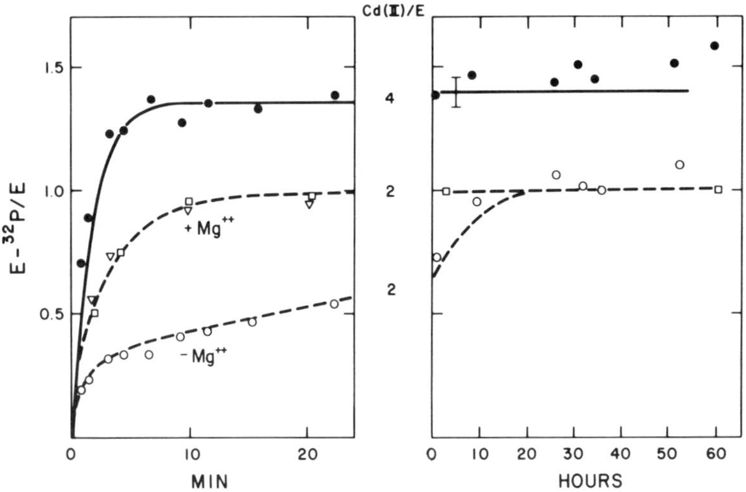

Figure 7 Effect of Mg(II) on the phosphorylation (E-^{32}P/E) of Cd(II) alkaline phosphatases. Conditions: 0.01*M* TrisHCl, 0.01*M* NaOAc, 0.1*M* NaCl, pH 6.5, 5.26 × 10^{-5}*M* alkaline phosphatase, 3.93 × 10^{-3}*M* H $^{32}PO_4^{2-}$. Metal ion stoichiometry: Cd(II)/alkaline phosphatase = 4.07 (●); Cd(II)/alkaline phosphatase = 2.03, Mg(II)/alkaline phosphatase = 2.04 (○); Cd(II)/alkaline phosphatase = 2.0, [Mg(II)] = 5 m*M* (▿); Cd(II)/alkaline phosphatase = 2.0, [Mg(II)] = 10m*M* (□). From ref. 78.

and one an orthorhombic form (space group *I*222) (17, 52, 80). The first space group contains one monomer per asymmetric unit and thus the two monomers are related by a crystallographic as well as a molecular twofold axis. The second space group contains one dimer per asymmetric unit and the molecular twofold axis relating the monomers does not coincide with crystallographic symmetry, thus twofold symmetry is not forced on the dimer. Nevertheless, the electron density map at 4.1 Å resolution shows no significant deviation from twofold symmetry in the dimer. While an electron density map at 2.8 Å resolution has been calculated for the $P3_121$ crystals, the quality of the maps from the *I*222 form is much better and the structural work is now being completed on this form. Currently the data are at 4.1 Å and the map for the nonphosphorylated enzyme has been refined to 3.1 Å.

7.1 Molecular Boundaries

The boundaries of individual dimers are very well resolved in the present electron density maps, but the monomer–monomer interface is closely interconnected and solvent appears to be pretty well excluded from this interface. These boundaries as well as other general features of the dimer are illustrated

in Figure 8, which is a silhouette of the dimer with the unresolved molecular twofold axis represented by the dotted line. At this stage of the analysis, the active center appeared to contain two metal-binding sites labeled A and B in Figure 8 (see below for modification). The α/β domain, which constitutes the core of the molecule, is illustrated by the six strands of β-pleated sheet shown by the black bars which are surrounded by a number of short stretches of α-helix. Only helices behind the sheets are illustrated in this view, hence to complete the helices of the lower monomer, the three helices of the monomer on the upper right should be rotated over in front of the lower monomer around the twofold axis. The α/β core accounts for approximately 27% of the 449 residues. The upper part of each monomer in Figure 8 constitutes a smaller domain that consists of much irregular secondary structure that is difficult to characterize at the present stage of refinement.

7.2 Metal-Binding Sites

Difference maps between the native and apoenzyme prepared from both crystal forms have readily identified a Zn peak at position A of the active site (Fig.

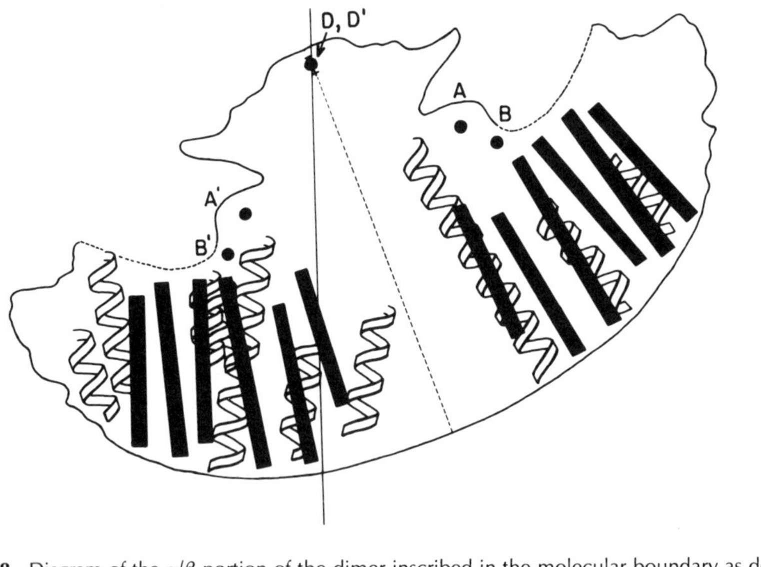

Figure 8 Diagram of the α/β portion of the dimer inscribed in the molecular boundary as defined from the 6 Å map. Black points indicate functional metal-binding sites, the broken line indicates a molecular boundary which is not well defined. Noncrystallographic two-fold axis (broken line) runs 19.5° off the z crystallographic direction (solid line). Only helices behind the sheets are shown. Helices in the upper monomer would thus be folded down in front of the lower β sheet by the two-fold axis. From ref. 17.

8). Such difference maps show that Zn removal from the active center is accompanied by significant movement of protein structure, side chains, and possibly backbone in the immediate vicinity of the active center. In contrast to Zn(II), both Co(II) and Cd(II) demonstrate the phenomenon known as anomalous dispersion which causes X-rays scattered from these ions to have a phase shift that is manifested in different intensities for the general pairs of reflections (*hkl* and *-h-k-l*). Electron density maps calculated from data on the Co(II)- and Cd(II)-substituted alkaline phosphatases using the anomalous dispersion from the metal centers unambiguously reveal the positions of the metal ions without interference from protein shifts. The 2.66*M* $(NH_4)_2SO_4$ at pH 7.5 used as the crystallization medium has a tendency to remove metal ions from the protein, and in order to compensate for this the crystals have been soaked in 10 to 50m*M* metal ions. Under these conditions, in the unphosphorylated enzyme, at relatively low resolution both Co(II) and Cd(II) appeared to occupy two sites at each active center ~4.9 Å apart, labeled A and B in Figure 8. Site A consists of a rather elongated region of density that was originally interpreted as fusion of metal with ligand density. Full occupancy of site B by Cd required 50m*M* Cd in the crystallization medium.

7.3 Three Potential Metal-Binding Sites at the Active Centers of Alkaline Phosphatase

A 3.4 Å electron density map of the phosphoryl Cd(II) enzyme at pH 7.5 has recently been obtained and difference maps based on the anomalous dispersion of the Cd centers show that the initial high occupancy A site (Fig. 8) actually consists of two closely spaced metal sites, A_1 and A_2, ~3.9 Å apart. These two sites and the original B site form a triangle at the active center with sides A_1—A_2 ≅ 3.9 Å, A_2—B ≅ 4.9 Å, and A_1—B ≅ 7 Å (81). The A_2 site appears to be stabilized by formation of E-P and probably is responsible for the ^{113}Cd resonance at 50–70 ppm (depending on conditions) observed in the NMR spectrum of the ^{113}Cd alkaline phosphatase, which arises from a phosphate-induced or phosphate-stabilized binding site (see Fig. 4) and which has been designated B in the ^{113}Cd NMR spectra (see below). The original "B" site in the electron density map of the unliganded enzyme and labeled *B* in Figure 8 appears to contain only carboxylate ligands (2 Asp and 2 Glu residues). The chemical shift of the most upfield of the three ^{113}Cd resonances observed in $^{113}Cd(II)_6$AP is at 1.5 ppm and corresponds most closely to the chemical shift expected from an exclusively oxygen-liganded ^{113}Cd(II) (see Sec. 10). Thus the original X-ray site B may correspond to the ^{113}Cd NMR signal designated as arising from site C. The Ser-102 residue is immediately adjacent to sites A_1 and A_2, about equidistant from both of them, and not close to the third site originally called B in the electron density map (Fig. 8). The A_1A_2B triad in one monomer is 32 Å from the same set in the opposite monomer. A fourth

pair of bound Cd ions is located near the top of the molecule in the unphosphorylated crystals soaked in 50m*M* Cd(II) ~12.5 Å off the twofold axis, and designated *D* in Figure 8.

In the ^{113}Cd NMR of the phosphorylated enzyme at 2m*M* concentration, pH 6, highly resolved homogeneous signals from only three pairs of ^{113}Cd sites are detected at ~150, 70, and 2 ppm designated A, B, and C (see Sec. 10). Sites A and B as defined by ^{113}Cd NMR probably correspond to sites A_1 and A_2 in the crystal structure. As outlined below, NMR signals from weakly bound Cd(II) can be broadened beyond detection by chemical exchange and this might be the case for one of the bound metals seen in the X-ray data. Exchange at the A_2 site might well be expected to result in conformational flux at site A_1, 3.9 Å away. Sites A_1 and A_2 are in the midst of a cluster of side chains including a number of histidyl residues and precise identification of the ligand structure in this complex site requires more refinement and higher resolution. The general folding of the polypeptide chain with approximate positions of the A_1, A_2 and B (spectroscopic A, B, and C) metal ion binding sites is shown in Figure 9.

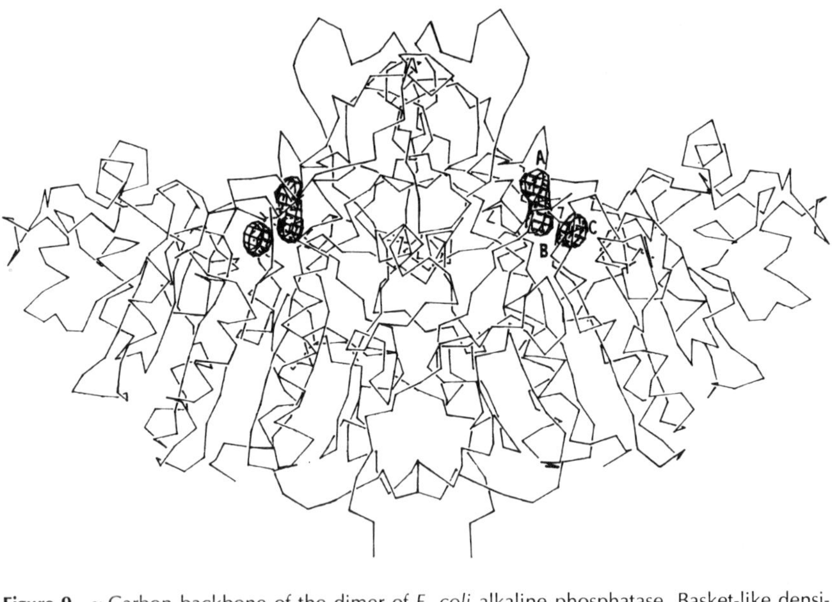

Figure 9 α-Carbon backbone of the dimer of *E. coli* alkaline phosphatase. Basket-like densities are the three metal-binding sites at each active center which can be occupied by Cd(II) ions as represented by the anomalous dispersion difference map of the Cd(II) derivative. Labels A, B, and C represent those expected as most likely to give rise to the chemical shifts of the three ^{113}Cd(II) NMR signals of the $Cd(II)_6AP$ as shown in Figure 12. Figure is kindly supplied by Professor Harold Wyckoff and summarizes the work of J. M. Sowadski, H.M.K. Murthy, M. Handschumacher, and H. W. Wyckoff (81).

8 OPTICAL SPECTROSCOPY OF ACTIVE-SITE METAL IONS

The derivatives of alkaline phosphatase in which Mn(II), Co(II), and Cu(II) have been substituted for the native Zn(II) have been extensively characterized in terms of optical and ESR transitions that arise from the *d-d* electronic transitions. The low-energy transitions giving rise to absorption in the visible spectral region have been extensively characterized only for the Co(II) enzyme (60, 61, 63, 82). Of the three pairs of metal-binding sites at the active center identified by crystallography (Sec. 7), or of the two tight-binding sites identified by ^{60}Co binding studies (Sec. 6), one pair when occupied by Co(II) gives rise to an unusual intense visible absorption spectrum characteristic of distorted tetrahedral or five-coordinate complexes of Co(II) (60, 61, 63, 82, 83). Whether the first two Co(II) ions added to the apoenzyme fully generate this spectrum is a sensitive function of the experimental conditions. Full generation often requires 4 or 6 Co(II) ions. A titration of the apoenzyme with Co(II) ions under experimental conditions where the first two Co(II) ions generate the spectrum is shown in Figure 10*A*. If 10m*M* Mg(II) is present, the first pair of Co(II) ions added always generates the full spectrum (43, 78, 82). Under many conditions, however, several pairs of Co(II) ions are required which appears to relate to the similar affinities of several pairs of binding sites for Co(II) such that the generation of heterogeneous species at low metal ion stoichiometries is the rule rather than the exception. This is especially true at the high protein concentrations ($\sim$m*M*) required for many spectroscopic studies. This phenomenon involving multiple competing metal ion binding sites for Cd(II) in alkaline phosphatase has recently been extensively documented by ^{113}Cd NMR (84).

The magnetically induced circular dichroism of this Co(II) site (Fig. 10*B*) has a magnitude of $\sim 10^4$ deg cm^2/dmol at a field of 50 kG which identifies the Co(II) site as one of basically tetrahedral rather than octahedral coordination (63, 83). The main bands relate to the ^4T_1P manifold expected from tetrahedral Co(II) complexes (63, 83). The distinction between distorted tetrahedral or five-coordinate complexes cannot be made from these optical spectra. Co(II)-substituted carbonic anhydrase shows a *d-d* absorption spectrum and MCD spectrum very similar to that observed for Co(II) alkaline phosphatase (63). In the former enzyme the ligands are the nitrogens of three histidyl residues and one (or possibly two) coordinated solvent molecules (85, 86).

Both phosphate and arsenate (as E·P or E·As) decrease the intensity of the two lowest energy visible absorption bands and result in an even more striking change in the natural CD associated with the Co(II) bands (63). Surprisingly E·P induces a large negative CD band, while E·As induces a large positive CD band, illustrating the extreme sensitivity of the natural CD bands associated with the *d-d* transitions of metalloproteins to small symmetry changes in the

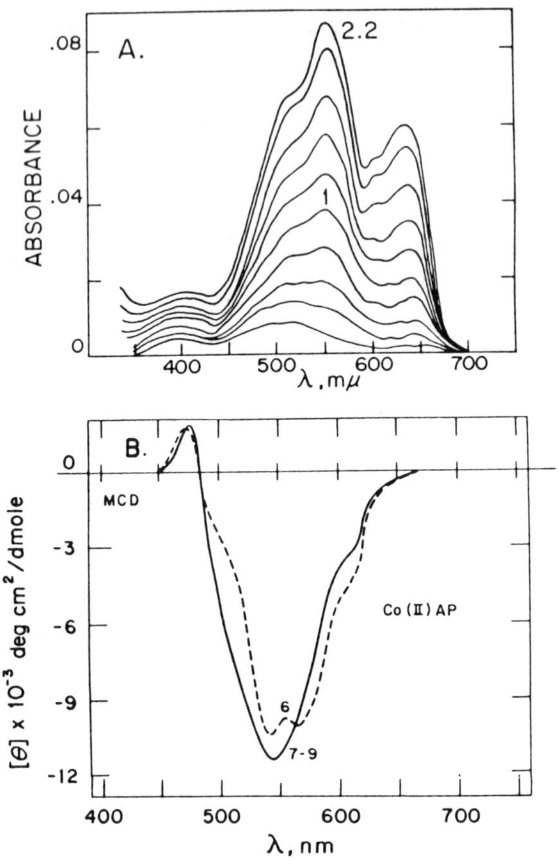

Figure 10 (*A*) Visible absorption spectrum of Co(II) alkaline phosphatase generated by adding two equivalents of Co(II)/dimer. From ref. 60. (*B*) Magneto circular dichroism through the Co(II) absorption bands in part *a*; (---) pH 6, (——) pH 7–9. From ref. 63.

surrounding potential field. That these changes do not reflect large changes in the coordination geometry around Co(II) is confirmed by the fact that formation of E·P or E·As produces very minor changes in the MCD bands shown in Figure 10*B* and practically no change in the magnitude of the major negative MCD band at 550 nm (63). In contrast to CD, MCD is largely a basic spectroscopic property of the immediate coordination sphere. As will be shown below by the observation of ^{113}Cd—O—^{31}P coupling in the ^{31}P NMR spectrum, phosphate in E·P coordinates the A- or catalytic-site metal ion at least in the Cd(II) enzyme (51, 84, 87). Most evidence indicates that it is the A site that gives rise to the unusual optical and ESR spectra rather typical of Zn(II)

metalloenzymes substituted with first-transition period metal ions (79, 88). Phosphate may replace a coordinated solvent molecule in the inner coordination sphere of the A-site metal ion without a major shift in coordination geometry, a conclusion also supported by the ESR signals from Mn(II) in the A sites (see below).

9 ELECTRON SPIN RESONANCE OF Mn(II), Co(II), AND Cu(II) ALKALINE PHOSPHATASES

An X-band ESR spectrum of the phosphoalkaline phosphatase containing two Mn(II) ions substituted homogeneously in the A sites [the adjacent B sites being filled by Mg(II)] is shown in Figure 11*A*. This spectrum, at −160°C, is unusual for solid-state ESR spectra of Mn(II) in that a series of low-field transitions are present almost down to zero field. Such spectra are characteristic of Mn(II) in coordination sites with significant rhombic distortions from cubic symmetry (79, 87, 88). While the ESR spectra of a d^5 ion cannot unambiguously distinguish between tetrahedral and octahedral complexes, the X-band Mn(II) ESR spectra of Mn(II)-substituted carbonic anhydrase, carboxypeptidase A, and thermolysin all show similar low-field transitions (88). In these three enzymes the crystal structures of the corresponding Zn(II) enzymes have shown the coordination geometry to consist of distorted tetrahedral symmetry (85, 90, 91). It can be shown that the unique ESR spectrum of Mn(II) occupying one pair of the three pairs of metal ion binding sites at the active center arises from sites normally occupied by Zn(II). When one Mn(II) was added to a homogeneous E·P complex of $Zn(II)_2Mg(II)_2AP$ prepared in the same manner as the $Mn(II)_2Mg(II)_2AP$, no low-field transitions are observed and the Mn(II) shows an ESR spectrum typical of Mn(II) in symmetrical environments (Fig. 11*B*). The *K*-band (35 GHz) ESR spectrum of the $Mn(II)_2Mg(II)_2AP$ is shown in Figure 11*C*. This spectrum, in contrast to the X-band spectra, can be computer simulated by a program based on perturbation theory (51). The analysis shows the D and E coefficients of the spin Hamiltonian, which characterize the axial and rhombic distortions from cubic symmetry, to be ~0.16 and ~0.02 cm^{-1}, respectively (79). Thus there are relatively small rhombic distortions in the complex. E/D is ~0.1 for the alkaline phosphatase A site, but can be as great as 0.33 in Mn(II) complexes with maximum rhombic distortions (89).

The Mn(II) ESR spectrum of the phosphoderivative of $Mn(II)_4AP$ has considerably less than twice the spin density observed for $Mn(II)_2Mg(II)_2AP$ and many of the low-field transitions are severely broadened (79). Thus the two pairs of Mn(II) ions occupy sites close enough for some spin–spin interaction to occur, but do not demonstrate classic antiferromagnetic coupling. If these two pairs of

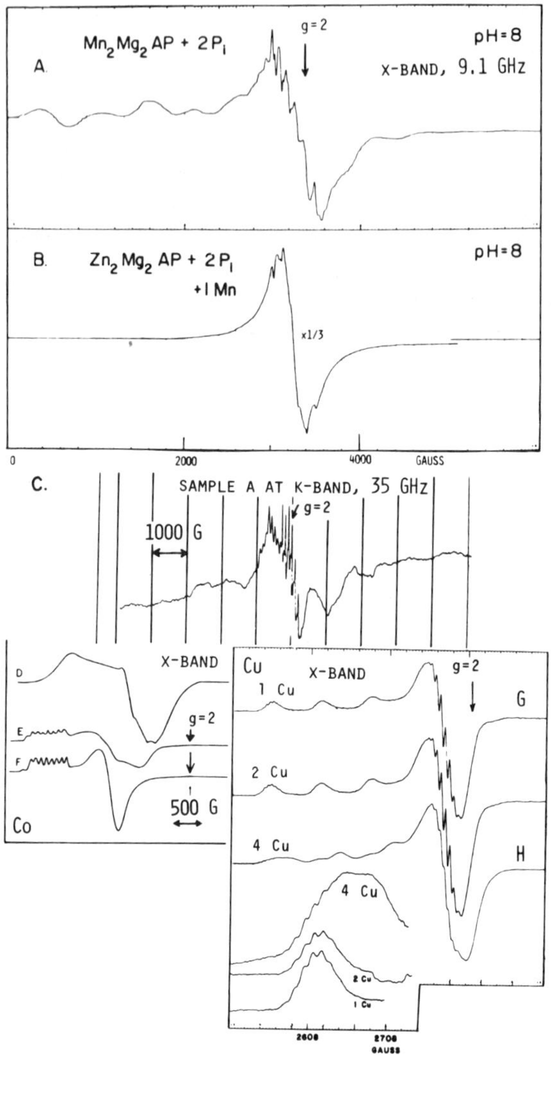

Figure 11 ESR spectra of transition-metal ion-substituted alkaline phosphatases. (*A*) X-band (9.1 GHz) spectrum of $Mn(II)_2Mg(II)_2AP$ plus 2 mol P_i/mol dimer, pH 8, −120°C. (*B*) X-band spectrum of $Zn(II)_2$ $Mg(II)_2AP$ plus 2 mol P_i/mol dimer plus one equivalent of Mn(II). Conditions as in part (*A*). (*C*) K band (35 GHz) of sample in part (*A*). (*D*) X-band spectrum of Co(II) thermolysin. (*E*) X-band spectrum of $Co(II)_2Zn(II)_2AP$. (*F*) X-band spectrum of $Co(II)_2Zn(II)_2AP$ plus P_i. Spectra in parts (*D*), (*E*), and (*F*) taken at 4–10°K. (*G*) X-band spectrum of $Cu(II)_1Mg(II)_2AP$ and $Cu(II)_2Mg(II)_2AP$. (*H*) X-band spectrum of $Cu(II)_2Cu(II)_2AP$, 111°K. *Insert:* Expansion of low-field copper hyperfine line. Data in parts (*A*), (*B*), and (*C*) from ref. 79; (*D*), (*E*), and (*F*) ref. 93, and (*G*) and (*H*) ref. 64.

sites correspond to two of the sites observed in the crystal structure, the slight spin–spin interaction suggests that no bridging ligand is present, since any bridged system in which the metal centers are 6 Å or less apart shows large antiferromagnetic coupling, 20–100 cm^{-1} (79).

ESR spectra of high-spin Co(II) complexes are relatively broad even at liquid helium temperatures and the few ESR spectra reported for Co(II)-substituted Zn(II) metalloenzymes are no exception (92–94). A typical example for Co(II) thermolysin is shown in Figure 11*D*. Alkaline phosphatase to which 2 Co(II) ions are added shows a broad featureless spectrum with a suggestion of hyperfine in the $g = 6$ region (95). Two other examples of Co(II)-substituted alkaline phosphatase are shown in Figures 11*E* and 11*F*. Both are $Co(II)_2Zn(II)_2$ hybrids, with and without phosphate (93). In general, they show the same broad features, but have low-field regions where Co(II) hyperfine structure is well resolved, more characteristic of octahedral Co(II) complexes. Recent NMR experiments have shown it to be very difficult to prepare hybrids with two species of metal [except when one is Mg(II)] in which each metal species is forced to occupy only one pair of the several metal-binding sites present. Accompanying phosphorylation there is often facile intersite metal exchange (51, 84). Thus some heterogeneity may exist and two types of Co(II) site may contribute to these spectra.

A homogeneous $Cu(II)_2Mg(II)_2AP$ has been prepared as described for the Mn(II) system above and the Cu(II) ESR spectrum of this species is shown in Figure 11*G* (64). This is a rather typical Cu(II) signal from an essentially axially symmetric (close to square planar) complex, perhaps with slight rhombic distortion. The lowest field copper hyperfine line clearly consists of seven hyperfine lines with a splitting of ~20 G indicating the presence of three magnetically equivalent nitrogen atoms as ligands (64). The most likely candidates would be three imidazole nitrogens from three histidyl ligands. The present analysis of the crystal structure shows that there are at least three histidyl residues in the active-center cavity, but their distribution between sites A_1 and A_2 is not clear as yet. The Cu(II) enzyme is not active and the axial symmetry may be forced by this ion. Thus precise analogies between the Cu(II) enzyme and those containing other transition and IIB metals ions as regards symmetry of the coordination sphere and even ligands must be made with caution.

Addition of a second pair of Cu(II) ions generates an ESR spectrum which consists of signals from two overlapping Cu(II) sites with different g and A values (Fig. 11*H*). From the detail of the low-field region made up of two overlapping Cu(II) hyperfine lines, the spectrum from the original site with the nitrogen hyperfine structure remains intact and a new signal of smaller $A_{\parallel}$ is superimposed. From the ESR spectrum of $Cu(II)_4AP$ there is no evidence of significant spin–spin interaction between the Cu(II) ions in the two

sites. It is possible that in contrast to Mn(II), Cu(II) cannot occupy the A_2 site (see Sec. 7).

10 ^{113}Cd AND ^{111}Cd NMR OF Cd(II) ALKALINE PHOSPHATASE

The substitution of Cd(II) for Zn(II) in zinc-containing metalloenzymes is an obvious and attractive means of opening up these systems to observation by the powerful techniques of nuclear magnetic resonance. The span of chemical shifts observed for ^{113}Cd(II) [or ^{111}Cd(II)] complexes with the potential donor groups expected to occur in proteins is ~800 ppm (96, 97). The softer thiol ligands found in metallothionein (97) are strongly deshielding with respect to nitrogen ligands (96, 98, 99) which are in turn deshielding with respect to oxygen ligands such as carboxyls (100–102). This large range of chemical shifts makes ^{113}Cd a potentially sensitive probe of the structure of metal-binding sites in enzymes as well as of conformational changes which may accompany binding of additional ligands. ^{113}Cd(II) alkaline phosphatase illustrates this sensitivity well, but also highlights the complexities associated with metal binding to multisubunit proteins with several potential metal-binding sites.

The affinity of apoalkaline phosphatase for Cd(II) is dependent on pH and the presence or absence of phosphate (see Sec. 6). At pH 6.5 only one pair of cadmium atoms binds tightly, although a stoichiometry of six Cd(II) ions per dimer (binding to A, B, and C sites) can be achieved at high [Cd(II)] or in the presence of phosphate. At pH 6.5, with only two sites occupied, ^{113}Cd(II)$_2$AP has an NMR spectrum consisting of only one sharp resonance at 169 ppm relative to cadmium perchlorate, arising from symmetrical binding of cadmium at both A sites (42) (Fig. 12*A*). The A site appears to correspond to the low-symmetry or catalytic Zn(II) site defined earlier by various spectroscopic techniques. The identical NMR signal from both Cd(II) sites is additional evidence that the dimer has twofold symmetry. This narrow resonance at 169 ppm diminishes linearly in intensity, without being broadened, upon titration of the enzyme with a second pair of cadmium ions (Fig. 13). Allowing for the decreased signal-to-noise, the ^{113}Cd NMR spectrum of ^{113}Cd(II)$_4$AP resulting from this titration can be collected and consists of a broad A-site resonance at 161 ppm of intensity 2, that is, only ~50% of the Cd(II) (Fig. 12*C*). The resonance of ^{113}Cd(II) at the B sites is apparently too broad to be detected (84).

One of the more interesting features of the alkaline phosphatase molecule revealed by ^{113}Cd NMR is the finding that it is not possible to treat the protein as a rigid ligand. The most striking phenomenon reflecting this is the chemical exchange modulation that must occur to account for the disap-

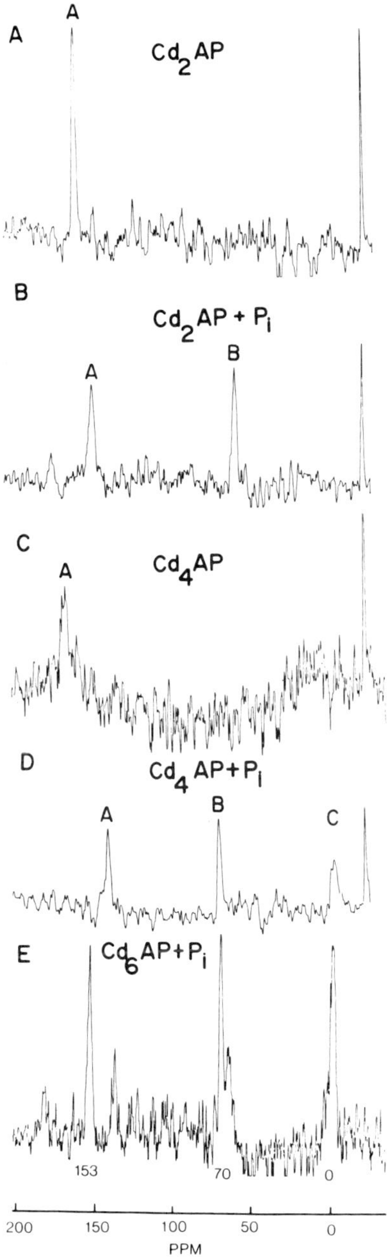

Figure 12 ^{113}Cd-NMR spectra of ^{113}Cd-substituted alkaline phosphatase as functions of Cd(II) and phosphate stoichiometry. (*A*) $^{113}Cd_2AP$, pH 6, no phosphate. (*B*) Sample A + phosphate. (*C*) $^{113}Cd_4AP$, pH 6, no phosphate. (*D*) Sample C + phosphate, pH 6.5 (*E*) $^{113}Cd_6AP$ + phosphate, pH 8.3 ^{113}Cd NMR spectra were obtained on a Bruker CXP-200 at 44.37 MHz and chemical shifts are expressed relative to $Cd(ClO_4)_2$. Standard at the far right is Cd acetate resonating under these conditions at −17 ppm. Data compiled from ref. 84.

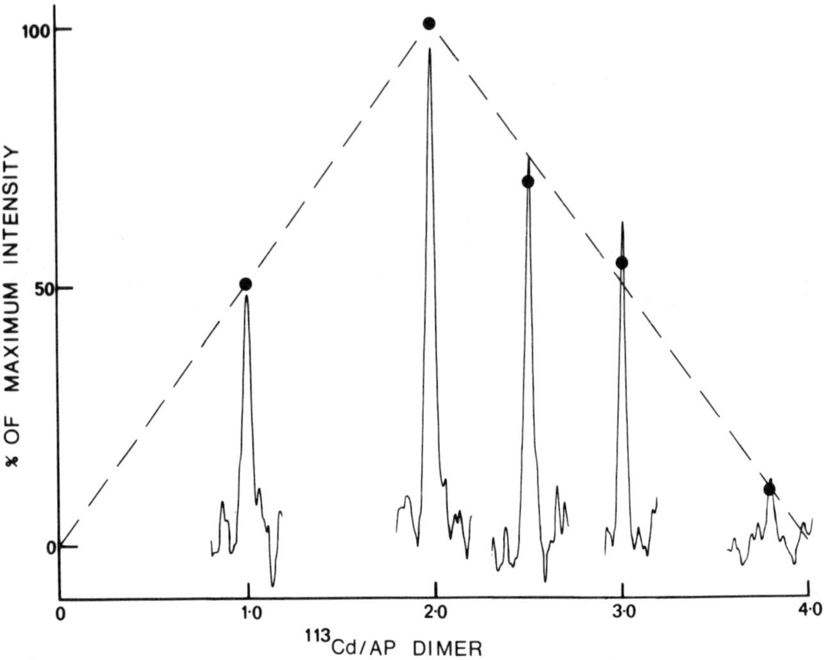

Figure 13 ^{113}Cd NMR of Cd(II)-alkaline phosphatase during progressive addition of 1 to 4 g at ^{113}Cd/mol of alkaline phosphatase dimer, pH 6.5, 0.01*M* Tris HCl-0.1*M* NaCl, 25°C. (●) Represents the integrated area of the resonance. Spectra were recorded at 19.96 MHz and are the average of 46K transients. From ref. 84.

pearance of the sharp A-site ^{113}Cd resonance of $^{113}Cd(II)_2AP$ as the second pair of ^{113}Cd(II) ions is added and its replacement by a severely broadened A-site resonance. To explain the lack of broadening of the sharp 169 ppm resonance of $^{113}Cd(II)_2AP$ as it diminishes in intensity, it is necessary to propose the existence of a high conformational barrier between the species that give rise to the sharp and broad A-site resonances, which are only 1 ppm apart. While the basic mechanism may be the exchange of the central metal ion at the B site (whose resonance has completely disappeared), the A-site metal must remain tightly bound (78). Thus the broadening of its signal must reflect conformational flux transmitted to the A site and sensed as a modulation of the chemical shift. The frequency of such a conformational flux required to result in the intermediate chemical exchange condition depends on the chemical shift difference between the species in flux. For a chemical shift difference of 25 ppm (500 Hz at 19.96 MHz) as estimated from the A-site resonance at 169 ppm when B site is empty and 144 ppm when B site is occupied (Fig. 12), the required rate would be between 3×10^2 and 3×10^4

sec^{-1}. As argued in Sec. 6, the exchange of the second pair of Cd(II) ions in the absence of phosphate ligand at pH 6.5 could be this rapid. This modulation is removed by phosphorylation of the $Cd(II)_4AP$ species attesting to the increased rigidity of the phosphoenzyme (Fig. 12*D*). Accompanying phosphorylation of $Cd(II)_4AP$, however, part of the Cd(II) migrates to a third or C site (Fig. 12*D*). All three sites are not fully occupied until a total of six Cd(II) ions are added to the phosphoryl enzyme (Fig. 12*E*). In contrast phosphorylation of $Cd(II)_2AP$ at pH 6.5 results in the replacement of the original A-site resonance by two new resonances of equal intensity (Fig. 12*B*). As will be shown below this results from the migration of Cd(II) from the A site of one monomer to occupy the B site of the phosphorylated monomer.

An increase in the pH of $Cd(II)_2$ alkaline phosphatase has the same effect as phosphorylation and results in a migration of cadmium from A site in one subunit to the B site of the other (as defined by the ^{113}Cd NMR), thus producing an asymmetric metal ion distribution of one monomer with two cadmium ions and the other devoid of metal. In the NMR spectrum this is seen as a gradual disappearance of the 169 ppm resonance and its replacement by two resonances of equal intensity at 144 and 52 ppm arising from A and B sites, respectively, as the pH rises (Fig. 14). The most stable species thus appears to be an active center in which both A and B sites are occupied. The reasons for a nonstatistical arrangement of these in which all dimers have one monomer with A and B sites filled, the other vacant are discussed below.

10.1 Phosphorylation of $^{113}Cd(II)_2$ Alkaline Phosphatase

Phosphorylation of the $^{113}Cd(II)_2AP$ at pH 6.5 induces a slow rearrangement of cadmium from the A site of one subunit to the B site of the other subunit (103). At enzyme concentrations of 2m*M* it is possible to follow the time course of formation of the phosphoseryl residue by ^{31}P NMR and the $^{113}Cd(II)$ migration by ^{113}Cd NMR as illustrated in Figure 15. Following 72 hr of incubation with phosphate, the single ^{113}Cd resonance of $^{113}Cd(II)_2AP$ has become two equal resonances characteristic of A- and B-site occupancy (Fig. 15, $t = 0$ and $t = 72$ hr) and the ^{31}P resonance shows the phosphoenzyme to be exclusively E-P (^{31}P signal at 8 ppm) (Fig. 15, $t = 66$ hr). In contrast, the ^{113}Cd spectrum taken at the intermediate time of 14 hr shows three closely spaced ^{113}Cd resonances near the chemical shift of the original A-site $^{113}Cd(II)$, two of approximately equal intensity at 163 and 152 ppm and the third, somewhat smaller resonance at 141 ppm (Fig. 15, $t = 14$ hr).

The ^{31}P spectra from the same sample can be recorded at more frequent intervals because of the greater sensitivity of ^{31}P NMR. At $t = 1$ hr the bound phosphate is present predominantly as the noncovalent complex (E·P) as

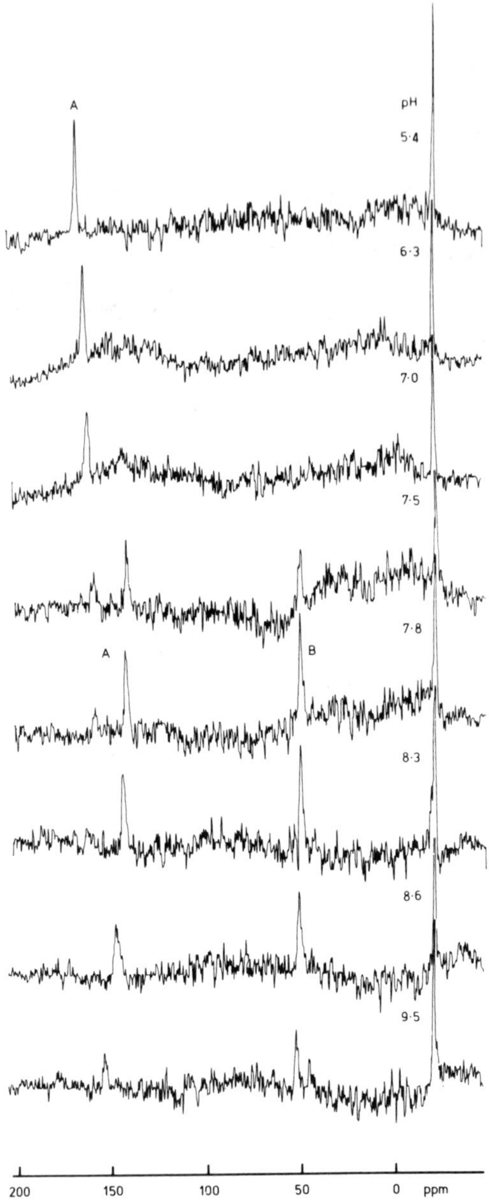

Figure 14 pH titration of $^{113}Cd(II)_2$ alkaline phosphatase (2.50m*M*) followed by ^{113}Cd NMR (19.96 MHz). Spectra are the average of 120,000 scans. Designations A and B correspond to the metal-binding site responsible for the resonance. From ref. 84.

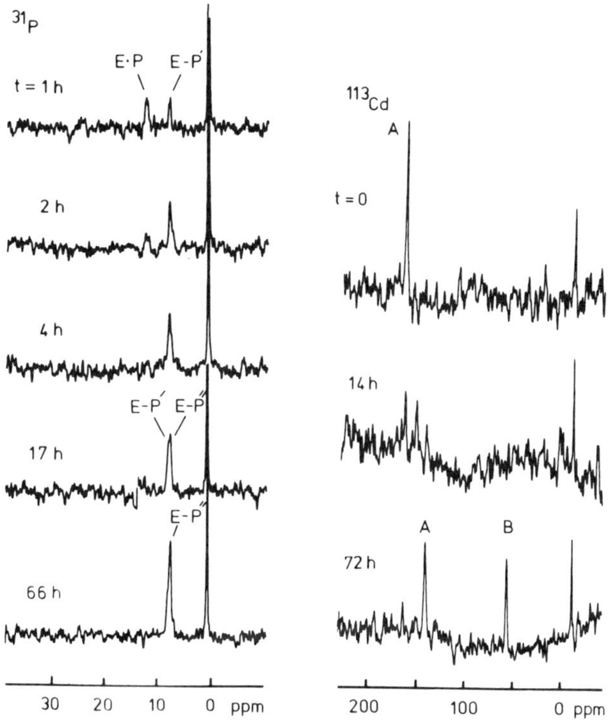

Figure 15 Time course of phosphorylation of $^{113}Cd(II)_2$ alkaline phosphatase (2.47m*M*, pH 6.5) followed by ^{31}P NMR (80.9 MHz) and ^{113}Cd NMR (44.37 MHz). At t = 0, two equivalents of inorganic phosphate were added and ^{31}P and ^{113}Cd spectra recorded at the times indicated. Each ^{31}P spectrum is the average of 1500 scans (30° pulse angle, 3 sec repeat time). ^{113}Cd spectra are the average of 60,000 scans. Extreme upfield cadmium resonance is an external $^{113}Cd(acetate)_2$ standard. Times are given for the midpoint of each accumulation. From ref. 84.

shown by its characteristic ^{31}P resonance at 12.79 ppm (19) (Fig. 15, t = 1 hr). The E·P resonance slowly disappears to be replaced by the resonance at 8.30 ppm characteristic of the E-P species (19). By t = 4 hr, 0.6 equivalents of phosphate are bound and by t = 17 hr the maximum formation of one equivalent of E-P has been completed.

The ^{31}P spectrum at 17 hr was obtained immediately after completion of the intermediate ^{113}Cd spectrum shown in Figure 15 at t = 14 hr. Hence phosphorylation has occurred before significant cadmium rearrangement has taken place. The E-P resonance at this stage consists of two overlapping resonances of slightly different chemical shift. One is the chemical shift observed for E-P at final equilibrium (72 hr) and the other, slightly downfield, is the chemical shift first observed for E-P at 1–4 hr (Fig. 15, t = 1, 2, and 4

hr). This suggests that two different conformations of E-P are present early in the phosphorylation. The three ^{113}Cd resonances in Figure 15 (t = 14 hr) would thus all appear to arise from A-site ^{113}Cd(II), the most downfield from unphosphorylated active centers, while the two upfield arise from active centers containing the two conformers of E-P. Over the next 48 hr the ^{113}Cd(II) rearranges and now gives rise to two ^{113}Cd resonances of equal intensity, one at the most upfield position (141 ppm) of the A-site group and a new one far upfield at 55 ppm which arises from the B-site ^{113}Cd(II). The total amount of E-P has not changed during this rearrangement, but the E-P resonance has now become homogeneous (Fig. 15, t = 66 hr) and the system has reached equilibrium.

From the time course of the changes in the ^{31}P and ^{113}Cd spectra during phosphorylation of ^{113}Cd(II)$_2$AP at pH 6.5 (Fig. 15), it is apparent that formation of E·P and phosphorylation of Ser-102 from this complex can occur with only the A-site metal ion in place. The conformation of the active center in the E-P state is such that the potential stability of the A + B site in the presence of E-P is so great as to cause the depopulation of half the A sites.

The sequence of phosphorylation of ^{113}Cd(II)$_2$ over a 4 hr period followed by migration of half the ^{113}Cd(II) from A site of the unphosphorylated monomer to the B site of the phosphorylated monomer requires a ligand-induced asymmetry in the dimeric protein. This asymmetry must be introduced to explain the phosphate-induced transition between the two most stable states; that is, the initial one with Cd(II) in two identical equally stable A sites and the final one with Cd(II) occupying both the A and B sites of the phosphorylated monomer. No metal is present in either the A or B sites of the unphosphorylated monomer. One could postulate that an asymmetry is initially present in the dimeric enzyme which favors phosphorylation of only one subunit and that presently available probes are not sensitive enough to pick up the inequivalence of the monomers. These probes, however, include ^{113}Cd NMR, which shows identical metal-binding sites (Fig. 15), ^{13}C NMR of the [γ-^{13}C]-labeled His residues, which shows one ^{13}C resonance for each pair of histidines (104), and the crystal structure, which shows a molecular two-fold axis to exist between the monomers of the dimer in the unliganded enzyme (17).

Phosphorylation of the first active site containing the A-site metal ion momentarily renders the system asymmetric by virtue of the bound phosphate, but if no ligand-induced conformational changes were propagated to the site on the other monomer, the second site should independently phosphorylate and symmetry would be restored. This would remove the driving force for Cd(II) migration and result in phosphorylation of both sites. That does not occur in Cd(II)$_2$AP (Fig. 15).

10.2 Slow Equilibria Associated with Phosphorylation of $Cd(II)_2$ Alkaline Phosphatase

In this section we show that the equilibrium percentage of E·P increases along a pH profile related to the activity profile and typical of each metal ion stoichiometry. For the $Cd(II)_2$ enzyme, where the E-P ⇌ E·P equilibrium is shifted far to alkaline pH, the phosphoenzyme, if formed at pH 6 and titrated to pH 9, still has a E-P:E·P ratio of ~85:15, even at pH 9.0. In contrast, if $^{113}Cd(II)_2AP$ formed at pH 6.5 is first taken to pH 9 [where migration of half the Cd(II) ions from the A site to the B site of the opposite monomer has occurred] and is then phosphorylated, the phosphoenzyme formed in the first hour is 85–90% E·P as shown by the ^{31}P NMR signal at 13.38 ppm (Fig. 16*A*). The ^{113}Cd spectrum of this sample (E·P complex) consists of two resonances at 168 ppm (A′) and 50 ppm (B′) (Fig. 16*A*). This is not the final equilibrium species, however. Over the next several days there is a gradual change in both the ^{31}P and ^{113}Cd spectra of the enzyme; a change which finally results in a complete reversal of the E·P:E-P ratio. After two days the E·P:E-P ratio has fallen from 90:10 to 50:50 as shown by the ^{31}P NMR, and the ^{113}Cd spectrum shows two sets of ^{113}Cd resonances with half of the original resonance intensity in each set (Fig. 16*B*). One set occurs at the original chemical shifts of 168 and 50 ppm (A′ and B′) and the new set at 144 and 57 ppm (A and B). Ten days after addition of phosphate the E·P:E-P ratio has fallen further to 10:90, and the ^{113}Cd spectrum has converted almost exclusively to the species with signals at 144 and 57 ppm (Fig. 16).

Thus for the $^{113}Cd(II)_2$ enzyme at pH 9.0, with the A and B sites of only one monomer occupied, formation of E·P is rapid, but phosphorylation of Ser-102 from this complex is very slow. The slow phosphorylation step allows the demonstration that the bonding around the Cd(II) ions in both A and B sites depends on which phosphoenzyme intermediate is present, E·P or E-P. A-site Cd(II) is more sensitive, the chemical shift is ~25 ppm further upfield in the E-P than in the E·P complex (Fig. 16). The ^{113}Cd NMR signal of the B site is 6 ppm further downfield for E-P than for E·P. The greater sensitivity of the A-site Cd(II) may reflect the direct coordination of the phosphate of E·P to this metal ion (see Sec. 12). The chemical shifts of the $^{113}Cd(II)$ ions in the $^{113}Cd(II)_6AP$ show the same sensitivity to the E-P ⇌ E·P equilibrium, except that for the A site the E-P resonance is now 16 ppm downfield (rather than upfield) of that for the E·P complex. While the ^{31}P spectrum is not shown, the small extra ^{113}Cd signals in Figure 12*E* represent the residual amount of E·P present in this sample (84). That the sign of the change in chemical shift of the A-site resonance induced by formation of E-P from E·P depends on the total stoichiometry of Cd(II) (two or six ions per dimer), despite the fact that both active centers contain A- and B-site Cd(II)

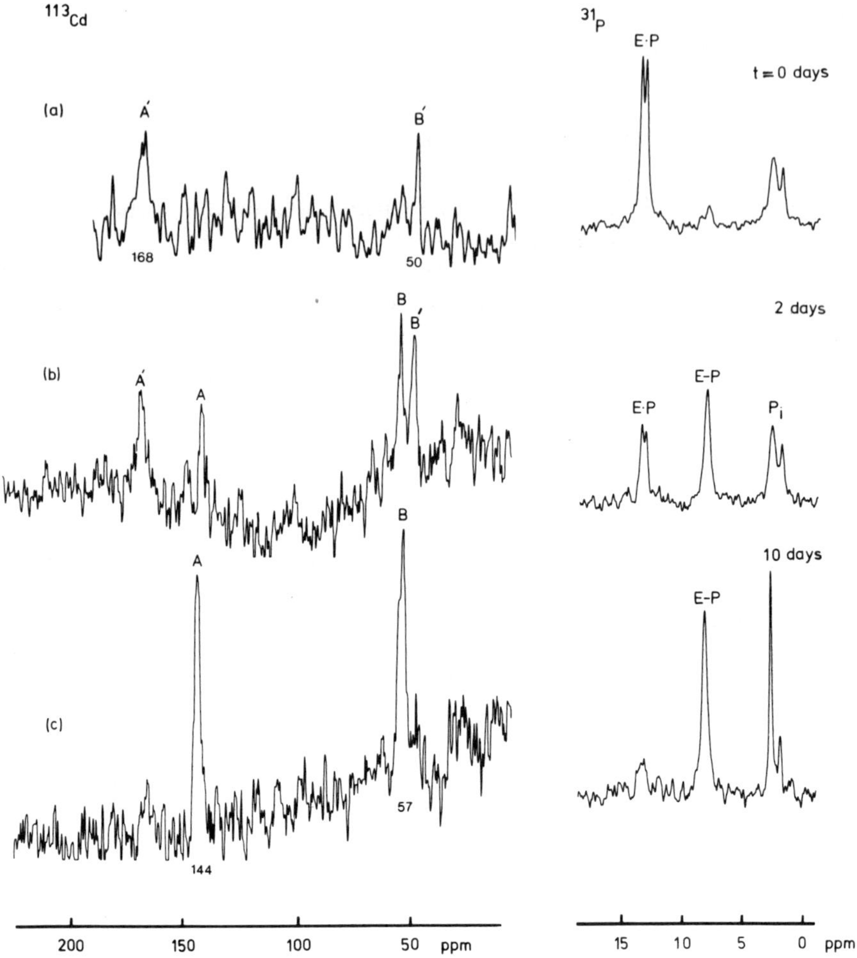

Figure 16 Time course of the slow phosphorylation of $^{113}Cd(II)_2$ alkaline phosphatase (2.76mM) containing two equivalents of P_i at pH 9.0. Spectra on the right are ^{31}P NMR (80.9 MHz) and on the left are ^{113}Cd NMR (44.3 MHz, *a*) and (19.96 MHz, *b* and *c*). Phosphate was added to $^{113}Cd(II)_2AP$, pH 9 at t = 0. For the ^{113}Cd spectra 37,000 (*a*), 116,000 (*b*), and 162,000 (*c*) scans were recorded. For the ^{31}P spectra 18,000 (*a*), 12,000 (*b*), and 6,000 (*c*) scans were recorded. From ref. 84.

ions, not only illustrates the great sensitivity of ^{113}Cd chemical shifts, but is additional evidence of subunit-subunit interactions in this enzyme.

11 ^{13}C, ^{19}F, AND ^{35}Cl NMR OF ALKALINE PHOSPHATASE

11.1 ^{13}C NMR Studies of β,β-[γ-^{13}C]Dideuterohistidine-Labeled Alkaline Phosphatase

Given the low natural abundance of ^{13}C (1.1%), carbon NMR studies of macromolecules as large as alkaline phosphatase require significant enrichment of ^{13}C label for satisfactory sensitivity. This is usually achieved by biosynthesis using a bacterial auxotroph for the [^{13}C]-labeled amino acid of interest. Thus ^{13}C resonances from one type of residue are selectively enhanced. If there are strong indications of which residues are critical to structure or function, this is a very useful spectral probe. For alkaline phosphatase the amino acids known or suspected to be of importance in the active center are histidine (64, 105), serine (34), and arginine (106). Of these only the role of histidine has been investigated by ^{13}C NMR, using an alkaline phosphatase biosynthetically labeled with β,β-[γ-^{13}C]dideuterohistidine (104, 107). At 25 MHz resonances from nine of the ten histidines in each subunit were resolved for the metal-containing enzyme and at 50 MHz it was possible to distinguish all ten resonances (Fig. 17). For preparations containing a symmetrical distribution of metal ions, each pair of histidines in the dimer gives rise to only a single resonance, indicating conformational equivalence of the monomers or rapid interconversion. In contrast to the narrow range of chemical shifts found previously for histidine C_γ resonances in diamagnetic proteins (~7 ppm) (108), the resonances in both zinc and cadmium alkaline phosphatases are spread over 13–14 ppm. Only four show the normal pH titration behavior of a solvated histidine side chain. The values of their chemical shifts and relaxation times (T_1) also indicate that they are relatively mobile probably surface residues. The remaining six residues must be in environments that prevent protonation or deprotonation over the pH range studied (pH 5–10) (107) and give rise to the observed spread of chemical shifts. Their T_1 values are two to eight times longer than for the surface histidines and suggest a much higher degree of immobilization due either to being buried or rigidly coordinated.

By constituting separate samples of the enzyme with $^{112}Cd(II)$ or $^{113}Cd(II)$ it was shown that there is direct ^{113}Cd—^{13}C three-bond coupling to three histidyl side chains. Coordination of two of them was assigned to the A-site metal (resonances at 128.6 and 127.8 ppm) and the third to B-site metal (resonance at 127 ppm) (Fig. 17). It was proposed that two further histidyls

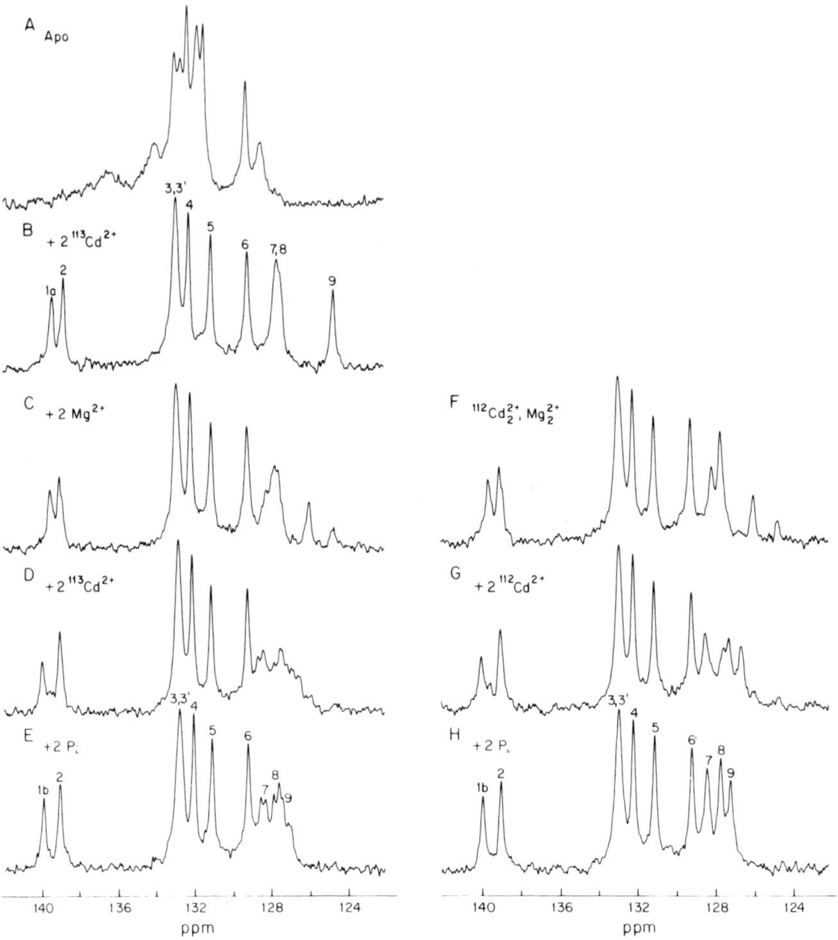

Figure 17 ^{13}C NMR spectra (50.3 MHz) of [γ-^{13}C]-histidyl-substituted alkaline phosphatase. (*A*) apoenzyme; (*B*) plus 2 ^{113}Cd(II); (*C*) *B* plus 2 Mg(II); (*D*) Cd(II)$_4$AP plus 2 Mg(II); (*E*) *D* plus 2 P_i. (*F*) ^{112}Cd(II)$_2$Mg(II)$_2$AP; (*G*) ^{112}Cd(II)$_4$AP plus 2 Mg(II); (*H*) *G* plus 2 P_i. From ref. 104.

whose resonances are shifted >3 ppm downfield upon binding of Cd(II) or Zn(II) to A site (resonances 1 and 2, Fig. 17), are also coordinated to the A-site metal (107). The downfield chemical shift suggests coordination through the π-nitrogen of the imidazole ring rather than the more usual τ-nitrogen coordination. For ^{13}C the resulting two-bond coupling constant is expected to be smaller than the three-bond coupling constant and hence the coupling may not be resolved.

When two equivalents of Mn(II) are bound to apoalkaline phosphatase

the two most downfield histidyl ^{13}C resonances as well as the other three shown from ^{113}Cd spin–spin coupling to be implicated in metal coordination are severely broadened (104), demonstrating that these five are all within ~10 Å of the Mn(II) ion, although it does not prove that all five are coordinated. The other five histidyls are definitely outside the immediate active-center region. The chemical shifts of the two most downfield resonances, however, are not affected by phosphate–metal coordination, which is known to occur to the A-site metal ion in the E·P complex (51, 87). In contrast, the ^{13}C resonances of the two histidyls coordinated to A-site cadmium (as shown by $^{113}Cd—^{13}C$ coupling) are significantly affected by phosphate binding, while the B-site-coordinated histidine resonance is shifted much less and the remaining residues not at all (107). This may raise some question about the direct coordination of the two histidyls responsible for the two downfield resonances. While the three upfield spin-coupled resonances are all sensitive to occupancy of both A and B sites by cadmium, only one of the two downfield resonances senses B-site occupation, which is somewhat unexpected if both residues are metal ligands since the A and B metal sites appear to be only ~3.9 Å apart (81). It was also shown that the major changes in chemical shift of the ^{13}C resonances produced by saturating A and B sites with metal were reproduced in the zinc enzyme. Thus the three resonances that shift by ~3 ppm upon metal binding are all similarly affected by zinc. At zinc : enzyme stoichiometries of less than 4 : 1 there is evidence for heterogeneity arising from the high affinities of both A and B sites for zinc, which results in a statistical distribution of metal among the available sites. Addition of a third pair of metals to $Zn(II)_4AP$, whether zinc or magnesium, neither alters the homogeneity of the spectrum nor appreciably alters the chemical shifts of the resonances, as would be expected for occupation of a site thought to consist solely of carboxylate ligands.

11.2 ^{19}F NMR of Alkaline Phosphatase Containing Fluorinated Aromatic Amino Acids

Fluorinated amino acids tend to have an inhibitory effect on cell growth, making it more difficult to use auxotrophic bacteria to incorporate these labeled residues into proteins of interest than it is to incorporate isotopically labeled amino acids. The biosynthesis of alkaline phosphatase from *E. coli* is maximal after all growth has leveled off and therefore, by using appropriate auxotrophs, incorporation of fluorinated amino acids into alkaline phosphatase can be obtained in good yield by adding the labeled amino acid only at the stage of greatest enzyme production. In this way *m*-fluorotyrosine (109), *p*-fluorophenylalanine (67), and 4-fluorotryptophan (110) have been separately incorporated into alkaline phosphatase.

There are 11 fluorotyrosine residues per subunit, each of which gives a resolvable fluorine resonance in the native enzyme (Fig. 18). As with the ^{13}C resonances of γ-^{13}C-enriched histidyl AP, the fluorine nuclei from each pair of residues in the dimer produce only one resonance, indicating either equivalent monomer structures or a time-averaged equivalence. Because of the potential of tyrosine for hydrogen bonding as well as for simple van der Waals interactions, the chemical shift range of fluorotyrosines in AP is relatively large (12 ppm), giving quite good separation of resonances and the potential for monitoring conformational changes at many points throughout the structure. In later work (111), a significant narrowing of the linewidths was achieved by incorporating perdeuterated fluorotyrosine. This is done in preference to simple proton decoupling, since the latter gives a near nulling of most of the ^{19}F signals due to unfavorable nuclear Overhauser enhancements in macromolecules where τ_c, the rotational correlation time, is long.

It was shown that while no tyrosine is directly involved in metal or phos-

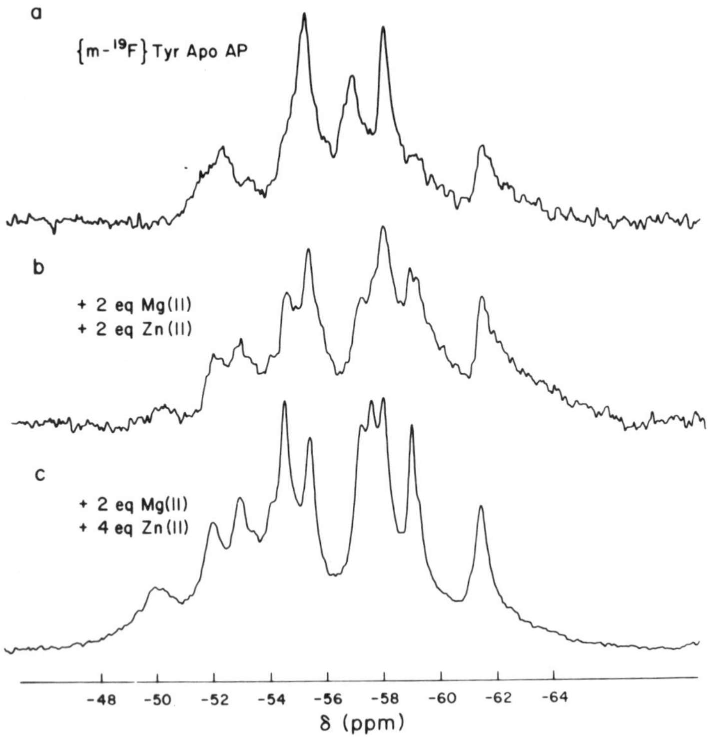

Figure 18 ^{19}F NMR spectra (84.6 MHz) of *m*-F-tyrosine-substituted alkaline phosphatase. From ref. 67.

phate binding, changes in ^{19}F chemical shift could be used to monitor conformational changes produced by zinc, cobalt, and phosphate binding (111). Zinc:enzyme stoichiometries of less than 4:1 produce complex ^{19}F spectra in keeping with the heterogeneity expected to arise from the high affinity of both metal-binding sites for zinc. In contrast to more direct observations to be presented below, it was inferred that the $Zn(II)_4$ enzyme binds two equivalents of phosphate with $K_D < 5 \times 10^{-6} M$ at pH 8.

Fluorophenylalanyl alkaline phosphatase proved to be of much less utility than the fluorotyrosyl enzyme. Apart from a relatively sharp resonance at −36 ppm attributed to denatured enzyme, the resonances of the eight pairs of phenylalanines were unresolved for the most part, covering a chemical shift range of only ~5 ppm (67).

The situation with tryptophan substitution is greatly simplified by there being only three residues per subunit (although it was believed at the time that there were four residues). In the fluorinated enzyme three distinct resonances can be seen, with a spread of 6 ppm (110). None occurs at the chemical shift of free 4-fluorotryptophan (45.6 ppm). Interestingly, the authors claim that the activity of 4-fluorotryptophan-containing AP is consistently 15% higher than for the normal enzyme. This is the only reported work on the enzyme.

11.3 ^{35}Cl NMR of Alkaline Phosphatase

^{35}Cl is a quadrupolar nucleus with narrow NMR signals only when placed in a symmetrical environment and when the correlation time, τ_c, associated with its motion is short. Cl^- bound to macromolecules fulfills neither of these conditions and produces linewidths in the kHz range. On the other hand, if fast exchange of Cl^- exists between aqueous and protein environments, perturbations of the free signal can be monitored for small percentages of macromolecule-bound Cl^- and the fully bound linewidth calculated. The method is therefore indirect, but can be used not only to monitor Cl^- binding, but other structural changes in the molecule which influence the bound linewidth.

^{35}Cl NMR studies on alkaline phosphatase have reported the effects of zinc:enzyme stoichiometry and phosphate binding on Cl^- linewidths (112–114). Cottam and Ward (112) reported that four Zn(II) ions were necessary to restore complete activity to apoalkaline phosphatase. These four ions had no effect upon the Cl^- linewidth. Zinc in excess of four equivalents resulted in a linear broadening. Norne et al. (113) countered this by claiming that only two equivalents of Zn(II) could be added without affecting the Cl^- linewidth, but later reported that four equivalents of Zn(II) produced no

broadening (114). The contradictory results were explained by postulating two fully active forms of alkaline phosphatase with different Zn(II) binding properties.

From ^{113}Cd NMR studies it is known that the A-site cadmium resonance is very sensitive to Cl^- concentration, changing its chemical shift by ~50 ppm between no NaCl and 0.1*M* NaCl present (98). This suggests a direct Cd(II)—Cl^- interaction. Work from this laboratory shows a substantial broadening of the Cl^- resonance by $Cd(II)_6AP$, approximately four times the magnitude of the effect shown by $Zn(II)_4AP$ (unpublished results).

The only reported effect of phosphate binding on the Cl^- resonance is the report that upon phosphate addition to $Zn(II)_2AP$ a reduction in the excess linewidth is observed for the first equivalent of phosphate only (113). In view of the evidence for the heterogeneity of AP samples at less than 4:1 zinc:enzyme stoichiometries, this result cannot be interpreted unambiguously. Recent work by Gettins and Coleman (unpublished results) suggests that for $Zn(II)_4AP$ at pH 8 only the first equivalent of phosphate has an effect on the broadening of the Cl^- resonance. Under the conditions of the experiment it is known from ^{31}P NMR that only about one equivalent of phosphate is tightly bound per AP dimer (51). At lower pH, in the region where phosphate is bound as a mixture of noncovalent E·P and covalent E-P, there appears to be a correlation between the net Cl^- excess linewidth and the percentage of E·P present. Formation of the E-P complex has little effect on the Cl^- linewidth, whereas formation of E·P reduces it substantially compared with the unphosphorylated zinc enzyme (our unpublished results). This suggests a competition between P_i and Cl^- for coordination at zinc.

12 ^{31}P NMR OF PHOSPHOENZYME INTERMEDIATES

The reactivities of the phosphoenzyme intermediates, E-P and E·P, the hydrolysis of the former and dissociation of the latter, are the key elements in the alkaline phosphatase mechanism. Studies of the exchange of ^{18}O from $H_2{}^{18}O$ into inorganic phosphate catalyzed by alkaline phosphatase (115, 116) and the demonstration using ^{16}O, ^{17}O, ^{18}O chiral phosphate esters that enzymatic hydrolysis proceeds by retention of configuration around phosphate (117) strongly suggest that E-P is an essential intermediate even at alkaline pH where equilibrium concentrations of this intermediate are small. The chemical shifts, the presence of metal-phosphorus coupling in E·P but not E-P, and the change in concentration of E-P and E·P as a function of pH observed by ^{31}P NMR provide much new information on the chemistry of these intermediates.

12.1 ^{31}P NMR Spectra of Homogeneous Zn(II)-Containing Alkaline Phosphatase Species

At zinc:enzyme stoichiometries of less than 4:1 the ^{31}P NMR spectrum resulting from phosphorylation is usually heterogeneous (118). This arises from the high affinities of both A and B as well as other sites for zinc (78, 119), so that binding is not sufficiently selective to give homogeneous $Zn(II)_2AP$ species when additions are made at the millimolar concentrations used for NMR experiments. Only by adding sufficient zinc to occupy both pairs of A and B sites fully can a homogeneous ^{31}P NMR spectrum be observed. Spectra for this species are shown in Figure 19 for the pH range 5.0–6.0. At pH 5.0 two resonances are seen, one at 8.60 ppm corresponding to E-P and a second at 4.23 ppm arising from E·P. The percentage of phosphate bound as E-P decreases, while that bound as E·P increases as the pH is raised. The midpoint ([E-P] = [E·P]) of the transition is ~pH 5. There is no variation in chemical shift for either E·P or E-P resonances over this pH range. At pH values above 6 no E-P is seen, while above pH 7 a growing resonance from dissociated P_i is visible. The chemical shift of E·P changes slightly at higher pH, moving from 4.16 ppm at pH 6.0 to 3.94 ppm at pH 9.1.

A complete pH titration of the ^{31}P NMR signals from the phosphoenzyme intermediates formed by $Zn(II)_4Mg(II)_2AP$ with 1 mol of phosphate bound at pH 4.9 is shown in Figure 20. In the low pH range, 4.9–6.9, the major equilibrium species shifts from E-P to E·P without dissociation of E·P. Above pH 7.0, however, K_d for E·P increases rapidly to ~mM with the appearance of a major signal at the chemical shift of P_i. The NMR findings suggest a larger K_d for phosphate in the alkaline pH range than estimates based on equilibrium dialysis at much lower concentrations of enzyme which have ranged between 10^{-6} and $10^{-5}M$ (52, 120). Several determinations of the kinetic inhibition constant, K_i, for phosphate at pH 8 have been made and the values range between 0.6×10^{-6} and $10^{-5}M$ (49, 121–124). K_i has generally been assumed to be relatively pH independent. With the recognition that at pH 7, where significant equilibrium concentrations of E-P are no longer present, the rate-limiting step has shifted from hydrolysis of E-P to dissociation of E·P (41, 87), one must infer that the dissociation rate of E·P increases dramatically between pH 7 and 10, since the enzyme activity rises dramatically over this pH range along a sigmoid pH-rate profile with a midpoint near pH 8 (45). The ^{31}P NMR in Figure 20 supports such a picture. The factors that influence the magnitude of K_d for E·P require further investigation.

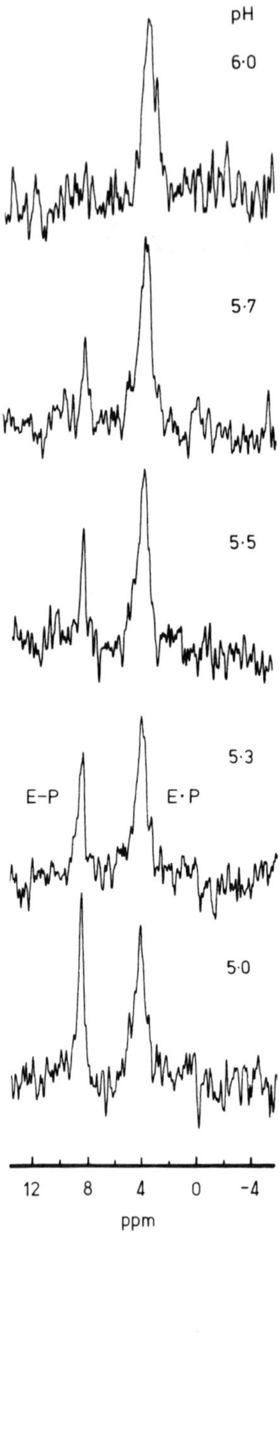

Figure 19 80.9 MHz ^{31}P NMR spectra of $Zn(II)_4$ alkaline phosphatase containing two equivalents of phosphate as a function of pH. Enzyme concentration was 1.72m*M*. Each spectrum represents the average of 9000 transients. From ref. 51.

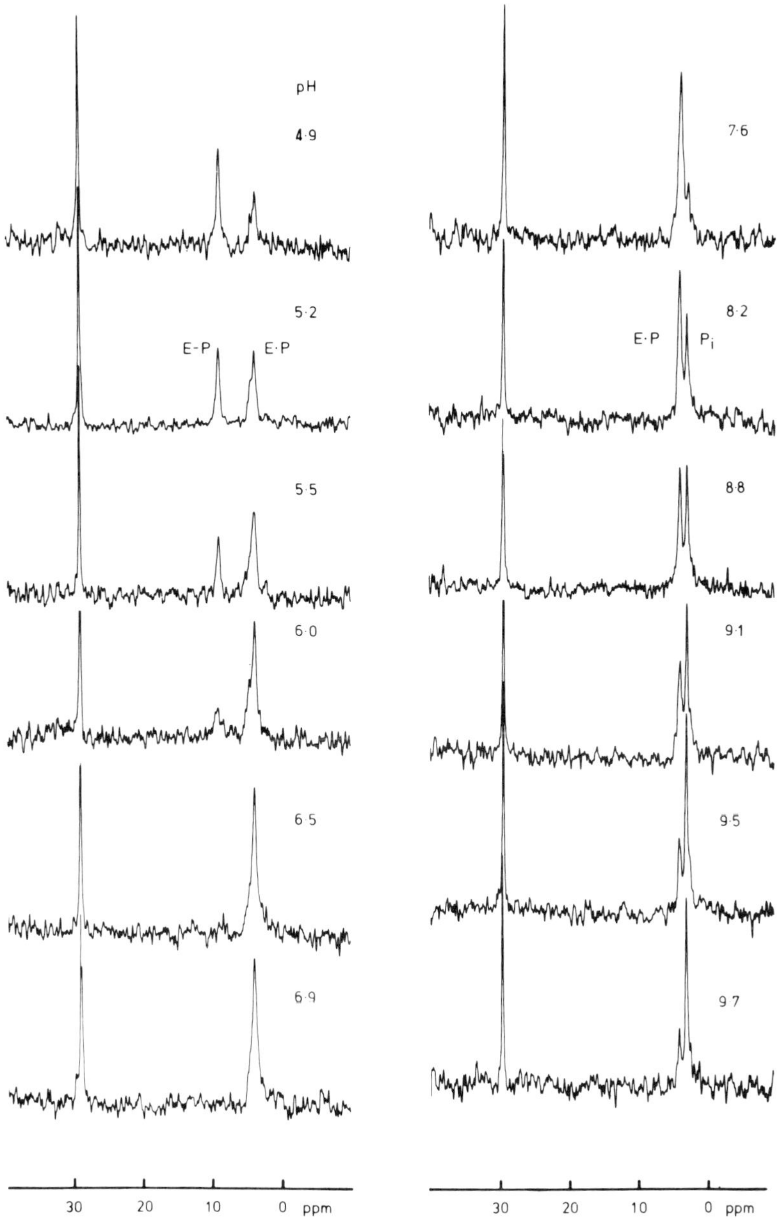

Figure 20 ^{31}P NMR spectra (80.9 MHz) of $Zn(II)_4Mg(II)_2AP$, containing one equivalent of phosphate, as a function of pH. Enzyme concentration was 2.13m*M*. Spectra are the average of 14,000 transients. Most downfield resonance is external methyl phosphonate. From ref. 51.

12.2 Homogeneous Cd(II)-Substituted Alkaline Phosphatase

The effects on the ^{31}P NMR spectrum of phosphorylated $^{113}Cd(II)_6AP$ (one equivalent of phosphate added) as the pH is raised are shown in Figure 21. At low pH, where the species is first formed, only a single resonance is seen, arising from the phosphoseryl residue of the E-P complex. Its chemical shift is 8.45 ppm and it appears to be a singlet in the proton-decoupled spectrum, although any coupling less than ~20 Hz would go undetected given the linewidth of ~43 Hz. By pH 8 the percentage of bound phosphate existing as E·P is sufficiently high (~10%) to make its resonance clearly discernible at 13.02 ppm. The ^{31}P resonance of E·P is a doublet with a coupling of ~33 Hz to a single $^{113}Cd(II)$ nucleus. The intrinsic linewidth of the resonance is ~40 Hz. Both the E·P and E-P resonances have comparatively narrow linewidths, which may reflect some rotational freedom about bonds to the phosphorus. In marked contrast to the Zn(II) enzyme, phosphate does not dissociate from the Cd(II) enzyme until above pH 9.0, when a resonance from free P_i begins to appear. A second equivalent of phosphate added to the enzyme is bound as E·P and E-P without altering their relative proportions.

$^{113}Cd(II)_2AP$ forms E-P and E·P intermediates whose ^{31}P chemical shifts are generally similar to those of the same intermediates formed by $^{113}Cd(II)_6AP$. The $Cd(II)_2AP$, however, will bind only one equivalent of phosphate, hence a prominent P_i resonance from the second equivalent of phosphate is apparent throughout the pH titration. As illustrated by the ^{113}Cd NMR in Section 10, phosphorylation of $Cd(II)_2AP$ results in migration of half the Cd(II) from A to B sites, hence the phosphoenzymes of both $Cd(II)_2$ and $Cd(II)_6$ species represent active sites with both A and B sites filled. The active center does respond to the absence of metal ions in the other monomer. The most striking example is a shift of the pH profile for the E-P ⇌ E·P equilibrium ~1.5 pH units to the alkaline for $Cd(II)_2AP$ (Fig. 22).

12.3 E-P ⇌ E·P Equilibrium

Integration of the E·P, E-P, and P_i resonances allows the determination of the percentage composition of the phosphointermediates. This has been done for $Zn(II)_4$, $Zn(II)_4Mg(II)_2$, $Cd(II)_6$, and $Cd(II)_2AP$ (Fig. 22). Of the three homogeneous species plotted, only the $Zn(II)_4$ and $Cd(II)_6$ enzymes exhibit the complete shift from E-P to E·P and finally dissociation of P_i over the pH range examined, 5–10. These two have several features in common. At low pH [below pH 5 for $Zn(II)_4AP$ and $Zn(II)_4Mg(II)_2AP$ and below pH 8 for $Cd(II)_6AP$] E-P predominates. As the pH is raised, there is a shift in the equilibrium position toward E·P, which occurs at the same rate for all three systems, so that at 1.5 pH units above the midpoint of the transition, E-P is

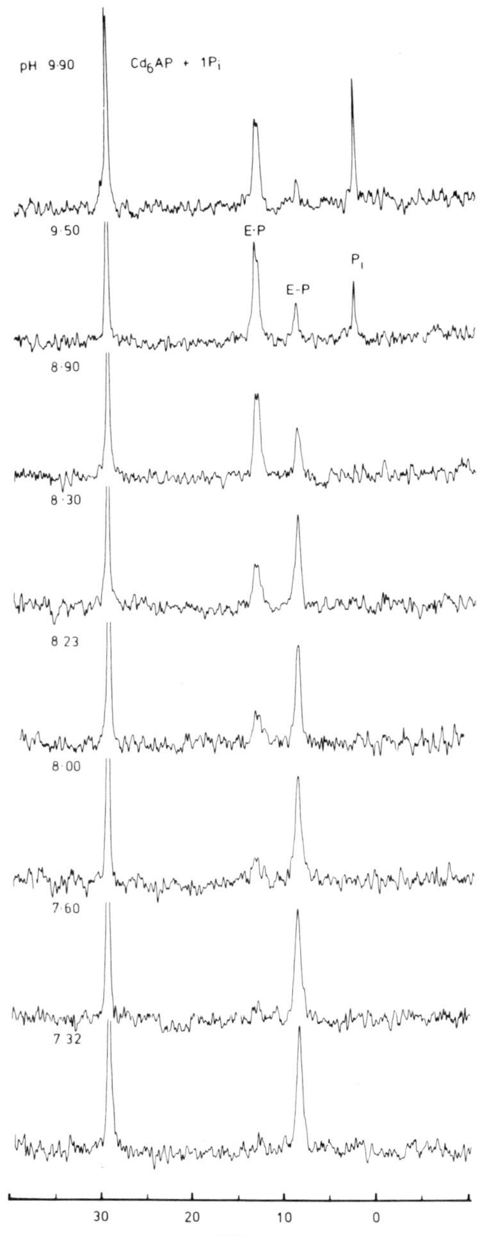

Figure 21 ^{31}P NMR spectra (80.9 MHz) of $^{113}Cd(II)_6AP$, containing one equivalent of phosphate, as a function of pH. Enzyme concentration was 2.0m*M*. Spectra are the average of 7000 transients. Most downfield resonance is external methyl phosphonate. From ref. 51.

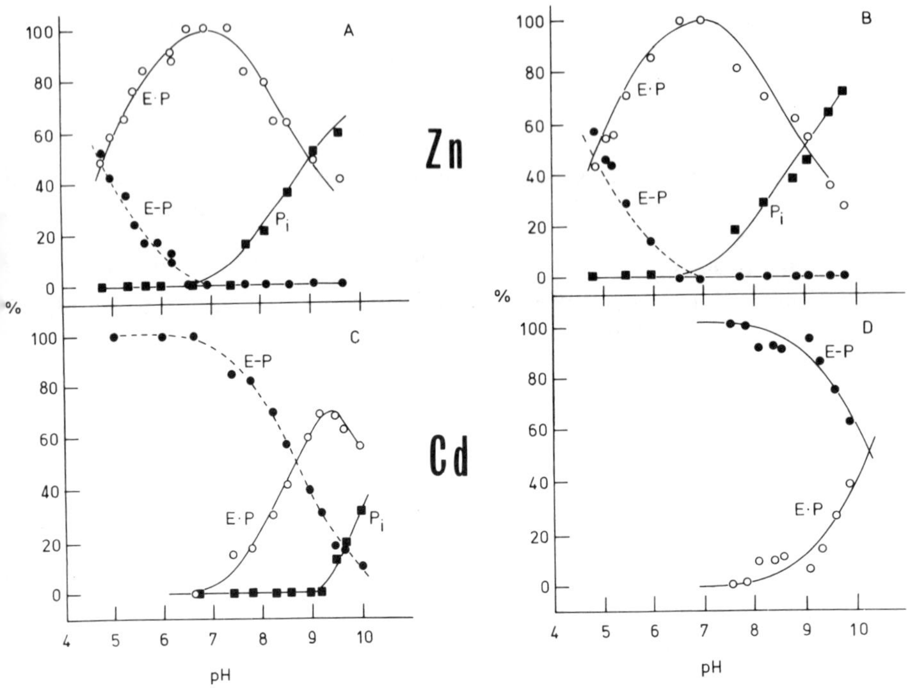

Figure 22 Relative equilibrium concentrations of phosphorus species for zinc- and cadmium-containing alkaline phosphatases as a function of pH. (○) E·P; (●) E-P; (■) P_i. *a*, 1.98m*M* Zn(II)$_4$AP plus one equivalent P_i; *b*, 2.13m*M* Zn(II)$_4$Mg(II)$_2$AP plus one equivalent P_i; *c*, 2.00m*M* Cd(II)$_6$AP plus two equivalents P_i; *d*, 1.83m*M* Cd(II)$_2$AP plus two equivalents of P_i. Phosphate was added at pH 6.5 and equilibrium attained before the pH was adjusted. Only one equivalent of phosphate remained bound throughout the titration of Cd(II)$_2$AP and therefore the percentages of E·P and E-P are normalized to this concentration rather than to two equivalents. Dashed lines represent E-P concentrations determined previously from ^{32}P measurements (52). From ref. 51.

no longer of measurable intensity. At still higher pH inorganic phosphate, which was previously bound, starts to dissociate. This is evident above pH 7 for Zn(II)$_4$AP and Zn(II)$_4$Mg(II)$_2$AP and above pH 9.5 for Cd(II)$_6$AP (Fig. 22). It should also be noted that the profiles for Zn(II)$_4$AP and Zn(II)$_4$Mg(II)$_2$AP are superimposable, despite the differences in chemical shift noted above for the ^{31}P resonances of their E·P and E-P intermediates.

12.4 Factors Affecting the Chemical Shift of E·P and E-P

Phosphoserine ionizes from the monoanionic to the dianionic state with a pK_a of 5.5 and with a concomitant change in chemical shift from −0.3 ppm

to 3.7 ppm (67). Inorganic phosphate behaves similarly and exhibits a downfield shift of 2.5 ppm (from 0.2 to 2.7 ppm) for the same ionization. In contrast the ^{31}P resonances of the E-P forms of alkaline phosphatase are in the 8.0–8.9 ppm range for the metalloenzymes [both Zn(II) and Cd(II)] and occur at ~6.2 ppm in apophosphorylalkaline phosphatase (124). Even more striking are the chemical shifts for the noncovalent E·P complexes which are between 3.3 and 4.3 ppm when Zn(II) occupies the A and B sites, but between 12.8 and 13.4 ppm when Cd(II) occupies the corresponding sites. In addition, unless the enzyme is denatured, there is no radical pH dependence of the chemical shift of the E-P resonance to indicate protonation at low pH (Fig. 19–21), so that it presumably exists solely as the dianion as low as pH 4 in both Zn(II) and Cd(II) enzymes. E·P is not formed as an equilibrium species below pH 7 except for the Zn(II)-containing enzyme. Over the observable pH range for this species (>pH 4.8) the chemical shift is also independent of pH, suggesting that it is also the dianion.

Compared with free phosphoserine there is a downfield shift of 2.5 ppm for the phosphoseryl group at the active center in the absence of metal ion (apophosphoryl enzyme) accompanied by a fall in the pK_a by at least 2.5 pH units (125). Binding of metal ions induces an additional downfield shift of 1.5–2.4 ppm (Table 3). The lowering of the pK_a presumably results from the existence of positive charges close to the phosphate moiety that stabilize the dianionic form. Since E·P is coordinated to one A-site metal ion, a divalent metal ion is clearly one of these groups.

The presence of proximal metal cations could stabilize both E·P and E-P and lower the pK_a of both intermediates. On the other hand, the pK_a of E-P in the apophosphoryl enzyme is also lowered implying the involvement of protein side chain(s) as well. This may be the arginine side chain, which has been implicated in the mechanism from chemical modification and protection studies (106, 126). A more important factor for alkaline phosphatase may be the conformation of the phosphate group. Changes in the torsional angles, ω and ω', of phosphate diesters and the associated alteration in O—P—O bond angles and lengths have been estimated to result in quite large chemical shift changes (127–129). A chemical shift difference of 4–6 ppm between *gauche-gauche* and *gauche-trans* conformations has been calculated for such esters (127).

The chemical shifts for the E·P complexes are much more sensitive to the change of metal ion species than is E-P. For the Zn(II)-containing enzyme the resonance of E·P occurs only 0.6–1.6 ppm downfield from HPO_4^{2-}, but for Cd(II)-substituted AP the chemical shift is ~10 ppm downfield from HPO_4^{2-}. Furthermore, the major source of the difference appears to be the nature of the metal at A site, since E·P in $Cd(II)_2AP$ at pH 6.5, prior to occupation of the B sites by cadmium migration, has a chemical shift of 12.79

Table 3 ^{31}P Chemical Shifts of E·P and E-P Phosphointermediates of Metalloalkaline Phosphatases Containing Cadmium or Zinc and of E-P of Apoalkaline Phosphatase

		Chemical Shifts (ppm)	
Enzyme	pH	E-P	E·P
apoAP	4-8	6.2	
Native AP [$Zn(II)_4Mg(II)_2$]	7.1		3.2
$Zn(II)_4AP$	6.2	8.63	4.26
	9.1		3.94
$Zn(II)_4Mg(II)_2AP$	6.0	8.60	3.47
	9.1		3.36
$Cd(II)_2AP$ A and B	6.1	7.9	
	9.1	8.18	13.43
$Cd(II)_2AP$[a] A and A′	6.5	8.30	12.79
$Cd(II)_4AP$	7.0	8.42	13.06
$Cd(II)_6AP$	6.6	8.45	12.8
	9.2	8.67	12.96
$Cd(II)_4Mg(II)_2AP$	9.0	8.89	12.89
$Cd(II)_2Mg(II)_2AP$	9.0	9.08	12.85

[a] Applies to phosphoenzyme prior to migration of Cd(II) from A to B site. ^{113}Cd signal is split into A and A′ (see Sec. 10).

ppm as described in Figure 15. It is unlikely that the difference between Zn(II) and Cd(II) arises from a difference in mechanism, since both metalloenzymes appear to proceed via the same intermediates. The most likely factor resulting in the large change in ^{31}P chemical shift of E·P on metal ion substitution is the difference in ionic radius between Zn(II) and Cd(II) (0.74–0.97 Å). It is unclear how this alters the properties of the phosphate group, but it may arise from a distortion of the group that is bound more tightly by the Cd(II) enzyme (see below).

The influence of the third pair of metal ions on the chemical shifts of the phosphointermediates shows no consistent pattern. For the $Zn(II)_4$ enzyme the presence of two magnesium ions has little effect on the E-P resonance, but a somewhat greater influence on the E·P resonance. Comparative data for the $Cd(II)_4$ system are limited by the heterogeneity induced by raising the pH, but Mg(II) does produce small changes in chemical shift of both E·P and E-P. It should be noted that there is no conclusive evidence to prove that when the third pair of metal ions is Mg(II), they occupy the same sites as a third pair of Cd(II) ions, the latter being defined to occupy C sites from ^{113}Cd NMR studies (84).

12.5 ^{113}Cd—O—^{31}P Coupling in E·P

As illustrated by the ^{31}P NMR spectra of E·P formed by all species of ^{113}Cd(II) alkaline phosphatase, the noncovalent phosphate complex shows a ~30 Hz coupling expected for a typical ^{113}Cd—O—^{31}P bond (Figs. 16 and 21). Thus the phosphate oxygen of E·P is coordinated to one, but not both, of the two Cd(II) ions, A and B, at each active center. In order to determine which Cd(II) ion is coordinated, we prepared ^{111}Cd(II)$_6$AP at pH 9.1 in which about equal concentrations of E·P and E-P are present (Fig. 21). The ^{111}Cd NMR spectrum of this species shows two resonances for site A separated by ~14 ppm, the upfield one corresponding to E·P, the downfield to the E-P species. Likewise two resonances occur for site B separated by 6 ppm, the upfield corresponding to E·P and the downfield to E-P. Using heteronuclear decoupling the set of ^{111}Cd signals was irradiated while observing the ^{31}P spectrum. Irradiation of the B-site ^{111}Cd signal corresponding to E·P causes no loss of coupling. Irradiation of the A-site signal corresponding to E·P causes partial collapse of the coupling on the ^{31}P signal of E·P (84). In contrast to the E·P resonance, if the downfield A-site ^{111}Cd signal, corresponding to the E-P species, is irradiated, no alteration of the coupling on the ^{31}P signal of E·P is observed. Moreover, no alteration in linewidth of the ^{31}P signal of E-P is observed either, confirming the impression that the phosphate of E-P is not coupled to either cadmium ion. Hence the phosphate in E·P appears to be coordinated to the A-site metal ion. Careful analysis of the line shape of the ^{31}P resonances for both E·P and E-P during the initial stages of phosphorylation of ^{113}Cd(II)$_2$AP, when both metal ions remain in A sites, tends to support the same conclusion (see Fig. 15). ^{31}P spectra over the first 2 h of phosphorylation were summed and show the resonance from E·P to be much broader than that for E-P and coupling appears to be present, although good resolution is not possible because of the transient appearance of E·P at pH 6.5.

12.6 Stoichiometry of Phosphate Binding

From the pH titrations of the homogeneous phosphoenzyme derivatives, it is apparent that phosphate binds more weakly at higher pH such that both phosphate binding sites may not be fully saturated at stoichiometric concentrations of inhibitor:enzyme. To quantitate this, phosphate was titrated into Cd(II)$_6$AP and Zn(II)$_4$AP at pH 6 and pH 9 or pH 8 and the fraction bound calculated. The data expressed as phosphate added versus phosphate bound (calculated from ^{31}P resonance area) are shown in Figure 23. For the cadmium derivative with all metal sites occupied the phosphate affinity constant characterizing both sites remains $>10^3 M^{-1}$ at both pH 6 and 9, so that

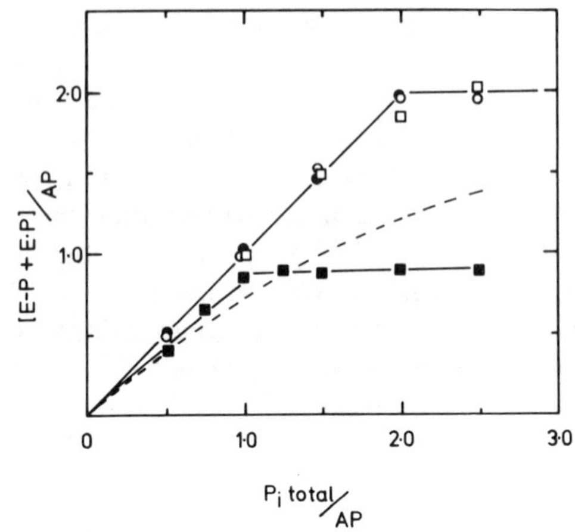

Figure 23 Equivalents of phosphate bound as the sum of E·P and E-P versus total phosphate added for Cd(II)$_6$AP and Zn(II)$_4$AP. Concentration of each species was determined from integration of the respective ^{31}P NMR resonance. (□) Zn(II)$_4$AP (1.29m*M*), pH 6.0; (■) Zn(II)$_4$AP (1.65m*M*), pH 8.0; (○) Cd(II)$_6$AP (1.11m*M*), pH 6.0; (●) Cd(II)$_6$AP (1.21m*M*), pH 9.0. Dashed curve is calculated for independent dissociation from two sites with a K_d of 1m*M*. From ref. 51.

phosphate in excess of two equivalents per dimer is required before a resonance due to free phosphate is seen. The same is true for the Zn enzyme at pH 6. The 2 mol of bound phosphate (Fig. 23) are distributed between E·P and E-P. In marked contrast, at pH 8 the zinc enzyme binds slightly less than one equivalent of phosphate. There is also an indication of a difference in binding constants for the two sites. After one equivalent of phosphate has been added (of which about 90% is bound) there is a relatively small increase in E·P concentration for each subsequent addition of phosphate (Fig. 23). It should be noted that pH 8 is approximately the midpoint of the pH profile for the sigmoid curve of catalytic activity of the native enzyme.

12.7 The E-P ⇌ E·P ⇌ P_i Equilibrium

The pH profiles describing the concentrations of the three species (E·P, E-P, and P_i) in equilibrium at the active center of alkaline phosphatase, as determined by integration of the ^{31}P NMR signals (Fig. 22), show an almost identical form for all species of Zn(II) and Cd(II) enzyme with one major difference. The pH where [E-P] = [E·P] varies from pH 5 to pH 10 depending on both the metal ion species and the metal:protein stoichiometry. The sigmoidal shape of the E-P → E·P transition suggests that an ionization in the

active center is responsible. The most remarkable change in this function is the shift from a midpoint of pH 5 to a midpoint of 8.7 on changing from the $Zn(II)_4$ to the $Cd(II)_6$ enzyme (Fig. 22). This implies an alkaline shift of 3.7 pH units in the pK_a of the group controlling the E-P ⇌ E·P equilibrium. The shift in E-P ⇌ E·P equilibrium on Cd(II) substitution is accompanied by some other significant changes at the active center: (a) the chemical shift of E·P shifts ~9 ppm downfield (Fig. 21), and (b) the binding affinity of phosphate becomes much higher in the alkaline pH range (Fig. 23).

An attractive candidate for the ionizing group influencing the E-P ⇌ E·P equilibrium is a metal-coordinated water, whose pK_a is lowered from 15.7 to 7.5–8 by the mixed enzyme–Zn(II) complex. Note that the actual value of the pK_a need not correspond to the midpoint of the E-P ⇌ E·P equilibrium, but may be several pH units higher depending on the absolute values of k_3 and k_{-3} (Scheme 1). Thus it may correspond to the apparent pK_a for enzymatic activity. Substitution of Cd(II) for Zn(II) could thus be interpreted to result in a far less effective lowering of the pK_a of the coordinated solvent. A similar effect has been observed for carbonic anhydrase, where there is evidence to support the existence of a Zn(II)—OH^--containing enzyme species as the active form (130, 131). There is also a shift of the pH-rate profile by two to three units toward alkaline pH upon substitution of Cd(II) for Zn(II) in this enzyme (69).

Cd(II) has a remarkable effect on the chemical shift of E·P, but little effect on the chemical shift of E-P (Fig. 21). Along with the absence of coupling of E-P to $^{113}Cd(II)$, this suggests that the phosphoseryl residue is less closely connected to the metal ion than the noncovalent complex. Perhaps the phosphate leaves the coordination sphere on formation of the ester. In contrast, the larger ionic radius of Cd(II) must have a significant effect on the geometry of the E·P complex.

12.8 Dissociation of E·P

The detailed examination of the E-P ⇌ E·P equilibrium, including the dissociation of P_i, made possible by ^{31}P NMR reveals significant aspects of the product dissociation reaction. At H^+ concentrations where significant equilibrium concentrations of E-P are present, both $Zn(II)_4$ and $Cd(II)_6AP$ bind phosphate sufficiently tightly to form 2 mol of phosphointermediate (E-P + E·P) at mM concentrations of enzyme and phosphate (Fig. 23). This is, of course, the pH region where dephosphorylation of E-P is expected to be the rate-limiting step as has been shown by many previous investigations of both Zn(II) and Cd(II) enzymes (45, 124, 131, 132). In contrast, when alkaline pH values are reached, where E·P becomes the exclusive equilibrium species, phosphate-binding affinity decreases dramatically, and for the Zn(II) en-

zyme K_d for at least one phosphate must be $>$mM, since 1 mol phosphate has completely dissociated by pH 9 (Fig. 20). The binding curve is non linear, not even hyperbolic (shown by the dashed curve, Fig. 23), suggesting that not only has K_d for phosphate increased, but two different binding constants may apply. One site has a $K_d \ll$ mM and the second has a $K_d \sim$ mM. Since dissociation of E·P is the rate-limiting step at alkaline pH, the increase in the dissociation rate of phosphate would appear to have functional significance. The appearance of a negative charge in the active center, for example, a metal-coordinated hydroxyl, could be responsible for reducing the phosphate-binding constant.

The same formulation could apply to the Cd(II)$_6$ enzyme except that the corresponding pH range is shifted ~2 pH units more alkaline (Fig. 22). On this basis one might argue that one of the reasons for the low activity of the Cd(II) enzyme is its much higher affinity for phosphate in the normal pH range. Activity might appear above pH 9.0 in the region where phosphate begins to dissociate. Another effect of Cd(II) substitution, however, may be primarily responsible for slow turnover and that is the slow phosphorylation rate associated with the distortion of E·P produced by Cd(II) substitution. As shown in Figure 16, phosphorylation of the seryl hydroxyl from E·P is very slow (hours to days) at both pH 6.5 and pH 9.0 (84). Thus the phosphorylation rate for HOP remains slower than P_i dissociation, although phosphorylation of the Cd(II) enzyme by ROP might be faster and allow significant turnover in the region of phosphate dissociation, ~pH 9.

12.9 Subunit–Subunit Interactions

The dimeric nature of alkaline phosphatase provides the opportunity for cooperativity of substrate binding. There have been both claims (6, 52, 133) and rebuttals (103, 111) that phosphate binding is negatively cooperative. We will now summarize results that are relevant both to negative cooperativity of phosphate binding and more generally to subunit–subunit interactions.

Cd(II)$_2$AP at alkaline pH has both A and B sites of one monomer occupied by cadmium, while both sites of the other monomer are vacant. Yet ^{113}Cd(II)$_4$AP, with A and B sites of both monomers filled, does not simply show the same ^{113}Cd NMR spectrum at twice the intensity, but a different character to the resonances. Not only are the precise chemical shifts different, but some type of chemical exchange modulation is present (84). This unexpected difference is further emphasized here by the surprising change in the pH where [E-P] = [E·P] between the Cd(II)$_2$ and Cd(II)$_6$ phosphoenzymes, from pH 10.2 in the former to pH 8.7 in the latter (Fig. 22). If changes in structure at the active center of one monomer do not influence the other, these changes are very difficult to explain.

Further support for the dissimilarity of $Cd(II)_2$-containing subunits in $Cd(II)_2AP$ vs. $Cd(II)_4AP$ comes from the ^{31}P chemical shift data on the phosphorylated species. The equilibrium E-P species of $Cd(II)_2AP$ formed at pH 6.5 has a ^{31}P chemical shift of 7.95 ppm compared with 8.42 ppm for E-P of the $Cd(II)_4AP$ at the same pH. Similarly the ^{113}Cd chemical shifts reflect the same difference; A- and B-site cadmium in phosphorylated $^{113}Cd(II)_2AP$ have resonances at 142 and 56 ppm compared with 155 and 70 ppm in $^{113}Cd(II)_4AP$ (84) (Fig. 12).

The most compelling evidence for negative cooperativity was provided by the sequential spectra during the phosphorylation of $^{113}Cd(II)_2AP$ at pH 6.5 (Fig. 15). Prior to addition of phosphate the subunits are identical with cadmium in both A sites. Phosphate binds, however, at only one active center. The subsequent migration of cadmium from the A site of the unphosphorylated monomer to the B site of the phosphorylated subunit is thus an effect and not a cause of the phosphate-induced asymmetry.

12.10 Mechanistic Implications

The completely analogous pH profiles for the formation of E·P and E-P by the Zn(II) and Cd(II) enzymes (Fig. 22) suggest that the latter is a reasonable model for the native enzyme. The latter has the great advantage of slow rate constants and the spin-1/2 isotopes of Cd(II) as NMR probes. The one striking difference between the two metal ions, the shift of the E-P ⇌ E·P equilibrium ~3 pH units to the alkaline could be explained, if not uniquely so, by the participation of a coordinated solvent whose ionized form is required for hydrolysis of E-P.

Formation of E·P by the Cd(II) enzymes appears to involve coordination to the A-site Cd(II) ion. Both formation of E·P and hydrolysis of E-P is metal ion dependent as shown by previous work (52) and can clearly occur slowly with only A site occupied as shown by the sequential ^{31}P and ^{113}Cd NMR spectra during phosphorylation of $^{113}Cd(II)_2AP$ illustrated in Figure 15. Coordination of the phosphate to A site would explain the metal ion dependence of phosphate binding. Such coordination must also place the phosphorus in a suitable position for nucleophilic attack by the hydroxyl oxygen of Ser-102. Formation of the seryl ester appears to move the phosphate out of the coordination sphere, since no coupling is observed on the ^{31}P signal from E-P and the chemical shift of E-P is relatively insensitive to metal ion species, in contrast to E·P.

The precise function of the B-site metal ion is less clear. Its presence clearly accelerates phosphorylation (Fig. 7). This could result from enhanced nucleophilic attack made possible by the additional neutralization by the B-site metal ion of the negative charge at the active site arising from the phosphate

dianion. The possible coordinated hydroxyl ion postulated to participate in hydrolysis of the phosphoseryl ester as outlined above could be associated with either the A- or B-site metal ion. Since phosphorylation of any Cd(II) enzyme species always induces occupancy of A and B sites at equilibrium, we do not know whether dephosphorylation can proceed in the presence of A-site metal ion alone. Since phosphate has apparently moved out of the inner coordination sphere of the A-site metal ion accompanying phosphorylation, there would appear to be an open coordination site on the A-site metal ion available for a coordinated solvent molecule that could create the nucleophile for the final step of the mechanism.

ABBREVIATIONS

$Me(II)_2AP$, $Me(II)_4AP$, or $Me(II)_2Mg(II)_2AP$ Alkaline phosphatase containing metal ions at A sites alone, the same metal at A and B sites or Me(II) at A sites and Mg(II) at B sites

Me(II) Mn(II), Co(II), Cu(II), Zn(II), or Cd(II)

E-P Phosphoryl enzyme

E·P Noncovalent complex with inorganic phosphate

REFERENCES

1. A. Boivin and L. Mesrobeanu, *Compt. Rend. Soc. Biol.*, **112,** 611 (1933).
2. T. Horiuchi, S. Horiuchi, and D. Mizuno, *Nature (London)*, **183,** 1529 (1959).
3. A. Torriani, *Biochim. Biophys. Acta*, **38,** 460 (1960).
4. D. J. Plocke, C. Levinthal, and B. L. Vallee, *Biochemistry*, **1,** 373 (1962).
5. D. J. Plocke and B. L. Vallee, *Biochemistry*, **1,** 1039 (1962).
6. T. W. Reid and I. B. Wilson, *The Enzymes*, Vol. 4, 3rd ed., 373 (1971).
7. J. E. Coleman and J. F. Chlebowski, *Adv. Inorg. Biochem.*, **1,** 1 (1979).
8. R. B. McComb, G. N. Bowers, and S. Posen, *Alkaline Phosphatase*, Plenum, New York, 1979.
9. R. W. Brockman and L. A. Heppel, *Biochemistry*, **7,** 2554 (1968).
10. M. H. Malamy and B. L. Horecker, *Biochemistry*, **3,** 1893 (1964).
11. H. Inouye, C. Pratt, J. Beckwith, and A. Torriani, *J. Mol. Biol.*, **110,** 75 (1977).
12. W. P. Smith, P.-C. Tai, R. C. Thompson, and B. D. Davis, *Proc. Nat. Acad. Sci. USA*, **74,** 2830 (1977).
13. C. N. Chang, H. Inouye, P. Model, and J. Beckwith, *J. Bacteriology*, **142,** 726 (1980).
14. R. A. Bradshaw, F. Cancedda, L. H. Ericsson, P. A. Neumann, S. P. Piccoli, M. J. Schlesinger, K. Shriefer, and K. A. Walsh, *Proc. Nat. Acad. Sci. USA*, **78,** 3473 (1981).

15. P. Y. Chou and G. D. Fasman, in *Adv. in Enzymology*, Vol. 47, A. Meister, Ed., Wiley, New York, 1978, p. 45.
16. J. A. Reynolds and M. J. Schlesinger, *Biochemistry*, **6,** 3552 (1967).
17. J. M. Sowadski, B. A. Foster, and H. W. Wyckoff, *J. Mol. Biol.*, **150,** 245 (1981).
18. D. R. Harkness, *Arch. Biochem. Biophys.*, **126,** 503 (1968).
19. N. K. Ghosh and W. H. Fishman, *Biochem. J.*, **108,** 779 (1968).
20. K. D. Gerbitz, H. J. Kolb, and O. H. Wieland, *Hoppe-Seyler's Z. Physiol. Chem.*, **358,** 435 (1977).
21. R. Felix and H. Fleisch, *Biochim. Biophys. Acta*, **350,** 84 (1974).
22. R. H. Eaton and D. W. Moss, *Enzymologia*, **35,** 31 (1968).
23. M. Fosset, D. Chappelet-Tordo, and M. Lazdunski, *Biochemistry*, **13,** 1783 (1974).
24. G. Cathala, C. Brunel, D. Chappelet-Tordo, and M. Lazdunski, *J. Biol. Chem.*, **250,** 6040 (1975).
25. A. Ohkubo, N. Langerman, and M. M. Kaplan, *J. Biol. Chem.*, **249,** 7174 (1974).
26. F. K. Yoshizumi and J. E. Coleman, *Arch. Biochem. Biophys.*, **160,** 255 (1974).
27. F. M. Hulett-Cowling and L. L. Campbell, *Biochemistry*, **10,** 1364 (1971).
28. R. N. Glew and E. C. Heath, *J. Biol. Chem.*, **246,** 1556 (1971).
29. S. Y. Ali, S. W. Sajdera, and H. C. Anderson, *Proc. Nat. Acad. Sci. USA*, **67,** 1513 (1970).
30. R. E. Wuthier, *Calcif. Tiss. Res.*, **23,** 125 (1977).
31. R. J. Majeska and R. E. Wuthier, *Biochim, Biophys. Acta*, **391,** 51 (1975).
32. Symposium on Matrix Vesicle Calcification. *Fed. Proc.*, **35,** 105 (1976).
33. D. Chappelet-Tordo, M. Fosset, M. Iwatsubo, C. Gache, and M. Lazdunski, *Biochemistry*, **13,** 1788 (1974).
34. L. Engström and G. Ågren, *Acta Chem. Scand.*, **12,** 357 (1958).
35. L. Engström, *Ark. Kemi.*, **19,** 129 (1962).
36. L. Engström, *Biochim. Biophys. Acta*, **92,** 79 (1964).
37. J. Schwartz, A. Crestfield, and F. Lipmann, *Proc. Nat. Acad. Sci. USA*, **49,** 722 (1963).
38. M. J. Melancon and H. F. DeLuca, *Biochemistry*, **9,** 1658 (1970).
39. E. S. Holdsworth, *J. Mem. Biol.*, **3,** 43 (1970).
40. L. V. Avioli and L. G. Raisz, in *Metabolic Control and Disease*, P. K. Bondy and L. E. Rosenberg, Eds., W. B. Saunders, Philadelphia, 1980, pp. 1709–1814.
41. W. E. Hull, S. E. Halford, H. Gutfreund, and B. D. Sykes, *Biochemistry*, **15,** 1547 (1976).
42. J. F. Chlebowski, I. M. Armitage, and J. E. Coleman, *J. Biol. Chem.*, **252,** 7053 (1977).
43. J. F. Chlebowski and J. E. Coleman (unpublished data).
44. H. Neumann, *J. Biol. Chem.*, **243,** 4671 (1968).
45. J. F. Chlebowski and J. E. Coleman, *J. Biol. Chem.*, **249,** 7192 (1974).
46. R. Breslow and I. Katz, *J. Am. Chem. Soc.*, **90,** 7376 (1968).
47. P. Mushak and J. E. Coleman, *Biochemistry*, **11,** 201 (1972).
48. I. B. Wilson, J. Dayan, and K. Cyr, *J. Biol. Chem.*, **239,** 4182 (1964).
49. H. N. Fernley and P. G. Walker, *Biochem. J.*, **111,** 187 (1969).
50. M. L. Applebury and J. E. Coleman, *J. Biol. Chem.*, **244,** 308 (1969).

51. P. Gettins and J. E. Coleman, *J. Biol. Chem.*, **258,** 408 (1983).
52. M. L. Applebury, B. P. Johnson, and J. E. Coleman, *J. Biol. Chem.*, **245,** 4968 (1970).
53. J. H. Schwartz, *Proc. Nat. Acad. Sci. USA*, **49,** 861 (1963).
54. J. L. Bock and M. Cohn, *J. Biol. Chem.*, **253,** 4082 (1978).
55. M. Caswell and M. Caplow, *Biochemistry*, **19,** 2907 (1980).
56. J. R. Knowles, *Ann. Rev. Biochem.*, **49,** 877 (1980).
57. D. R. Trentham and H. Gutfreund, *Biochem. J.*, **106,** 455 (1968).
58. M. Cocivera, J. McManaman, and I. B. Wilson, *Biochemistry*, **19,** 2901 (1980).
59. J. R. Bale, *Fed. Proc.*, **37,** 1287 (1978).
60. M. L. Applebury and J. E. Coleman, *J. Biol. Chem.*, **244,** 709 (1969).
61. R. T. Simpson and B. L. Vallee, *Biochemistry*, **7,** 4343 (1968).
62. W. F. Bosron, R. A. Anderson, M. C. Falk, F. S. Kennedy, and B. L. Vallee, *Biochemistry*, **16,** 610 (1977).
63. J. S. Taylor, C. Y. Lau, M. L. Applebury, and J. E. Coleman, *J. Biol. Chem.*, **248,** 6216 (1973).
64. J. S. Taylor and J. E. Coleman, *Proc. Nat. Acad. Sci. USA*, **69,** 859 (1972).
65. F. S. Kennedy, H. A. O. Hill, T. A. Kaden, and B. L. Vallee, *Biochem. Biophys. Res. Commun.*, **48,** 1533 (1972).
66. R. E. Weiner, J. F. Chlebowski, P. H. Haffner, and J. E. Coleman, *J. Biol. Chem.*, **254,** 9739 (1979).
67. J. E. Coleman, I. M. Armitage, J. F. Chlebowski, J. D. Otvos, and A. J. M. Schoot Uiterkamp, in *Biological Applications of Magnetic Resonance*, R. G. Shulman, Ed., Academic, New York, 1979, p. 345.
68. R. Bauer, P. Limkilde, and J. T. Johansen, *Biochemistry*, **15,** 334 (1976).
69. J. E. Coleman, in *Biophysics and Physiology of Carbon Dioxide*, C. Bauer, G. Gros, and H. Bartels, Eds., Springer-Verlag, Berlin, 1980, pp. 133–150.
70. J. L. Bock and M. Cohn, *J. Biol. Chem.*, **253,** 4082 (1978).
71. M. L. Harris and J. E. Coleman, *J. Biol. Chem.*, **243,** 5063 (1968).
72. W. F. Bosron, F. S. Kennedy, and B. L. Vallee, *Biochemistry*, **14,** 2275 (1975).
73. J. A. Reynolds and M. J. Schlesinger, *Biochemistry*, **8,** 588 (1969).
74. C. Petitclerc, C. Lazdunski, D. Chappelet, A. Moulin, and M. Lazdunski, *Eur. J. Biochem.*, **14,** 301 (1970).
75. C. N. A. Trotman and C. Greenwood, *Biochem. J.*, **124,** 25 (1971).
76. H. Csopak and H. Szajn, *Arch. Biochem. Biophys.*, **157,** 374 (1973).
77. R. A. Anderson, F. S. Kennedy, and B. L. Vallee, *Biochemistry*, **15,** 3710 (1976).
78. J. E. Coleman, K.-I. Nakamura, and J. F. Chlebowski, *J. Biol. Chem.*, **258,** 386 (1983).
79. C. Schulz and J. E. Coleman, (unpublished data).
80. J. R. Knox and H. W. Wyckoff, *J. Mol. Biol.*, **74,** 533 (1973).
81. J. M. Sowadski, H. M. K. Murthy, M. Handschumacher, and H. W. Wyckoff, *Nature*, in press.
82. R. A. Anderson, F. S. Kennedy, and B. L. Vallee, *Biochemistry*, **15,** 3710 (1976).
83. J. E. Coleman and R. V. Coleman, *J. Biol. Chem.*, **247,** 4718 (1972).
84. P. Gettins and J. E. Coleman, *J. Biol. Chem.*, **258,** 396 (1983).

85. S. Lindskog, L. E. Henderson, K. K. Kannan, A. Liljas, P. O. Nyman, and B. Strandberg, *The Enzymes*, Vol. 5, 1971, pp. 587–665.

86. K. K. Kannan, in *Biophysics and Physiology of Carbon Dioxide*, C. Bauer, G. Gros, and H. Bartels, Eds., Springer-Verlag, Berlin, 1980, p. 184.

87. J. D. Otvos, J. R. Alger, J. E. Coleman, and I. M. Armitage, *J. Biol. Chem.*, **254,** 1778 (1979).

88. P. H. Haffner, F. Goodsaid-Zalduondo, and J. E. Coleman, *J. Biol. Chem.*, **249,** 6693 (1974).

89. R. E. Weiner, J. F. Chlebowski, P. H. Haffner, and J. E. Coleman, *J. Biol. Chem.*, **254,** 9739 (1979).

90. D. C. Rees and W. N. Lipscomb, *Proc. Nat. Acad. Sci. USA*, **78,** 5455 (1981).

91. W. R. Kester and B. W. Matthews, *J. Biol. Chem.*, **252,** 7704 (1977).

92. E. Grell and R. C. Bray, *Biochim. Biophys. Acta*, **236,** 503 (1971).

93. F. S. Kennedy, H. A. O. Hill, T. A. Kaden, and B. L. Vallee, *Biochem. Biophys. Res. Comm.*, **48,** 1533 (1972).

94. P. H. Haffner and J. E. Coleman, *J. Biol. Chem.*, **248,** 6630 (1973).

95. R. A. Anderson, W. F. Bosron, F. S. Kennedy, and B. L. Vallee, *Proc. Nat. Acad. Sci. USA*, **72,** 2989 (1975).

96. D. B. Bailey, P. D. Ellis, A. D. Cardin, and W. D. Behnke, *J. Am. Chem. Soc.*, **100,** 5236 (1978).

97. J. D. Otvos and I. M. Armitage, *J. Am. Chem. Soc.*, **109,** 7734 (1979).

98. I. M. Armitage, A. J. M. Schoot Uiterkamp, J. F. Chlebowski, and J. E. Coleman, *J. Mag. Res.*, **29,** 375 (1978).

99. A. J. M. Schoot Uiterkamp, I. M. Armitage, and J. E. Coleman, *J. Biol. Chem.*, **255,** 3911 (1980).

100. D. B. Bailey, P. D. Ellis, and J. A. Fee, *Biochemistry*, **19,** 591 (1980).

101. T. Drakenberg, B. Lindman, A. Cave, and J. Parello, *FEBS Lett.*, **92,** 346 (1978).

102. S. Forsén, E. Thulin, T. Drakenberg, J. Krebs, and K. Seamon, *FEBS Lett.*, **117,** 189 (1980).

103. J. D. Otvos and I. M. Armitage, *Biochemistry*, **19,** 4031 (1980).

104. J. D. Otvos and I. M. Armitage, *Biochemistry*, **19,** 4021 (1980).

105. G. H. Tait and B. L. Vallee, *Proc. Nat. Acad. Sci. USA*, **56,** 1247 (1966).

106. J. M. Daeman and J. F. Riordan, *Biochemistry*, **13,** 2865 (1974).

107. J. D. Otvos and D. T. Browne, *Biochemistry*, **19,** 4011.

108. E. Oldfield, R. S. Norton, and A. Allerhand, *J. Biol. Chem.*, **250,** 6368 (1975).

109. W. E. Hull and B. D. Sykes, *Biochemistry*, **13,** 3431 (1974).

110. D. T. Browne and J. D. Otvos, *Biochem. Biophys. Res. Commun.*, **68,** 907 (1976).

111. W. E. Hull and B. D. Sykes, *Biochemistry*, **15,** 1535 (1976).

112. G. L. Cottam and R. L. Ward, *Arch. Biochem. Biophys.*, **141,** 768 (1970).

113. J.-E. Norne, H. Csopak, and B. Lindman, *Arch. Biochem. Biophys.*, **162,** 552 (1974).

114. J.-E. Norne, H. Szajn, H. Csopak, P. Reimarsson, and B. Lindman, *Arch. Biochem. Biophys.*, **196,** 552 (1979).

115. D. H. Eargle, V. Licko, and G. L. Kenyon, *Anal. Biochem.*, **81,** 186 (1977).

116. J. L. Bock and M. Cohn, *J. Biol. Chem.*, **253,** 4082 (1978).

117. S. R. Jones, L. A. Kindman, and J. R. Knowles, *Nature*, **257,** 564 (1978).
118. J. L. Bock and A. Kowalsky, *Biochem. Biophys. Acta*, **526,** 135 (1978).
119. S. McCracken and E. Meighen, *J. Biol. Chem.*, **255,** 2396 (1980).
120. W. Bloch and D. Bickar, *J. Biol. Chem.*, **253,** 6211 (1978).
121. C. Lazdunski and M. Lazdunski, *Biochim. Biophys. Acta*, **113,** 551 (1966).
122. M. Gottesman, R. Simpson, and B. L. Vallee, *Biochemistry*, **8,** 3776 (1969).
123. D. Levine, T. W. Reid, and I. B. Wilson, *Biochemistry*, **8,** 2374 (1969).
124. J. D. Otvos, I. M. Armitage, J. F. Chlebowski, and J. E. Coleman, *J. Biol. Chem.*, **254,** 4707 (1979).
125. J. F. Chlebowski, I. M. Armitage, P. P. Tusa, and J. E. Coleman, *J. Biol. Chem.*, **251,** 1207 (1976).
126. B. W. Augustus and D. W. Hutchinson, *Biochem. J.*, **177,** 377 (1979).
127. F. R. Prado, C. Giessner-Prettre, B. Pullman, and J.-P. Daudey, *J. Am. Chem. Soc.*, **101,** 1737 (1979).
128. D. G. Gorenstein, *Ann. Rev. Biophys. Bioeng.*, **10,** 355 (1981).
129. D. Perahia and B. Pullman, *Biochim. Biophys. Acta*, **435,** 282 (1976).
130. S. Lindskog and J. E. Coleman, *Proc. Nat. Acad. Sci. USA*, **70,** 2505 (1973).
131. S. H. D. Ko and F. J. Kezdy, *J. Am. Chem. Soc.*, **89,** 7139 (1967).
132. R. T. Simpson and B. L. Vallee, *Biochemistry*, **9,** 953 (1970).

CHAPTER 6
The Role of Zn(II) in RNA and DNA Polymerases

JOSEPH E. COLEMAN

Department of Molecular Biophysics and Biochemistry
Yale University
New Haven, Connecticut

CONTENTS

1 INTRODUCTION, 221

2 RNA POLYMERASES—MOLECULAR PROPERTIES AND MECHANISM, 221

3 PROKARYOTIC RNA POLYMERASES, 228

3.1 *Escherichia coli* RNA Polymerase, 229
3.2 *Bacillus subtilis* RNA Polymerase, 232
3.3 Bacteriophage RNA Polymerases, 233

4 EUKARYOTIC RNA POLYMERASES, 236

4.1 RNA Polymerases from *Euglena gracilis*, 236
4.2 RNA Polymerase II from Wheat Germ, 237
4.3 RNA Polymerases from Yeast, 238

5 PHYSIOLOGICAL EVIDENCE FOR A ROLE FOR Zn(II) IN TRANSCRIPTION-TRANSLATION, 239

6 REVERSE TRANSCRIPTASES, 241

7 DNA POLYMERASES, 243

8 MOLECULAR ASPECTS OF Zn(II) FUNCTION, 245

ACKNOWLEDGMENTS, 248

ABBREVIATIONS, 248

REFERENCES, 249

1 INTRODUCTION

DNA-dependent RNA polymerase from *Escherichia coli* was first reported to contain 2 g atom Zn(II)/mol [molecular weight (MW) of 370,000] by Scrutton et al. (1) in 1971. At approximately the same time Slater et al. (2) reported that the DNA polymerase I from *E. coli* also contained ~2 (1.7 to 2.4) g atom of tightly bound Zn(II)/mol of MW 109,000. This was later revised downward to 1.0 ± .15 g atom Zn(II)/mol (3). Both enzymes were inhibited by the chelating agent 1,10-phenanthroline, while the nonchelating 1,7- phenanthroline showed much weaker inhibition. Since 1971, many further studies on the *E. coli* RNA polymerase as well as other RNA polymerases from both prokaryotes and eukaryotes have amply confirmed the initial observations. Work on the DNA polymerases has been less extensive. Tables 1 and 2 summarize the currently available Zn(II) content and subunit structure of the RNA polymerases. With the exception of DNA polymerase I and the bacteriophage RNA polymerase all of these enzymes are large multisubunit proteins and elucidation of the precise role of Zn(II) in the structure and/or function of the enzyme has been difficult to achieve. Even in the one subunit enzymes, the role of Zn(II) is far from clear.

Since a significant nondialyzable Zn(II) content has been found in every DNA-dependent RNA polymerase examined thus far the circumstantial evidence for some role for Zn(II) in the enzyme seems compelling. Indeed earlier studies of Zn(II) deficiency in the simple eukaryote, *Euglena gracilis*, showed a major alteration in nucleic acid synthesis to be induced by Zn(II) deficiency suggesting a role of zinc in the enzymes of DNA and RNA biosynthesis (4, 5). Since structural and functional studies have not progressed to the point where even the skeleton of a unified structural or functional model can be presented, the particular findings on the best studied pure proteins will be presented below.

Several excellent recent reviews have appeared on the general properties of the RNA and DNA polymerases (6–15) and the purpose of this chapter will be to focus on the studies relating to the role of Zn(II).

2 POLYMERASES—MOLECULAR PROPERTIES AND MECHANISM

The enzyme from *E. coli* is the prototype for the prokaryotic RNA polymerases and the "core" enzyme has the subunit structure $\alpha_2\beta\beta'$ of ~380,000 Daltons (Table 1) (7). The holoenzyme contains an additional subunit, σ, of 613AA (MW 70,263) required for specific promoter recognition (7, 16). A single organism, for example, *Bacillus subtilis*, can contain several types of σ subunit

Table 1 Zinc Content of Prokaryotic RNA Polymerases

Enzyme Source	Molecular Weight (MW)	Subunit Structure	Subunit MW	Zn(II) g atom/mol	Reference
E. coli (gram negative)	453.5K	$\alpha_2\beta\beta'\sigma$		2	1, 6, 7, 39–43
		α	36.5K	0	
		β'^{a}	160 K	1–2	
		β	150.5K	?1	
		σ	70 K	0	
Bacillus subtilis (gram positive)	~440K	$\alpha_2\beta\beta'\sigma$		2	6, 7, 50, 55
		α	43 K	0	
		β^{a}	154–155 K (150 K)	0	
		β'^{a}	155 K (?145 K)	2	
		σ	56 K	0	
T7 Bacteriophage	98K[b]	Monomer	98 K	2	64, 66

[a]Precise molecular weight values for the RNA polymerases have been difficult to achieve. The values given for the *E. coli* subunits are calculated from the gene sequences (or amino acid sequence) except for β'. β' is slightly larger than β. In contrast, for the *B. subtilis* enzyme β is larger than β' and when run on the same gel with the *E. coli* enzyme, both are smaller than β of *E. coli*. The estimates in parentheses are thus adjusted relative to the true molecular weight of the *E. coli* subunits. On SDS gels, σ of *E. coli* has consistently calibrated (86–95 K) as considerably larger than the true molecular weight determined from the gene sequences.
[b]Molecular weight of the T7 RNA polymerase is from the gene sequence (66).

Table 2 Zinc Content of Eukaryotic RNA Polymerases

	MW and Subunit Structure						Zinc Content (g atom/mol)			
	Total MW			Enzyme Type			Enzyme Type			
Organism	I	II	III	I[a]	II	III	I	II	III	Reference
Saccharomyces cerevisiae	610 K	550 K	588 K		220 K		2.4	1.0	4.3	9, 70–72
				190 K						
						160 K				
					150 K					
				135 K						
						128 K				
						82 K				
				49 K						
					44.5 K					
				43 K						
				40 K		40 K				
				34.5 K		37 K				
					32 K					
						30 K				
				27 K	27 K	27 K[b]				
				23 K	23 K	23 K				
				19 K		19 K				
					16.5 K					
				14.5 K	14.5 K	14.5 K				
				14.0 K						
					12.6 K					
				12.2 K		11 K				
				10 K	10 K					

Table 2 *(Continued)*

Organism	MW and Subunit Structure						Zinc Content (g atom/mol)			Reference
	Total MW			Enzyme Type			Enzyme Type			
	I	II	III	I[a]	II	III	I	II	III	
Euglena gracilis	624 K	700 K					2.2	2.2		75, 76, 90
					208 K					
				162 K						
					160 K					
				148 K						
					133 K					
					83 K					
				74 K						
					64 K					
				36 K						
				34 K						
				28 K						
				24 K						
				18 K						
				3 K						
Wheat Germ	418 K	640 K	639 K		220 K			~7		9, 73, 79
				200 K						
						150 K				
					140 K					

	125 K		130 K	
			94 K	
			55 K	
		40 K		
	38 K		38 K	
			28 K	
		27 K		
		25 K	25 K	
			24.5 K	
		24 K		
		21 K	20.5 K	
	20 K	20 K	20 K	
			19.5 K	
	17.8 K	17.8 K	17.8 K	
	17.0 K	17.0 K	17.0 K	
		16.3 K		
		16.0 K		
		14.0 K		
Sea Urchin	I	II (Inhibited by 1,10-phenanthroline)	III	116
Rat Liver	I	II (Inhibited by 1,10-phenanthroline)		116

[a]The precise subunit structure of the eukaryotic RNA polymerases has varied somewhat between reports for at least two reasons. The affinity of some of the associated polypeptide chains is weak enough to cause removal by harsh methods and unless care is taken in extraction partial proteolysis of chains can occur (9).

[b]Some subunits are common to two or three types of polymerase in one cell as indicated by underlining.

which apparently are present to control promoter selection (7, 17–20). For summaries of the many structural and enzymatic studies of the prokaryotic RNA polymerases, recent reviews should be consulted (7–9). The *E. coli* enzyme as well as all other RNA polymerases catalyze polymerization using a DNA template and base pairing between the incoming ribonucleoside triphosphates and one DNA strand (coding strand) of the DNA template to form a transient RNA-DNA hybrid at the active center. The Mg^{2+} or Mn^{2+} complexes of the substrates are required. The reaction taking place at the RNA start site of the DNA, a site determined by the recognition of a specific DNA sequence (promoter sequence) by a specific RNA polymerase, is illustrated schematically in Figure 1 using the most restricted promoter sequence known, the one for T7 RNA polymerase (21–23). All T7 promoters have this same sequence with the exception of one or two base changes (22–24).

The RNA polymerase reaction can be divided into the following four stages:

1. *Promoter location and melting* of the double strand to form an open complex as pictured in Figure 1. The enzyme binds to the DNA and locates the specific promoter sequence, perhaps by a one-dimensional diffusion. In the latter process it is not clear what the degree of specificity is at each stage; that is, does the polymerase recognize the specific sequence from the "outside" of the double helix and then melt or is it able to transiently melt a num-

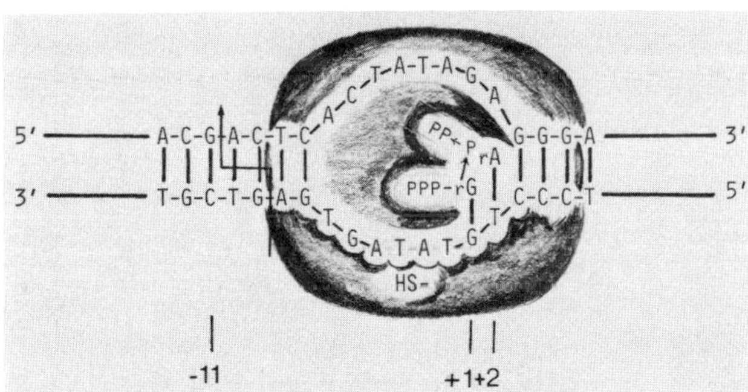

Figure 1 Diagrammatic representation of the open promoter-T7 RNA polymerase complex with the initiating (+1) and elongating (+2) ribonucleoside triphosphates in place. Adjacent to the TATA box is an essential —SH group whose carboxymethylation abolishes the proper recognition of the promoter by the polymerase (21, 22). Sequence from +1 to −11 is the unique sequence required for promoter recognition by T7 RNA polymerase (22). Arrow indicates a Hinf restriction site, which is present in all T7 promoters sequenced thus far. While cleavage at this site abolishes proper recognition and initiation, this site is not protected in the open complex and hence the polymerase molecule does not totally surround this sequence.

ber of (or all) sequences, forming a complex of long lifetime only with the specific sequence. There is some evidence for the latter mechanism in the case of the T7 RNA polymerase, since the enzyme is able to transiently melt double-strand DNA of any sequence (22, 25), but forms a melted or open complex of long lifetime only with the specific promoter sequence (Fig. 1).

2. *Initiation of the RNA chain*. The enzyme in the open complex has two adjacent binding sites for the nucleoside triphosphates which base pair with the first two bases of the coding strand corresponding to the start of the RNA (positions +1 and +2, Fig. 1). The first or initiating nucleotide (usually a purine) retains the 5′-triphosphate. The 3′-OH of this nucleotide is the nucleophile that attacks the α-phosphate of the second nucleotide, forming the first 3′-5′ diester with ejection of pyrophosphate.

The phosphorothioate analog of ATP, ATPαS, has two possible stereochemical arrangements around phosphorus, $A(S_p)$ and $B(R_p)$. Using the separated isomers, Yee et al. (26) have shown that the S_p isomer is preferred as the initiating nucleotide by the *E. coli* enzyme, although the R_p will work. On the other hand, using UTPαS as the substrate inserting at the second position, the stereochemical discrimination is absolute (26–29). Only the A(or S_p) configuration is utilized. The R_p diastereoisomer is a weak inhibitor. The product, $pppAp(S)U$, is rapidly hydrolyzed by snake venom phosphodiesterase under conditions where this enzyme only hydrolyzes the R_p isomer, hence the configuration around phosphorus has been inverted from S_p to R_p by formation of the diester bond. This proves that formation of the first diester is a single in-line nucleophilic attack by the 3′-OH with inversion of configuration around phosphorus and no covalent enzyme intermediates are involved (30).

3. *Elongation of the RNA chains*. Successive nucleotides are added to the 3′-OH terminus of the growing chain as the enzyme processes along the template, melting it as it goes and forming diesters as described above for the first diester bond. This general elongation reaction has also been shown to be absolutely stereospecific for the S_p or A isomers of the incoming nucleotide triphosphates (27, 28). This mechanism requires that RNA chains are always synthesized in the direction 5′ → 3′.

The description of polymerases as processive simply means that the enzyme does not dissociate from the enzyme–template–product complex before repeating the catalytic reaction with the next substrate in line. The one-dimensional movement along the template could be associated with the polymerization step itself. There is some evidence that movements of the enzyme along the template can be stimulated by the nucleotide binding step without formation of the diester (25).

4. *Termination of RNA chains*. When the elongating RNA polymerase encounters certain sequences known as terminators, the enzyme pauses fol-

lowed by the release of the RNA chain and dissociation of the polymerase. In the case of the bacterial polymerases these sequences generally have a G—C-rich segment (perhaps difficult for the enzyme to melt) followed by a string of T residues. The complementary RNA structure can usually form a stem and loop structure suggesting that formation of secondary structure in the emerging RNA may help terminate the transcription. The termination process has been studied most extensively for the tryptophan operon (31–34) and the terminator sequence is given in Figure 2. In the case of multisubunit RNA polymerases, recognition of certain sequences as terminators may require additional associated polypeptides, for example, the polypeptide known as ρ in the case of *E. coli* RNA polymerase (6). On the other hand, some *E. coli* terminators are ρ independent.

3 PROKARYOTIC RNA POLYMERASES

Bacterial, bacteriophage, and eukaryotic cell RNA polymerases comprise a large group of remarkably different molecular species catalyzing a basically identical chemical reaction. The differences in the gross topology of the catalytic molecules would appear therefore to be directed at various features required for the control of gene expression such as recognition of start sites (species specific, e.g., viral vs. host), modulation of the spectrum of promoter recognition during cell development, or as a feature of metabolic control. The regulation of the efficiency of promoter recognition and termination may control the amount of a given RNA synthesized. Despite large differences in molec-

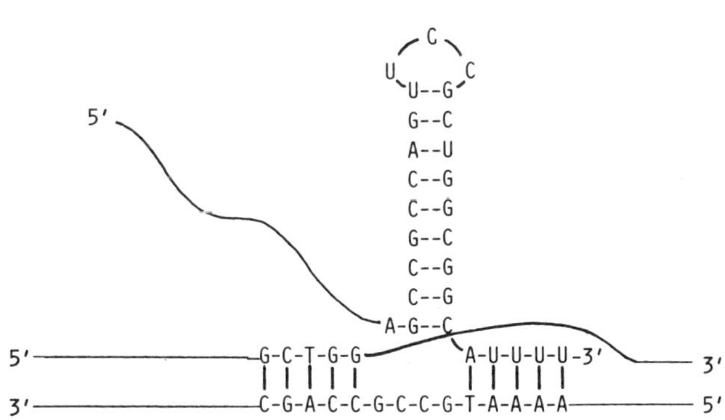

Figure 2 Base sequence at the termination region of the *trp* operon. mRNA can form a secondary structure in this region consisting of a stem and loop structure as shown. mRNA ends at the fourth U in a string of U's. Secondary structure of the mRNA may be involved in the termination event (see ref. 106 for discussion.)

ular properties, all RNA polymerases appear to share certain basic properties relating to the initiation of an RNA chain including the TATA box characteristic of promoter sequences and frequently called the Pribnow (prokaryotes) (35) or Hogness (eukaryotes) box (36). Some aspects of phage, bacterial, and eukaryotic promoters are schematically pictured in Figure 3 to emphasize certain common features. The phage T7 polymerase, a monomer, seems to differ from the others in that promoter efficiency is not affected by sequence in the immediate upstream region of the template.

3.1 Escherichia coli RNA Polymerase

This enzyme was the first RNA polymerase shown to contain Zn(II), ~2 g atom/mol of core enzyme ($\beta'\beta\alpha_2$) (1) (Table 1). The genes for these subunits have been mapped, cloned, and completely or partially sequenced. The *rpo B* (β gene) gives a sequence of 1342 AA (MW = 150,619) (39, 40), *rpo D* (σ gene) gives a sequence of 613 AA (MW = 70,263) (41), the *rpo A* (α gene) is partially sequenced (42) and the complete amino acid sequence determined earlier is 329 AA (MW = 36,512) (43). The α gene is cotranscribed with several ribosomal protein genes (44) and the β and β' genes are separated from the preceding ribosomal protein genes by a 321 *bp* intercistronic region (45).

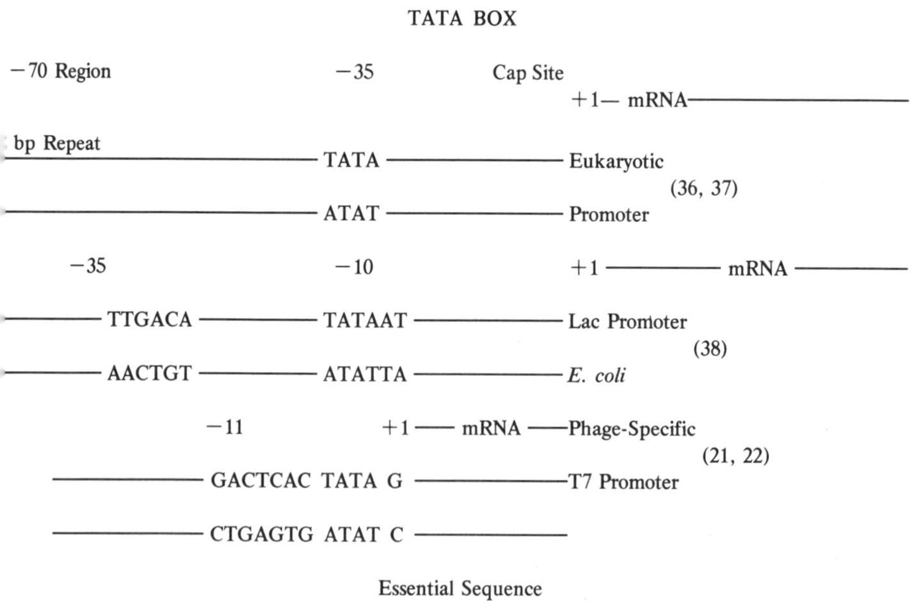

Figure 3. Features of the sequence in the region of promoters for RNA polymerases.

While β and β' are cotranscribed with genes for the ribosomal proteins L10 and L7/L12 (46, 47), the control is different under some circumstances suggesting that regulatory sequences are present in the 321 *bp* region separating the genes (42).

The Zn(II) ions cannot be removed from either the holo or the core enzyme by EDTA or 1,10-phenanthroline. After treatment with 7*M* urea, the individual subunits can be isolated by chromatography on a phosphocellulose column in 7*M* urea (48, 49). Elution with potassium phosphate gives three well-separated protein peaks, α, β, and β', none of which contains Zn(II). Thus 7*M* urea denatures the Zn-binding sites sufficiently to dissociate the metal ions. The potential for Zn(II) binding by these isolated subunits was tested by renaturing them by dialysis against buffers containing $10^{-5}M$ Zn(II), followed by dialysis against 10m*M* EDTA to remove loosely bound Zn(II). Under these conditions the α subunit binds <0.1 g atom Zn(II)/mol, the β subunit 0.3–0.9 g atom Zn(II)/mol, while β' binds 0.9–1.9 g atom Zn(II)/mol (49). If the $\alpha_2\beta$ complex is renatured, it will bind 0.4–1.0 g atom Zn(II)/mol. The large variance in these numbers makes it difficult to decide whether the original binding sites for the two Zn(II) ions in the native enzyme are located exclusively on the β' subunit or one on β' and one on β. As will be described below passage of the *B. subtilis* RNA polymerase over a blue dextran–Sepharose column in 6.5*M* urea retains the β' subunit on the column and this subunit retains both of the Zn(II) ions found in the native enzyme (50). In contrast, the same procedure applied to *E. coli* enzymes, while it results in differential binding and isolation of the β' subunit, produces isolated β' subunits and $\alpha_2\beta$ complexes that contain no Zn(II) (49).

Complete restoration of polymerase activity by renaturation–reassociation of the urea-dissociated subunits has not been possible. Only 30–50% of the native activity can be restored by dialyzing the proper stoichiometry of subunits against renaturation buffer (49). Surprisingly, addition of Zn(II) to the buffer did not have much effect on the level of restored activity, which the investigators speculate was due to the difficulty of removing contaminating Zn(II) from the sulfhydryl-containing buffers (49). The fact that the subunit location of the Zn(II) in *E. coli* RNA polymerase has to be assessed by the rebinding to initially denatured subunits is an unsatisfactory situation, since the rebound stoichiometry is variable. From the data quoted above it seems safe to conclude that one of the Zn(II) ions is located in β'. Whether the second one is also located in β', in β, or on the $\beta'\beta$ interface appears uncertain.

By growing *E. coli* on a Zn(II)-depleted, Co(II)-supplemented medium, Speckhard et al. (51) were able to isolate a biosynthesized Co(II) RNA polymerase with an enzymatic activity identical to that of the native Zn(II) enzyme. The Co(II) enzyme contained 1.8–2.2 g atom Co(II)/mol and showed a distinctive visible absorption spectrum with absorption maxima at 584 nm

($\epsilon \cong 250$) and 703 nm ($\epsilon \cong 340$) (Fig. 4). The energies and intensities of these bands suggest that a tetrahedral or 5-coordinate geometry surrounds the Co(II) ions. The addition of nucleoside triphosphates and the template analog, $d(pT)_{10}$, to the Co(II) enzyme modified both absorption peaks moderately (largely accounted for by a 20–30% drop in intensity, without much band shift). While the spectrum suggests that Co(II) has been placed at highly specific sites, the absorption changes accompanying substrate or template binding are too general to define whether any alteration in ligands at the Co(II) sites occur on substrate or template binding.

An attempt was made to better localize the metal ions to specific subunits by oxidizing the Co(II) to Co(III) with H_2O_2 to produce the substitution-inert Co(III) complexes (49). Interpretation of such data requires the assumption that the cobalt has not migrated during oxidation, an assumption that has been documented for several other metalloenzymes (52–54). Following this oxidation, which destroyed activity, one of the Co(III) ions was found with the urea-denatured β' subunit bound to the blue dextran–Sepharose column (49), thus supporting the conclusion that one of the specific metal-binding sites is in β'. Oxidation by H_2O_2 inactivates the Zn(II) enzyme unless the —SH groups are protected by prior treatment with potassium tetrathionate,

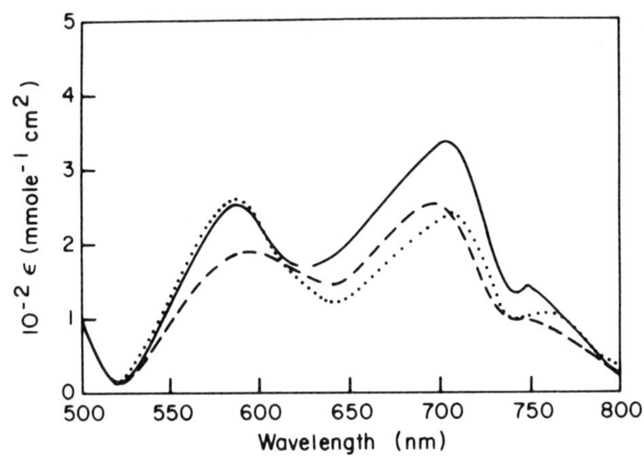

Figure 4 Difference spectra between Co- and Zn-RNA polymerase. Sample was 10 mg/ml Co-RNA polymerase in 0.05*M* tris-HCl buffer (pH 8) containing 0.05*M* KCl, 0.01*M* $MgCl_2$, 5% glycerol, 0.1m*M* EDTA, and 0.1m*M* dithiothreitol. Reference was the same concentration of Zn-RNA polymerase in the same buffer. Spectrum was measured in a 0.5 ml masked quartz cuvet using a Cary 118C spectrophotometer equipped with a scatter-transmission accessory. Both the sample and reference cells were thermostated at 15°C: (——) no substrate or template added; (---) 0.4m*M* ATP or GTP added; (···) $10^{-5}M\ d(pT)_{10}$ added. Molar absorptivity was calculated from the measured absorbance with a path length of 1 cm and the molar concentration of cobalt ion (from ref. 51 with permission.)

in which case the enzyme can be reactivated with β-mercaptoethanol after oxidation. If the Co(II) enzyme is protected in analogous fashion during oxidation to Co(III), the Co(III) enzyme cannot be reactivated. While this might suggest a direct function for Co(II) in the activity, secondary changes in structure cannot be ruled out.

3.2 Bacillus subtilis RNA Polymerase

The RNA polymerase from *B. subtilis* (a gram-positive organism) has the typical bacterial enzyme subunit structure $\beta'\beta\alpha_2\sigma$, but differs from the *E. coli* enzyme in that the β subunit (MW $\cong$ 155K) is larger than the β' subunit (MW $\cong$ ~150K) (50, 55). The larger subunit was initially believed therefore to correspond to β', the designation for the largest subunit of the *E. coli* enzyme. The larger subunit, however, was found to determine the rifampicin-resistant phenotype in a mutant polymerase (55), hence the larger polypeptide corresponds to β of the *E. coli* enzyme. In contrast the β subunit is the smaller of the $\beta\beta'$ pair in the *E. coli* enzyme and determines the rifampicin phenotype.

The Zn(II) content and subunit distribution of the metal ions are probably more certain for the *B. subtilis* enzyme than for any other RNA polymerase. Elution of the enzyme from a DNA–cellulose column with KCl results in two peaks of RNA polymerase activity. The first corresponds to core enzyme, while the later peak contains holoenzyme (50). Both contain ~2 g atom Zn(II)/mol, hence σ does not contain Zn(II). Dialysis of both core and holoenzymes against 1m*M* EDTA for 7 days fails to remove the Zn(II) (50). On the other hand, dialysis against 10m*M* EDTA for 7 days removes 50% of the Zn(II), accompanied by loss of 80% of the activity (50).

After dissociation of the subunits in 6.5*M* urea, passage of the enzyme through a blue dextran–Sepharose column results in binding of only β' (the smaller subunit) to the column, which can then be eluted with 0.6*M* KCl. After dialysis of the separated subunits against 1m*M* EDTA for a short time to remove nonspecifically bound metal ions, Zn(II) analysis showed the β' subunit to contain ~2 g atom Zn(II)/mol, while the α and β subunits contained negligible amounts (50).

Dialysis of the isolated β' subunit against 1m*M* EDTA for nine days resulted in loss of ~50% of the Zn(II), suggesting as in the case of the core enzyme, that one Zn(II) is bound less tightly than the other. This Zn(II) is, however, more labile in the isolated subunit, since 1m*M* EDTA will remove it. Inhibition studies with the phenanthrolines show that the chelating 1,10-phenanthroline inhibits the enzyme by 40% at 1m*M* concentration, while 4m*M* will inhibit it by 96%. In contrast, 1–2m*M* concentrations of 1,7-phenanthroline (nonchelating) inhibit only ~20% (50). A summary of the com-

parative properties of β and β′ subunits of the *B. subtilis* and *E. coli* RNA polymerases is shown in Figure 5 from the paper of Halling et al. (50).

3.3 Bacteriophage RNA Polymerases

A number of bacteriophage genomes have genes which code for RNA polymerases (8). These enzymes recognize specific promoters on the phage DNA and are responsible for phage-specific transcription which ensues a few minutes after infection following initial transcription of the phage genome, including the RNA polymerase gene, by the host polymerase. Phage polymerases from bacteriophages T7 (56, 57), T3 (58, 59), PBS2 (60, 61), and N4 (62, 63) have received considerable study and their properties have been recently reviewed (6, 8). The role of Zn(II) in the relatively small T-phage monomeric RNA polymerases has been investigated only for the enzyme from the coliphage T7 (64). T7 RNA polymerase is the product of T7 gene 1, which extends from 7.93 to 14.55% of the linear T7 genome of 40,000 base pairs (24) and is transcribed early in infection by the *E. coli* RNA polymerase which terminates at 18.9% of the genome (65). Gene 1 has been cloned and sequenced which shows the single polypeptide to contain 883 amino acids giving a total molecular weight of 98,092 (66).

A relatively efficient scheme for preparation of the T7 polymerase from phage-infected cells has been developed (56, 67, 68), and the physicochemical and enzymatic properties of the purified enzyme are summarized in Table 3. Average preparations of this enzyme contain from 2–4 g atom Zn(II)/mol based on an original MW of 107,000 (64). The most highly purified preparations, following dialysis to remove less tightly bound Zn(II), approach a Zn(II) content of 2 g atom/mol without loss of activity. Dialysis of the enzyme against $5 \times 10^{-3} M$ 1,10-phenanthroline, 4°C, results in loss of all Zn(II) and all activity by 96 hr, while control samples dialyzed against metal-free buffer lose a

Property	E. coli	B. subtilis
RIFAMPICIN SENSITIVE	β	β
ELUTION FROM PC	β	β
CONTAIN ZINC	β'	β'
LOW SALT AGGREGATION	β'	β'
PROTEOLYZED EASILY	β'	β'

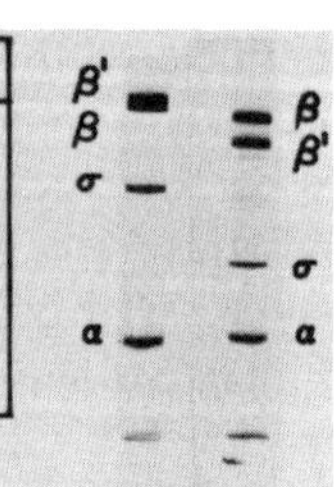

Figure 5 Summary of the properties of the large polypeptides of *E. coli* and *B. subtilis* RNA polymerase when the polypeptides in the urea–SDS–polyacrylamide gels are labeled as shown (from ref. 55 with permission.)

Table 3 Physicochemical Properties of T7 RNA Polymerase

Property	Comments
Molecular weight	98,092 (from gene sequence)
Number of amino acids	883 (from gene sequence)
Polymerization activity	Steep sigmoid pH profile; midpoint pH 6.5, plateau[a] pH 7–9
DNA binding (filter assay)	Highly temperature dependent; correlates with temperature dependence of activity, 10°C (10%), 37°C (100%) Sigmoid pH dependency, $\sim pK_a = 7$[a]
Single titrable SH in native enzyme	Reacts with 1 mol ^{14}C-iodoacetamide/mol protein[b]; reaction blocked by Nbs_2; reaction completely inactivates the enzymes; SH has pK_a of 7.8; alkylated polymerase still binds to DNA template, but is not centered on the TATA box[c]
Nitration of tyrosine	Abolishes DNA binding[b]
UV circular dichroism	Contains under 20% α-helix[b]
Zinc	Contains 2 Zn ions/mol; inactivated by 1,10-phenanthroline[d]

[a] Reference 105.
[b] Reference 68.
[c] Reference 25.
[d] Reference 25.

maximum of 30% of their activity. Unfortunately it has not been possible thus far to restore activity to the Zn-free enzyme by the readdition of Zn(II) (64). With a number of preparations of the polymerase, however, it has been possible to specifically activate the originally isolated enzyme two- to five-fold by the addition of $10^{-4}M$ Zn(II) (64), an activation not shown by similar concentrations of Mn(II), Co(II), or Cd(II).

There are a number of potential complications to the interpretation of metal ion effects on the catalytic activity of an RNA polymerase. The prototype *E. coli* enzyme requires Mg(II), Mn(II), or Co(II), since it uses the metal complexes of the nucleoside triphosphates as substrates and possibly also forms a substrate–metal–enzyme bridge complex (see Sec. 8). At concentrations below 3mM, Mn(II) and Co(II) are better activators than Mg(II), but inhibit at higher concentrations (29). Zn(II) and Cd(II) activate slightly, but maximally show less than 10% the activating activity of Mg(II) at concentrations of 1mM. Above 1mM Zn(II) and Cd(II) show potent inhibition (29).

The T7 RNA polymerase is also subject to metal ion inhibition even at relatively low concentrations of metal ions as illustrated in Figure 6 (69).

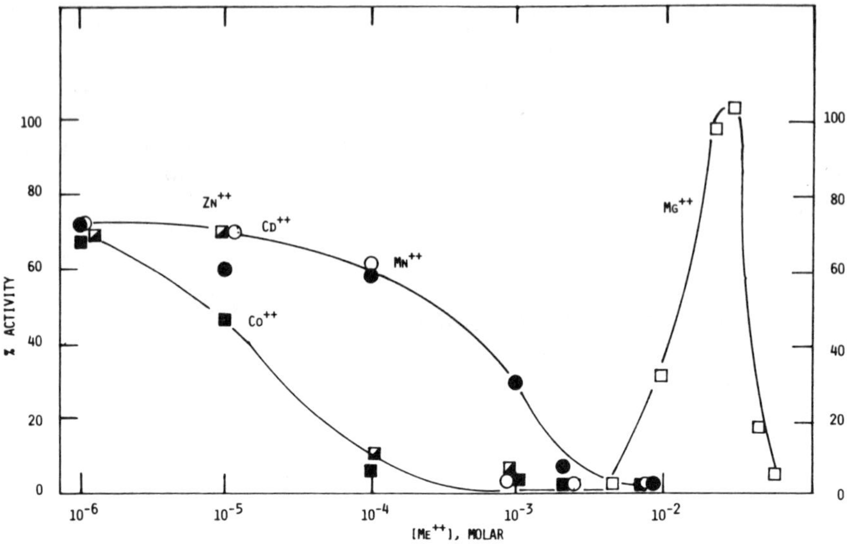

Figure 6 Inhibition of T7 RNA polymerase by transition- and IIB-metal ions as a function of metal ion concentration, $[Me^{2+}]$. Activation by Mg^{2+} versus concentration. Stock enzyme was incubated with the metal ion for 3 hr prior to assay by a 10-fold dilution into standard assay medium containing 40m*M* Tris HCl, 30m*M* $MgCl_2$, 10m*M* β-mercaptoethanol, 0.4m*M* NTP's (^{3}H-ATP), 12.5 μg/ml rifampicin, and 50 μg/ml T7 DNA, pH 8, 37°C. (◪) Zn^{2+}; (○) Cd^{2+}; (●) Mn^{2+}; (■) Co^{2+}; (□) Mg^{2+} (69).

These potent inhibitions by first-transition and IIB metal ions occur at concentrations one to two orders of magnitude (10^{-4}–$10^{-3}M$) lower than maximal Mg(II) activation ($10^{-2}M$). Thus Zn(II) can inhibit as well as activate depending on as yet poorly identified variables. Metal ion inhibition is similar whether the stock polymerase is incubated with the metal ions prior to a 1:10 dilution into the assay mix or if the metal ions are added directly to the reaction mixture. Mg(II) does not compete out this inhibition, suggesting that there are inhibitory metal-binding sites other than those normally forming the enzymatically competent Mg(II) complexes.

These features of metal ion modulation of RNA polymerase action using the simplest of polymerases illustrate the difficulty of interpreting simple metal addition experiments in these complex template-directed reactions. Until it is possible to separate the effects of substitution of metal ions in the nucleotide substrate complexes from the effects of metal ion binding to specific catalytic or structural sites on the enzymes and these in turn from metal ion binding to inhibitory sites, metal ion activation experiments must be interpreted with caution. Observed effects may be the result of two actions. A given metal nucleotide may serve as substrate, but the same metal may have

inhibitory sites on the protein. RNA polymerases in general contain a number of free SH groups. In fact, the T7 RNA polymerase is unique in this regard, since in the native conformation it possesses only a single titratable SH group essential for activity, located at the template binding site and apparently involved in the correct binding of the promoter sequence (68) (see Fig. 1). Binding of metal ions to this SH group may inhibit and obscure functional effects at other sites.

4 EUKARYOTIC RNA POLYMERASES

The RNA polymerases from nucleated cells are an order of magnitude more complex than the prokaryotic enzymes. The most highly purified preparations contain between 9 and 14 separate polypeptide chains when the best preparations are dissociated on SDS polyacrylamide gels (9). Nucleated cells contain three separate RNA polymerases: I (or*A*), responsible for ribosomal RNA synthesis and localized to the nucleoli; II (or*B*), responsible for general mRNA synthesis and selectively inhibited by the fungal toxin, α-amanitin; and III (or *C*), responsible for transcription of 5S and tRNA genes (9). As indicated in Table 2, where more details of subunit structure are given, significant amounts of Zn(II) are found associated with these polymerases. The reported contents show a significant variation, that is, RNA polymerases I (70), II (71), and III (72) of yeast are reported to contain 2.4, 1.0, and 4.3 g atom Zn(II)/mol enzyme, respectively, while RNA polymerase from wheat germ is reported to contain ~7 g atom Zn/mol (73). Since there are enzymes with molecular weights of between 500 and 700K for the undissociated multipolypeptide complex, reasonably large amounts of protein still contain relatively small amounts of zinc. While elegant assay techniques, such as microwave spectroscopy (74), can measure metal content in the picogram range with precision, the small Zn(II)-to-protein mass ratio makes the stoichiometry inherently subject to greater variation than data for the smaller monomeric Zn(II) metalloenzymes.

4.1 RNA Polymerases from *Euglena gracilis*

Among the most carefully controlled zinc analyses on the eukaryotic RNA polymerases are those of the RNA polymerases I and II from the simple unicellular dinoflagellate, *Euglena gracilis*. The RNA polymerase I from this organism has a molecular weight of 624,000 and consists of 10 subunits based on the SDS gels of the most highly purified preparations. Zn(II) content of

the enzyme averages 2.2 g atom/mol (75). RNA polymerase I of *Euglena* is considerably more sensitive to OP than RNA polymerase II (see below). The enzyme is inhibited completely by $10^{-4}M$ 1,10-phenanthroline and is 50% inhibited by $10^{-5}M$. The nonchelating 4,7-phenanthroline is not an inhibitor.

The RNA polymerase II from *Euglena* has a molecular weight of ~ 700,000 and dissociates into five separate bands on SDS gels. The bands appear to represent six subunits, two of which have the same molecular weight (76). Zn(II) analyses again show an average of 2.2 g atom/mol (76). The chelating agents 1,10-phenanthroline, 8-hydroxyquinoline, 8-hydroxyquinoline-5-sulfonic acid, and the metal-complexing antibiotic, lomofungin, inhibit the enzyme, while the nonchelating phenanthrolines, 1,7 and 4,7, do not inhibit. In contrast to the RNA polymerase I, $10^{-3}M$ 1,10-phenanthroline is required before complete inhibition is observed. The inhibition is instantaneous and is fully reversible on dilution, which suggests that a mixed enzyme–metal–OP complex forms. Reversible inhibition of this type by 1,10-phenanthroline due to formation of mixed enzyme–metal–chelator complexes was first described for alcohol dehydrogenase and detailed analysis of this type of inhibition is available (77, 78). A time-dependent inhibition of alcohol dehydrogenase accompanying Zn(II) removal can also be induced by OP and this type of inhibition appears to apply for some eukaryotic RNA polymerases (see below).

4.2 RNA Polymerase II from Wheat Germ

Large quantities of homogeneous RNA polymerase II can be obtained from wheat germ by the method of Jendrisak and Burgess (79). Samples of this protein used for zinc anaysis were, therefore, more concentrated than usually employed for polymerases and thus should yield relatively accurate values. The unusual aspect of this polymerase is that it contains reproducibly large amounts of Zn(II), ~ 7 g atom (6.2–7.7 range)/mol enzyme, MW 550,000 (73). The molecular weight has since been revised upward [639,000, 14 polypeptide chains (9)], but this value still indicates ~ 6 g atom Zn(II)/mol. The Zn(II) does not appear to be loosely bound, since extended dialysis and gel filtration against relatively metal-free buffers did not reduce the Zn(II) content. In addition, Zn(II) was not removed by dialysis for 7 days against a 1000-fold volume excess of 0.05M Tris, 0.1mM EDTA, 1mM dithiothreitol, 25% glycerol, pH 7.9, 4°C. 1,10-Phenanthroline (1–2.5mM) does not produce instantaneous inhibition of the wheat germ enzyme (73). On the other hand, if the enzyme is dialyzed against 2.5mM OP, 25°C, pH 7.9, for seven days, the Zn(II) is totally removed and the activity is abolished. Attempts to reactivate the apoenzyme with Zn(II) have failed thus far (73).

4.3 RNA Polymerases from Yeast

Yeast RNA polymerase I is another eukaryotic polymerase in which the relationship of Zn(II) to function has been examined. Using a molecular weight of 650,000 [the sum of recent subunit molecular weights is 610K (9)], the holoenzyme contains 2.4 g atom Zn(II)/mol (70). The inhibition by 1,10-phenanthroline is of the instantaneous type like that shown by the *Euglena* enzymes. In this case estimates of the number of chelator molecules bound per mole of enzyme, for both OP and bathocuproin, gave values ≅ 1:1 suggesting the formation of enzyme–Zn–chelator complexes (70). As observed for the other RNA polymerases, 1,7-phenanthroline was not an inhibitor.

A single RNA polymerase II can be isolated in milligram quantities from *Saccharomyces cerevisiae* (71). The preparation gives a single major band on polyacrylamide gels and from densitometry appears to have only minor contamination. SDS-gel electrophoresis revealed the following subunits in 1:1 stoichiometry: 170K, 150K, 41K, 28K, 23K, 19K, plus a small chain of 2.3K. This differs somewhat from the most recent data on yeast RNA polymerase II (Table 2). The Zn(II) content paralleled the RNA polymerase activity at each stage of the purification and the final Zn(II) content of eight preparations averaged 0.98 ± 0.4 g atom Zn(II)/mol of MW 460,000 (71). If a 10-fold molar excess of Zn(II) was added to the enzyme, the excess could easily be removed by gel filtration and the stoichiometry returned toward the value of 1:1. 1-10-Phenanthroline, 0.6mM, inhibits the enzyme activity by 50%, while 7,8-benzoquinoline (one N of OP replaced by C) shows no inhibition. OP, 3mM, completely inhibits the enzyme, but this inhibition can be reversed by adding excess Zn(II), but not by excess Mn(II). Mn(II) is used as the activating metal ion in the assay. Treatment of the RNA polymerase II with 1000-fold excess of OP will slowly remove the Zn(II), but results in irreversible denaturation of the enzyme (71).

Yeast RNA polymerase III has been purified to the point where polyacrylamide gels run under nondenaturing conditions show a single protein band (72). Analysis of the protein by atomic absorption showed the presence of 0.7 g atom Zn(II)/100,000 g protein or 4.3 g atom Zn(II)/mol protein assuming a MW of 620,000. The latter molecular weight is close to that of the sum of the molecular weights of the subunits (588K, Table 1). The Zn(II) cannot be removed from yeast RNA polymerase III by passage over a column of chelex resin (72). The enzyme is inhibited by the addition of 1,10-phenanthroline, 50% at 0.1mM, 100% at 1mM. Excess Mn(II), equimolar to the 1,10-phenanthroline, was added as activator without any effect on the inhibition which was interpreted to mean the OP was inhibiting via a Zn(II) complex rather than via the "activating" metal ion, Mn(II) or Mg(II) (72). It is not yet known

whether the OP inhibition is reversible or whether OP can remove Zn(II) from the yeast RNA polymerase III.

Analyses for other metal ions have been performed in most of the studies of eukaryotic and prokaryotic RNA polymerases and indicate the presence of less than 0.1 g atom/mol of Mg, Cr, Mn, Fe, Ni, Co, and Cu, although not all of this series has been analyzed in all instances.

5 PHYSIOLOGICAL EVIDENCE OF A ROLE FOR Zn(II) IN TRANSCRIPTION–TRANSLATION

Zinc has long been recognized as an essential nutrient for most organisms (80, 81). It is, however, extremely difficult to make the prokaryote, *E. coli*, zinc deficient (82), probably related to the low concentrations of zinc required coupled with efficient uptake mechanisms. These factors render usual metal extraction techniques insufficient to reduce the zinc concentration of media far enough (82). On the other hand, zinc-deficiency syndromes in eukaryotic organisms are not difficult to produce and occur naturally in several animals (83). A requirement of Zn(II) for cell division was first reported in 1869 (84). The most elegant biochemical characterization of eukaryotic zinc deficiency is provided by the series of studies by Vallee and co-workers on zinc deficiency in the single-celled eukaryote, *Euglena gracilis* (4, 5, 85, 86). Zinc deficiency in this organism is characterized by inhibition of cell proliferation and striking changes in cell morphology, which can be detected by both light and electron microscopy (85, 86). The cells increase in size by 14-fold on a dry weight basis along with a remarkable accumulation of cytoplasmic paramylon (a β-1,3-glucan) which appears as large storage granules in electron micrographs (86).

The first fractionations of the macromolecular components of the Zn-deficient cells showed that the DNA content was nearly double that of a normal cell, while both RNA and protein content were decreased on a dry weight or total nitrogen basis, findings which suggested a premitotic block (5). More detailed studies of the DNA content of individual cells made possible by laser-induced cytofluorimetry show the DNA content of individual Zn-deficient cells to be as much as twice that of normal Zn-sufficient cells (87). The majority of normal cells are in the *G*-1 phase of the cell cycle during the log phase of growth. By using alternating periods of dark and light, *Euglena* cells can be synchronized and the histograms for the relative DNA contents of synchronized cells compared to nonsynchronized cells in log phase can be used to plot relative DNA contents in cells at the *G*-1, *S*, and *G*-2 phases of the cell cycle. While normal *Euglena* cells in stationary phase show a histogram typical of cells in *G*-1, nonsynchronized Zn-deficient cells, after they

have stopped growing, show a histogram typical of the DNA content of S/G-2 phase cells (81). This suggests that as zinc is exhausted, cells in S do not continue to G-2, while those in G-2 do not proceed to mitosis.

The above findings might suggest that Zn primarily affects steps following DNA replication. On the other hand, a small fraction of cells present in the growth-arrested Zn-deficient cultures are in G-1, suggesting that there may also be a block in processes preceding DNA replication. The latter conclusion is further supported by resuspending stationary Zn-sufficient cells (most in G-1 phase) in Zn-deficient medium. Resuspension initiates a 25% increase in cell number before growth arrests a second time, but the histogram shows most cells are still in G-1. That the absence of zinc was responsible for the cessation of growth was shown by readdition of Zn(II) which led to rapid division of cells such that cell density equal to that of a Zn-sufficient culture was achieved at stationary phase.

These findings were and continue to be strong circumstantial evidence implicating zinc as a required cofactor in the fundamental processes of DNA replication and transcription and possibly protein synthesis, hypotheses further strengthened by the analytical demonstration of tightly bound Zn(II) in various DNA and RNA polymerases, particularly those of the test organism itself. While the biochemical phenomena associated with eukaryotic Zn deficiency are compatible with the idea that malfunctioning DNA and RNA polymerases are at least partly responsible for the biochemical lesions, the data do not prove it uniquely. Considering the known participation of Zn as an essential catalytic or structural component of enzymes involved in many aspects of metabolism (88, 89), blocks in a number of pathways could accompany Zn deficiency and produce many secondary anomalies. Zn-deficient *Euglena* accumulate much higher than normal levels of Mg, Cr, Mn, Fe, and Ni (90), which might also cause some functional alterations, Mn particularly in the function of nucleotidyl transferases (see below).

More recently in the continuing series of studies on the derangements of nucleic acid biosynthesis in Zn-deficient *Euglena*, Vallee and co-workers have examined RNA synthesis in more detail (90). Factoring out the increase in cell size by normalizing to cell number, it becomes apparent that the Zn-deficient cells contain approximately the same total amount of RNA as Zn-sufficient cells, but the distribution between ribosomal, transfer, and mRNA is altered. There is almost a doubling of the percentage represented by mRNA with a fall in the percentage of RNA. Moreover, there were easily detectable alterations in base composition of the RNA, a significant increase in the proportion of G + C plus some unidentified abnormal bases emerging as separate peaks on high-pressure liquid chromatography. An unexpected finding is that the Zn-deficient cells have as their major RNA polymerase a new single

species which appears to replace the normal three RNA polymerases, I, II, and III (90). While the new polymerase is a zinc enzyme, it has a different α-amanitin sensitivity. Since Zn-deficient cells accumulate 35-fold greater amounts of Mn(II) and five-fold greater amounts of Mg(II) than normal cells, the effects of Mn(II) on normal *Euglena* RNA polymerase function were tested (90). Increasing the Mn(II) concentration from 1 to 10mM significantly decreased the ratio of UMP/CMP in the resulting RNA. The same effect was observed in the new RNA polymerase species from Zn-deficient cells. Thus the abnormal metal ion profile may be a significant factor in addition to the absolute lack of Zn(II). Addition of unusual metal ions such as Cd(II) in concert with Zn deficiency can have additional striking synergistic effects on cell metabolism, and nucleic acid biosynthesis and mitosis (86). Original references should be consulted (5, 85, 90).

6 REVERSE TRANSCRIPTASES

Type *C* oncogenic RNA viruses contain an RNA-dependent DNA polymerase which synthesizes the DNA complement of the RNA genome, hence the name reverse transcriptase. The enzyme has been most extensively studied in the case of avian myeloblastosis virus and "highly purified" enzyme derived from the purified virus can be obtained reproducibly (91–95). The enzyme is generally assayed on poly(rA) with $(dT)_{12\text{-}18}$ as primer. The molecular weight of the enzyme has been determined as (1.6–1.8) $\times 10^5$ (92–94). With the use of G-100 Sephadex column as the final purification step, Auld et al. (74) first showed that the high specific activity fractions emerging from this column show a constant Zn:protein ratio of 1.8 to 2.0 g atom Zn(II)/mol enzyme. This enzyme is obtainable only in limited quantities and zinc assays had to be performed on 3 μg or less of protein. Auld et al. (74) developed a technique utilizing microwave-induced emission spectrometry, a method that can measure transition and IIB metals at levels as low as 10^{-14} g atom with precision. Zn(II) analyses of the enzyme fractions were on the order of 10^{-11} g atoms. Mn, Fe, and Cu were present at 10^{-13}–10^{-14} g atom/mol (74).

Reverse transcriptase is instantaneously and reversibly inhibited by 1,10-phenanthroline, $7 \times 10^{-5}M$ producing 50% inhibition, while $5 \times 10^{-4}M$ causes complete inhibition (74). The standard assays of this enzyme utilize 0.2mM Mn(II) as activator and under these conditions 0.1mM OP causes 60% inhibition. It seems doubtful that enough Mn(II) is removed by OP complexation to account for this inhibition. A complex with the Zn(II) seems the more likely mechanism of inhibition. EDTA, 8-hydroxyquinoline-5-sulfonate, 8-hydroxyquinoline, and EGTA at concentrations from 0.1 to 0.3mM

also inhibit polymerase activity by ~60%. If the enzyme is incubated with OP at 37°C prior to assay, there is a slow time-dependent irreversible inhibition suggesting that OP may slowly remove Zn(II) from the enzyme producing irreversible changes (see below). Neither 1,7 nor 4,7-phenanthroline are effective inhibitors, suggesting that inhibition requires metal chelation (74).

The most highly purified preparations of the AMV reverse transcriptase contain two subunits, an active α-subunit (MW = 60,000–70,000) and a β-subunit (MW 105,000–115,000) of unknown function (94-96). Using an isoelectric focusing method to separate the AMV reverse transcriptase, Poiesz et al. (97, 98) showed by atomic absorption that the peak fraction (containing both subunits) at a concentration of $6.5\mu M$ contained $8\mu M$ Zn(II) for a stoichiometry of 1.3 g atom Zn(II)/mol enzyme. The α subunit alone was shown to contain 1.2 g atom Zn(II)/mol. If the enzyme is incubated with excess ^{65}Zn(II), the enzyme retains ^{65}Zn after extensive dialysis against metal-free buffers and reisolation by isoelectric focusing. Zn(II) content remains similar to that of the initial preparation suggesting that the intrinsic Zn(II) can be exchanged (97).

The 1,10-phenanthroline inhibition studies presented by Poiesz et al. (97) followed a different protocol than those described above and involve the dialysis of the ^{65}Zn(II) exchange-labeled enzyme against $0.8mM$ OP. Under these conditions activity fell more rapidly than ^{65}Zn removal. By half an hour activity was at the 25% level, while ~60% of the ^{65}Zn was still bound. While the authors suggest that the discrepancy is caused by the slow time scale for the dialysis of the Zn–OP complexes, a combination of mixed enzyme–Zn–OP complexes and actual Zn removal is possible. These authors, however, failed to observe reversal by dilution as described above. Precise OP inhibition mechanisms will require further investigation. By 4 hr both Zn and activity were at the 20% level. With this preparation redialysis against $0.2\mu M$ Zn restored activity, while dialysis against metal-free buffer did not. If the apoenzyme was allowed to stand for more than 2 hr, however, it became progressively inactive and could not be reactivated by dialysis against Zn(II) under nitrogen. Restoration of activity by dialysis was specific for Zn(II). The metal ions Mg(II), Mn(II), Fe(II), Co(II), Ni(II), Cd(II), and Hg(II) did not restore activity under the same conditions (97). While detailed analyses have not been made, the reverse transcriptase associated with Rauscher murine leukemia virus is also inactivated by OP on a similar time scale (97). A further study by Auld et al. (99) has shown the reverse transcriptases from murine, simian, and feline RD-114 RNA tumor viruses to all be inhibited by OP. Moreover, the most highly purified fractions of the murine and woolly monkey viruses were shown to contain Zn(II). While precise stoichiometries require more enzyme, the estimated values were 1.4 and 1.0 g atom/mol enzyme for the murine and woolly monkey enzymes, respectively (99).

7 DNA POLYMERASES

In contrast to RNA polymerases, DNA polymerases cannot start a chain, but must add nucleotides to the 3′-OH of a preexisting primer, either RNA or DNA (10, 11). Otherwise the general features of the synthesis of the deoxynucleotide polymer in the 5′ → 3′ direction on a DNA template involving attack of the 3′-OH of the primer on the α phosphate of an incoming base-paired nucleoside triphosphate are similar to those that apply to RNA polymerases. DNA polymerases are in addition endowed with a 3′-5′ endonuclease function for error correction which can remove a mispaired base from the 3′ end of the growing chain. This implies that the 3′-end-base must be correctly paired before the subsequent diester can be formed.

The DNA polymerase I from *E. coli* was the first DNA polymerase described, since it normally is the most abundant DNA polymerase in the *E. coli* cell. Thus the molecular properties of this DNA polymerase have been the most extensively investigated (12,100). Despite its prototype role, *pol* I is primarily involved in DNA repair and chromosomal replication is carried out by the much more complex DNA polymerase III, the "core" enzyme of which is composed of three subunits, α, ϵ, and θ, with molecular weights of 140,000, 25,000, and 10,000, respectively (13). DNA polymerases vary in complexity from the 40,000 dalton single polypeptide of mammalian polymerase β (101) to the seven subunit complex of ~500,000 daltons which comprises the *pol* III holoenzyme of *E. coli* (13). In addition, it should be borne in mind that DNA polymerases do not appear in general to function as isolated enzymes, but as part of a complex replication apparatus involving single-stranded DNA binding proteins, helicase enzymes, as well as other polypeptides, an aggregate for which Kornberg has suggested the designation replisome (11). Except for *pol* I, DNA polymerases are present in a small copy number per cell, are difficult to purify, and are often unstable enzymes. This may be part of the reason that even analytical data on Zn(II) in DNA polymerases are far less extensive than for RNA polymerases. The existing data are given in Table 4.

Other than the data listed in Table 4 investigations relating Zn(II) to DNA polymerase structure or function have not been carried out except for *pol* I of *E. coli*. The 1 g atom/mol of tightly bound Zn(II) is nondialyzable, but will exchange with ^{65}Zn(II) and can be easily removed with 10m*M* 1,10-phenanthroline. The "apoenzyme" containing 0.06 g atom Zn(II)/mol after removal of the OP by dialysis against metal-free buffer is inactive, but can be completely reactivated by dialysis against 0.1m*M* Zn(II) (3).

The original picture of inhibition by 1,10-phenanthroline was clouded somewhat by the finding that the 2:1 cuprous complex of OP [$(OP)_2Cu(I)$] is a potent inhibitor of DNA polymerases (102–104). This complex at levels of

Table 4 Zinc Content of DNA Polymerases

Source	Molecular Weight	Zn (g atom/mol)	Reference
E. coli, pol I	109,000	1	2, 3[a]
Phage T4	114,000	1	3
Sea urchin	150,000	4	2

[a]The Klenow fragment of DNA polymerase I, isolated and crystallized from E. coli cells transformed with a plasmid carrying the gene for the Klenow fragment (Joyce, C. M. and Grindley, N.D.F., *Proc. Nat. Acad. Sci. U.S. 80*, 1830, 1983) contains <0.05 gm at Zn/mole, but has full polymerizing activity. (C. M. Joyce, N.D.F. Grindley, T. A. Steitz and J. E. Coleman, unpublished data).

$0.1\mu M$ catalyzes a rapid oxygen-dependent strand scission of both the poly (*d*A—*d*T) template and primer (104). The products of the clevage are potent inhibitors of *pol* I. Since most thiol reagents are contaminated with the required level of copper, the cleavage occurs readily in the usual polymerase assay mixtures (104). The DNA cleavage activity is not a property of the Cu(I) complexes of 2,2′-bipyridine, 2,2′,2″-terpyridine, or 2,9-dimethyl-1,10-phenanthroline and these complexes do not inhibit *pol* I. While this phenomenon may complicate the interpretation of the OP inhibition of *pol* I, it cannot be extrapolated readily to RNA polymerase data. The RNA polymerases are potently inhibited by a large variety of chelating agents including chelex resin and such metal binding anions as CN^- and ^-SH whose metal complexes are not likely to generate oxygen-assisted cleavage of DNA. In addition, the pattern of OP inhibition of various RNA polymerases is quite variable. There are those that show instantaneous reversible inhibition like the *Euglena* enzymes. In contrast, those from which Zn(II) can be removed by OP, like the wheat germ and T7 RNA polymerases, show little instantaneous inhibition, and require reasonably long time courses over which zinc is finally removed. Activity falls with Zn(II) content despite the fact that each sample is exposed to template for the same short time during assay. Contrary to statements in reference 104, conditions have not been found under which the Zn-free enzymes prepared from T7 and wheat germ RNA polymerase can be reactivated by the readdition of Zn(II). These patterns of OP inhibition are difficult to accommodate by a common mechanism involving destruction of the template and generation of inhibitory products.

The observation that 1,10-phenanthroline is an inhibitor of terminal deoxynucleotidyl transferase (107) led to the speculation that Zn(II) was essential for the function of the enzyme and triggered the analysis of DNA polymerases for Zn(II) content (3). Terminal deoxynucleotidyltransferase, however, is not

a polymerase by definition, since it assembles a random chain without utilizing a template.

8 MOLECULAR ASPECTS OF Zn(II) FUNCTION

Despite the considerable body of analytical data on the Zn(II) content of RNA and DNA polymerases, experiments to pinpoint the precise role of Zn(II) in the structure or catalytic mechanism of these enzymes have been difficult to devise. This is in part due to the difficulty of removing the Zn(II) reversibly, particularly in the case of the RNA polymerases. Several spectroscopic probes of the metal-binding sites have been applied to the *E. coli* DNA and RNA polymerase and these are briefly summarized here.

Zn(II) and its coordination compounds have a large effect on the nuclear quadrupolar relaxation of the ^{79}Br nucleus if the Br^- ion enters the inner coordination sphere, due to distortion of the symmetric distribution of electrons about the Br^-. A mixed complex between Br^-, Zn(II), and a macromolecular ligand of slow rotational correlation time may be particularly effective (108, 109). The enhanced relaxation of the ^{79}Br nucleus results in broadening of the ^{79}Br NMR signal of the bulk bromide in solution which must be in rapid exchange with the coordinated $^{79}Br^-$ (109). DNA polymerase ($\sim 50\mu M$) increases the linewidth of $0.2M$ $^{79}Br^-$ from 612 $\pm$ 14 Hz to 1344 $\pm$ 44 Hz, while free Zn(II) ($2.5mM$) increases the line only to 709 $\pm$ 23 Hz (3). If Zn(II) was removed from the enzyme with $10mM$ OP, the "Zn-free" enzyme broadened the $^{79}Br^-$ signal to $\sim$1150 Hz (based on the same protein concentration), suggesting that there is a $\sim$200 Hz contribution to the braodening from the Zn(II) ion. The latter is some 80 times greater than the contribution of ionic Zn(II) of the same concentration to $^{79}Br^-$ line broadening (3). This suggests that the Zn(II) of DNA polymerase has an open coordination site or at least one occupied by solvent which Br^- can replace. If $(dT)_{6\text{-}9}$ was added to the Zn(II) enzyme, the ^{79}Br linewidth narrowed to 1164 $\pm$ 27 Hz, a value similar to that observed for the apoenzyme. The large remaining contribution to broadening is likely due to asymmetric charges on the protein surface which interact transiently with Br^-. The decrease in line broadening as a function of increasing $d(pT)_{6\text{-}9}$-to-enzyme ratio suggested that a 1:1 complex between the oligonucleotide and DNA polymerase was responsible for the displacement of Br^-. In contrast, mononucleotides like *d*TTP showed no effect. In agreement with this picture $(dT)_{6\text{-}9}$ did not further narrow the ^{79}Br linewidth in a solution of the apoenzyme.

These results are compatible, but not uniquely so, with the postulate that Zn(II) coordinates the 3′-hydroxyl group of the primer as pictured in Figure 7. The small contribution to line broadening by Zn(II) compared to the much

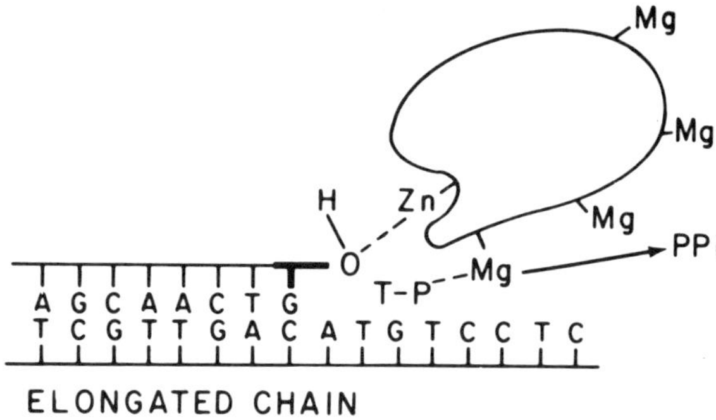

Figure 7 Schematic representation of the possible role of Zn(II) in the DNA polymerase reaction. (Reprinted from ref. 110 with permission.)

larger contribution of nonspecific protein interactions to ^{79}Br line broadening, makes interpretation of the data as showing direct Zn(II) coordination by $d(p\mathrm{T})_{6-9}$ highly speculative. It should be pointed out that the actual NMR experiments did not contain the DNA template, hence it must be assumed that the template does not significantly alter binding of the olignucleotide or Zn(II). A Zn—3′-OH complex could potentiate the attack of the 3′-oxygen on the incoming α-phosphorus of the nucleoside triphosphate. Such a complex might be even more useful in the RNA polymerase reaction because the electron-withdrawing property of the adjacent 2′-OH tends to make it more difficult to polarize the 3′-OH than in the 2′-deoxynucleotide.

A series of NMR studies has been applied to the interaction of the enzyme-bound Mn(II) of both RNA and DNA polymerase with the nucleoside triphosphates. These studies have determined spatial relationships between nucleotide and metal based on distances derived from the inverse r^6 dependence of relaxation of the ^{1}H and ^{31}P nuclei of the nucleotide by the paramagnetic center. The findings have recently been reviewed in detail (110) and will be given only briefly here. The conclusion in the case of *pol* I is that multiple Mn(II) [or Mg(II)] binding sites are present as indicated in Figure 7. In contrast to the original postulate that coordination to the α-phosphoryl group occurred in ternary enzyme–Mn–nucleotide complexes (111), the most recent studies (110, 112) suggest that an enzyme–Mn–nucleotide complex forms between a single tightly bound Mn(II) and the γ phosphate of the incoming nucleoside triphosphate and thereby assists the departure of the leaving pyrophosphate group (110). There is no indication from the NMR studies that Mn(II) participates in DNA binding (110).

Studies of Mn(II) binding to *E. coli* RNA polymerase by EPR and NMR relaxation show the presence of one tight Mn(II) binding site per molecule and ~6 weaker sites of ~100-fold less affinity (110, 113, 114). The tight-

binding site appears to be at the active site and the Mn(II) functions in chain elongation, since the K_d for this site is the same order of magnitude as the K_m for Mn UTP in the elongation reaction (113). The distance measurements between Mn(II) and initiator nucleotides such as A*p*A and A*p*U show a considerably longer distance (~10 Å) than between Mn(II) and the elongation nucleotides, UTP and ATP (~5 Å). This suggests that Mn(II) is bound at the elongation site and forms a bridge complex with the elongation substrate. The initiating nucleotides, A*p*A or A*p*U, however, cause the elongating nucleotides to displace water molecules from the Mn(II) suggesting that conformational changes are induced at the Mn(II) site by the cooperative binding of initiating and elongating nucleotides (110). The paramagnetic Cr(III) ATP can also be bound at the initiating site and can be incorporated into the RNA, although easily displaced by A*p*U. The Cr(III) provides a second paramagnetic reference point at the initiation site which confirms the conclusions drawn from the Mn(II) data (110, 115). Precise models of the geometries of the nucleotides bound at the initiator and elongation sites have been constructed from the NMR distance measurements and these have been discussed in detail recently (110) and are summarized in Figure 8 (114). While the model indicates rather normal nucleotide conformations, appropriate for base pairing with B-DNA, as well as a bridge complex between enzyme-bound Mn(II) and the γ phosphate of the elongating nucleotide, all of these com-

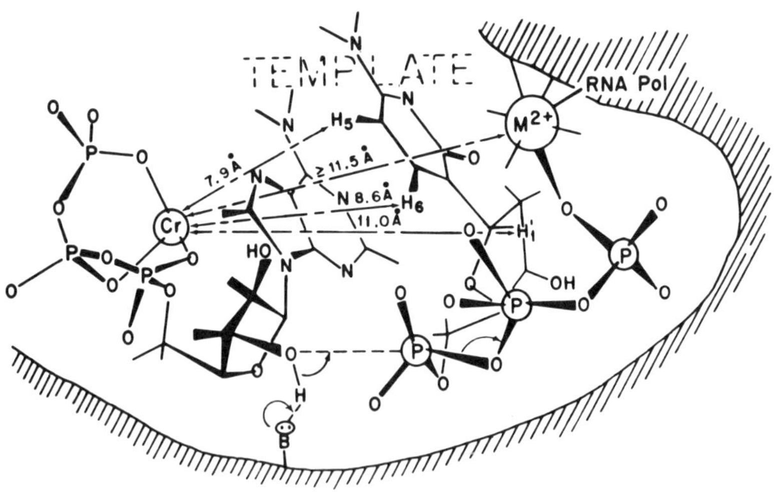

Figure 8 Arrangement of the initiator, CrATP, the elongation substrate, CTP, and the divalent activator, Mn(II), on RNA polymerase consistent with the indicated distances and with those from Mn(II) to the initiator and substrate as determined by NMR methods (see text). (Reprinted from ref. 115 with permission.)

ABBREVIATIONS

Tris Tris-hydroxymethyl amino methane
EDTA Ethylenediaminetetraacetic acid
OP 1,10-Phenanthroline
Nbs_2 5,5′-dithiobis(2-nitrobenzoic acid)

plexes are formed in the absence of the DNA template. Thus it must be assumed that binding of template does not introduce significant additional conformational changes in the enzyme active center.

The additional role of Zn(II) in this picture can only be speculative at the present time. Zn(II) could play an exclusively structural role in RNA polymerase and not be bound in the immediate vicinity of the active center. Both $d(pT)_{10}$ and purine nucleotides modify the visible absorption spectrum of the Co(II) enzyme (51) (Fig. 4). Whether these reflect conformational changes propagated to the cobalt site or insertion or exchange of a ligand cannot be determined at present. Clearly Co(II) substitutes for Zn(II) at an unusual metalloenzyme site and maintains normal catalytic activity (51). Since *in vitro* methods of metal replacement lead to denaturation of *E. coli* RNA polymerase, production of large quantities of the biosynthesized Co(II) enzyme would probably provide the most potentially useful probe of the role of the metal ion in RNA polymerase action.

ACKNOWLEDGMENT

Original research in the author's laboratory was supported by NIH grants AM09070 and GM21919 and NSF grant PCM76 82231.

REFERENCES

1. M. C. Scrutton, C. W. Wu, and D. A. Goldthwait, *Proc. Natl. Acad. Sci. USA*, **68,** 2497 (1971).
2. J. P. Slater, A. S. Mildvan, and L. A. Loeb, *Biochem. Biophys. Res. Commun.*, **44,** 37 (1971).
3. C. F. Springate, A. S. Mildvan, R. Abramson, J. L. Engle, and L. A. Loeb, *J. Biol. Chem.*, **248,** 5987 (1973).
4. C. A. Price and B. L. Vallee, *Plant Physiol.*, **37,** 428 (1962).

5. W. E. C. Wacker, *Biochemistry*, **1,** 859 (1962).
6. R. Losick and M. Chamberlin, *RNA Polymerase*, Cold Spring Harbor Laboratory Monograph (1976).
7. M. J. Chamberlin, *The Enzymes* (3rd ed) *15, part B*, 61–86 (1982).
8. M. J. Chamberlin and T. Ryan, *The Enzymes* (3rd ed) *15, part B*, 87–108 (1982).
9. M. K. Lewis and R. R. Burgess, *The Enzymes* (3rd ed) *14, part B*, 109–153 (1982).
10. A. Kornberg, *DNA Replication*, Freeman, San Francisco, 1980.
11. A. Kornberg, *The Enzymes* (3rd ed) *15, part A*, 3–13 (1982).
12. R. I. Lehman, *The Enzymes* (3rd ed) *15, part A*, 15–37 (1982).
13. C. McHenry and A. Kornberg, *The Enzymes 15, part A*, 39–50 (1982).
14. R. I. Lehman, *The Enzymes* (3rd ed) *15, part A*, 51–65 (1982).
15. A. Weissbach, *The Enzymes* (3rd ed) *15, part A*, 67–86 (1982).
16. Z. Burton, R. Burgess, J. Lin, D. Moore, S. Holder, and C. Gross, *Nucleic Acids Res.*, **9,** 2889 (1981).
17. R. R. Burgess, A. A. Travers, J. Dunn, and E. K. F. Bautz, *Nature (London)*, **221,** 43 (1969).
18. W. G. Haldenwag and R. Losick, *Proc. Natl. Acad. Sci. USA*, **77,** 7000 (1980).
19. J. L. Wiggs, M. Z. Gilman, and M. J. Chamberlin, *Proc. Natl. Acad. Sci. USA*, **78,** 2762 (1981).
20. W. G. Haldenwag, N. Lang, and R. Losick, *Cell*, **23,** 615 (1981).
21. J. L. Oakley and J. E. Coleman, *Proc. Natl. Acad. Sci. USA*, **74,** 4266 (1977).
22. H. L. Osterman and J. E. Coleman, *Biochemistry*, **20,** 4884 (1981).
23. M. Rosa, *Cell*, **16,** 815 (1979).
24. J. J. Dunn and F. W. Studier, *J. Mol. Biol.*, **148,** 303 (1981).
25. R. E. Strothkamp, J. L. Oakley, and J. E. Coleman, *Biochemistry*, **19,** 1074 (1980).
26. D. Yee, V. W. Armstrong, and F. Eckstein, *Biochemistry*, **18,** 4116 (1979).
27. F. Eckstein, V. W. Armstrong, and H. Sternbach, *Proc. Natl. Acad. Sci. USA*, **73,** 2987 (1976).
28. P. M. J. Burgers and F. Eckstein, *Proc. Natl. Acad. Sci. USA*, **75,** 4798 (1978).
29. V. W. Armstrong, D. Yee, and F. Eckstein, *Biochemistry*, **18,** 4129 (1979).
30. J. R. Knowles, *Ann. Rev. Biochem.*, **49,** 877.
31. K. Bertrand, L. Korn, F. Lee, T. Platt, C. L. Squires, C. Squires, and C. Yanofsky, *Science*, **189,** 22 (1975).
32. K. Bertrand, L. J. Korn, F. Lile, and C. Yanofsky, *J. Mol. Biol.*, **117,** 227 (1977).
33. I. P. Crawford and G. V. Stauffer, *Ann. Rev. Biochem.*, **49,** 163 (1980).
34. A. M. Wu and T. Platt, *Proc. Natl. Acad. Sci. USA*, **75,** 5442 (1978).
35. D. Pribnow, *J. Mol. Biol.*, **99,** 419 (1975).
36. M. L. Goldberg, Ph. D. Thesis, Stanford University, Palo Alto, Calif., 1979.
37. R. G. Roeder and W. S. M. Wold, *Proc. Natl. Acad. Sci. USA*, **79,** 41 (1982).
38. R. D. Simpson, *Cell*, **18,** 277 (1979).
39. Y. Ovchinnikov, G. Monastryskaia, V. Gubanov, S. Guriev, O. Chertov, N. Modyanov, V. Grinkevich, I. Markova, T. Marchenko, I. Polvonikova, V. Lipkin, and E. Sverdlov, *Dokl. Acad. Nauk. USSR*, **253,** 994 (1980).

40. Y. Ovchinnikov, G. Monstryskaia, V. Gubanov, S. Guriev, O. Chertov, N. Modyanov, V. Grinkevich, I. Markova, T. Marchenko, I. Polovnikova, V. Lipkin, and E. Sverdlov, *European J. Biochem.*, **116,** 621 (1981).

41. Z. Burton, R. Burgess, J. Lin, D. Moore, S. Holder, and C. Gross, *Nucleic Acids Res.*, **9,** 2889 (1981).

42. L. Post and M. Nomura, *J. Biol. Chem.*, **254,** 10604 (1979).

43. Y. Ovchinnikov, V. A. Lipkin, N. Modyanov, O. Chertov, and Y. Smirnov, *FEBS Lett.*, **76,** 1081 (1977).

44. S. R. Jaskunas, R. R. Burgess, and M. Nomura, *Proc. Natl. Acad. Sci. USA*, **72,** 5036 (1975).

45. L. E. Post, G. D. Strycharz, M. Nomura, H. Lewis, and P. P. Dennis, *Proc. Natl. Acad. Sci. USA*, **76,** 1697 (1979).

46. M. Yamamoto and M. Nomura, *Proc. Natl. Acad. Sci. USA*, **75,** 3891 (1978).

47. T. Linn and J. Scaife, *Nature*, **276,** 33 (1978).

48. L. R. Yarbrough and J. Hurwitz, *J. Biol. Chem.*, **249,** 5400 (1974).

49. C.-W. Wu, F. Y.-H Wu, and D. C. Speckhard, *Biochemistry*, **16,** 5449 (1977).

50. S. M. Halling, F. J. Sanchez-Anzaldo, R. Fukuda, R. H. Doi, and C. F. Meares, *Biochemistry*, **16,** 2880 (1977).

51. D. C. Speckhard, F. Y.-H. Wu, and C.-W Wu, *Biochemistry*, **16,** 5228 (1977).

52. R. A. Anderson and B. L. Vallee, *Proc. Natl. Acad. Sci. USA*, **72,** 394 (1975).

53. H. Shinar and G. Navon, *Biochim. Biophys. Acta*, **334,** 471 (1974).

54. E. P. Kang, C. B. Storm, and F. W. Carson, *Biochem. Biophys. Res. Commun.*, **49,** 621 (1972).

55. S. M. Halling, K. C. Burtis, and R. H. Doi, *J. Biol. Chem.*, **252,** 9024 (1977).

56. M. Chamberlin and J. Ring, *J. Biol. Chem.*, **248,** 2235 (1973).

57. M. Chamberlin and J. Ring, *J. Biol. Chem.*, **243,** 2245 (1973).

58. J. J. Dunn, W. T. McAllister, and E. K. F. Bautz, *Eur. J. Biochem.*, **29,** 500 (1972).

59. H. A. Küpper, W. T. McAllister, and E. K. F. Bautz, *Eur. J. Biochem.*, **38,** 581 (1973).

60. S. Clark, R. Losick, and J. Pero, *Nature (London)*, **252,** 21 (1974).

61. S. Clark, *J. Virol.*, **25,** 224 (1978).

62. S. Falco, K. Vanderlaan, and L. Rothman-Denes, *Proc. Natl. Acad. Sci. USA*, **74,** 520 (1977).

63. S. Falco, W. Zehring, and L. Rothman-Denes, *J. Biol. Chem.*, **255,** 4339 (1980).

64. J. E. Coleman, *Biochem. Biophys. Res. Commun.*, **60,** 641 (1974).

65. J. J. Dunn and F. W. Studier, *Nucleic Acids Res.*, **8,** 2119 (1980).

66. S. J. Stahl and K. Zinn, *J. Mol. Biol.*, **148,** 481 (1981).

67. E. G. Niles, S. W. Conlon, and W. C. Summers, *Biochemistry*, **13,** 3904 (1974).

68. J. L. Oakley, J. A. Pascale, and J. E. Coleman, *Biochemistry*, **14,** 4684 (1975).

69. J. E. Coleman, unpublished.

70. D. S. Auld and I. Atsuya, *Biochem. Biophys. Res. Commun.*, **69,** 548 (1976).

71. H. Lattke and U. Weser, *FEBS Lett.*, **65,** 288 (1976).

72. T. M. Wandzilak and R. W. Benson, *Biochem. Biophys. Res. Commun.*, **76,** 247 (1977).

73. P. Petranyl, J. J. Jendrisak, and R. R. Burgess, *Biochem. Biophys. Res. Commun.*, **74,** 1031 (1977).

74. D. S. Auld, H. Kawaguchi, D. Livingston, and B. L. Vallee, *Proc. Natl. Acad. Sci. USA*, **71,** 2091 (1974).

75. K. H. Falchuk, L. Ulpino, B. Mazus, and B. L. Vallee, *Biochem. Biophys. Res. Commun.*, **74,** 1206 (1977).

76. K. H. Falchuk, B. Mazus, L. Ulpino, and B. L. Vallee, *Biochemistry*, **15,** 4468 (1976).

77. B. L. Vallee and F. L. Hoch, *J. Biol. Chem.*, **225,** 185 (1957).

78. A. Sytkowski and B. L. Vallee, *Proc. Natl. Acad. Sci. USA*, **73,** 344 (1976).

79. J. Jendrisak and R. R. Burgess, *Biochemistry*, **14,** 4639 (1975).

80. B. L. Vallee, *Physiol. Rev.*, **39,** 443 (1959).

81. B. L. Vallee in *Cancer Enzymology*, J. Schultz and F. Ahmad, Eds., Academic, New York, 1976, p. 159.

82. M. I. Harris and J. E. Coleman, *J. Biol. Chem.*, **243,** 5063 (1968).

83. "Zinc", Report of the Committee on Medical and Biologic Effects of Environmental Pollutants, Division of Medical Sciences, Assembly of Life Sciences, National Research Council, University Park Press, Baltimore, 1979.

84. J. Raulin, *Ann. Sci. Natl. Botan. Biol. Vegetale*, **11,** 93 (1869).

85. K. H. Falchuk, D. W. Fawcett, and B. L. Vallee, *J. Cell Sci.*, **17,** 57 (1975).

86. K. H. Falchuk, D. W. Fawcett, and B. L. Vallee, *J. Submicr. Cytol.*, **7,** 139 (1975).

87. K. H. Falchuk, A. Krishan, and B. L. Vallee, *Biochemistry*, **14,** 3439 (1975).

88. B. L. Vallee and W. E. C. Wacker, *The Proteins*, 2nd ed., H. Neurath, Ed., Academic, New York, 1970.

89. J. F. Chlebowski and J. E. Coleman in Metal Ions in Biological Systems, Vol. 6, H. Sigel, Ed., Dekker, New York, 1976, p. 1.

90. K. H. Falchuk, C. H. Hardy, L. Ulpino, and B. L. Vallee, *Proc. Natl. Acad. Sci. USA*, **75,** 4175 (1978).

91. J. Ross, E. M. Scolnick, G. J. Todaro, and S. A Aaronson, *Nature New Biol.*, **231,** 163 (1971).

92. D. L. Kacian, K. F. Watson, A. Burney, and S. Spiegelman, *Biophys. Biochem. Acta*, **246,** 365 (1971).

93. J. Hurwitz and J. P. Lies, *J. Virol.*, **9,** 116 (1972).

94. D. P. Grandgenett, G. F. Gerard, and M. Green, *Proc. Natl. Acad. Sci. USA*, **70,** 230 (1973).

95. N. Battula and L. A. Loeb, *J. Biol. Chem.*, **249,** 4086 (1974).

96. D. L. Kacian and S. Spiegelman, *Methods in Enzymology*, **32,** 150 (1974).

97. B. J. Poiesz, G. Seal, and L. A. Loeb, *Proc. Natl. Acad. Sci. USA*, **71,** 4892 (1974).

98. B. J. Poiesz, N. Battula, and L. A. Loeb, *Biochem. Biophys. Res. Commun.*, **56,** 959 (1974).

99. D. S. Auld, H. Kawaguchi, D. M. Livingston, and B. L. Vallee, *Biochem. Biophys. Res. Commun.*, **62,** 296 (1975).

100. T. Kornberg and A. Kornberg, *The Enzymes*, Vol. 10, 3rd ed. 119, (1974).

101. D. Korn, P. A. Fisher, J. Battey, and T. S.-F. Wang, *Cold Spring Harbor Sym. Quant. Biol.*, **43,** 613 (1978).

102. V. D'Aurora, A. M. Stern, and D. S. Sigman, *Biochem. Biophys. Res. Commun.*, **78,** 170 (1977).

103. V. D'Aurora, A. M. Stern, and D. S. Sigman, *Biochem. Biophys. Res. Commun.*, **80,** 1025 (1978).

104. D. S. Sigman, D. R. Graham, V. D'Aurora, and A. M. Stern, *J. Biol. Chem.*, **254,** 12269 (1979).

105. J. L. Oakley, R. E. Strothkamp, A. H. Sarris, and J. E. Coleman, *Biochemistry*, **18,** 528 (1979).

106. T. Platt, *Cell*, **24,** 10 (1981).

107. L. M. S. Chang and J. E. Bollum, *Proc. Natl. Acad. Sci. USA*, **65,** 1041 (1970).

108. T. E. Bull, J. Andrasko, E. Chiancone, and S. Forsen, *J. Mol. Biol.*, **73,** 251 (1973).

109. R. L. Ward, *Biochemistry*, **8,** 1879 (1969).

110. A. S. Mildvan and L. A. Loeb, *CRC Critical Reviews in Biochemistry*, **6,** 219 (1979).

111. J. P. Slater, I. Tamir, L. A. Loeb, and A. S. Mildvan, *J. Biol. Chem.*, **247,** 6784 (1972).

112. D. L. Sloan, L. A. Loeb, A. S. Mildvan, and R. J. Feldmann, *J. Biol. Chem.*, **250,** 8913 (1975).

113. R. Koren and A. S. Mildvan, *Biochemistry*, **16,** 241 (1977).

114. B. L. Bean, R. Koren, and A. S. Mildvan, *Biochemistry*, **16,** 3322 (1977).

115. P. J. Stein and A. S. Mildvan, *Biochemistry*, **17,** 2675 (1978).

116. P. Valenzuela, R. W. Mooris, A. Faras, W. Levinson, and W. J. Rutter, *Biochem. Biophys. Res. Commun.*, **53,** 1036 (1973).

117. E. T. Butler and M. J. Chamberlin, *J. Biol. Chem.*, **257,** 5772 (1982).

118. G. A. Kassavetis, E. T. Butler, D. Roulland, and M. J. Chamberlin, *J. Biol. Chem.*, **257,** 5779 (1982).

CHAPTER 7

Roles for the Metal Ion in Reactions of Coordinated Substrates and in Some Metalloenzymes

NICHOLAS E. DIXON and ALAN M. SARGESON

Research School of Chemistry
The Australian National University
Canberra, Australia

CONTENTS

1 INTRODUCTION, 255

1.1 Metal Ions as Lewis Acids, 256

2 INTER- AND INTRAMOLECULAR REACTIONS OF COORDINATED NITRILES, 261

3 COORDINATED LIGANDS AS NUCLEOPHILES IN INTERMOLECULAR REACTIONS—CARBONIC ANHYDRASE, 266

4 INTERCONVERSIONS OF COMPLEXES OF CARBAMATE, CYANATE, AND UREA–UREASE, 274

5 HYDROLYSIS OF AMINO ACID ESTERS AND AMIDES—CARBOXYPEPTIDASE, PROTEASES, AND ESTERASES, 288

5.1 Intermolecular Hydrolysis of Chelated Derivatives, 291
5.2 Intramolecular Hydrolysis of Coordinated Nucleophiles, 296
5.3 Intermolecular Hydrolysis of *N*-Monodentate Derivatives, 301
5.4 Coordination of the Leaving Group, 302

6 METAL ION-PROMOTED REACTIONS INVOLVING OLEFINS—ACONITASE, AMMONIA LYASES, 304

6.1 Hydration, 306
6.2 Aconitase Mechanism, 311
6.3 Amination and Deamination, 313

7 INTER- AND INTRAMOLECULAR REACTIONS OF COORDINATED IMINES—SERINE SULFOHYDROLASE, 318

7.1 Transamination and Deamination of Amino Acids, 318

8 PHOSPHATE ESTER CHEMISTRY—ALKALINE AND ACID PHOSPHATASES, PHOSPHORYL TRANSFERASES, 323

ACKNOWLEDGMENTS, 345

REFERENCES, 346

1 INTRODUCTION

Enzymes have evolved to catalyze biochemical reactions with efficiencies that are seldom approached in the analogous chemical systems under such mild conditions. It is the origin of these enormous catalytic efficiencies which is a central puzzle in enzymology and, in a more specific sense, is the fundamental problem that we address in this chapter. A large proportion of the known enzymes rely on nonprotein cofactors to assist them in carrying out their highly efficient chemical transformations, and in many of these the cofactor is or contains a transition-metal ion.

Metal ions in metalloenzymes are generally involved in either one or the other of two types of reactions. They may exert their ability to exist in a number of distinct oxidation states. Those in this category catalyze reactions involving the oxidation or reduction of their substrates and often have active-site Fe(II/III), Cu(I/II), Mn(II)-(IV), or Co(I)-(III) ions to assist them in that task. Alternatively, the active-site metal ion(s) may activate a coordinated substrate involved in a solvolytic or substitution reaction. In metalloenzymes of this second class, the nucleophilic substitution reactions may occur within the coordination sphere of the metal ion, which itself remains in the same formal oxidation state throughout the process. Since most of the known zinc metalloenzymes fall into this latter category, we shall confine our discussions to mechanisms by which metal ions in metalloenzymes activate substrates toward these types of reactions. Since the details of mechanisms of action of specific metalloenzymes are reviewed in greater detail in other chapters in this volume and elsewhere, we shall restrict ourselves to an examination of ways in which metal ions in simple coordination complexes promote nucleophilic substitution reactions, applying the principles to specific metalloenzyme systems as the opportunity presents itself.

Our approach to understanding enzyme chemistry in terms of the chemistry of model complexes rests on a number of presumptions that we shall defend later in the discussions. The first is that the mechanisms used by enzymes are always chemically reasonable. That is, the chemistry used by an enzyme conforms to chemical principles that are either established or reasonably predicted. The importance of the model chemistry in establishing chemical principles is evident. While the detailed mechanisms of enzymes generally appear to be considerably more complex than those in even the most sophisticated model systems, the chemist is able to model individual steps in the enzyme mechanism with some success, and is thus able to comment on their relative contributions to the overall efficiency of the reaction.

The second assumption, which appears to be true for zinc and other first-row transition-metal ions that promote substitution reactions in metalloenzymes, is that the effect of coordination of a metal ion to a substrate parallels the effect of protonation of the substrate at the donor atom. This generaliza-

tion that the major role of the metal ion is to act as a Lewis acid was recognized early by Westheimer (1) who coined the term "superacid catalysis" to describe the ability of metal ions to catalyze acid-promoted reactions at pH values where substantial concentrations of the protonated substrates cannot exist.

The third assumption follows. If the predominant role of the metal ion in any of these systems is to act merely as a Lewis acid, then the particular metal ion used ought to have little effect on the mechanisms of reactions of the coordinated ligands, although the extent of activation observed will of course be dependent on the Lewis acidity of the metal ion as well as the electronic structure of the coordinated ligand. In metalloenzyme systems in which several different metal ions are catalytically efficient, this appears to be generally true (2). Although, for reasons detailed later, the majority of model systems described in this chapter involve activation of substrates by coordination to substitution-inert metal ions such as Co(III), the chemistry observed is directly relevant to discussion of the mechanisms of reactions of complexes containing labile metal ions and therefore to the action of metalloenzymes which necessarily contain only metal centers that are labile to substitution [e.g., Zn(II), Ni(II), Fe(II), etc.].

A second factor that contributes enormously to the apparent reactivity of ligands in metal complexes is the ability of the metal ion to organize the chemistry about itself. Since a metal ion contains multiple coordination sites (usually four to six), it may simultaneously coordinate both substrates of a normally intermolecular nucleophilic reaction. The resulting intramolecular reaction within the coordination sphere of the metal ion often occurs many orders of magnitude faster than its intermolecular counterpart. The use of this template effect for the synthesis of macrocyclic (3) and macropolycyclic (4) ligands is well documented, and relevant examples of simpler reactions are given later. In spite of the facility of these reactions in the model systems, there appears to be no extant example of a metalloenzyme in which both substrates are coordinated to a single metal ion, perhaps because most of the coordination sites on the active-site metal ion are engaged in binding to the protein ligands. Nevertheless, an enzyme has other means to bind one of the substrates via interactions with amino acid side chains, or can deliver intramolecular nucleophiles or general acid or base catalysts in the proximity of the coordinated substrate. By using model complexes containing both substrates, we are thus able to mimic the proximity effects apparent in intramolecular reactions at the active sites of the enzymes.

1.1 Metal Ions as Lewis Acids

Before considering in detail the consequences of coordination of substrates to metal centers, it is instructive to compare some of the properties of metal ions with those of the proton.

1. In spite of the fact that a proton is limited to a single positive charge, it has an extremely high charge density, and therefore enormous polarizing power. On the other hand, metal ions carry variable charge that depends not only on their particular oxidation state but also on the charges on the ligands. They are always larger than a proton, and generally bear a much lesser charge density and correspondingly less capacity to polarize bonds in the coordinated substrate. The charge density on Pt(IV) may approach that of H^+, and this is reflected in the properties of complexes of this metal ion (e.g., pK_a values and reactivity of coordinated amines and H_2O) (5).

2. The coordination number of the proton is usually one; but it can be greater if hydrogen bonding is considered. The larger coordination numbers of the transition-metal ions allow them more flexibility in grouping reagents for intramolecular reactions.

3. The concentration of protons (and of many protonated substrates) is obviously limited at neutral pH. On the other hand, analogous metal complexes are often stable over a wide pH range often including the physiological pH, and may exist in the presence of appreciable concentrations of OH^- or other nucleophiles.

The properties of many transition-metal complexes are nevertheless directly related to the chemistry of the free ligands protonated at the same donor atom, although the effects are mitigated somewhat by the lesser charge density, reflected in the lesser Lewis acid strength of the metal ion in comparison with the proton. This generalization allows ready rationalization of the properties of complexes of Co(III), and of Zn(II) at the active sites of metalloenzymes, and will be used frequently to predict and explain the chemistry of model complexes in the discussion that follows. The properties on which the Lewis acid character of the metal ion are most obviously dependent are the reactivity and pK_a values of the coordinated ligands.

In cases where the substrate is activated toward reaction by protonation, coordination to a metal ion may also produce activation, although generally to a much lesser extent. If on protonation the substrate becomes less reactive, the effect of the metal ion in the complex will also be to protect it against reaction. This concept that the major role of the metal ion is merely to act as a Lewis acid is certainly true for those complexes in which the metal–ligand bond has substantial σ character. The activating or protective effect of coordination derives from withdrawal of electron density from the ligand by the metal ion. In complexes in which there is also substantial π character in the metal–ligand bond, back donation of electron density from filled d orbitals on the metal ion into the π^*-antibonding ligand orbitals may result in little or no net electron withdrawal, or in some instances a net gain (5a), for example, Os^{II}-pyrazine. The predicted effect is that the effective Lewis acid character

of the metal center will appear to be diminished or nonexistent, and the effect of coordination on the reactivity of the ligand will be minimal (5b). With the exception of those complexes that fall into the latter class, and none of these includes zinc complexes in metalloenzymes or apparently the Co(III) model complexes, the most significant role of the metal ion is to act as a substitute for the proton. The Lewis acid strength of a particular metal ion will, in simple terms, be a function of the charge density on the metal center in the complex. The expectation is, therefore, that a complex of a metal ion like Zn(II) or Co(III) ought to display properties intermediate between those of the free ligand and of its conjugate acid. A simple example is the effect of coordination to Co(III) on the rate constants for bromination of aniline (Table 1) (6). In this case, coordination protects aniline against substitution by an electrophile, although to a lesser extent than does protonation. Similarly, coordination of a carbonyl group through a lone pair of electrons on its oxygen atom, or of a nitrile through its nitrogen atom, results in withdrawal of electrons from the bond and an increase in the susceptibility of the carbon atom to attack by nucleophiles. The metal complex will be inherently less reactive than the protonated substrate, but can exist at high pH values where the concentration of protonated substrate would be negligible. Under these circumstances the reaction with OH^- is readily observed. Examples of these types of reactions are considered in detail later in this chapter.

The second property of complexes that also reflects the Lewis acidity of the metal center is the effect of coordination on the pK_a' values of amphoteric ligands. Consider water as the simplest example. Coordination of OH_2 will result in electron withdrawal from the ligand by the metal ion, and two extreme resonance structures for the complex may be envisaged (I↔II) (7). In I, no formal bond exists between water and the metal ion and the complex will have a pK_a' approaching that of the free ligand (i.e., 15.74). In II, where

Table 1 Second-Order Rate Constants for the Bromination of Aniline[a]

Substrate	k (M^{-1} sec^{-1})
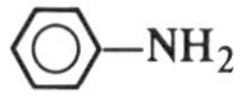	3×10^{10}
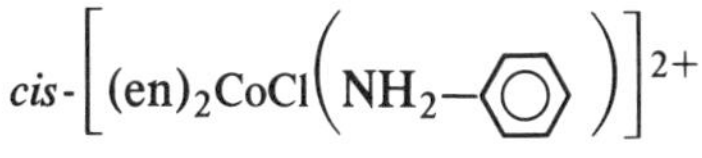	1.4×10^{-1}
	3×10^{7}

[a]Reference 6.

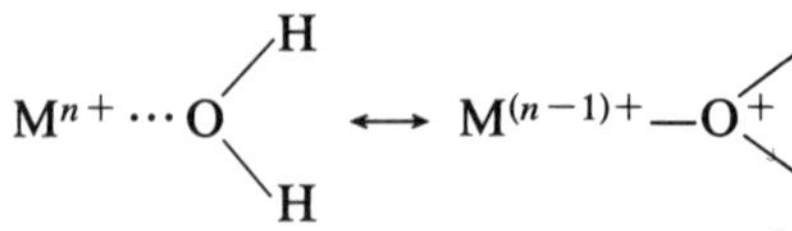

I II

charge separation is complete, the complex will resemble the hydronium ion and will have a pK_a' near -1.7. The extent to which the actual degree of charge separation resembles either of the extremes will determine: (a) the pK_a' of the complex which for Co(III) and Zn(II) will always be intermediate between -1.7 and 15.7; (b) the strength (and lability) of the metal–ligand bond, and

Table 2 Selected pK_a' Values of Co(III) Complexes

Acid	pK_a'	Reference[a]
H_3O^+	-1.74	
H_2O	15.74	
cis-$Co(en)_2(NH_3)(OH_2)^{3+}$	6.1	8
cis-$Co(en)_2(OH_2)_2^{3+}$	6.1	8
cis-$Co(en)_2(OH)(OH_2)^{2+}$	8.2	8
cis-$Co(en)_2Cl(OH_2)^{2+}$	7.1	9
cis-$Co(NH_3)_4Cl(OH_2)^{2+}$	6.6	9
trans-$Co(en)_2(NH_3)(OH_2)^{3+}$	5.8	8
trans-$Co(en)_2(OH_2)_2^{3+}$	4.5	8
trans-$Co(en)_2(OH)(OH_2)^{2+}$	7.9	8
trans-$Co(en)_2Cl(OH_2)^{2+}$	6.1	9
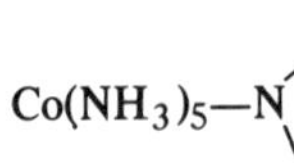	7.0	
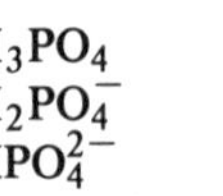	14.5	
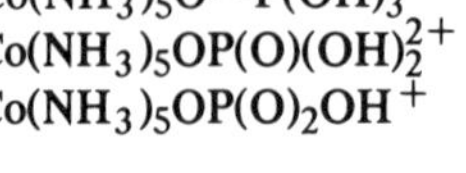	10.0	10
H_3PO_4	2.1	
$H_2PO_4^-$	7.2	
HPO_4^{2-}	12.3	
$Co(NH_3)_5O{=}P(OH)_3^{3+}$	-0.2	11
$Co(NH_3)_5OP(O)(OH)_2^{2+}$	3.6	11
$Co(NH_3)_5OP(O)_2OH^+$	8.5	11

Table 2 *(Continued)*

Acid	pK_a'	Reference[a]
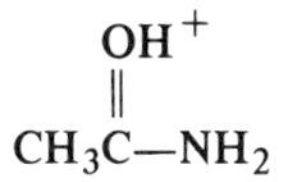	−0.5	
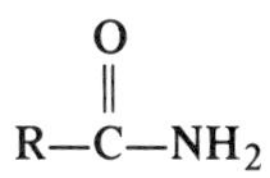	>15	
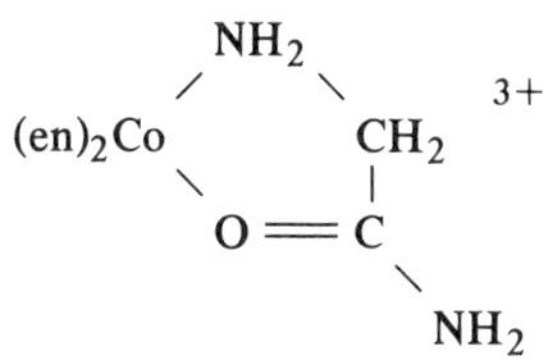	11.2	12
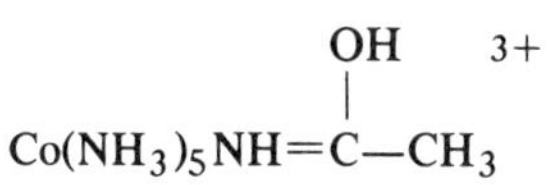	3.0	13
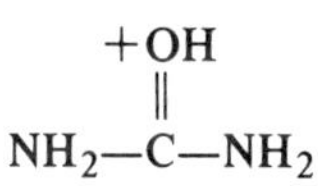	0.2	
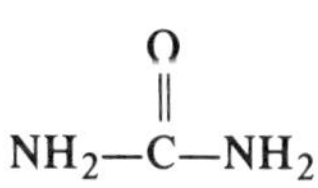	~14	
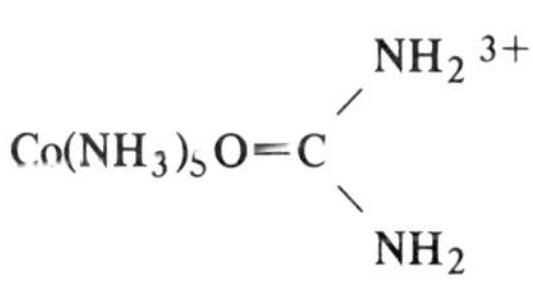	13.2	14
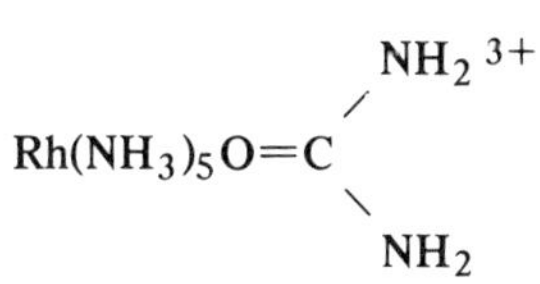	13.7	15
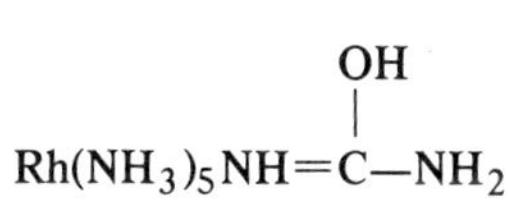	3.9	15

[a] Values for the free ligands are taken from ref. 16.

(c) the extent to which bonds in the ligand are polarized on coordination (often reflected in the vibrational spectra of the ligand or in the electronic spectra of the complex). Examination of selected pK_a' values of complexes of water and other ligands with Co(III) (Table 2) illustrates their dependence both on the effective charge at the metal center [e.g., $Co(en)_2(NH_3)(OH_2)^{3+}$ vs. $Co(en)_2Cl(OH_2)^{2+}$] and on the electronic structure of the ligand [e.g., $Co(NH_3)_5OC(NH_2)_2^{3+}$ vs. $(en)_2\overset{|}{Co}\bar{N}H_2CH_2CONH_2^{3+}$, *vide infra*]. The higher values of the pK_a' of water coordinated in octahedral complexes of divalent metals ions such as Zn^{2+}, Co^{2+}, Fe^{2+}, and Ni^{2+} (ca. 8–10) (16) reflects the lesser charge densities on the metal centers in comparison with Co(III).

We now turn to an examination of the effects of coordination on the reactivity of ligands including nitriles, hydroxide and amide ions, carboxamides, carboxylic and phosphate esters, imines, olefins, and urea derivatives. The observations are related to the general principles outlined above and to the mechanisms of some specific metalloenzymes.

2 INTER- AND INTRAMOLECULAR REACTIONS OF COORDINATED NITRILES

The rate of hydration of 2-cyano-1,10-phenanthroline by intermolecular attack of OH^- to form the corresponding carboxamide shows a first-order dependence on $[OH^-]$. The second-order rate constant k_{OH} is about 10^7 times larger in the presence of a saturating concentration of Ni^{2+} than in its absence (17). This activation was interpreted in terms of Mechanism A (Scheme 1), in which the metal center acts as an intermolecular "general acid" catalyst to stabilize the developing carboxamide anion in the transition state. The temperature dependence of the reactions showed that activation was essentially completely accounted for by a large increase in $\Delta S^\dagger$, and this was interpreted as being due to this interaction (charge neutralization), perhaps coupled with the displacement of a water molecule from the coordination sphere of the metal ion. The alternative Mechanism B (Scheme 1) involving intramolecular attack of coordinated OH^- to give the O-coordinated amide was argued to be unlikely (17) on the basis of the effects of nucleophiles other than OH^-, but conclusive evidence was not obtained since the product complexes were not sufficiently inert to enable isolation. Substitutions of various nucleophiles at nitrile carbon atoms have since been observed to be promoted by Lewis acids such as Zn^{2+}, Ni^{2+}, and other labile metal ions, and these reactions have proven useful in organic synthesis (18). However, these studies give little further indication of the mechanism responsible for activation since it is not possible to assign unequivocally the composition of the active complexes, and kinetic studies are often complicated by ligand dissociation equilibria.

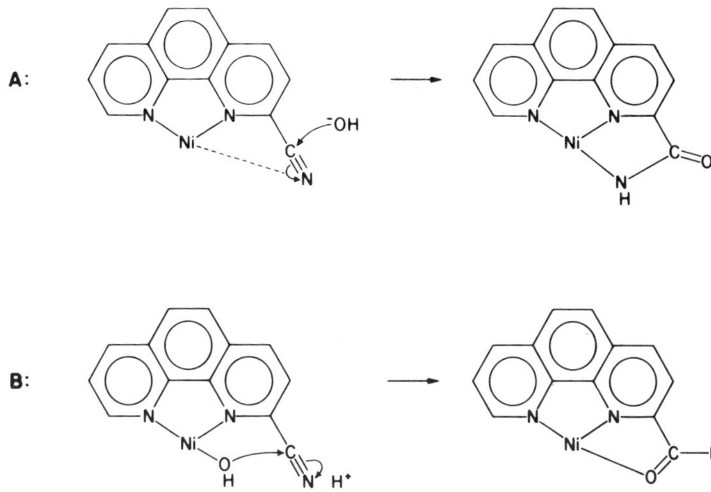

Scheme 1

Securely characterized complexes of N-bound nitriles with inert metal ions have been examined, and do comment further on the mechanisms (13, 19, 20).

$$\underset{\substack{\uparrow \\ ^{-}OH}}{(NH_3)_5Co\,N{\equiv}C{-}R^{3+}} \rightarrow \underset{\text{“imidol”}}{(NH_3)_5CoN{=}\underset{\substack{| \\ OH}}{C}{-}R^{2+}} \rightarrow (NH_3)_5CoNH{-}\underset{\substack{\| \\ O}}{C}{-}R^{2+}$$

Reaction of the Co(III)–acetonitrile (R $=$ CH_3) complex with hydroxide ion gives the deprotonated N-bound acetamido complex quantitatively, with $k_{OH} = 3.4M^{-1}\ sec^{-1}$ ($\mu = 1.0M$, 25°C) (13). The benzonitrile (R $=$ C_6H_5) complex gives coordinated benzamide with $k_{OH} = 18.8M^{-1}\ sec^{-1}$ (19) (Table 3). In each case, the acceleration in rate over that for the OH^--dependent hydration of the uncoordinated nitrile is $\sim(2 \times 10^6)$-fold. In terms of our concepts of the essential function of the metal ion in such reactions, the substantial activiation of the coordinated nitrile derives from polarization of the —C≡N bond by electron withdrawal to the metal center, resulting in the formation of an incipient carbonium ion. The expectation is that the free energy of the ground state of the substrate would be raised relative to that of the uncomplexed substrate, and that activation would be reflected in a lower $\Delta H^{\dagger}$ for the reaction. However, as has been frequently observed for reactions of these types at metal centers, the activation derives almost completely from a large increase in the entropy term (Table 3) (19, 20, 23). The reasons for these effects are in no instance yet fully understood,

Table 3 Representative Second-Order Rate Constants (k_{OH}, 25°C) and Activation Parameters for the Hydration of Nitriles in Base

Substrate	k_{OH} (M^{-1} sec^{-1})	$\Delta H^{\dagger}$ (kcal mol^{-1})	$\Delta S^{\dagger}$ (cal deg^{-1} mol^{-1})	References
$N{\equiv}C-CH_3$	1.60×10^{-6}	18.5	−23	13, 20
$N{\equiv}C-C_6H_5$	8.2×10^{-6}	19.3	−15	19
$N{\equiv}C-C_6H_4-Cl$	7.2×10^{-6}	17.4	−19	19
$(NH_3)_5CoN{\equiv}C-CH_3^{3+}$	3.40	18.0	+4	13, 20
$(NH_3)_5RhN{\equiv}C-CH_3^{3+}$	1.0			21
$(NH_3)_5RuN{\equiv}C-CH_3^{3+}$	2.2×10^2			21
$(NH_3)_5RuN{\equiv}C-CH_3^{2+}$	$<6 \times 10^{-5}$			21
$(NH_3)_5CoN{\equiv}C-C_6H_5^{3+}$	18.8	16.5	+3	19
$(NH_3)_5RuN{\equiv}C-C_6H_5^{3+}$	2.0×10^3			21
$(NH_3)_5Co-N{\equiv}C-C_6H_4-O^{2+}$	0.2	16	−7	19
$(NH_3)_5Co-N{\equiv}C-C_6H_4(o\text{-}O)^{2+}$	3.6	15	−6	19
$(NH_3)_5Co-N{\equiv}C-CH{=}CH_2^{3+}$	35			20
$(NH_3)_5Co-N{\equiv}C-N(CH_3)_2^{3+}$	3.06			22

although it is unlikely that they result simply from differential effects of solvation in the transition state. The argument that they may reflect charge neutralization by the metal center in the transition state (in the absence of solvent reorganization) is also tenuous since, for these complexes in which the substrate remains N-coordinated throughout, such an effect is a corollary to electron withdrawal by the metal center in the ground state and ought also to be reflected in an energy rather than a probability term. More likely the problem devolves into a consideration of the differences in mechanism between the two reactions, and it may, in general, be more instructive to make comparisons of the metal ion-promoted reactions with the corresponding H^+-catalyzed processes. In addition, both the catalyzed and uncatalyzed reactions necessarily involve two steps: OH^- addition to produce an "imidol" species followed by tautomeric proton transfers. A simple consideration of the effects of the cationic metal center on the pK_a's of the donor NH in the amide complex and of the more remote imidol OH indicates that the imidol ought to be significantly stabilized by coordination to Co(III), and the

second proton transfer (tautomerization) step may in fact become rate determining in the overall reaction. For these reasons, the differences among the activation parameters given in Table 3 may not be amenable to simple interpretation.

The OH^--promoted hydration reaction has been investigated for a range of Co(III)-nitrile complexes (24) and, for the aceto- and benzonitrile complexes, with several other metal ions (Table 3). Other nucleophiles such as BH_4^- (25) and CN^- (26, 27) also add at the nitrile carbon of Co(III) complexes at rates that are $\geq 10^4$ times those of the corresponding reactions of the free ligands. Especially interesting cases are CO_3^{2-} catalysis of the hydration of $(NH_3)_5RuNCCH_3^{3+}$ (21) and of the acrylonitrile complex $(NH_3)_5CoNC{-}CH{=}CH_2^{3+}$ (20). In the latter complex, a direct nucleophilic path results in incorporation of oxygen from CO_3^{2-} into the amide product (20).

As predicted by considerations earlier, coordination also increases the acidity of CH protons α to the nitrile. With $(NH_3)_5CoNC{-}CH_3^{3+}$, methyl proton exchange occurs in base at a rate comparable with that of the hydration reaction (13), while the malonodinitrile complex $(NH_3)_5CoNC{-}CH_2{-}CN^{3+}$ has a pK_a of 5.7 which may be compared with 11.3 for the uncoordinated ligand (20). The carbanion produced on deprotonation is also an efficient nucleophile and reacts readily with CH_3I, Br_2, and methyl pyruvate (20).

The data in Table 3 show that, with the exception of the Ru(II) complex, a variety of metal ions activate the hydration reaction to a remarkably consistent extent (10^6–10^8-fold). The constancy of these numbers and the anomaly of the Ru(II) complex are consistent with the proposition that it is the Lewis acidity of the metal ion which is essential to activation. In the Ru(II) complex, considerable metal–ligand π bonding is expected, resulting in back donation of electron density from the metal center into the $C{\equiv}N$ bond, and this is simply confirmed by measurements of the $C{\equiv}N$ stretching frequency of free nitriles and of their complexes with Ru(II) and the other metal ions (28). The result is little or no net polarization of the nitrile by the metal ion, and the effect on reactivity is correspondingly small.

Dixon et al. (2), on the basis of evidence derived from studies of enzyme systems, have argued that it is the Lewis acid character of the metal ion which is of predominant importance in activation of substrates by coordination at the active sites of metalloenzymes which catalyze nucleophilic substitution reactions. The present observations derived from coordination chemistry are consistent with that view, and show that similar chemical principles may be applied in relating Co(III) chemistry to the chemistry of the Zn(II) metalloprotein systems.

The coordination chemist chooses Co(III) while an enzyme chooses Zn(II) (or other metal ion) to accomplish the same chemistry for related reasons which do not, however, impinge on the essential aspects of mechanism and

sources of activation. Co(III) is used for much of the model chemistry because it forms easily characterized octahedral complexes that are generally inert to ligand substitution reactions. Thus, kinetic behavior may be interpreted without the complications of ligand dissociation equilibria and the bonding mode and structure of the substrate and product complexes are usually readily established. The reactions are generally quantitative rather than catalytic, and this greatly simplifies the analysis of the separate factors responsible for the activating or protective effect of coordination in individual steps of the reactions.

On the other hand, a metalloenzyme would not employ a metal ion which was inert to substitution since it requires rapid turnover of substrate and product complexes at its active site in order to maximize its catalytic efficiency. Rate constants for water exchange at Co(III) centers are typically orders of magnitude smaller than the turnover number of a typical enzyme (10–$10^5\ sec^{-1}$). Water exchange at labile centers such as Zn^{2+} ($\sim 10^7\ sec^{-1}$), Mn^{2+} ($\sim 10^7\ sec^{-1}$), Fe^{2+} ($\sim 10^6\ sec^{-1}$), Mg^{2+} ($\sim 10^5\ sec^{-1}$), and Ni^{2+} ($\sim 10^4\ sec^{-1}$) (29) occurs much more rapidly and at rates in excess of most enzymatic processes. The particular choice of labile metal ion in a metalloenzyme would therefore not necessarily be determined by the mechanism of the reaction it catalyzes. Evolutionary pressures relating to the preferences of particular metal ions for the available donor atoms and donor atom geometry, and the availability of the metal ion in the environment at the critical point in evolutionary time may be significant.

The uncertainty involved in modeling enzymatic reactions using labile metal complexes is illustrated above in the case of hydration of the 2-cyano-1,10-phenanthroline–Ni(II) complex where either mechanism (A or B, Scheme 1) is consistent with the rate law and the observation of the formation of the (uncoordinated) amide as product. Similar examples in ester and amide hydrolysis will be given later. In the present instance, the efficacy of mechanisms related to A has been established above using Co(III) complexes. That mechanism B (intramolecular attack of coordinated OH^-) is also plausible is implied by the facile intramolecular hydration of the uncoordinated nitrile in the *cis*-[hydroxo(aminoacetonitrile)bis(ethylenediamine)Co(III)]$^{2+}$ ion to produce the chelated amide product (23, 30). At values of pH sufficiently far

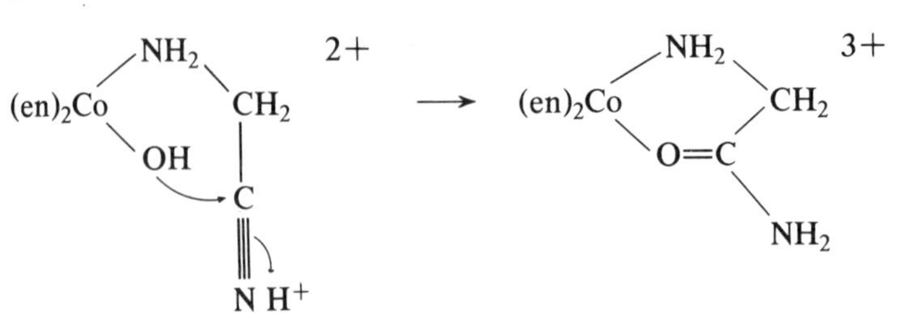

above the pK_a' (5–6) of the *cis*-aqua complex, and observed first-order rate constants are pH independent, with $k = 1.15 \times 10^{-2}\ \text{sec}^{-1}$. At pH 7, the reaction occurs $\sim 10^{11}$ times more rapidly than the intramolecular hydration of free acetonitrile by OH^- ($k_{OH} = 1.6 \times 10^{-6} M^{-1}\ \text{sec}^{-1}$) (23). In part this activation at neutral pH derives from the inherent strength of coordinated OH^- as a nucleophile (see Sec. 3), but mostly from the proximity effect of its coordination in a position which favors the intramolecular process. At this pH, the corresponding intramolecular reaction with *cis*-coordinated aminopropionitrile to produce a six-membered chelated amide product takes place at least 10^5 times more slowly (23). These considerations relate to proximity effects thought to be important in reactions at the active sites of enzymes; the activations produced in the model chemistry are no less dramatic than those seen in many enzymatic systems.

Further enhancement of the rates of the reactions of both the aminoacetonitrile and aminopropionitrile complexes by factors in excess of 10^6 is observed in the presence of $1M$ Hg^{2+} ion, and has been interpreted in terms of a mechanism involving additional coordination of the nitrile to the labile metal ion (23, 30). The additional activation is also observed in the presence of Ag^+, Zn^{2+}, and Cd^{2+}, and apparently derives from a mechanism analogous to Co(III)—promotion of the intermolecular hydration of the $[(NH_3)_5CoN{\equiv}C{-}R]^{3+}$ ions in base. These studies indicate how the concerted action of two metal centers, one to provide the intramolecular nucleophile, the other acting as an electrophile to assist the nucleophile, can be used to produce extraordinarily efficient catalysis of such reactions. These factors may be relevant to the chemistry of those enzymes thought to use more than one metal ion at their active sites [e.g., jack bean urease (7)].

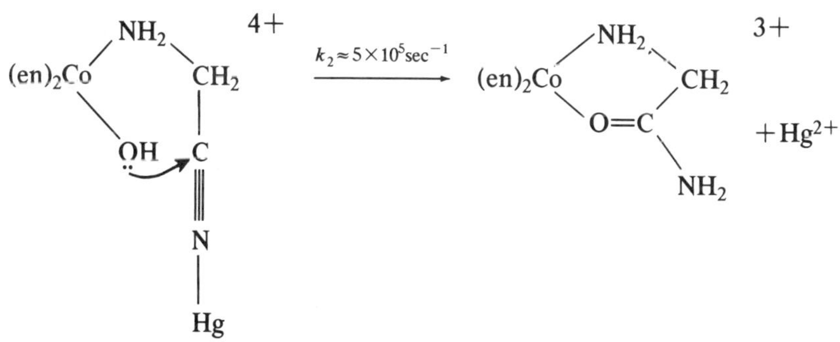

3 COORDINATED LIGANDS AS NUCLEOPHILES IN INTERMOLECULAR REACTIONS—CARBONIC ANHYDRASE

In Section 2, a reaction in which coordinated OH^- behaves as a nucleophile in an intramolecular reaction was discussed, and further examples will be

given later. Hydroxide ion coordinated to Co(III) has also been observed to function as a nucleophile in intermolecular reactions, and it is pertinent to examine this chemistry in order to determine the extent to which the rate enhancements observed in the intramolecular processes derive from the inherent nucleophilic strength of coordinated OH^- as opposed to the proximity effect ascribable to loss of the translational entropy of one of the reagents. Much of this chemistry has been reported and reviewed (31).

Reactions of the $[(NH_3)_5CoOH]^{2+}$ or $[(NH_3)_5CoOH_2]^{3+}$ ions with electron-deficient species including NO^+ (32), NCO^- (33), CO_2 (34), SO_2 (35), carboxylic acid esters and anhydrides (10, 31), acetylacetone (36), $HSeO_3^-$ (37), WO_4^{2-} (38), MoO_4^{2-} (39), $H_2AsO_4^-/HAsO_4^{2-}$ (40), and HIO_3 (41) take place at rates orders of magnitude faster than the rates of normal substitution or water exchange reactions of the Co(III) complex ions. In those cases where ^{18}O-tracer experiments have been carried out, it has been shown that the donor oxygen of the product derives from the coordinated water. It appears therefore that the product complex is formed in each case by reactions involving coordinated OH^- acting as a nucleophile, rather than by substitution at the Co(III) center, for example,

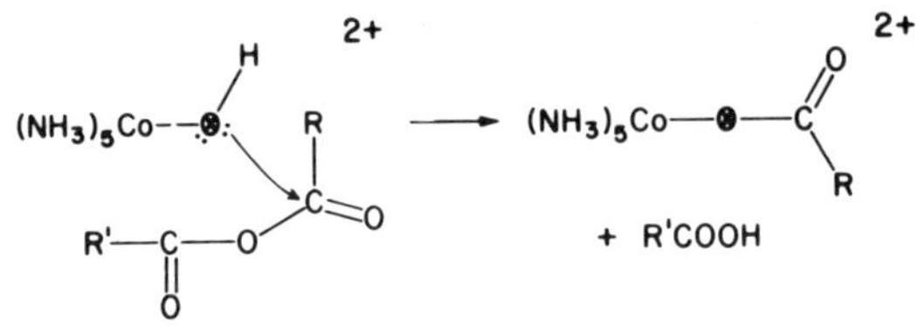

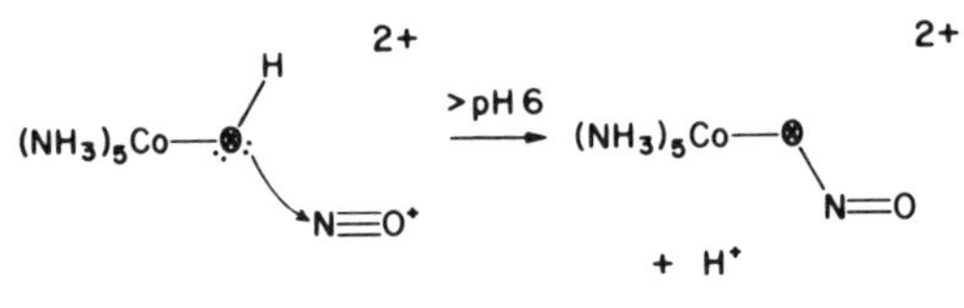

That coordinated OH^- might be a good nucleophile is qualitatively predicted by arguments based on the Lewis acidity of the metal center. The nucleophilicity of a related series of bases is generally related to the pK_a of their conjugate acids (Brønsted relationship). Since metal ions are usually weaker Lewis acids than H^+, their aqua complexes frequently have a pK_a intermediate between those of H_3^+O and H_2O. The $M\text{-}OH^{(n-1)+}$ complex is therefore predicted to be intermediate between H_2O and OH^- in its strength as a nucleophile.

The kinetics of acylation of a series of $M\text{-}OH^{(n-1)+}$ complexes by propionic anhydride and carboxylic esters containing good leaving groups have been examined. The inert hydroxo complexes used were $(NH_3)_5CoOH^{2+}$, $(NH_3)_5CrOH^{2+}$, $(NH_3)_5RuOH^{2+}$, $(NH_3)_5IrOH^{2+}$, $(H_2O)_5CrOH^{2+}$, *trans*-$(NH_3)_4Co(NO_2)(OH)^+$, and $(CN)_5CoOH^{3-}$. The pK_a of the corresponding aqua complexes varied between ~4 [for $(H_2O)_6Cr^{3+}$] and ~10 [for $(CN)_5CoOH_2^{2-}$], while the overall charge on the hydroxo complexes varied between 2+ and 3−. Each of the reactions was followed under pseudo-first-order conditions, and second-order rate constants were obtained over a range of pH including the pK_a of the $M\text{-}OH_2^{n+}$ ion. For each, the dependence of the second-order rate constants on pH described a titration curve giving limiting values for reactions of $M\text{-}OH_2^{n+}$ (k_{MOH_2}) at low pH and of $M\text{-}OH^{(n-1)+}$ at high pH. The ratio of k_{MOH_2} to k_{MOH} was always $<10^{-3}$, indicating the hydroxo complex to be the reactive species (31).

For reactions of $M\text{-}OH^{(n-1)+}$ with propionic anhydride, the Brønsted plot of log k_{MOH} versus the pK_a of $M\text{-}OH_2^{n+}$ (Fig. 1) described a smooth curve including values for reaction with H_2O and OH^-.

If a line is drawn between points for H_2O and points for $M\text{-}OH^{(n-1)+}$ spe-

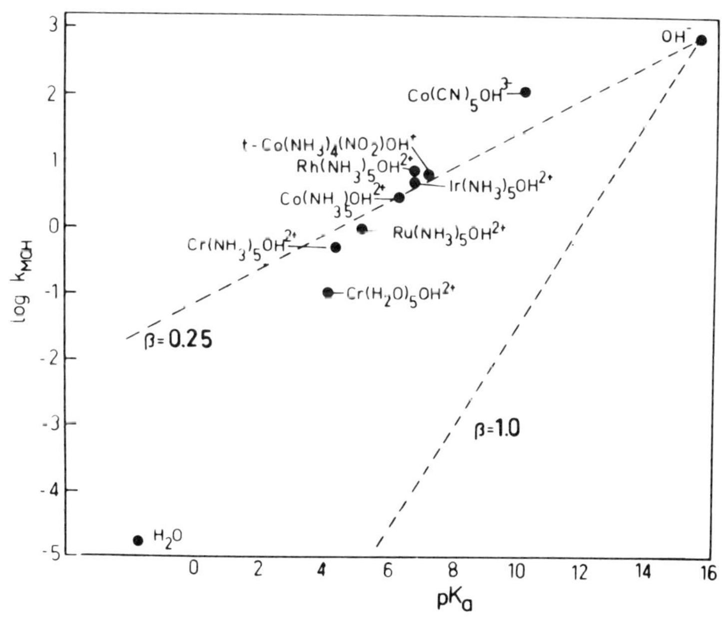

Figure 1 Plot of log k_{MOH} versus pK_a (of MOH_2^{n+}) for the reaction of various metal hydroxide complexes with propionic anhydride at 25°C and $\mu = 1.0$ ($NaClO_4$) (31). (Reproduced with permission from ref. 31.)

cies in Figure 1, a good fit of data to a Brønsted β of ~ 0.6 is obtained. This relationship for complexes of a number of different metal ions whose overall charge varied from -3 to $+2$ suggests that it is the Lewis acidity of the metal center itself in comparison with H^+, mediated somewhat by the charges of other ligands in the complex, which controls both the reactivity of the M-$OH^{(n-1)+}$ species and the pK_a of its conjugate acid. Thus a comparison of rates of nucleophilic reactions of M-$OH^{(n-1)+}$ with corresponding reactions of water (at the same concentration) at pH values where the complex is completely ionized will give rate enhancements which depend on the pK_a of the M-OH_2^{n+} species varying in this case from about 10^4 (p$K_a' \sim 4$) to 10^7 (p$K_a' \sim 10$). Taking account of incomplete ionization of the complexes with higher pK_a's at pH 7, the enhancements at neutral pH are all in the range of 10^4 to 10^5.

On the other hand, if the line is drawn as in Figure 1 to include the point for k_{OH}, a Brønsted β of ~ 0.25 is obtained. Thus, for example, although k_{MOH} for $(NH_3)_5CoOH^{2+}$ ($3M^{-1}$ sec^{-1}) is about 10^3-fold less than k_{OH}, its reaction will compete favorably at neutral pH with the OH^--promoted hydrolysis. At pH 7, where the Co(III) complex exists almost completely as the M-OH^{2+} species, the first-order rate constants for nucleophilic attack by OH^- would be $\sim 10^{-4}$ sec^{-1}. A $1M$ solution of $(NH_3)_5CoOH^{2+}$ would produce $k_{obs} \sim 2.5$ sec^{-1}, an increase in rate greater than 10^4-fold. This factor could be much greater in an intramolecular reaction where the effective concentration of the nucleophile in the vicinity of the other substrate may be greatly in excess of $1M$. The role of the metal ion in such reactions is thus to deliver an effective concentration of an efficient nucleophile at an acceptably low pH.

Brønsted plots similar to those given in Figure 1 have been constructed for other reactions of M-$OH^{(n-1)+}$ species with substrates including 4-nitrophenyl acetate, 2,4-dinitrophenyl acetate, CO_2, and acetaldehyde (31).

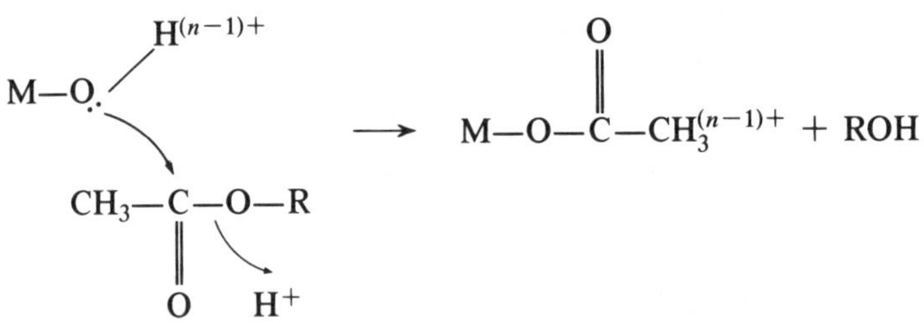

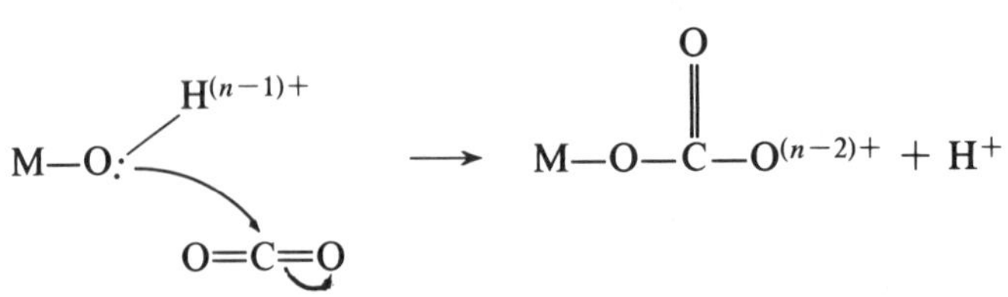

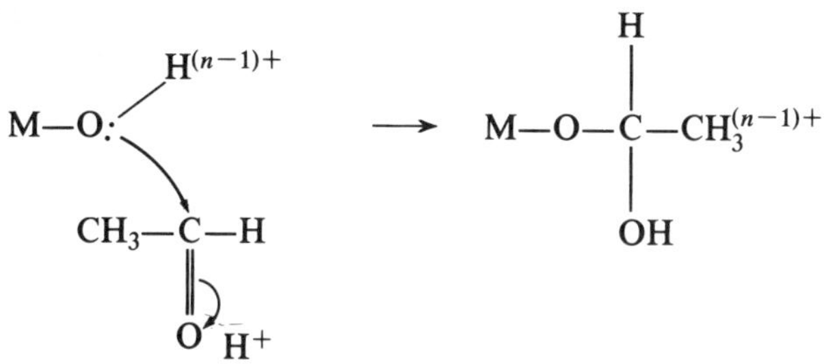

The Brønsted β values (through k_{OH}) of 0.4, 0.33, 0.18, and 0.40, respectively (31), each showed a general insensitivity to the pK_a of the conjugate acid of the nucleophile, and this once again establishes the superiority of M-$OH^{(n-1)+}$ species over OH^- and H_2O as nucleophiles (in a rate sense) at neutral pH. These latter plots also included data for hydroxo complexes of labile metal ions including Zn^{2+} and Cu^{2+} (31). In each case, the plots were similar and points for the Zn^{2+} and Cu^{2+} complexes lay on the same lines obtained with the inert complexes, indicating there to be no essential difference between the mechanisms used by labile and inert metal centers to accomplish the same chemistry.

The carbonic anhydrases catalyze reversibly the hydration of CO_2 and aldehydes, and the hydrolysis of esters containing good leaving groups. They contain active-site zinc ion, and there is evidence in support of zinc-bound hydroxide ion acting as a nucleophile in their mechanisms. Much of the enzymology and its relationship to the model chemistry discussed above and elsewhere in this article has been recently reviewed (31, 42, 43), and the reader is referred to those articles for detailed discussions.

Models that relate to alternative mechanisms for carbonic anhydrase, where zinc-coordinated histidine is the active nucleophile (44) are discussed below, but first let us consider some further implications of the Brønsted relationships (e.g., Fig. 1), which potentially impinge on the chemistry of those metalloenzymes which use M-$OH^{(n-1)+}$ or other coordinated nucleophiles.

The data in Figure 1 indicate that the strongest M-$OH^{(n-1)+}$ nucleophiles are those whose conjugate acids have the highest pK_a's. Although enzymes usually function best near neutral pH where the concentration of the (deprotonated) nucleophilic form of these more reactive complexes is low, their increased nucleophilic strength partially compensates for their decreased concentration. In general, therefore, so long as the pK_a of M-OH_2^{n+} is greater than about 6, the observed rate *at neutral pH* will be insensitive (to within a factor of < 10) to the actual values of pK_a. Further, any change (with pH, for example) which results in a slight decrease in the pK_a and a consequent increase in the concentration of the nucleophile (e.g., an increase in the overall charge in the vicinity of the metal ion consequent to protonation of another

ligand or nearby base) will necessarily be offset by a diminution of its reactivity. This balance between the two effects of the Lewis acid character of the metal (i.e., lowering of ligand pK_a's and increasing reactivity) becomes apparent in many aspects of the chemistry of metalloenzymes and model systems and must be carefully evaluated in considering the efficacy of proposals of mechanism.

Thus, arguments against the zinc–histidine mechanism for carbonic anhydrase based on the presumption that the high pK_a of Zn^{2+}-coordinated imidazole (pK_a > 10, Table 2) (10) would produce a negligible concentration of its conjugate base at neutral pH would not necessarily be significant if its high pK_a led to its conjugate base being a correspondingly more efficient nucleophile. Studies of the reaction of the $(NH_3)_5$Co-imidazolato^{2+} ion with p-nitrophenyl acetate offer some support for these arguments. Although it is a nucleophile of comparable reactivity to $(NH_3)_5CoOH^{2+}$ in water at neutral pH, it would actually be poorer than free imidazole under these conditions (Table 4) (10, 31). In dimethylsulfoxide, the $(NH_3)_5CoOH^{2+}$ ion becomes a very powerful base and nucleophile, while the imidazolato complex is only a little more reactive than in water (10). If these observations have any relevance to the situation at the active site of carbonic anhydrase where the local dielectric constant could be much less than that of the bulk solvent, the model chemistry offers more support for the Zn–OH^+ mechanism than for the Zn–histidine mechanism. A further difficulty with the latter proposal arises when reaction of the enzyme with CO_2 is considered. Nucleophilic attack by Zn–His^+ on CO_2 would produce a carbamate complex which, on the basis of established model chemistry (see Sec. 4), ought to undergo a homolytic decarboxylation rather than hydrolysis to give HCO_3^-.

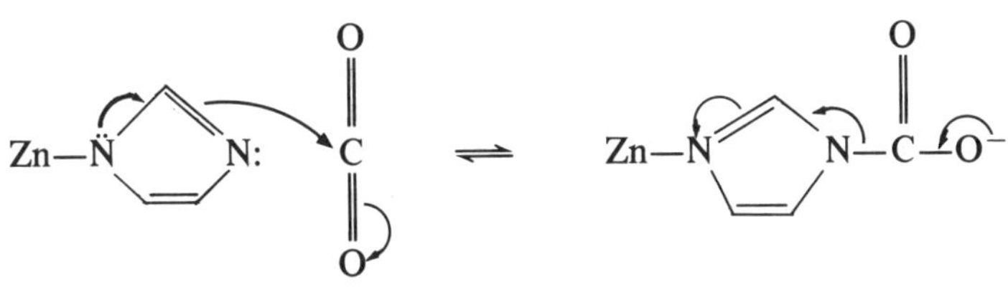

After consideration of the Co(III) model chemistry outlined above, Buckingham (31) proposed an alternative two-step mechanism for carbonic anhydrase. The (uncoordinated) active-site histidine, acting as a nucleophile, could attack CO_2 or an activated ester to produce an N-acyl derivative (carbamate or amide). Hydroxide ion coordinated to the active-site zinc ion could then act as a nucleophile in the hydrolysis of the acyl enzyme. In a model study, the $(NH_3)_5CoOH^{2+}$ ion was found to react rapidly with N-acetylimidazole (k = 14.9M^{-1} sec^{-1} at 25°C) to produce the stable O-acetylated

Table 4 Second-Order Rate Constants for Reactions of Nucleophiles with *p*-Nitrophenyl Acetate (25°C)[a]

Nucleophile	k (M^{-1} sec^{-1})	pK_a' (Conjugate Acid)
OH^-	9.5	15.7
OH^-/H_2O (pH 7)	10^{-6}	
$(NH_3)_5CoOH^{2+}$	1.52×10^{-3}	6.3
$(NH_3)_5CoOH^{2+}/(NH_3)_5CoOH_2^{3+}$ (pH 7)	1.3×10^{-3}	
	3160	14.2
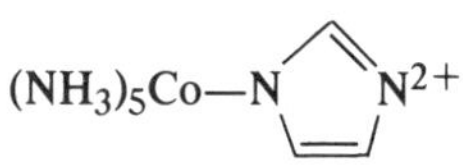	9	10.0
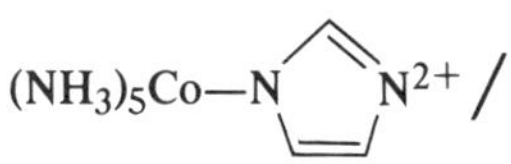 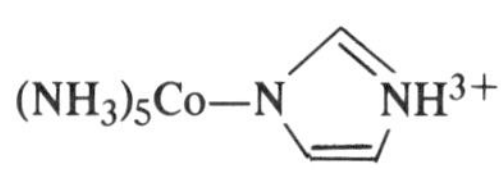 (pH 7)	9×10^{-3}	
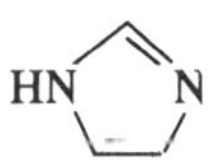	0.6	7.2
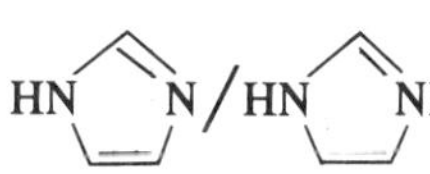 (pH 7)	0.23	

[a] Data taken from refs. 10 and 31.

derivative [the $(NH_3)_5CoO_2CCH_3^{2+}$ ion] (31). Further examples of intramolecular hydrolysis of amides by Co(III)-coordinated OH^- are given in Section 5.

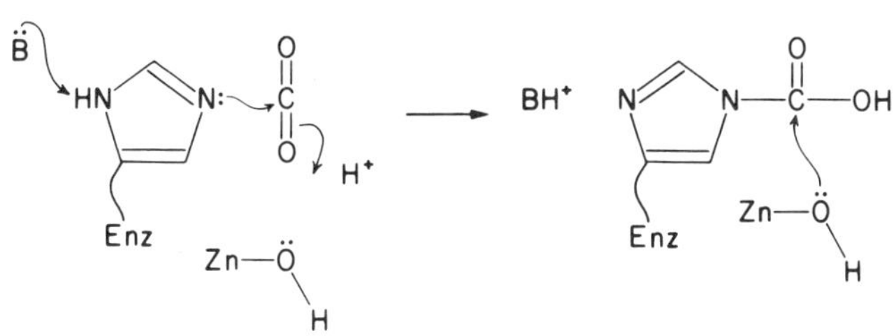

The last mechanism, which appears to be generally consistent with the enzymology, involves the zinc-hydroxide nucleophile in catalysis of deacyla-

tion. This may be a rather general mechanism in the Zn^{2+}-dependent hydrolases since there is a suggestion based on low-temperature kinetic studies that active-site Zn–OH^+ may be involved in deacylation of the intermediate anhydride in carboxypeptidase A-catalyzed ester hydrolysis (45) (see Sec. 5) and models (see Sec. 8) related to alkaline phosphatase indicate that its dephosphorylation might also involve Zn–OH^+.

We note in passing that a role for Zn–OH^+ in hydrolysis of an intermediate *N*-carboxyhistidine in carbonic anhydrase bears some analogy to a reasonable proposal for the involvement of metal-bound nucleophiles in reactions with *N*-carboxybiotin at the second subsite of biotin-dependent carboxylases (Scheme 2). In a sense, aspects of this mechanism have been modeled in the carboxylation of ketones by methylmagnesium carbonate (Scheme 3) (46).

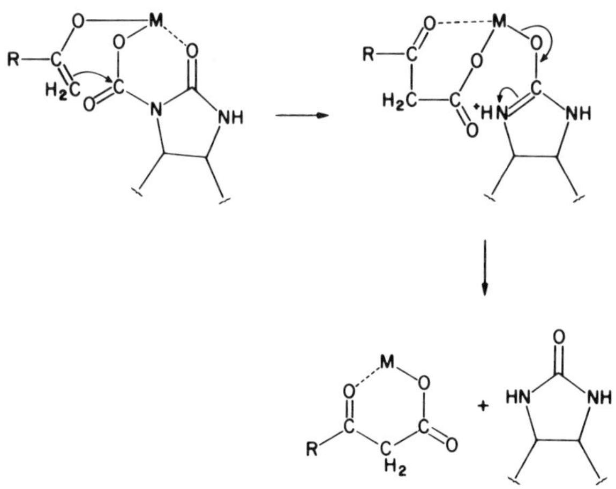

Scheme 2

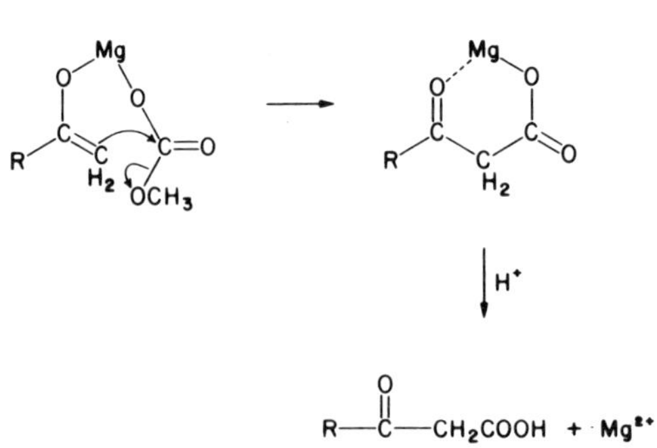

Scheme 3

4 INTERCONVERSIONS OF COMPLEXES OF CARBAMATE, CYANATE, AND UREA–UREASE

The synthesis of urea from ammonium cyanate and the reverse reaction which occurs very slowly in water presented chemists with some of their earliest problem in kinetics and mechanism.

$$NH_3 + H{-}N{=}C{=}O \longrightarrow H_2N{-}\overset{\overset{\displaystyle O}{\|}}{C}{-}NH_2$$

By and large, the mechanism still remains to be elucidated (47). Complexes of inert metal ions [Co(III) and Rh(III)] with cyanate, carbamate, and urea should mimic the role of H^+ in the reactions and allow comment on feasible mechanisms. They may also provide some insight into the chemistry of jack bean urease, the Ni(II) metalloenzyme (7, 48), which *hydrolyzes* urea to produce (initially) ammonia and carbamic acid (49). Studies on the carbamate

$$H_2N{-}\overset{\overset{\displaystyle O}{\|}}{C}{-}NH_2 + H_2O \xrightarrow{\text{urease}} H_2N{-}\overset{\overset{\displaystyle O}{\|}}{C}{-}OH + NH_3$$

complexes are also relevant to those enzymes that produce carbamates or use them as intermediates in carboxylation reactions (Sec. 3). As will be shown, much of the metal complex chemistry is complicated by linkage isomerization reactions. These ligands contain both nitrogen and oxygen as potential donor atoms, and both N-to-O and O-to-N isomerizations of their complexes occur. While the precise mechanism of linkage isomerization is not known, it is clear that it occurs entirely within the coordination sphere of the metal ion and not by dissociation and reformation of the complexes (50). Its relevance to events at the active site of an enzyme (e.g., urease) is uncertain, but the possibility of its occurrence at one or another stage of the enzymatic reactions ought not to be ignored.

The $(NH_3)_5CoNCO^{2+}$ ion (51) and the corresponding Rh(III) complex (52) are produced by reaction of $(NH_3)_5MOH_2^{3+}$ with urea at $>130°C$, perhaps by the deamination of an intermediate N-bonded urea complex (51). The corresponding O-bound cyanato complex is so far unknown in Co(III) chemistry (51). On treatment under acidic conditions, the Co(III) (51, 53), Rh(III) (52), and Ru(III) (52) complexes all undergo sequential reactions with the ultimate formation of $(NH_3)_6M^{3+}$ ions. The first reaction to occur is a H^+-catalyzed hydration to the N-bonded carbamate complex,

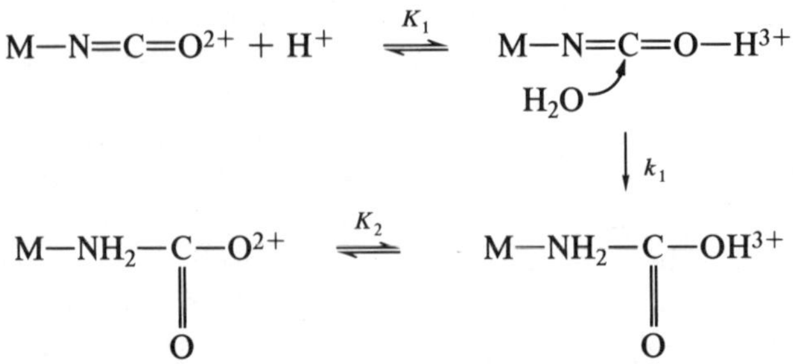

For the Co(III) complex, the rate law has the form $k_{obs} = K[H^+]/([H^+] + K_1)$, where the dissociation constant $K_1 > 20$ ($pK_1 < -1.3$) (53). Under conditions where $[H^+] \ll K_1$, a plot of k_{obs} versus $[H^+]$ was linear, giving a slope $k_a = k_1/K_1 = 0.17M^{-1}\ sec^{-1}$ ($\mu = 1M$, 25°C). For M = Rh(III) and Ru(III), $k_a =$ 0.62 and $0.06M^{-1}\ sec^{-1}$, respectively (52), while for the corresponding reaction where $M{=}H^+$, $k_a = 0.12M^{-1}\ sec^{-1}$ ($\mu \simeq 0.2M$) (54). The near equivalence of the values of k_a for HNCO and the three M(III) complexes offers support for the proposition earlier (Sec. 1) that the predominant role of the metal ion is to act as a poor substitute for a proton. The pK_a of $H_2(NCO)^+$ ought to be at least three units lower than that for $(NH_3)_5CoNCOH^{3+}$ (*viz.* < -4) giving $k_1 > 10^3\ sec^{-1}$ for $M = H^+$. If for the Co(III) complex, $K_1 \sim 20$ then $k_1 \sim 3\ sec^{-1}$. The diminished reactivity of the $(NH_3)_5CoNCOH^{3+}$ ion is essentially completely offset by its higher pK_a (under conditions where $[H^+] \ll K_1$) and it shows similar reactivity to $HNCO + H^+$ at any particular $[H^+] < K_1$.

The final product of reaction of the $(NH_3)_5RhNCO^{2+}$ ion in strong acid was isolated and argued to be the protonated N-bound carbamate complex $(NH_3)_5RhNH_2COOH^{3+}$. In acid solution ($[H^+] > 0.2M$) it decarboxylates to produce $(NH_3)_6Rh^{3+}$ and CO_2 with a rate law $k_{obs} = k_0 + k_1/[H^+]$, where $k_0 = 2.1 \times 10^{-3}\ sec^{-1}$ and $k_1 = 9.6 \times 10^{-2}M\ sec^{-1}$ (25°C, $\mu = 1.0M$) (52). This is consistent with the mechanism shown for which the rate law has the form: $k_{obs} = [k_2[H^+] + k_3K_a/(1 + K)]/(K_a + [H^+])$. When $[H^+] \gg K_a$, this reduces to the required form where $k_0 = k_2$ and $k_1 = k_3K_a/(1 + K)$.

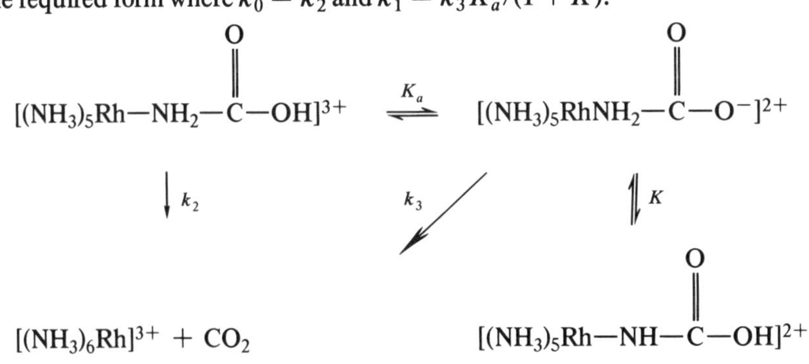

The Co(III) chemistry also requires the intermediacy of an N-bonded carbamate complex. With $(NH_3)_5CoNCO^{2+}$ in acid, the consecutive reactions are observable spectrophotometrically, and the first-order rate constants for the second process yield a rate law identical to that for the Rh(III) complex, with $k_2 \sim 10^{-2}\ sec^{-1}$, $k_3/(1+K) = 0.38\ sec^{-1}$, and $K_a = 0.42M$ ($\mu = 1.0M$, 25°C). However, product distributions and ^{18}O-tracer experiments indicate the mechanism to be complicated by linkage isomerization and aquation of the $(NH_3)_5CoNH_2COOH^{3+}$ ion (Scheme 4, $k_2 = k_2' + k_2''$) (53). The linkage isomerization is paralleled exactly by the similar N-to-O rearrangement of an N-bonded urea complex which occurs on a similar time scale and is discussed below.

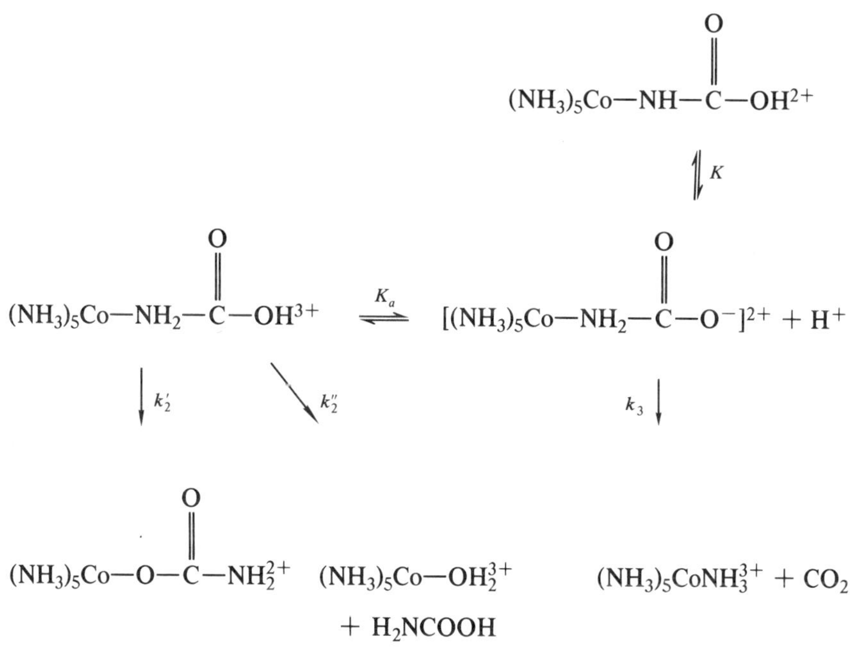

Scheme 4

Comparisons of the rate of decarboxylation of the $(NH_3)_5Co{-}NHCOOH^{2+}$ ion with that of H_2NCOOH in acid are difficult since no detailed kinetics for reactions of the free ligand have been reported, and uncertainty regarding the position of the equilibrium between the free acids and their more-reactive

zwitterions further complicates the issue. Roughton (55) has reported $k_4/(1 + K_c) \sim 80\ \text{sec}^{-1}$ in acid at 0°C.

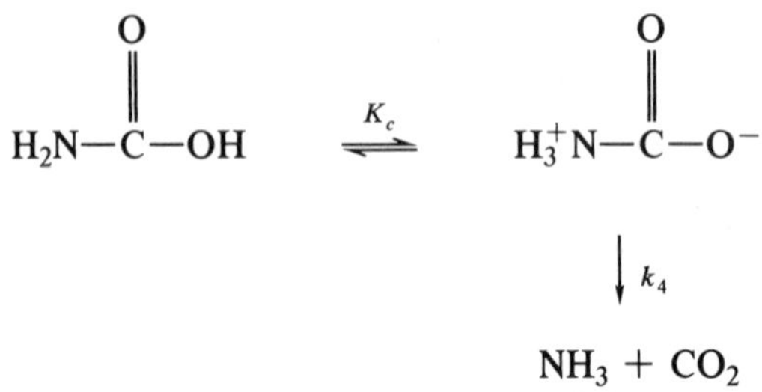

On the basis of predicted pK_a's for the two groups, K_c would be about 10^7. If the zwitterion is the only reactive form, then $k_4 \sim 10^9\ \text{sec}^{-1}$. Based on measured values of pK_a for N-coordinated amides bound to Co(III) (Table 2), K would be predicted to be about unity for the deprotonated carbamate complex (Scheme 4), giving $k_3 \sim 10^0\ \text{sec}^{-1}$. It is apparent from published data (49, 55) that the $H_2N{-}C({=}O){-}O^-$ ion does not decarboxylate at any appreciable rate ($k < 10^{-3}\ \text{sec}^{-1}$), the slow decomposition of carbamate even at high pH being due essentially to the $k_4/(1 + K_c)$ pathway. The above analysis establishes one further facit of our concepts of the role of the metal ion as a poor substitute for a proton. As predicted, $(NH_3)_5Co{-}NH_2^{2+}{-}$ is a much poorer leaving group than $H_3^+N{-}$, but is still considerably better in this regard than is $H_2N{-}$. Similarly, with oxygen donors, the leaving group capacity ought to decrease in the order $H_2^+O{-} > M{-}OH{-}^{n+} > HO{-} > M{-}O{-}^{(n-1)+} > {}^-O{-}$, and examples of this behavior will be given later.

The $(NH_3)_5CoO{-}C({=}O){-}NH_2^{2+}$ ion may be simply prepared by reaction of $(NH_3)_5CoOH^{2+}$ with HNCO (Sec. 3) (33, 56).

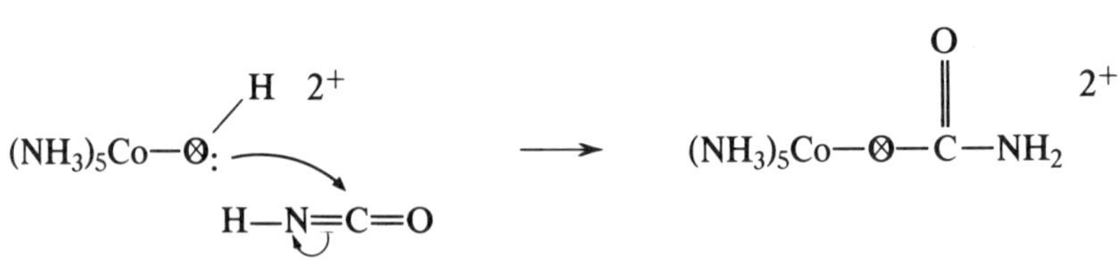

On the basis of predictions similar to those above, it ought to be much less reactive than ${}^-O{-}C({=}O){-}NH_2$ toward decarboxylation, and it has been observed to be stable at neutral and alkaline pH (14, 53). Under acidic condi-

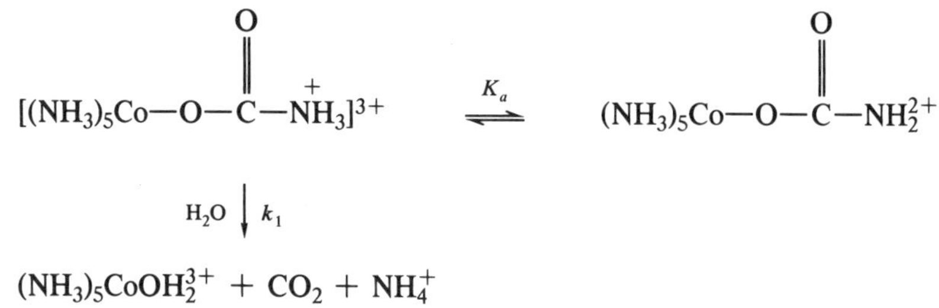

Scheme 5

tions, it undergoes a relatively slow reaction to produce $(NH_3)_5CoOH_2^{3+}$, NH_4^+, and CO_2 (Scheme 5, $K_a = 0.14$, $k_1 = 2.2 \times 10^{-5}$ sec^{-1}, only the protonated species being reactive) (57). Once again, it is predicted that $[(NH_3)_5CoOCONH_3]^{3+}$ ought to be protected relative to $H_2N—COOH$ (or $^+H_3N—COO^-$) by coordination to Co(III) and ought to show reactivity between that of $H_2N—COOH$ and $H_2N—COO^-$. It might be expected to be of similar reactivity to the $(NH_3)_5Co—NH_2—C(=O)—OH^{3+}$ ion, in which the roles of the two Lewis acids [Co(III) and H^+] are reversed. Unfortunately, with this latter complex (Scheme 4), the occurrence of competing reactions precluded the observation of decarboxylation (53). The alternative to Scheme 5, involving an S_N1 dissociation of O-protonated $(NH_3)_5CoO_2CNH_2^{2+}$ cannot be excluded by the kinetic and product distribution studies (57).

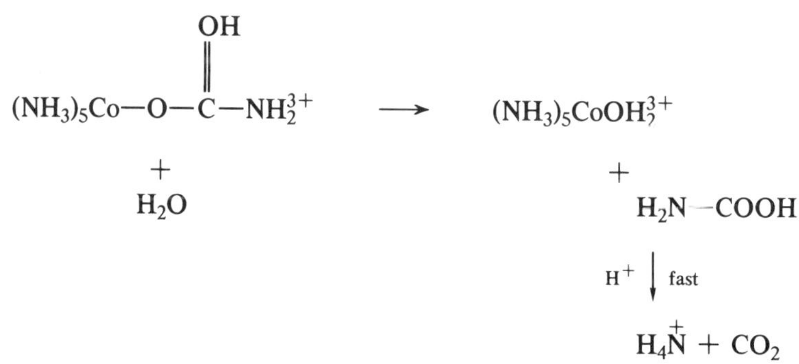

The reaction of the *cis*-$(en)_2Co(OH_2)_2^{3+}$ ion with cyanate has also been investigated (33). The ^{18}O-tracer studies and product analysis are consistent with the mechanism given in Scheme 6, which involves a chelated O,O′-bound carbamato complex as an intermediate which hydrolyzes to produce chelated $(en)_2CoCO_3^+$. Kinetic studies were not carried out, although it was noted that the carbonato complex is formed with $t_{1/2} \sim 14$ min at 25°C in 1*M*

NaNCO (33). Thus k (Scheme 6) $\geq 8 \times 10^{-4}$ sec^{-1}, depending on which step is rate determining.

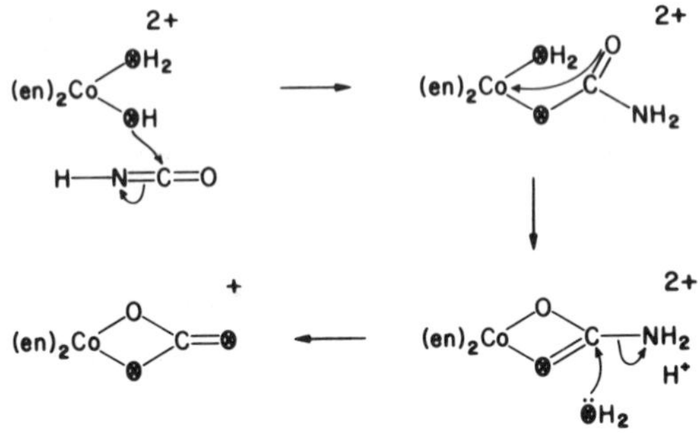

Scheme 6

The chemistry of N- and O-coordinated Co(III)–urea complexes is related in many respects to that of the carbamate complexes. The O-bonded isomer, the $(NH_3)_5CoOC(NH_2)_2^{3+}$ ion is readily prepared by exchange of urea for the labile $CF_3SO_3^-$ ligand in $(NH_3)_5CoOSO_2CF_3^{2+}$ (58). It was anticipated that coordination of urea to the metal center might polarize the carbonyl sufficiently to allow nucleophilic attack by external OH^- leading to its hydrolysis.

Such a reaction has been observed with the $(NH_3)_5CoOCH{\cdot}N(CH_3)_2^{3+}$ ion (59), which hydrolyzes to formatopentaammine-Co(III) and dimethylamine under basic conditions in competition with a minor pathway involving cobalt-oxygen bond breakage by the S_N1CB mechanism familiar in the base hydrolysis of many Co(III) amine complexes (14, 50, 60) (Scheme 7). Kinetic analysis gave a strict first-order dependence on $[OH^-]$, $k_{OH} = 1.3M^{-1}$ sec^{-1}. Product distributions gave values for $k_3 = 0.17\ M^{-1}$ sec^{-1} and the second-order rate constant for the dimethylformamide hydrolysis, $k'_{OH} = 1.1M^{-1}$ sec^{-1} at 25°C ($\mu = 1.0M$). The value of k'_{OH} is at least 10^4-fold greater than that for the OH^--catalyzed hydrolysis of free dimethylformamide. As with Co(III)-promoted nitrile hydration (Sec. 2), the increase in rate on coordination derives almost entirely from an increase in $\Delta S^\dagger$ (59), and once again this is not amenable to simple interpretation since the reactions may not be directly comparable.

The hydrolysis of uncoordinated dimethylformamide, in common with normal amide hydrolysis (61) would be expected to proceed through a tetrahedral intermediate which would be expected to collapse *via* two paths, one (k_{-1}) leading to oxygen exchange, the other giving hydrolysis (k_2):

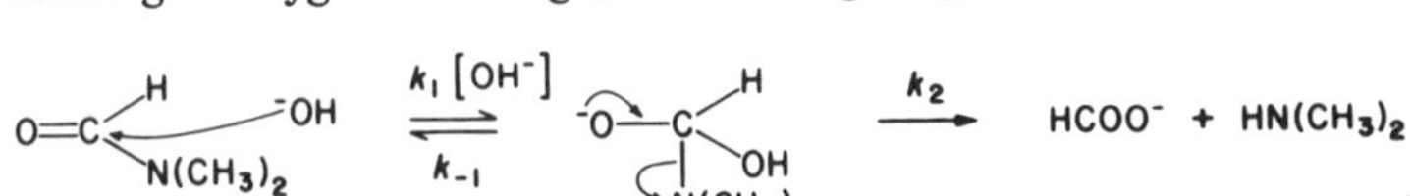

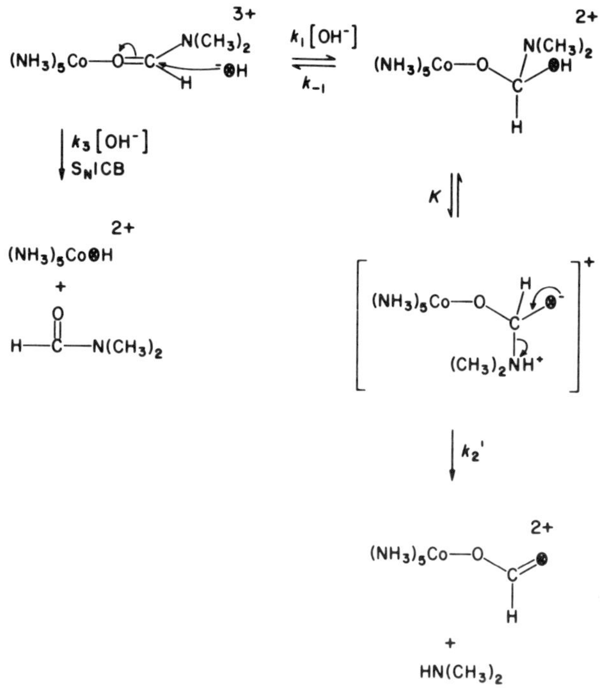

Scheme 7

Since —OH is a better leaving group than the amine from such an intermediate (61), $k_{-1} > k_2$, and breakdown of the intermediate is rate determining.

In the case of the Co(III)-promoted reaction, solvent exchange from the intermediate is not possible, and a proton transfer is required to enable it to collapse to products. The intermediate (Scheme 7) could collapse with expulsion of either $(NH_3)_5CoOH^{2+}$ or $HN(CH_3)_2$. The pK_a of $(NH_3)_5CoOH^{2+}$ is estimated to be about 20–25 and that of $HN(CH_3)_2$ would have a similar value; both are therefore very poor leaving groups. Tracer studies with ^{18}O showed that the hydroxopentaammine-Co(III) product ($\sim 12\%$ of total products at 25°C) derived at least 80% of its oxygen from solvent by an S_N1CB process and $<20\%$ from the ligand oxygen (by carbon–oxygen cleavage) (59). The amine therefore appears in this case to be the better of the two leaving groups by a factor greater than 40.

A simplified rate law, derived assuming a steady-state concentration of the intermediate, for both the unpromoted and Co(III)-promoted hydrolysis reactions gives $k'_{OH} = k_1k_2/(k_{-1} + k_2)$, where $k_2 = k'_2K$ (Scheme 7). It is apparent that the rate enhancement observed with the Co(III) complex could

derive from a greater value of either k_1 or the ratio k_2/k_{-1}. A greater value of k_1 could arise from polarization of the carbonyl group, making it more susceptible to attack by the nucleophile, while k_{-1} would be expected to show a corresponding diminution due to charge neutralization by the coordinated metal ion in the intermediate. On the other hand, the value of k_2 is governed by an unfavorable equilibrium in the Co(III) complex (i.e., $k_2 = k_2' \cdot K$), and this may offset the advantage gained through the decrease in k_{-1}.

The $(NH_3)_5CoOC(NH_2)_2^{3+}$ ion reacted under basic conditions to produce $(NH_3)_5CoOH^{2+}$ as the only Co(III)-containing product. The reaction was first order in $[OH^-]$ at pH < 11, $k_{OH} = 15.3M^{-1}\ sec^{-1}$ (25°C, $\mu = 1.0M$), but approached a limiting rate at high pH as the complex dissociated to its inactive conjugate base. Tracer studies (^{18}O) showed that 97% of the base-catalyzed reaction occurred with Co—O bond cleavage by the usual dissociative conjugate-base mechanism (14, 60). The remaining 3% reaction led to carbon-oxygen bond rupture via a tetrahedral Co(III)-urea-OH^- adduct (Scheme 8). The pK_a of the coordinated urea was determined kinetically to be 13.26, which compares with ~14 for the free ligand (Table 2) (14).

The reactions of the $[(NH_3)_5CoOC(NH_2)_2]^{3+}$ ion in base bear some anal-

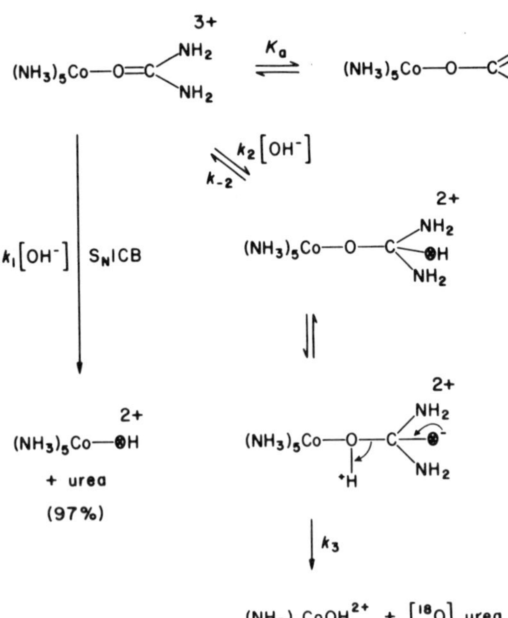

Scheme 8

ogy to those of the $(NH_3)_5CoOCH \cdot N(CH_3)_2^{3+}$ ion. It is apparent that the Co—O bond in the urea complex is somewhat more labile than in the dimethylformamide complex ($k_1 = 14.8M^{-1}$ sec^{-1}, *cf.* $0.17M^{-1}$ sec^{-1}) toward dissociation, and that this reaction competes more substantially with hydroxide-ion attack at the carbonyl carbon (k_{OH} for this pathway is $0.44M^{-1}$ sec^{-1}, *cf.* $1.1M^{-1}$ sec^{-1}). Apparently, with the urea complex $(NH_3)_5CoOH^{2+}$ is a better leaving group from the tetrahedral intermediate than is NH_3 since no $(NH_3)_5CoO_2CNH_2^{2+}$ ion was produced, and hydrolysis of urea was not observed (14). It is evident that coordination to a metal ion through its carbonyl oxygen does not of itself provide sufficient activation of urea to enable it to undergo hydrolysis. The stability of coordinated urea derives from the same source as the stability of free urea, and is attributable to its unusual resonance stabilization. That the urea carbonyl is not polarized to the same extent as the O-coordinated amide is therefore not unreasonable, and is confirmed by the physical properties of the complex.

As argued previously, the extent to which the ligand carbonyl is polarized by coordination ought to be reflected in the extent to which its acid dissociation constant is lowered (7). With urea, the effect is minimal (~1 unit) (14). An analogous case with a coordinated amide is the $(en)_2\overline{Co{-}NH_2{-}CH_2{-}CO}NH_2^{3+}$ ion, whose pK_a is lowered to ~11.2 from ~15 for the free ligand. It, like the

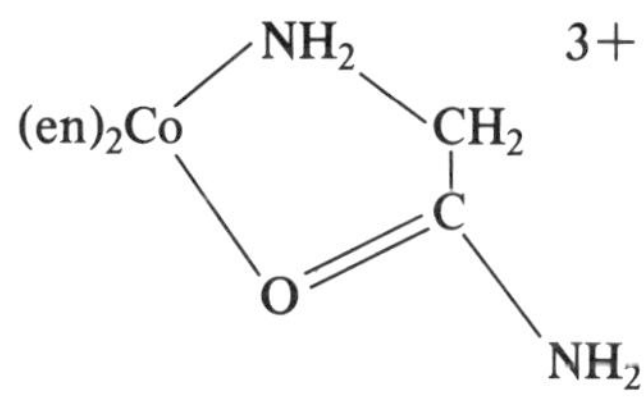

dimethylformamide complex, is susceptible to nucleophilic attack by OH^-, and the amide is hydrolyzed ~10^4-fold faster than uncoordinated glycine amide (Sec. 5) (12). It is therefore apparent that, in the $(NH_3)_5Co{-}O{=}C(NH_2)_2^{3+}$ ion, the carbonyl bond of urea is not polarized to the same extent as it is in the O-coordinated amides, and is therefore less susceptible to hydrolysis. That is, the situation more closely resembles structure A where the ligand is

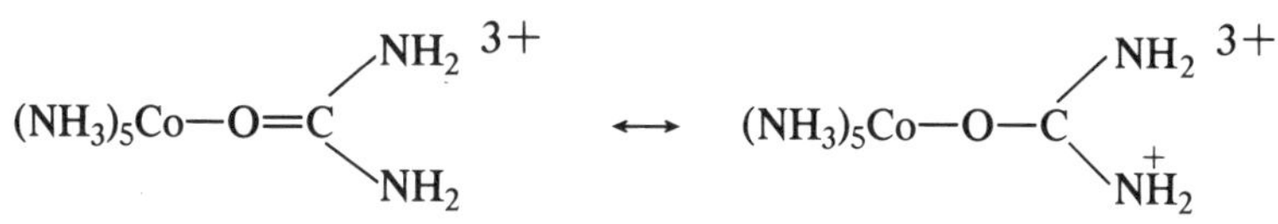

p$K_a \approx 14$ (less reactive) A

p$K_a \approx 1$ (more reactive) B

coordinated through a carbonyl group than B where the donor atom is an imidol hydroxyl. Although such correlations are often tenuous, this is also reflected both in the increased lability of the Co—O bond to solvolysis (58, 59), and in the wavelength of the first ligand-field band in the electronic spectra of the complexes. The urea complex absorbs maximally at 518 nm (14) (more typical of coordinated O=C<) while the dimethylformamide complex has λ_{max} = 506 nm (58) (more typical of coordinated $^{-}$O—C≤).

A molecule of urea is at least 10^{14}-fold more reactive toward hydrolysis when coordinated to Ni(II) at the active site of urease than it is in water at neutral pH (7). The model system raises two points of relevance to the chemistry of the enzyme. Firstly, simple coordination of the substrate to Ni^{2+} through its carbonyl oxygen would not of itself provide sufficient activation of the carbonyl oxygen to enable hydrolysis to be observed. In the mechanism proposed for the enzyme (7) (Scheme 9, reproduced with permission from ref. 7), the substrate is further polarized by the proximity of a nearby anion (a carboxylate). Although this concept has yet to be tested in a model system, it would be expected to lead to enhanced susceptibility to nucleophilic attack by OH^{-}. Secondly, the enzyme would need to provide a mechanism to ensure that ammonia, rather than Ni-OH^{+} would leave preferentially from the tetrahedral intermediate. In Zerner's mechanism (7), a general acid catalyst

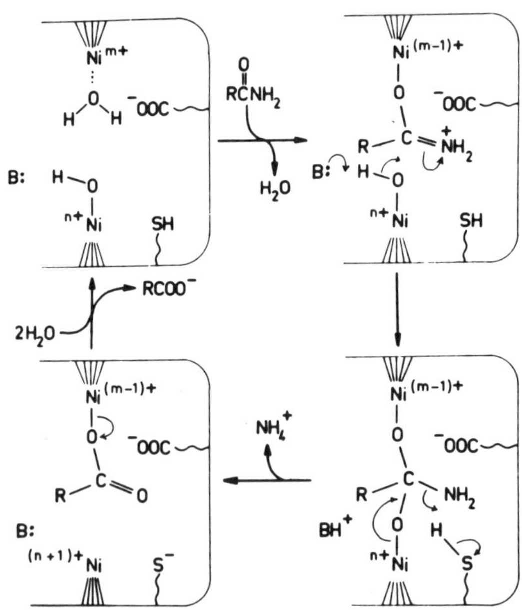

Scheme 9

(cysteine-SH) fulfils this role. In the proposed mechanism, the 10^{14}-fold rate enhancement is achieved by the additional use of hydroxide ion coordinated to the second Ni(II) center as an intramolecular nucleophile, and this role is also consistent with the enzyme studies (7). As discussed previously (Sec. 2), the use of two metal centers acting in concert in an analogous fashion leads to rate accelerations at neutral pH of up to 10^{18}-fold in the Hg^{2+}-catalyzed intramolecular hydration of the nitrile in the *cis*-$(en)_2Co(OH)(NH_2CH_2C{\equiv}N)^{2+}$ ion (23, 30). Although the Co(III) model systems yet examined do not provide a system that hydrolyzes urea, they have contributed some understanding of the subtleties of the enzyme mechanism.

The metal–ligand bonds in Rh(III) complexes are generally less susceptible by a factor of $\sim 10^2$ to substitution than in the Co(III) complexes. We examined therefore the reactions of the $(NH_3)_5RhOC(NH_2)_2^{3+}$ ion in base in the hope that urea hydrolysis might be competitive with ligand dissociation. The kinetics of reaction of the Rh(III) complex were analogous to those of the Co(III) complex, showing $k_{OH} = 0.17M^{-1}\,\text{sec}^{-1}$, $pK_a = 13.63$, but product distribution studies demonstrated a major product to be $(NH_3)_5RhNHCONH_2^{2+}$, produced as a result of base-catalyzed linkage isomerization (Scheme 10). No $(NH_3)_5RhO_2CNH_2^{2+}$ was produced (15).

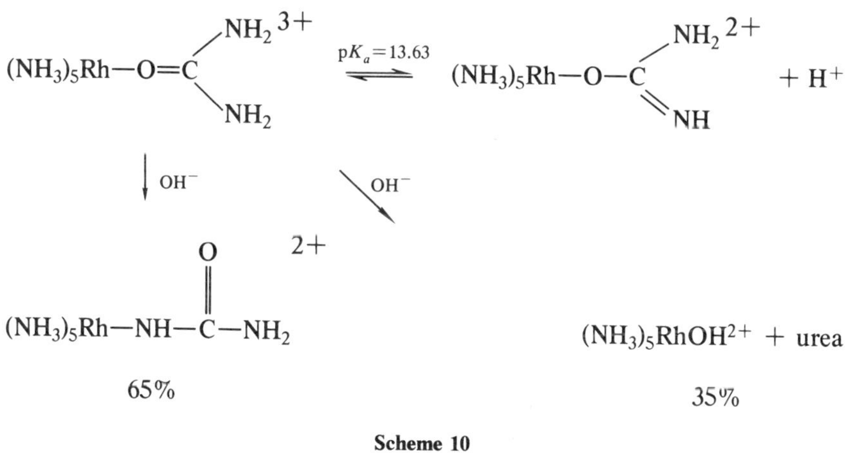

Scheme 10

The deprotonated N-bound urea complex is completely stable to further reaction under basic conditions, but undergoes reaction in acid to produce $(NH_3)_6Rh^{3+}$ and $(NH_3)_5RhOH_2^{3+}$ as products. Absorbance–time curves at suitable wavelengths show two exponential processes. The same two products are also produced in different proportions from $(NH_3)_5RhOC(NH_2)_2^{3+}$ under the same conditions. In contrast to the single first-order reactions observed with the O-bound isomer in base, the kinetics of its decomposition in acid

also show two exponential processes, the two exponential parameters being identical to within experimental error to those for reaction of the N-bonded isomer under the same conditions. The two exponential parameters are invariant with pH at pH < 2 and describe a titration curve at higher pH. The data are consistent with the mechanism shown in Scheme 11, and the combination of kinetic data and results of product distributions allow values of the pK_a of the $(NH_3)_5Rh{-}NH_2{-}C(=O){-}NH_2^{3+}$ ion and the five rate constants to be evaluated; $pK_a = 3.9$, $k_1 = 6.3 \times 10^{-5}\ sec^{-1}$, $k_{-1} = 6.4 \times 10^{-5}\ sec^{-1}$, $k_2 = 2.1 \times 10^{-5}\ sec^{-1}$, $k_3 = 3.9 \times 10^{-5}\ sec^{-1}$, $k_4 = 0.96M^{-1}\ sec^{-1}$ ($\mu = 1.0M$, 25°C).

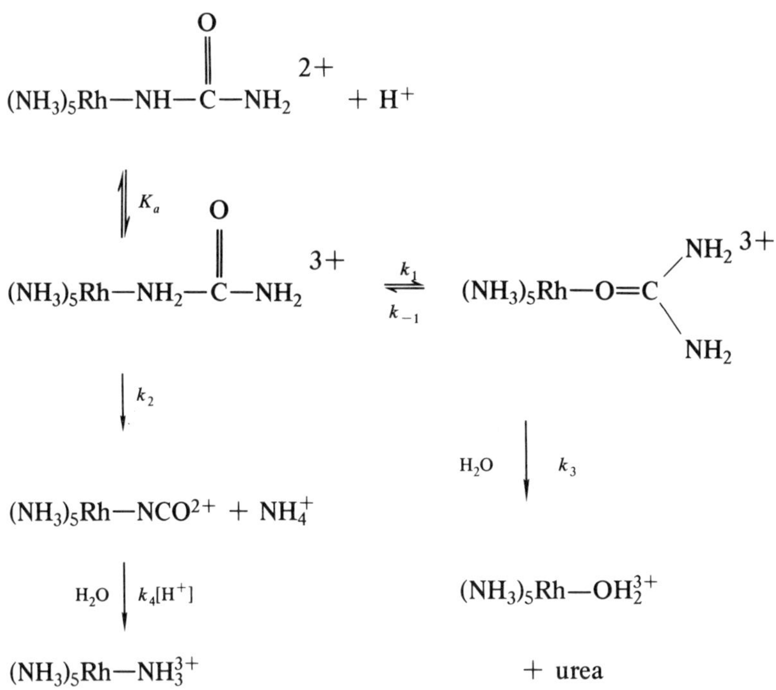

Scheme 11

The rate constants k_{-1} and k_1 describe a slow linkage interconversion of the O-bonded and protonated N-bonded forms of the urea complexes, while k_3 relates to dissociation of urea from the more labile O-bonded isomer. The reaction of direct relevance to the enzyme chemistry is the k_2 path which involves Rh(III)-promotion of the decomposition of coordinated urea. There

are four possible mechanisms which might be responsible, two via a hydrolytic path,

$$(NH_3)_5Rh{-}NH_2{-}\overset{\overset{\displaystyle O}{\|}}{C}{-}NH_2^{3+} + H_2O \rightarrow (NH_3)_5RhNH_3^{3+} + HO{-}\overset{\overset{\displaystyle O}{\|}}{C}{-}NH_2 \quad (1)$$

$$(NH_3)_5Rh{-}NH_2{-}\overset{\overset{\displaystyle O}{\|}}{C}{-}NH_2^{3+} + H_2O \rightarrow \left[(NH_3)_5Rh{-}NH{-}\overset{\overset{\displaystyle O}{\|}}{C}{-}OH\right]^{2+} + NH_4^+ \quad (2)$$

and two by a path analogous to the homolytic decomposition of the free ligand:

$$\underset{\text{(or its imidol form)}}{(NH_3)_5Rh{-}NH_2{-}\overset{\overset{\displaystyle O}{\|}}{C}{-}NH_2^{3+}} \longrightarrow (NH_3)_5Rh{-}N{=}C{=}O^{2+} + NH_4^+ \quad (3)$$

$$(NH_3)_5Rh{-}NH_2{-}\overset{\overset{\displaystyle O}{\|}}{C}{-}NH_2^{3+} \longrightarrow (NH_3)_5RhNH_3^{3+} + H{-}N{=}C{=}O \quad (4)$$

The isocyanato complex has been observed at pH 3–4 so an appreciable part of the reaction occurs by path (3) and possibly all of it. Both the isocyanato and N-carbamato complexes [eqs. (2) and (3)] are known to decompose to $Rh(NH_3)_6^{3+}$ more rapidly than the reactions of the urea complexes in acid (pH < 3) and so are difficult to detect as intermediates (15).

An N-bonded Co(III)-urea complex has been prepared by hydration of the dimethylcyanamide complex in base (Sec. 3), $k_{OH} = 3.1M^{-1}\ \text{sec}^{-1}$ (25°C, $\mu = 1.0M$) (22).

$$(NH_3)_5Co{-}N{\equiv}C{-}N(CH_3)_2^{3+} \xrightarrow{k_{OH}[OH^-]} (NH_3)_5Co{-}NH{-}\overset{\overset{\displaystyle O}{\|}}{C}{-}N(CH_3)_2^{2+}$$

Like the N-bonded Rh(III) complex it is stable under basic conditions, but its conjugate acid undergoes a rapid first-order linkage isomerization to the O-bonded isomer, $pK_a = 2.92$, $k = 0.016\ \text{sec}^{-1}$ ($\mu = 1.0M$, 25°C) (22).

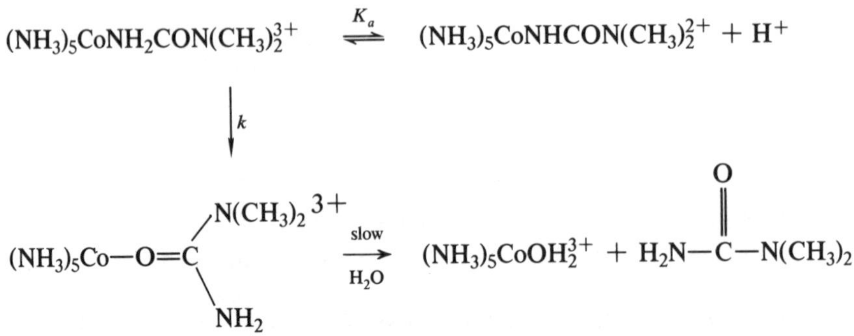

The inactivity of N-coordinated amides and ureas toward nucleophilic attack by HO^- is consistent with the Lewis acid properties of the metal center. Since the metal ions are poor substitutes for H^+, it follows that the ordering of the capacity of groups to leave from a tetrahedral intermediate is $^+NH_3—$ > $M—NH_2—^{n+}$ > $NH_2—$ > $M—NH—^{(n-1)+}$. Metal ions could promote the hydrolysis of N-bonded amides relative to uncoordinated amides only under conditions where the complex existed as the protonated species in the presence of appreciable concentrations of the nucleophile. Since N-coordination of an amide or urea to Co(III) or Rh(III) lowers its pK_a' by 11 or more units (Table 2) to ~4 or less, the protonated species ($M—NH_2—^{n+}$) cannot exist in the presence of $>10^{-10}M$ OH^-, and hydrolysis has been too slow to be observed in competition with solvolysis or linkage isomerization reactions (13, 22, 62). Because the Rh(III)-urea complexes undergo linkage isomerization and ligand dissociation only slowly in acid, and because the pK_a' of $(NH_3)_5RhNH_2CONH_2^{3+}$ is a little higher than that of the analogous Co(III) complex, it may yet provide the only case where the potential of $M—NH_2—^{n+}$ as a leaving group is realizable. It may also be that it would function as a good leaving group in reactions of N-coordinated amides with nucleophiles other than OH^- (e.g., NH_2OH), but this proposition is yet to be tested experimentally.

At the active site of an enzyme that used Lewis acids weaker than Co(III) or Rh(III) (e.g., Zn^{2+}, Ni^{2+}), the pK_a of an N-coordinated amide would be higher (perhaps 7–9), and its potential toward attack by an intramolecular nucleophile could be exploited. The arguments for the mechanism of urease (Scheme 9, based on enzymology) apply equally well for an N-coordinated urea-Ni^{2+} complex. It would merely require a reversal of the roles of the two Lewis acids (Ni^{2+} and H^+) and is equally reasonable from a chemical standpoint.

In aqueous ethanol in the presence of excess Ni^{2+}, *N*-(2-pyridylmethyl) urea has been reported to slowly undergo ethanolysis to give *N*-(2-pyridylmethyl) carbamate and hydrolysis to give *N*-(2-pyridylmethyl)amine (63).

The reaction, which occurs $>(7 \times 10^4)$-fold more rapidly than decomposition of the urea in the absence of metal ions, is proposed to involve activation of the urea toward solvolysis by O-coordination to Ni(II). A number of alternative mechanisms have not yet been rigorously excluded (15).

5 HYDROLYSIS OF AMINO ACID ESTERS AND AMIDES—CARBOXYPEPTIDASES, PROTEASES, AND ESTERASES

A number of Zn(II)-dependent peptidases, including most notably the carboxypeptidases A and B, thermolysin, and leucine amidopeptidase, catalyze the hydrolysis of peptide bonds in their natural substrates and also of some simple amino acid amides and esters *in vitro*. In the first potential model system, Kroll (64) in 1952 observed substantial increases in the rates of hydrolysis of a series of amino acid methyl esters in the presence of several divalent metal ions (Cu^{2+}, Co^{2+}, Ni^{2+}). Subsequent investigations of catalysis by labile metal ions of the alkaline hydrolysis of amino acid esters (65–70) and amides (67–71), and of other potentially chelating esters (72–74) and amides (75), have confirmed and extended these observations. This work is the subject of an extensive review (69). Although rate accelerations (at particular pH values) in excess of 10^5-fold have been observed in some of these systems, they have generally proved unrewarding as useful models for the enzymatic processes. The kinetic studies have been complicated by ligand dissociation equilibria, while the lability of the substrate and product complexes in solution has rendered assignment of structures to the reactive species and mechanistic inferences equivocal. At least four general mechanisms which might lead to rate acceleration are conceivable:

1. *Intermolecular attack of solvent on the N,O-chelated substrate:*

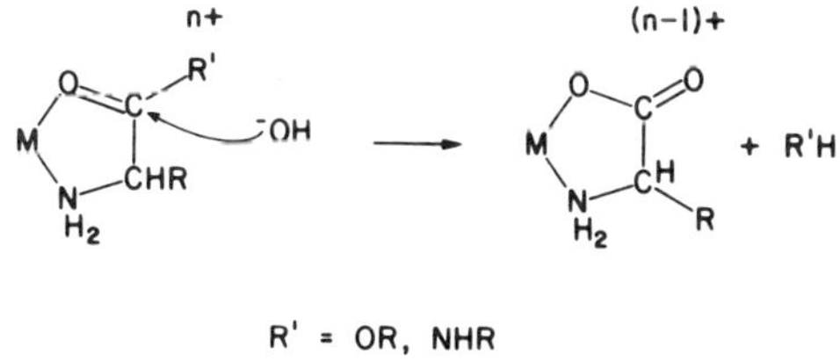

The mechanism is analogous to that discussed (Sec. 4) for the hydroxide ion-promoted hydrolysis of the [(dimethylformamide)pentaammine-Co(III)]$^{3+}$ ion. Coordination through the oxygen atom polarizes the carbonyl bond, producing an incipient carbonium ion much more susceptible to attack by the external nucleophile.

2. *Intramolecular attack of coordinated* OH^- *on the N-bound monodentate substrate:*

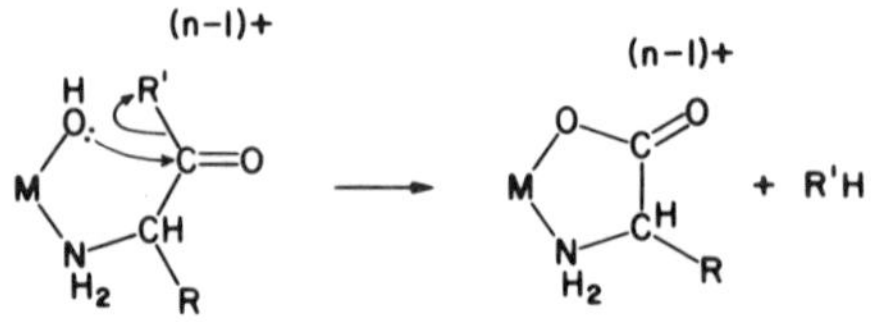

As discussed previously (Sec. 2), hydroxide ion when coordinated is still an efficient nucleophile, and an example was given (Sec. 2) of its involvement in the intramolecular hydration of aminoacetonitrile coordinated to Co(III) through its NH_2 group. In these reactions, the rate acceleration derives principally from the proximity effect of having both substrates coordinated in positions suitable for reaction; this effect offsets the decrease in reactivity of coordinated versus free OH^-, and leads to substantial rate increases.

3. *Intermolecular attack of solvent on the N-bound monodentate substrate:*

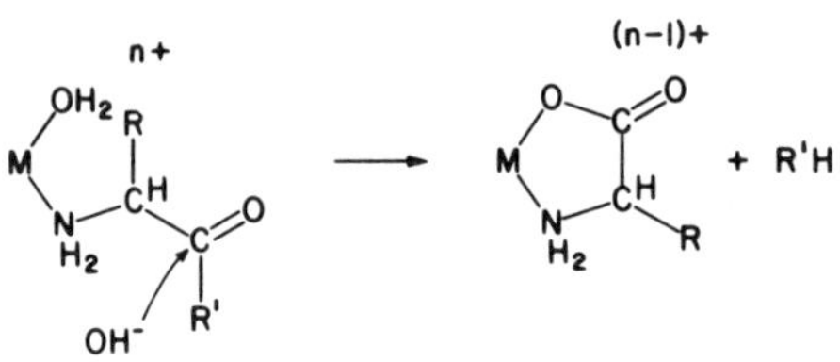

The rate enhancement here could derive only from electrostatic interaction between the cationic center and the incoming nucleophile. Since the metal ions concerned are weaker Lewis acids than H^+, their complexes would be expected to show reactivity intermediate between those of $R'CO{-}CHR{-}NH_2$ and $R'CO{-}CHR{-}NH_3^+$. The metal complex might achieve some selective advantage at high pH where the amino acid ester (or amide) is uncharged, but the effect ought to be minimal since k_{OH} (25.0°C, $\mu = 0.16M$) for the OH^--catalyzed hydrolysis of $EtO{-}CO{-}CH_2{-}NH_3^+$ is only 41-fold greater than that for $EtO{-}CO{-}CH_2{-}NH_2$ (67). That a complex formed with Cu^{2+} (saturating $[M^{n+}]$) was a further 3200-fold more reactive under the same conditions suggests that this mechanism is not involved, and favors either of the alternative mechanisms 1 or 2.

4. *Coordination of the leaving group:*

According to the Lewis acid concept as applied to these systems (Sec. 1), the reactivity of amines as leaving groups ought to decrease in the order $R{-}\overset{+}{N}H_2{-} >$

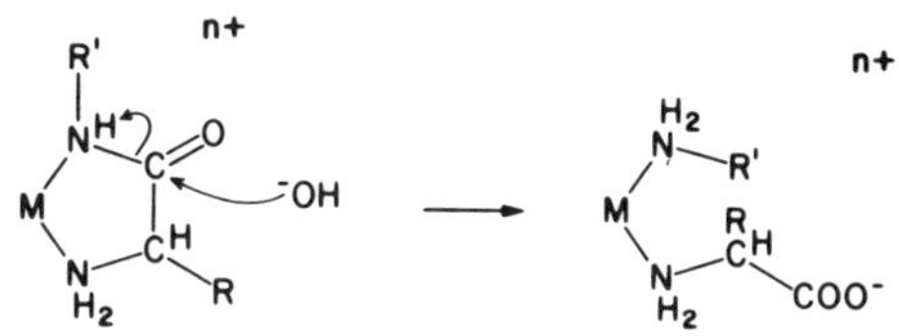

M—NH(R)—$^{n+}$ > R—NH— > M—N(R)—$^{(n-1)}$ and a related series may be constructed for the alcohols. Thus, at a pH where a coordinated amide remains protonated (perhaps as high as 7 for a Zn^{2+} complex), it would be more reactive toward a nucleophile than the predominant unprotonated form of the free amide (R—C(=O)—NH_2). At higher pH, the N-coordinated amide is deprotonated giving M—NH—C(=O)—R$^{(n-1)+}$, in which the leaving group is poorer than that in R—C(=O)—NH_2, and the amide ought to be protected from reaction. Such behavior has been discussed above in the cases of the N-bound urea complexes (Sec. 4) and will not be labored further here.

The suggestion that N-coordination of an amide ought to activate it toward nucleophiles at low pH is valid whether the amide is protonated at the carbonyl oxygen or nitrogen in the metal complex, since both structures are strongly activated in comparison with the free unprotonated amide: (A) by

M—NH=C(—OH)—R (A)

M—NH_2—C(=O)—R (B)

polarization of the C=N bond (cf. mechanism 1) and (B) by the provision of an excellent leaving group at high pH. This problem of the site of protonation bears some analogy to the yet unresolved question of the mechanistically significant site of protonation in the acid hydrolysis of normal carboxamides (2). Although amides are believed to be predominantly O-protonated in acid (76, 77), this does not necessarily imply that it is the predominant O-protonated form rather than the minor N-protonated species which is the most reactive.

All four mechanisms described above are consistent with the general observation made on the labile metal ion systems and, provided pH < pK_a of the coordinated H_2O in mechanism 2, are also consistent with the observed

rate law, $k_{obs} = k_{OH}[OH^-]$ (when $[M^{n+}]$ is saturating). With the exception of the last alternative 4, each of these mechanisms has been probed with substitution-inert complexes of Co(III), and we are able to make some comments on the relative efficiency of the three processes.

The use of the $[(trien)Co(OH_2)(OH)]^{2+}$ ion as a reagent to sequentially cleave N-terminal amino acids from peptides was first reported in 1963 (78). Subsequent study extended to other amino acid and peptide esters and amides (79), and to the use of the analogous *cis*-$[(en)_2Co(OH_2)(OH)]^{2+}$ ion to accomplish the same types of reactions, albeit less efficiently (80). Two groups reported the use of nucleophiles other than OH^- in organic solvents to produce peptides from amino acid esters in the presence of *cis*-β-[(trien) $CoCl_2]^+$ (81, 82) and $[(en)_2CoGlyOR]^{3+}$ (83). More recently, a number of papers have appeared describing the use of specific Co(III) complexes for the hydrolysis of the N-terminal bonds of peptides, with a view to using such reactions for peptide sequencing (84), but these studies have not of themselves proved mechanistically revealing. In every case, the reaction is stoichiometric, resulting in the incorporation of only the N-terminal amino acid as a chelate into the coordination sphere of the metal ion; such a product could arise by any of mechanisms 1–3 above. A number of detailed studies of the mechanisms of reactions of these types have, however, been carried out, and are discussed in turn below.

5.1 Intermolecular Hydrolysis of Chelated Derivatives

Alexander and Busch (85) discussed the Hg^{2+}-induced hydrolyses of *cis*-$[(en)_2Co(GlyOR)X]^{2+}$ ions [X = Cl,Br; R = CH_3,CH_2CH_3,$CH(CH_3)_2$] to

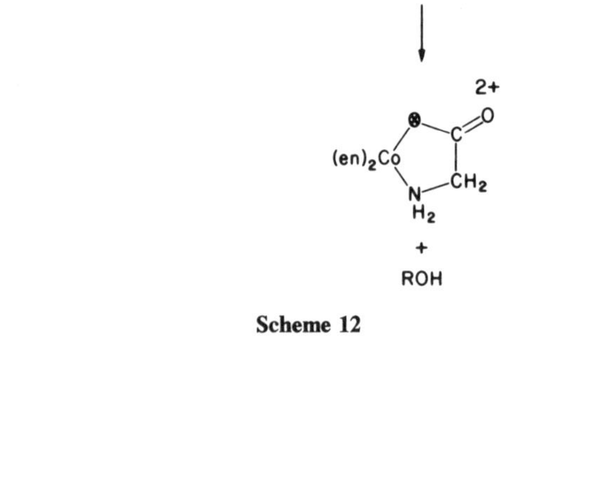

Scheme 12

give $[(en)_2CoGly]^{2+}$, which occur by consecutive first-order reactions corresponding to Hg^{2+}-assisted removal of coordinated X^- followed by ester hydrolysis (Scheme 12). It was proposed that removal of the coordinated halide led to formation of the chelated glycine ester, and that this species was activated toward ester hydrolysis by polarization of the coordinated carbonyl. Evidence in support of such an intermediate was a change in carbonyl stretching frequency as the monodentate ester (1735 cm^{-1}) formed the chelate (1610 cm^{-1}) and then hydrolyzed to the chelated acid (1640 cm^{-1}), but no direct evidence was presented to exclude the intramolecular pathway.

Once a series of *chelated* glycine ester, amide and dipeptide ester complexes had been isolated (83) and securely characterized (86), it was possible

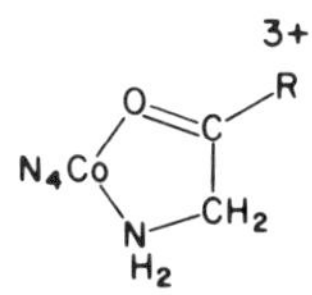

N_4 = $(en)_2$, trien
R = OR′, NHR

to confirm directly the mechanism proposed by Alexander and Busch (85). The kinetics of hydrolysis of the authentic $[(en)_2Co(GlyOCH(CH_3)_2)]^{3+}$ ion at acid pH ($k = 1.1 \times 10^{-3}$ sec^{-1} at 25°C) were identical to those for the second reaction in the Hg^{2+}-induced hydrolysis of *cis*-$[(en)_2$-$CoBr(GlyOCH(CH_3)_2)]^{2+}$ (87). That the ester hydrolysis proceeded in each case by solvent attack at the coordinated carbonyl was shown by ^{18}O-tracer experiments. It was found that $[^{18}O]$-$[(en)_2CoGly]^{2+}$ formed by Hg^{2+}-induced hydrolysis of carbonyl-labeled $[^{18}O]$-*cis*-$[(en)_2CoBr(GlyOCH_3)]^{2+}$ in unlabeled solvent showed no exchange of label after 16 days whereas labeled $[(en)_2CoGly]^{2+}$ produced by similar hydrolysis of unlabeled chelated ester in labeled solvent completely exchanged its label with $k \approx 10^{-6}$ sec^{-1} under the same conditions (87). The two rates were ascribed to the two inequivalent glycinate oxygen atoms. The positions of the label in each case, together with the result that in all cases one solvent oxygen was incorporated into product, demand that the reaction proceed via the carbonyl-oxygen chelated intermediate and that hydrolysis occur without opening of the chelate ring. A mechanism in which coordination of alkoxide ion facilitates its leaving is not permitted.

The rates of hydrolysis of the chelated esters are independent of pH in the range 0–4 (85, 87, 88). This is anticipated since coordination to Co(III) prevents protonation of the carbonyl oxygen and consequent acid hydrolysis. The corresponding fully protonated ester (89) is estimated to be ~10^5-fold more reactive to nucleophilic attack by H_2O than is the $[(en)_2CoNH_2CH_2COOCH(CH_3)_2]^{3+}$ ion, and this is due to Co(III) being a weaker Lewis acid than is H^+. The metal complex has the advantage of existence at pH's where the concentra-

tion of the protonated substrate would be minute. The net increase in rate over that of the amino-protonated ester substrate at pH ~ 3 may be as high as 10^7-fold (88).

Moreover, the Co(III) complexes are stable at much higher pH values where nucleophilic reactions with OH^- and other oxygen or nitrogen bases become possible. In buffers over the pH range 0 to 7, first-order rate constants for hydrolysis of the $[(en)_2CoNH_2CH_2COOCH(CH_3)_2]^{3+}$ ion obeyed the rate law:

$$k_{obs} = k_{H_2O} + k_{OH}[OH^-] + k_\beta[\beta]$$

where β is the buffer base. Product analyses showed that nitrogen bases acted as nucleophiles to form stable chelated amides or peptides (88), while competition between acetate and aminoacetonitrile established that oxygen bases also acted by a nucleophilic path, although the anhydride generated in those cases was rapidly hydrolyzed (90). It was apparent that the aminolysis reaction, at least, proceeded by way of a relatively stable tetrahedral amine-alcohol intermediate (88, 90), a situation anticipated for such reactions of highly acyl-activated esters with poor leaving groups (91). Such a species was observed directly as the intermediate product of sequential first-order processes in the reaction of $NH_2{-}CH_2{-}CO_2CH_2CH_3$ as nucleophile with the $[(en)_2CoNH_2CH_2COOCH(CH_3)_2]^{3+}$ ion in dimethylsulfoxide (Scheme 13), and was characterized as lacking a normal carbonyl resonance in its infrared spectrum (92).

The extraordinary stability of the intermediate is of some relevance. It is anticipated that polarization of the carbonyl by the metal center would promote nucleophilic attack, and this appears in general to produce an ~10^6-fold

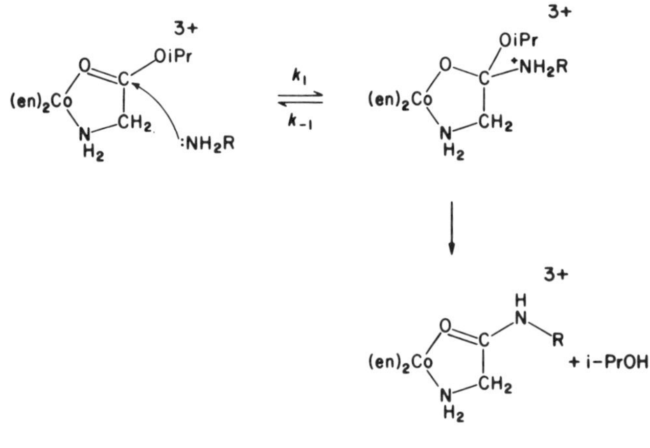

Scheme 13

enhancement in rates of reactions of the chelated esters with both O and N nucleophiles (86, 88). Axiomatically, it should be more difficult for a group to leave with its electron pair, and this might produce little net rate acceleration were the process concerted (92). The pronounced stabilization of the tetrahedral adduct (Fig. 2) over that in the unpromoted equivalent apparently derives from neutralization of the negative charge on the carbonyl oxygen, making it less able to donate electron density back into the C—O bond as the leaving group departs. However, such comparisons may be compli-

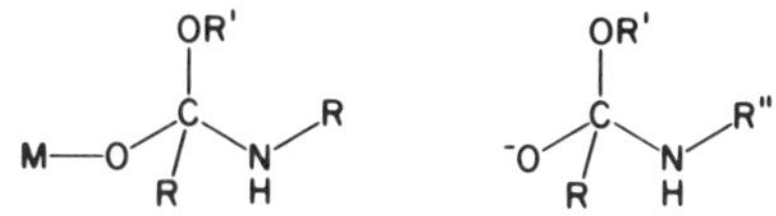

cated by the states of protonation of the two species at any particular pH (e.g., H^+ would neutralize the charge more effectively than Co^{3+}), and more work is required adequately to define pK_a values for such intermediates.

The second-order rate constant for the hydroxide ion-dependent pathway in the hydrolysis of the $[(en)_2CoNH_2CH_2COOCH(CH_3)_2]^{3+}$ ion ($k_{OH} = 1.5 \times 10^6 M^{-1}\ sec^{-1}$ at 25°C) is about six orders of magnitude greater than

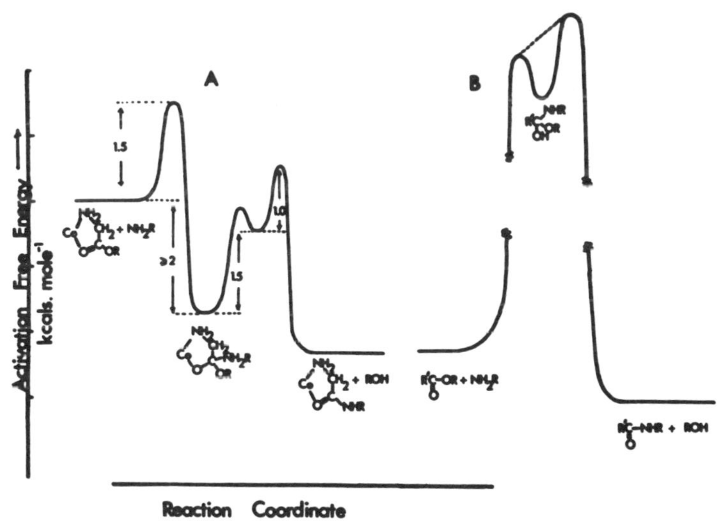

Figure 2 Comparison of reaction coordinate diagrams for metal-ion-assisted (*A*), and unassisted (*B*) aminolysis of an organic ester. Compared to (*B*) the tetrahedral intermediate (*A*) is substantially stabilized with respect to the ester and this leads to more rapid addition of the amine. However, loss of alcohol in (*A*) is slow compared to (*B*), the latter process being either subsequent to (——) or concurrent with (—·—) amine addition (92).

that for hydrolysis of the uncoordinated ester (88). The enhancement is entirely attributable to a larger increase in $\Delta S^{\dagger}$ for the chelated ester hydrolysis (88, 93), and is presumed to be related to the factors responsible for stabilization of the tetrahedral adduct.

The alkaline hydrolysis of a series of N,O-chelated glycine amides, glycylglycine and glycylglycine esters, $[N_4CoNH_2CH_2CONRR']^{3+}$ [N_4 = trien or $(en)_2$], has also been studied (12), and the rate data were found to be consistent with the mechanism shown in Scheme 14 for $[(en)_2CoNH_2CH_2CONHR]^{3+}$. With the exception of the dimethyl amide complex, which could not deprotonate and so showed a reaction first order in $[OH^-]$ even at high pH, the rate profiles (Fig. 3) showed a first-order dependence on $[OH^-]$ to pH ≈ 10, but curved off depending on pK_a's of 11–12 at higher pH as the unreactive chelated imide predominated. The limiting rate constants were in each case 0.01 to 0.05 sec^{-1} at 25°C in spite of the differences in pK_a'. The constancy of the limiting rates derives from the complementarity of k_{OH} and pK_a' predicted earlier by arguments based on Lewis acidity—increased polarization of the carbonyl leads to an increased value of k_{OH} which is offset by a corresponding decrease in pK_a of the coordinated amide. Coordination through the carbonyl oxygen in metal ion-assisted amide hydrolysis therefore will not produce substantial further enhancement of the limiting rate at high pH without an additional cooperative means of enforcing protonation of the leaving group in the transition state. A general acid catalyst at the active site of an enzyme could fulfill this role, as discussed in detail above in consideration of models for urease.

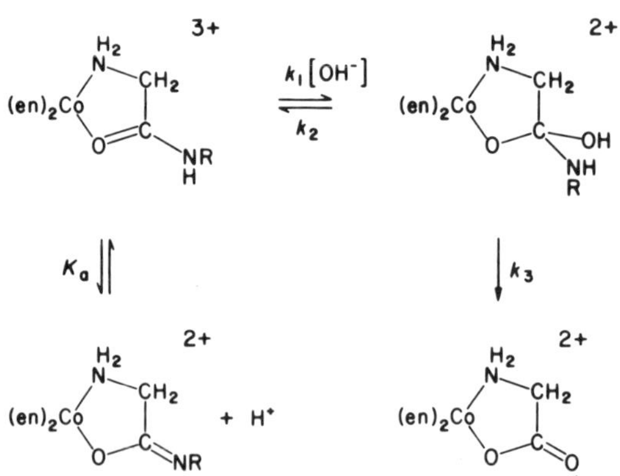

$$k_{OH} = k_1k_3/(k_2 + k_3)$$

Scheme 14

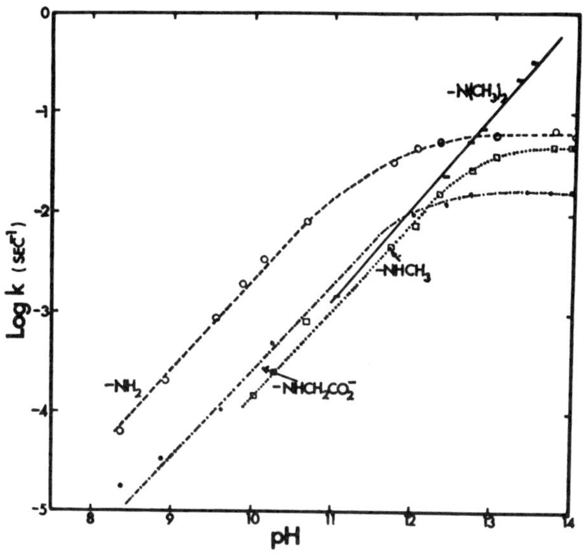

Figure 3 pH profile of first-order rate constants observed in the base hydrolysis of $[Co(en)_2GlyR]^{3+,2+}$ ions [R = NH_2, $NHCH_3$, $N(CH_3)_2$, $NHCH_2CO_2^-$]; 25°C, μ = 1.0*M* (12).

The chelated glycine amides and peptides hydrolyze with second-order rate constants about 10^4- to 10^6-fold less than those for the analogous chelated esters; but such is also true for the respective uncoordinated substrates (12). Carbonyl coordination here, as in the case of similarly coordinated monodentate dimethylformamide (59) (Sec. 4), increases the reactivity of the amide or ester substrate toward nucleophiles by factors of 10^4 to 10^6; and this range provides a reasonable estimate of the contribution that carbonyl coordination to a metal center might make toward the efficiency of a reaction at the active site of a metalloenzyme.

5.2 Intramolecular Hydrolysis by Coordinated Nucleophiles

In addition to their Hg^{2+}-induced hydrolysis described above and shown to occur by way of the N,O-chelated ester under acidic conditions, the complex ions of the type *cis*-$[N_4CoX(NH_2CH_2CO_2R)]^{2+}$ [N_4 = $(en)_2$; X = Cl,Br; R = CH_3,CH_2CH_3,$CH(CH_3)_2$,$(CH_2)_3CH_3$,$C(CH_3)_3$ (94); or N_4 = trien; X = Cl; R = CH_2CH_3 (95)] also undergo hydrolysis under basic conditions. In the bis(ethylene diamine) systems, numerous minor side products were observed, but the major product was $[(en)_2\overline{CoNH_2CH_2CO_2}]^{2+}$ which appeared to be produced very rapidly following rate-limiting loss of X^- from the coordination sphere, and concurrently with the production of ROH [R = $CH(CH_3)_2$,

X = Br] (94). The loss of halide occurs via the usual dissociative conjugate-base pathway and leads to the production of a five-coordinate intermediate (14, 60) which either captures the ester carbonyl (Scheme 15, path A) or solvent (path B).

The carbonyl-coordinated species base hydrolyzes as described earlier to product chelated glycine in which the coordinated oxygen is derived from the ester carbonyl (Scheme 12). The two carboxylate oxygens in the glycinato

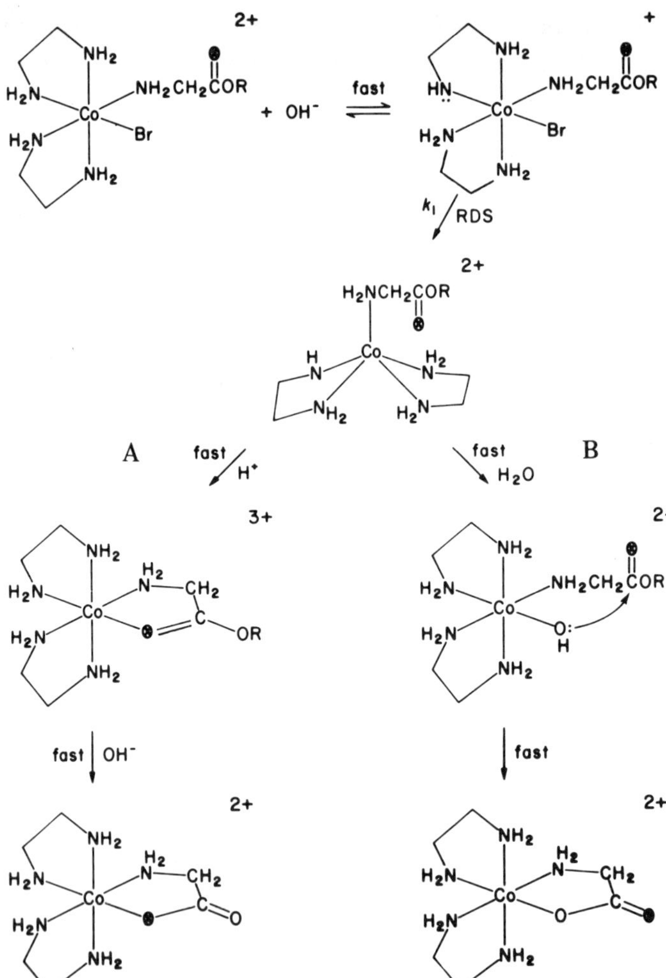

Scheme 15

complex are distinguishable by their exchange rates (87), and ^{18}O studies show that only ~50% of the single atom of ^{18}O incorporated from solvent into $[(en)_2\overline{CoNH_2CH_2CO_2}]^{2+}$ derived from *cis*-$[(en)_2CoBr(NH_2CH_2CO_2CH(CH_3)_2)]^{2+}$ resided in the *exo* (uncoordinated) position (Scheme 15). The remainder, in the more slowly exchanging bridging position, could only have derived from *cis*-coordinated OH^- incorporated by addition of solvent to the five-coordinate intermediate (Scheme 15) (94). This work suggested for the first time that coordinated OH^- could act as a very efficient intramolecular nucleophile, and showed this to be a viable path for ester and amide hydrolysis at the active sites of metalloenzymes. The reaction of the β_2-$[(trien)CoCl(NH_2CH_2CO_2CH_2CH_3)]^{2+}$ ion in base occurred ~3000 times more rapidly and produced chelated glycine in much better yield. Once again the rate of removal of halide appeared to limit the overall rate, and ^{18}O-labeling established a portioning between paths A (84%) and B (16%) (95).

Similar ^{18}O-tracer studies on $[(en)_2\overline{CoNH_2CH_2CO_2}]^{2+}$ produced in the base hydrolysis of the monodentate glycineamide complex $[(en)_2CoBr(NH_2CH_2CONH_2)]^{2+}$ revealed a similar distribution (~1:1) of label between the products of the two mechanisms analogous to A and B (Scheme 15), but in this case the two paths were chemically separable since the chelated glycineamide complex was relatively stable to further hydrolysis (96, 97). As with the esters, the only process observable kinetically was the loss of halide. The faster subsequent intramolecular reaction was estimated to occur at least 10^7 times faster than the hydrolysis of uncoordinated glycineamide at pH 9 (96). This estimate was refined following isolation in solution of the intermediate *cis*-hydroxoamide complexes from the *cis*-bromo ions after HOCl oxidation or controlled base hydrolysis, and separate study of the ensuing intramolecular reactions (98, 99). The pH dependences of the reactions were not simple, and have been interpreted in terms of the mechanism of Scheme 16 involving a tetrahedral adduct T, for which the addition reactions (k_2, k_2', and k_3') are rate determining, and the k_2' reaction is general base catalyzed. Equilibria among the various prototropic forms of T are established rapidly, but are influenced by the buffers used. Since Co—O^- and R—NH^- are poor leaving groups even in comparison with HO^-, the amine must be protonated ($T\pm$ or $T+$) in order to leave from the intermediate to give amide hydrolysis. In the form T, the best leaving group is —OH, and the collapse of T leads to formation of the chelated amide (99).

A further interesting aspect of this work is the greater apparent reactivity of the *cis*-aqua complex in comparison with the *cis*-hydroxo species (98, 99). Since M-OH_2^{n+} ought to be a poorer nucleophile than H_2O (but better than H_3^+O!), it may not be the predominant form A but the kinetically equivalent species B which is reactive.

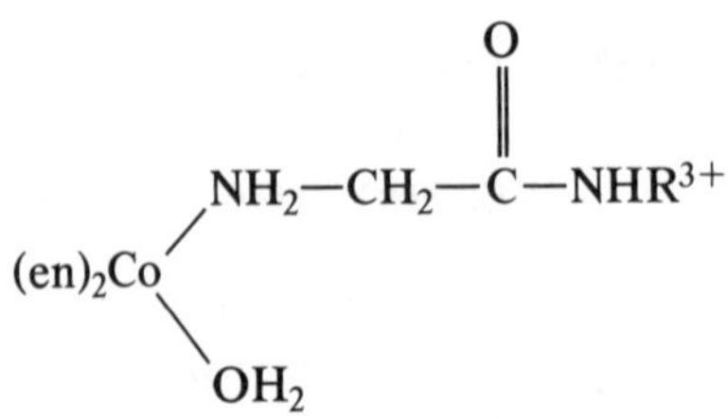

A

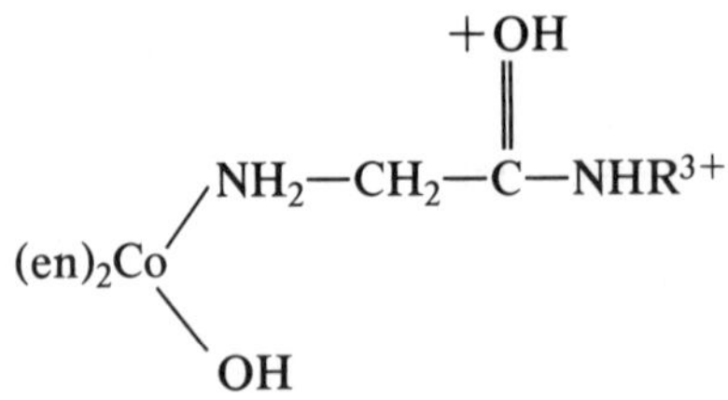

B

Boreham et al. (99) have argued an alternative possibility involving a general acid-general base mediated proton transfer from $Co-OH_2$ to the amide nitrogen in the transition state (proton tunnelling). Such a mechanism is supported by the lack of a kinetic H/D isotope effect in the cyclization reaction, and its catalysis by acetate ion, but the arguments are not compelling.

For a path involving B (whose concentration must be minute) to be significant in comparison with *cis*-hydroxide attack on the unprotonated monoden-

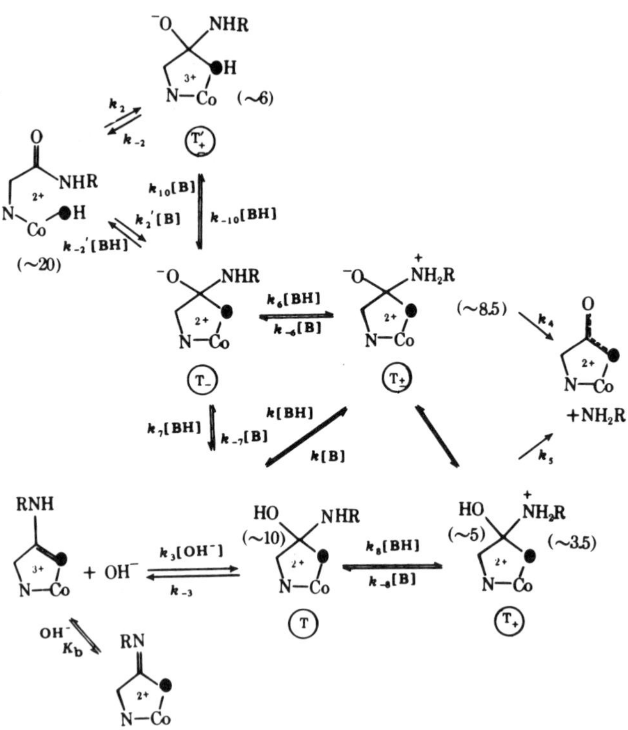

Scheme 16

tate amide (Scheme 16) at pH < 5, it would necessarily be a very efficient process [$k \approx 10^8\ sec^{-1}$, calculated by Boreham et al. (99)]. The use of carbonyl coordination (or enforced protonation by a nearby general acid catalyst) in conjunction with coordinated OH^- as a nucleophile would model such a reactive species (cf. B). It would provide a very attractive mechanism for an enzymatic reaction and could produce enormous rate accelerations (cf. urease, Sec. 4). In the present cases, the intramolecular paths involving coordinated OH_2 (formally) and OH lead to specific rates 10^{11}- and 10^7-fold greater than the rates of hydrolysis of free glycylglycine at pH 5 and 8, respectively, and a further acceleration of the hydroxide reaction is observed in the presence of buffer catalysts (99).

A further instance of an intramolecular nucleophilic addition was observed in the base hydrolysis of the $[(NH_3)_5CoNH_2CH_2COOC_2H_5]^{3+}$ ion (100), for which the rate law has the form $k_{obs} = k_1[OH^-] + k_2[OH^-]^2$. The k_1 path represents the hydroxide ion-promoted hydrolysis of the monodentate ester to $[(NH_3)_5CoNH_2CH_2COO^-]^{2+}$, while the k_2 pathway produces N,N'-chelated glycineimide by nucleophilic attack of a *cis*-aminate ion at the carbonyl carbon (Scheme 17). At pH < pK_1 (pK_1 ~ 15–16) such reactions have a first-order dependence on $[OH^-]$. In the present instance, the term in

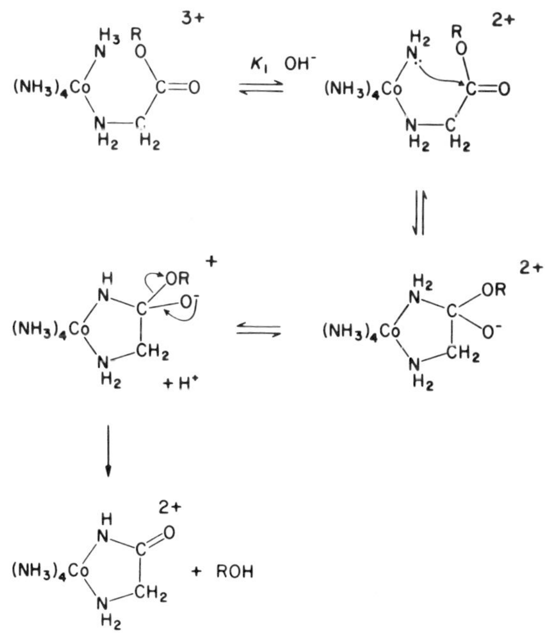

Scheme 17

$[OH^-]^2$ arises from the additional requirement to deprotonate the tetrahedral intermediate so that RO^- can leave in preference to $M—NH_2—^{(n-1)+}$.

In summary, intramolecular hydrolysis of N-monodentate glycine derivatives by *cis*-coordinated OH^- at Co(III) centers occurs $>10^7$-fold more rapidly than the intramolecular base hydrolysis of the free substrates at near-neutral pH, and deprotonated coordinated ammonia (and other amines) is also a good intramolecular nucleophile.

Wells and Bruice (74) have presented a detailed study of a reaction believed to involve Ni^{2+}- or Co^{2+}-coordinated OH^- as nucleophile in an intramolecular ester hydrolysis. The reactions occurred 10^3- to 10^5-fold more rapidly than reaction of the uncomplexed substrate with OH^- at pH 7–8, and demonstrate that chemistry similar to that in the Co(III) systems may be used

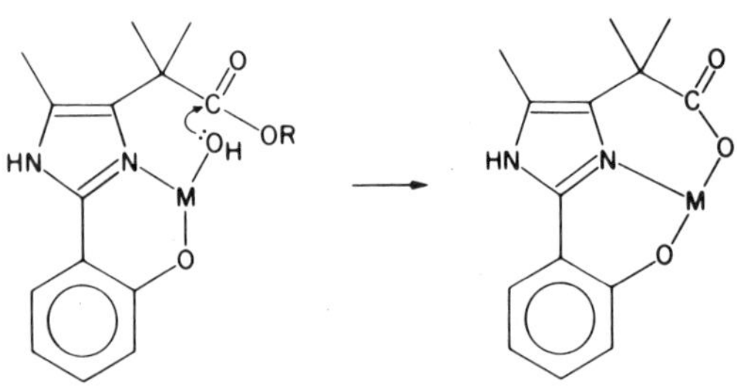

M = Ni^{2+}, Co^{2+}

to similar advantage with the labile metal ions. The reactions proceed as expected through a preequilibrium formation of the tetrahedral adduct. This intermediate is believed to collapse to produce initially a coordinated carboxylato species ("carboxyl–metal anhydride") which is susceptible to trapping by good nucleophiles. The chemical structure of this species is uncertain, although Bruice has presented some uncompelling speculations (74). It is clear from the Co(III) chemistry that the coordinated carboxylate in the final product ought to be essentially unreactive toward nucleophiles (101), as would be predicted for reactivity intermediate between that of $R—COO^-$ and R—COOH (Sec. 1).

5.3 Intermolecular Hydrolysis of N-Monodentate Derivatives

The second-order rate constant for the formation of the N-monodentate glycinate complex during the base hydrolysis of the [ethyl glycinatopentaammine-Co(III)]$^{3+}$ ion ($\sim 50M^{-1}$ sec^{-1}) (100) is only twice that of the ethyl glycinium ion ($k_{OH} = 24M^{-1}$ sec^{-1}) (67), suggesting that the effect of the nearby $(NH_3)_5Co^{3+}$ moiety on reactivity is comparable to that of a proton when each is coordinated to the amino group. The rate enhancement over that of free

unprotonated ethyl glycinate ($k_{OH} = 0.58M^{-1}\ sec^{-1}$) (67) is less than 100-fold.

$$[(NH_3)_5\,Co{-}NH_2{-}CH_2{-}\overset{\overset{\displaystyle O}{\|}}{C}{-}OEt]^{3+} \xrightarrow{OH^-} [(NH_3)_5\,Co{-}NH_2{-}CH_2{-}COO^-]^{2+} + EtOH$$

5.4 Coordination of the Leaving Group

For reasons of inaccessibility of suitable substrate complexes for study of reactions at sufficiently high $[OH^-]$, there are no extant examples of increased reactivity of Co(III) complexes of N-coordinated amides or alkoxy-O-coordinated esters in base hydrolysis. The protonated N-coordinated amides are accessible at low pH but are insufficiently reactive under these conditions to enable hydrolysis to be observed (see Sec. 4); reactions with nucleophiles other than OH^- and H_2O have yet to be studied.

Fife and co-workers (102–104) have examined the effects of divalent (labile) metal ions on the hydrolysis of amides and esters constrained as part of large chelating ligands which also (in some instances) incorporated a dangling carboxyl group placed as a potential general base or nucleophile. The enhancements of second-order rate constants as large as 10^9-fold which were observed in the metal ion-promoted OH^--catalyzed reactions at ester and amide carbonyl groups (102–104) derived from mechanisms akin to those considered above (i.e., carboxyl coordination and/or use of coordinated OH^- as an intramolecular nucleophile).

In more recent work from the same group, divalent metal ions (Co^{2+}, Ni^{2+}, Zn^{2+}) have been shown to be very effective catalysts of the hydrolysis of acetals (105) and esters (106) of the types A and B, respectively.

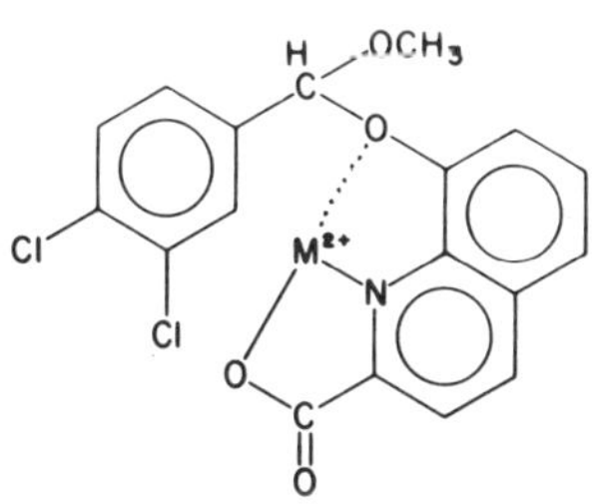

A

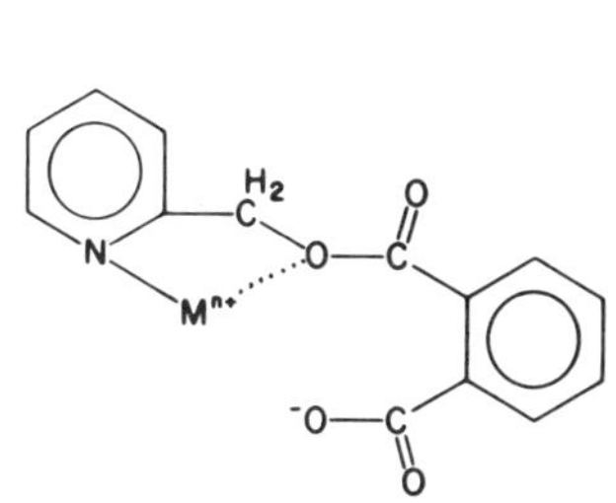

B

In each case, Fife has interpreted the results in terms of the structures shown in which the metal ion coordinates the leaving group (or, at least, stabilizes the developing negative charge on the alkoxide oxygen in the transition state). He has described the processes as "transition-state effects" to take account of such mechanisms of catalysis. Unfortunately, the mechanisms rest on arguments which are not always compelling, for example, carbonyl coordination in B might equally well be consistent with the data (c.f. ref. 106), and the lability of the systems often renders structural inferences equivocal. Unfortunately, the systems are not readily amenable to study using Co(III) since the alkoxide oxygens of the substrates are likely to be exceedingly poor donors and would as such be readily displaced by H_2O in any but the most constrained chelate systems.

Thus, models that establish unequivocally the efficacy of coordination of the leaving groups in ester and amide hydrolyses are yet to be forthcoming. Such mechanisms are nonetheless soundly based on considerations of the Lewis acidity of the metal ions, and their possible relevance to the chemistry of the zinc-dependent peptidases and esterases has been recognized (2, 106). The foregoing studies allow some evaluation as to the relative likelihood of mechanisms 1–4 operating in the divalent metal ion-catalyzed hydrolyses of amino acid esters and amides, and some speculation on possible modes of activation in the Zn^{2+}-dependent peptidases.

Mechanism 3, intermolecular hydrolysis of the N-monodentate complex, is excluded for model systems. The $(NH_3)_5Co^{3+}$ moiety attached to the amino nitrogen atom of ethyl glycinate is little more effective than H^+, similarly bound, in promoting its hydrolysis, and the enhancement of the second-order rate constant over that for the unprotonated ester is less than 10^2.

Both mechanisms 1 (carbonyl activation) and 2 (coordinated OH^- as nucleophile) result in substantial activations of esters and amide substrates toward hydrolysis, and either could account for model divalent metal ion catalysis. The 10^4- to 10^6-fold activation (as assessed by comparisons of second-order rate constants) derived as a result of carbonyl coordination is alone insufficient to account for the efficiency of a typical metallohydrolase (perhaps 10^9–10^{14}), but the ability of the enzyme to deliver intramolecular nucleophiles or general acid/base catalysts at its active site could account for much of the disparity. The models using coordinated OH^- as an intramolecular nucleophile produce specific increases in rate near neutral pH of at least 10^7-fold (possibly as much as 10^{12}-fold). However, the models already account for the intramolecularity of the reaction, and allow the enzymes much less scope in further activating the systems.

Mechanism 4, in which the metal ion assists the leaving group to depart in a concerted reaction, is yet to be thoroughly assessed in model systems. The

work by Fife and co-workers (105, 106) suggests that the rate enhancements derived by this mechanism in appropriate systems may be no less substantial.

Thus, while the Co(III) chemistry confirms that mechanisms akin to 1 and 2 [and possibly (4)] would be chemically reasonable models for events at the active site of carboxypeptidase A, they do not provide a basis for distinguishing between the possibilities. All three mechanisms have been argued for the enzyme system at various times, and it may indeed be that the particular mechanism varies depending on whether the substrate is an ester or an amide (42, 107, 108). In recent studies at low temperature, an acyl enzyme was identified during the hydrolysis of an ester substrate, and shown by trapping experiments to be a mixed anhydride resulting from attack of Glu-270 at the carbonyl of the substrate (107). The role of the zinc ion in such a mechanism could be to polarize the carbonyl of the coordinated substrate, a situation modeled closely in the CH_3COO^--catalyzed hydrolysis of the $[(en)_2CoNH_2CH_2COOCH(CH_3)_2]^{3+}$ ion (90).

More recently (45), there has been a suggestion, also based on low-temperature studies, that the role of the zinc ion during hydrolysis of the same ester substrate is to deliver coordinated hydroxide as nucleophile in the hydrolysis of the acyl-enzyme (anhydride). Such a process has been modeled in reactions of Co(III)-coordinated hydroxide ion described above and in Section 3 (109).

Wherever these and other studies on the enzyme systems lead toward defining the mechanisms responsible for catalysis, it is clear that the model chemistry is playing a significant role in defining just which processes are chemically reasonable, and just how much activation these processes are capable of producing.

6 METAL ION-PROMOTED REACTIONS INVOLVING OLEFINS—ACONITASE, AMMONIA LYASES

Nucleophilic additions at olefins are not usually accomplished all that easily in organic chemistry. Often fairly stringent conditions are required such as high concentrations of a strong base and/or high temperatures (110). This situation is in contrast with a variety of biological reactions involving olefins that are rapidly hydrated or aminated in near neutral conditions and ambient temperatures. Some of these enzymatic systems involve metal ions and it is therefore of some interest to consider the role of the metal ion in such cases. Understanding this role not only elucidates the biological reaction but it may lead to procedures for nonenzymatic catalysis of olefin chemistry.

One such enzyme is aconitase or aconitic acid hydratase which catalyses the isomerisation of citric acid to isocitric acid (111) through the intermediate *cis*-aconitic acid (Scheme 18).

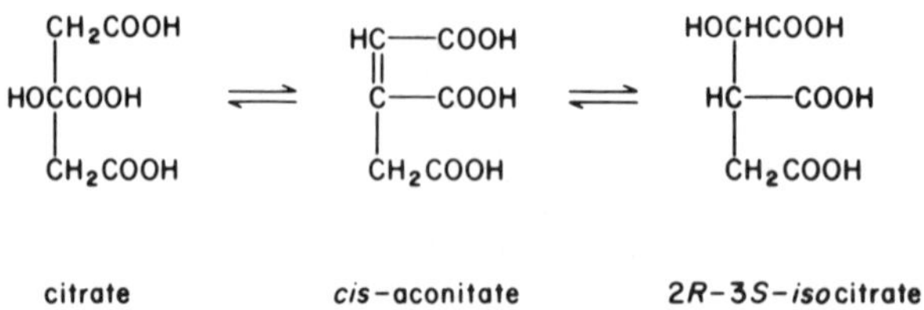

Scheme 18

The enzyme is part of the respiratory cycle; isocitric acid leads on to α-ketoglutarate while citric acid is available for fatty acid synthesis and feed-back regulation of glycolysis. The enzyme is activated by Fe^{2+} usually administered with a reducing agent such as cysteine or ascorbic acid. It has a molecular weight of ~90,000 and ~3 Fe atoms per molecule (111). An iron sulfur cluster (112) is implicated in the enzyme but its structure is not yet clear. NMR studies (113) have placed the substrates in the vicinity of the metal ion (within 4 Å) and it is presumed that they are coordinated to an Fe center. No sequence is available at the time of writing but several elegant studies (114–116) have been carried out to determine some stereochemical and mechanistic aspects of the substrate reactions. Of the two carboxymethyl groups in citric acid, only that derived from oxaloacetate is involved in the conversion to *cis*-aconitic and isocitric acid, namely the *pro-R* acetate residue. Moreover, only one proton of the methylene group is involved and only the 2*R*,3*S* isomer of isocitrate is a substrate. *Trans*-aconitic acid is a competitive inhibitor for this enzyme. Rose and O'Connell (116) showed that the proton abstracted from C(2) of citric acid, presumably intramolecularly by a base on the enzyme, is retained by the enzyme and restored to the C(3) atom of isocitrate without exchange with solvent protons. The oxygen atom involved in the hydration, however, does exchange with solvent oxygen. *Trans* addition of H^+ and OH^- stereospecifically to *cis*-aconitic acid accommodates the conversion of the prochiral *R* acetate group to 2(*R*),3(*S*)-isocitric acid (Fig. 4).

The body of evidence indicates that the general enzyme chemistry is essentially acid–base chemistry and not redox chemistry, even though an iron–sulfur cluster of some type appears to be involved. A number of mechanistic proposals involving Fe(II) as an active center have been offered to accommodate the results described above (111, 113). They are based on structures of metal-citrate complexes and NMR studies of Fe(II)- and Mn(II)-aconitase. All involve binding of citrate and isocitrate in a tridentate manner, and aconitate through the two carboxyl groups attached to the olefin. An attractive feature of the proposals is their use of a coordinated OH^- ion or H_2O to add to the olefin system when the olefin is *cis* to the bound nucleophile. If the proposal is valid, then there is a prospect of observing such reactions in model systems. The models

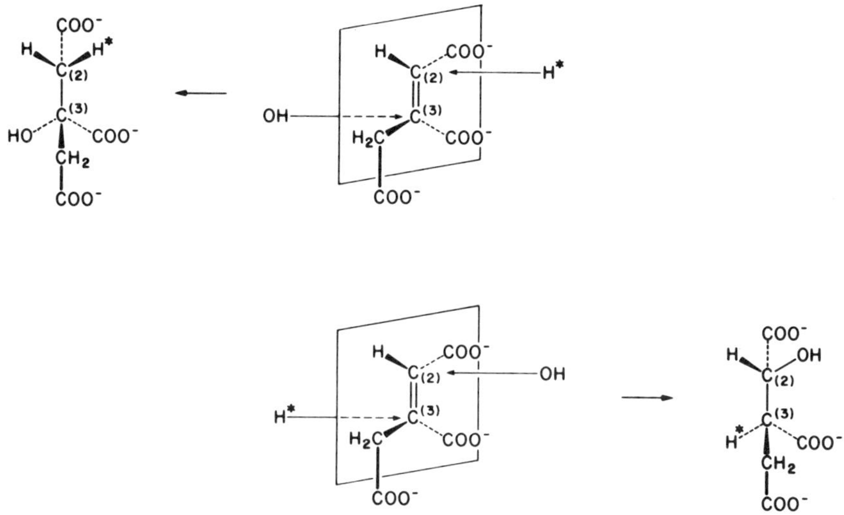

Figure 4 Stereochemistry of the conversion of *cis*-aconitate to citrate and isocitrate by aconitase.

could also indicate the efficiency of doing the chemistry in this way, whether there is a five- or six-membered chelate formed, the stereochemistry of the addition, and the electronic factors which govern the cyclization.

6.1 Hydration

Some simple carboxylato-olefin complexes have been constructed to examine these questions using the thesis that the metal ion holds the substrate and nucleophile together for an intramolecular reaction and that redox chemistry is not involved.

The simple systems chosen (117) are depicted in Figure 5. They are Co(III)-1,2-ethanediamine complexes of maleic and fumaric acid half esters containing a water molecule *cis* to the bound olefin substrate.

The Co(III) complexes were chosen because of the inertness of the ligands to dissociation and because bound OH^- ion can be generated near neutral pH. In this way the reaction of the ligand can be observed free of complicating ligand dissociation processes and the structure of the reactive species may be determined. The monoesters have been exploited since at neutral pH the free acids would exist with a dangling anionic $—COO^-$ group which would deactivate the olefin to nucleophilic attack. In the enzyme the carboxyl centers, if not coordinated, must be either H-bonded to some relatively acidic site or protonated in order to have the olefin sufficiently activated.

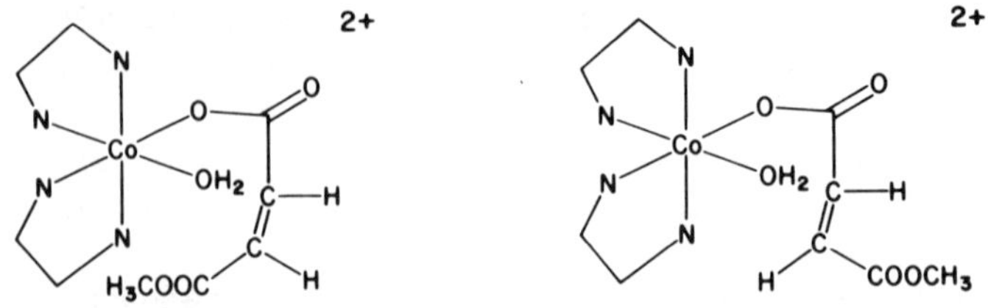

methyl maleate methyl fumarate

Figure 5 *Cis*-aqua bis (ethylenediamine)-Co(III) complexes of methyl maleate and methyl fumarate.

With the maleic acid complex, generation of OH^- bound to the Co(III) center leads to a rapid cyclization to give the chelated malate half ester in two diastereoisomeric forms (Scheme 19). The crystal structure of the ΛR isomer has been determined (118). In $0.2M$ imidazole buffer at pH 8 the rate constant for cyclization is $\sim 0.1\ \text{sec}^{-1}$ (25°C), which is more than 10^7-fold greater than that for hydration of the uncoordinated half ester under the same conditions. The cyclization gives exclusively the five-membered chelate. The rate is independent of pH in the range 8–11; thereafter it becomes first order in OH^-. Up to pH 8, the rate is also first order in OH^-, and this arises as a consequence of the degree of deprotonation of the coordinated water molecule (Sec. 3).

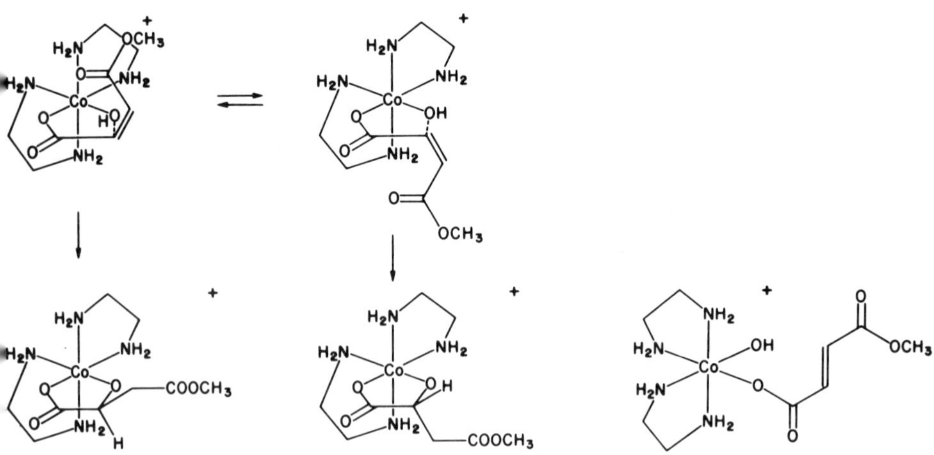

ΛS ΛR

Scheme 19

The two diastereoisomers arise from the orientation of the olefin plane relative to the coordinated OH^- nucleophile in relation to the chiral $Co(en)_2$ center (Scheme 19). In imidazole buffers and under pH-stat conditions at pH 8–11, the ratio of diastereoisomers is $\Lambda R/\Lambda S \sim 2$, but at high $[OH^-]$ this increases to 9. Tracer studies also show that bound $^{18}OH^-$ is retained in the chelated malate.

The other especially interesting features of the cyclization are its dependence on the buffer and buffer concentration in the pH range 8–11 and the appearance of the isomerized reactant, the *cis*-$[Co(en)_2(OH)$methylfumarato$]^+$ ion under pH-stat conditions (76%) and at low buffer concentrations. The reaction is clearly general acid catalyzed (Fig. 6). Under conditions where the coordinated hydroxide ion remains deprotonated, the best catalysts are the strongest acids. Increase in the concentration of imidazolium ion at constant pH leads not only to an increase in the rate of cyclization, but also to a decrease in the proportion of the fumarato product. It should be mentioned at this point that the *cis*-$[Co(en)_2(OH)$methylfumarato$]^{2+}$ ion reacts about 1000-fold more slowly than the maleato analog and is thus a stable product under the conditions of the experiment.

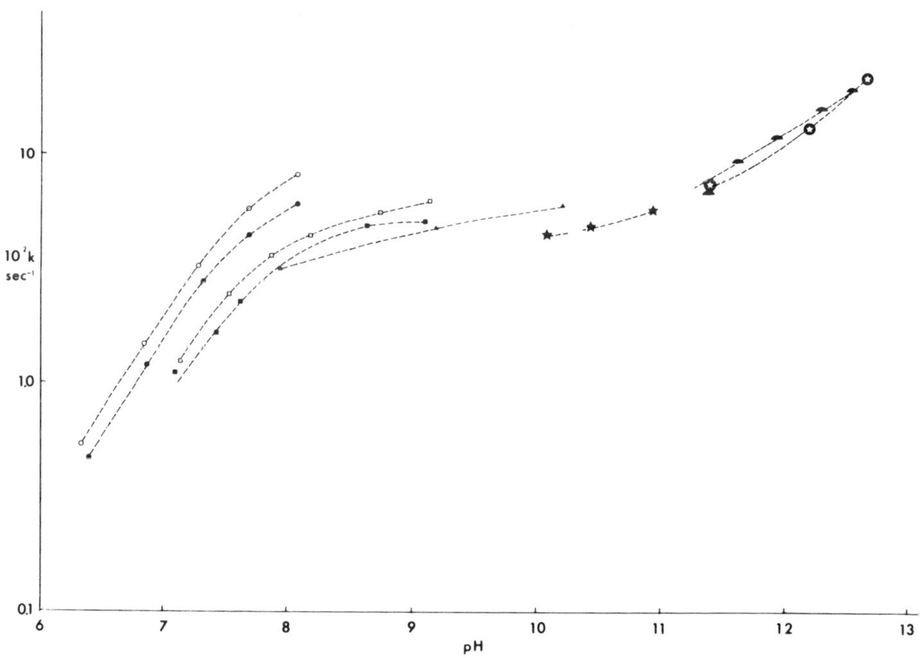

Figure 6 Intramolecular hydration of *cis*-$[Co(en)_2(H_2O)$ methyl maleato$]^{2+}$ ion at 25°C, μ = 1.0*M*. Imidazole: 0.05*M* (□); 0.1*M* (●), 0.2*M* (○), Diethanolamine: 0.2*M* (▲); pH Stat (■): Diethylamine: 0.05*M* (★) OH^- (**o**), Guanidine: 0.05*M* (⬬).

A mechanism consistent with these results is depicted in Scheme 20. In the range where the general acids are active (pH 8–11), the rate-determining step is protonation of the developing carbanion. Addition of the bound OH^- is faster than protonation, and rupture of the chelate is competitive with protonation at low BH^+ concentrations.

After OH^- adds to the olefin, rotation about the olefin C—C bond ensues. Loss of Co(III)-OH then leads to regeneration of starting material and also the production of the fumarate isomer. As argued previously, the cationic Co(III)-OH group is a better leaving group than is OH^- (Secs. 4 and 5). As BH^+ increases in concentration, the proportion of fumarate formed diminishes as protonation of the carbanion becomes more efficient.

The high base region clearly involves a different path with more compression in the transition state for addition of the bound nucleophile and consequently more specificity in the diastereoisomer distribution. This is an interesting mechanistic problem in its own right but not especially relevant here.

The implications from the models so far are that very efficient paths for hydration of olefins using intramolecular nucleophiles exist, and lead to the exclu-

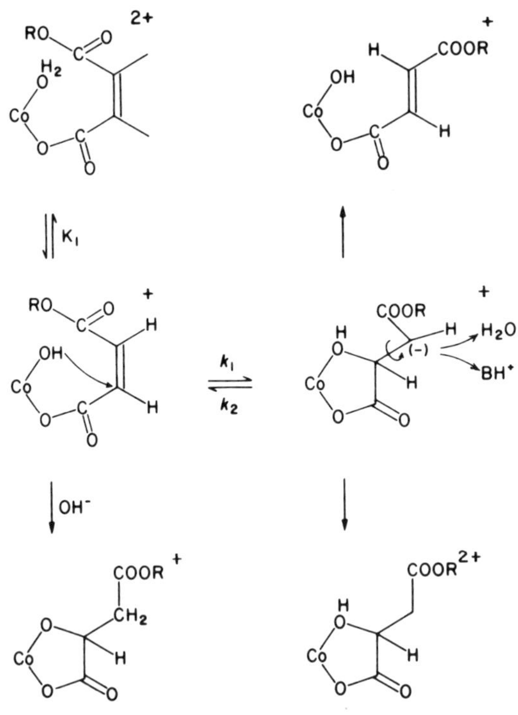

Scheme 20

sive formation of five-membered ring systems. The details of the mechanism are especially germane to features observed in the enzyme. The dechelation before protonation could mimic closely events in the enzyme after deprotonation of the citric acid at the methylene group by the intramolecular base (Scheme 20). At this point the bound OH^- can leave the carbanion to generate the olefin. Clearly that process occurs relatively rapidly (less than seconds).

The metal ion-promoted reactions occur very efficiently; the rate constant for cyclization is within two orders of magnitude of the turnover number of aconitase.

The electronic effects for the cyclization are interesting. Superficially, there is an expectation that coordinated acrylic acid would hydrate since it has an ester-like quality once coordinated and nucleophilic addition at the β-C atom could be expected to be so activated. Experimentally hydration is not observed by addition of a coordinated *cis*-OH^- ion (Scheme 21). However, both the 2- and 3-chloroacrylato complexes are hydrated by this intramolecular process (Scheme 21). The latter gives a five-membered chelate while the former gives a six-membered chelate. Although both form more slowly than the malate complex, the formation of the six-membered chelate ring is not forbidden. In fact, it is formed more rapidly than the five-mem-

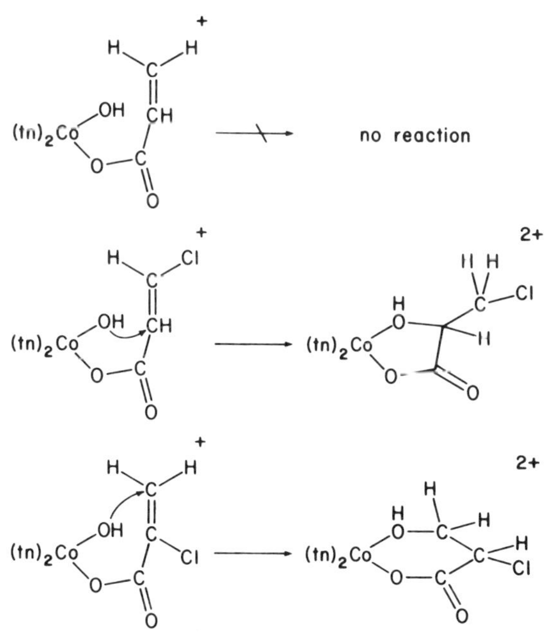

Scheme 21

bered chelate (~10-fold) and the reason for this is not clear at present. It appears from these examples that the cyclization is governed by an activating group *exo* to the ring and that a coordinated carboxylate ion alone is ineffective in promoting addition of a coordinated nucleophile to the β-carbon of an olefin.

The reason proffered for the lack of effect of the coordinated carboxylate ion is the stereochemistry required for the activated complex during the addition of the bound OH^-.

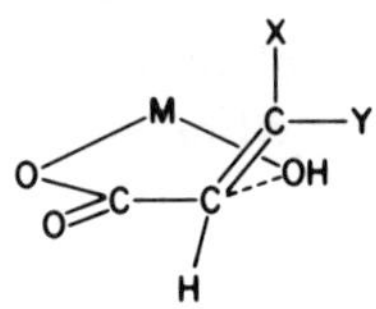

Essentially, the plane of the olefin moiety and that of the coordinated carboxylate moiety have to be orthogonal. This minimizes the interaction between their π systems and consequently minimizes the activating effect of the metal ion–carboxylato entity.

6.2 Aconitase Mechanism

The Glusker (111) proposals are depicted in Scheme 22. The enzyme abstracts a proton with a base from C(2) leaving the carbanion. This allows the

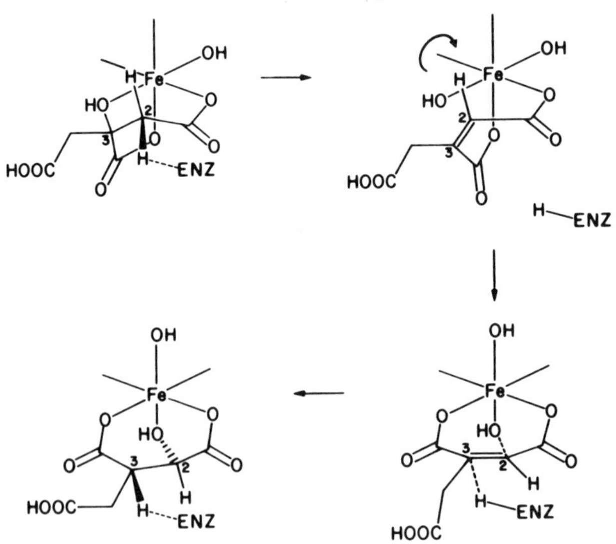

Scheme 22

coordinated OH^- to leave C(3) and generate *cis*-aconitate, coordinated in a bidentate manner. Following rotation about one of the Fe–ligand axes (shown) and a conformational change in the coordinated aconitate, another coordinated OH^- is added now at C(2). The enzyme then replaces the *original proton* on C(3) to give isocitric acid. In this process the OH^- removed from one face of the "olefin" is complemented by the addition of another OH^- at the opposite face.

The results from the models would require that in the chelated aconitic acid one coordinated carboxylate would need to remain almost coplanar with the olefin plane to assist addition of the nucleophile. Later we will see that this is a feasible process since chelated maleate can react in this manner. The conformational change for the olefin seems feasible and is likely to be fast enough. However, there must also be a linked enzyme conformational change which places the protonated base near the appropriate C atom.

An alternative mechanism that incorporates the implications of the models, the known biochemistry, the stereochemistry, and accommodates the orientation of the intramolecular base is shown in Scheme 23. It requires only monodentate coordination of the aconitic acid; different carboxyl groups are bound to the metal ion to yield the two diastereoisomeric products. The normally rapid exchange of ligands at Fe^{2+} ion would accommodate this requirement. Otherwise the process is very simple, involving the established rapid five-membered chelate formation. The proposal also meets the stereochemical requirement that OH^- add to the different faces of the olefin to produce

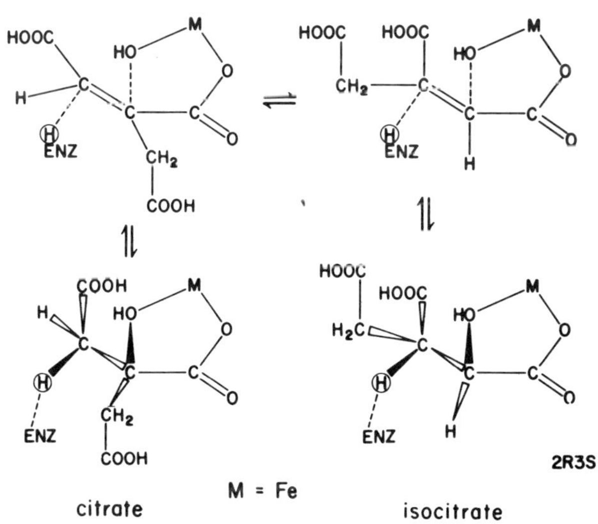

Scheme 23

the two diastereoisomers. In each instance there is a —COOH group in an appropriate position to activate addition of the bound OH^- at the β-carbon atom of the olefin. Moreover, one attractive feature of the proposal is that the orientation required of the intramolecular acid to protonate the carbanion is the same for the two orientations of the bound aconitic acid, that is, no conformational change in the enzyme is required to reorient the acid for the protonation step.

The role of the presumed Fe—S cluster (112) in the enzyme still needs elaboration. Possibly the enzyme is turned on and off by a redox process while the hydration is carried out by an Fe^{2+} ion not in the cluster. Possibly both processes are integrated and an Fe atom in the redox-active cluster also provides sites to bind the substrate and the nucleophile. These aspects are obviously speculative but at present the chemistry observed for the enzyme is more consistent with acid–base chemistry than with redox and radical chemistry.

6.3 Amination and Deamination

A consequence of the addition of coordinated OH^- to olefins is that other nucleophiles, notably a coordinated aminate ion (Sec. 5) should act analogously. This chemistry was first observed with the maleate ester complex $[(NH_3)_5CoOCOCH{=}CH{-}COO(t{-}Bu)]^{2+}$ in aqueous base (Scheme 24)

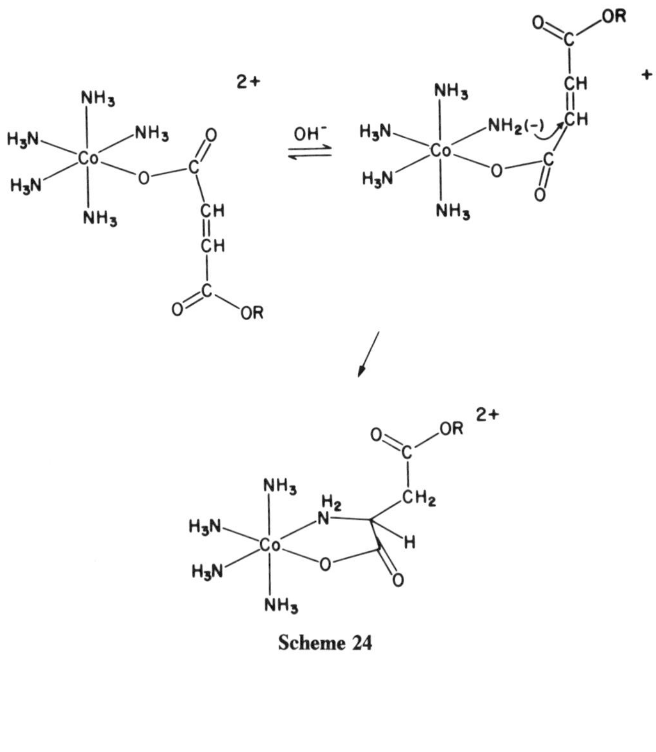

Scheme 24

(119a), and has been recently exploited in the synthesis of β-carboxyaspartic acid (119b).

The reaction of the *t*-butyl maleate complex occurs to the extent of ~50% and is complicated by hydrolysis of the maleate ester and some decomposition of the Co(III) complexes involved. Analogous systems, however, show more resistance to decomposition and display the intramolecular cyclization completely (Scheme 25) (120). NMR studies indicate that the reaction takes place as shown. Regiospecific attack by the deprotonated ethylenediamine gives two diastereoisomeric products containing *N*-2-aminoethyl-(*R*) or (*S*)-aspartic acid methyl ester in a 4:1 ratio. No attack by the deprotonated ammonia was observed and only one of the three possible ethylenediamine sites was effective. The second-order rate constant was $\sim 0.7M^{-1}\ \text{sec}^{-1}$ at 25°C. After the pK_a of the coordinated ethylenediamine (~16) is accounted for, the first-order rate constant for the attack of the coordinated deprotonated amine at the olefin site is estimated to be $\sim 70\ \text{sec}^{-1}$ at 25°C.

No general base effects were observed with the aminate nucleophiles and it is presumed that the rate-determining step here is attack by the nucleophile and not protonation of the carbanion. The ring closure certainly yields a five-membered chelate. The ultimate product of the intramolecular reaction shown in Scheme 24 after hydrolysis of the ester is the same as that synthesized from $[Co(NH_3)_4(OH)(OH_2)]^{2+}$ and authentic aspartic acid. Syntheses of this latter kind are known to give five-membered chelates of the type claimed for the intramolecular chelations.

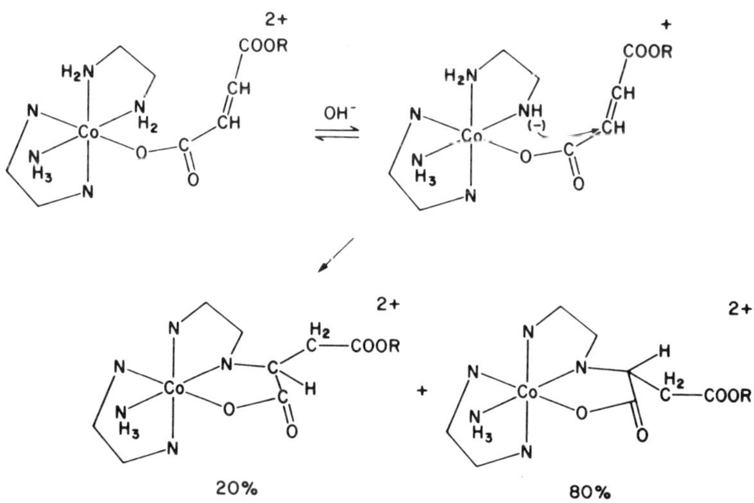

Scheme 25

The amination process is the reverse of the reactions catalyzed by several enzymes, notably the aspartate (121), 3-methylaspartate (121, 122), histidine (121, 123, 124), and phenylalanine (121, 125) ammonia lyases. These enzymatic reactions are not dependent on pyridoxal phosphate, and may involve transition-metal ions (2).

A role for a metal ion follows from the chemistry just described provided the enzyme supplies an intramolecular base for the removal of a proton. The proposal is outlined in Scheme 26 typified by deamination of (*S*),(*S*)-3-methylaspartate. Chelation of the amino acid is followed by removal of a proton from the β-carbon atom by an enzyme base. The carbanion so produced allows the coordinated aminate anion to leave readily and generate the olefin (2, 120). In this respect the chemistry is analogous to that observed for coordinated hydroxide as a nucleophile. The M—NH_2^- ion is a much better leaving group than NH_2^- (Secs. 4 and 5).

In the model chemistry only the addition of NH_2^- is observed since this is the rate-determining step (120). The models do not supply the intramolecular base to remove the β-proton at this stage of their development and protonation at the β-carbon is apparently more rapid than addition of NH_2^-. The model compounds have utilized, as substrate, a *cis* olefin which is likely to be more reactive than the *trans* isomer. Even so, the relative increase in rate between uncoordinated and coordinated substrate should be retained and the model compounds may mimic to a degree one direction for the enzymatic process. The enzymatic reactions are reversible; the equilibrium constants for reactions catalyzed by the 3-methylaspartate (0.19, pH 7–9), histidine (3.0, pH 9), and phenylalanine (4.1, pH 8.5) ammonia lyases (121) are not far from unity. Their turnover numbers are ~300, ~200, and ~3 sec^{-1}, respectively,

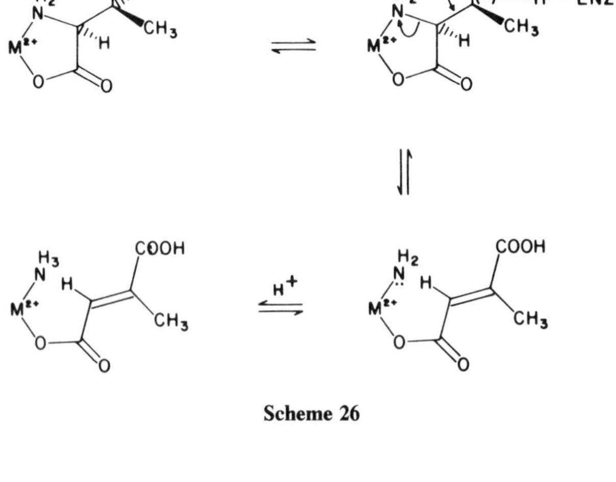

Scheme 26

at 25°C (121). Clearly, the model chemistry occurs at rates not much less than the enzymatic reactions, and a substantial part of the reactivity of the enzymes could be accommodated by such a path. The reverse reaction would rely on an intramolecular base for its efficiency. Once the β-proton is removed, the aminate ion should leave readily in much the same way as OH^- did in the malate chemistry described earlier.

There are several other features of the enzyme systems which are interesting. Ammonia bound to a metal ion clearly has to be deprotonated to be effective as a nucleophile. The pK_a of such a metal–ammonia group is likely to exceed 14 for most divalent metal ions bound to protein-type ligands. An intramolecular base would then need to be available to assist the nucleophile and a degree of concertedness would be implied. However, it has been shown that nitromethane inactivates both histidine (126) and phenylalanine (127) ammonia lyases. Experiments with $[^3H]BH_4^-$ also show inactivation of both enzymes. The results have been interpreted as a requirement for an active-site dehydroalanine moiety, but are also consistent with the involvement of an imine. Chemically this makes good sense because a coordinated imine becomes much more acidic than a coordinated ammonia molecule and access to the nucleophile is much easier.

A system has been constructed to look at this aspect of the chemistry (120), using the Co(III) ion to hold the pieces together (Scheme 27). The reactant consists of an imine moiety (tautomer of dehydroalanine) bound to the Co(III) ion *cis* to a methyl maleate ion. The remaining three positions on the Co(III) are tied by a diethylenetriamine tridentate.

The pK_a of the imine is 9.2 and the rate of addition of the imine N atom to the olefin becomes independent of pH just above this pK_a. The rate constant observed is 77 sec^{-1} at 25°C, which is close to that estimated for attack by the aminate ion in the previous model. The major advantage of using the imine as a nucleophile derives from the dramatic lowering of the pK_a of the ammonia on conversion to an imine. It is conceivable that the enzyme uses the same strategy to gain ready access to a N-nucleophile for the amination process via a rearranged dehydroalanine moiety. At least there is presumptive evidence for such a proposal from inhibition by nitromethane and BH_4^- (126–128). Imine chelates of this type also readily add nitromethane and are reduced to amino acids by BH_4^- (129).

This chemistry has been extended somewhat further to examine the reactivity of maleic acid when both carboxylate ions are coordinated to the same metal ion. The $[(en)_2Co(maleato)]^+$ ion was synthesized in 1921 by Duff (130) and it has the structure (131) shown in Scheme 28.

This complex reacts rapidly in base ($\sim 0.1 M^{-1}$ sec^{-1} at 25°C) to produce the *N*-2-aminoethylaspartate complex in Scheme 28 (132). Ring opening of the maleate chelate also occurs, but that is not relevant to this chapter. The

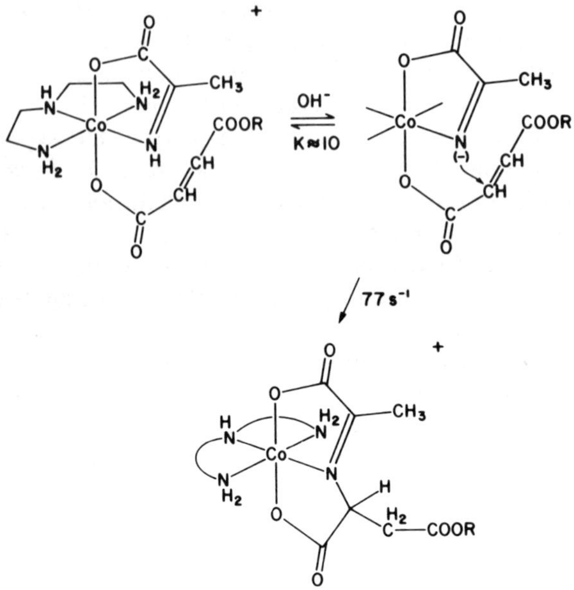

Scheme 27

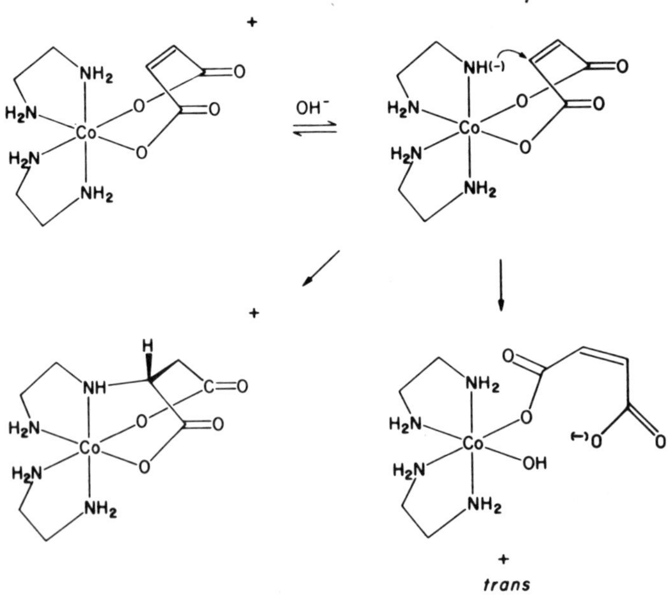

Scheme 28

reaction to produce the amino acid is stereospecific. Only the isomer that generates two five-membered rings *trans* to each other is produced and the estimated rate constant for the attack of the deprotonated amine at the olefin center is $\sim 100\ sec^{-1}$. The estimated rate constant is similar to that observed for cyclization of the imine complex described earlier. The experiment indicates that coordination of both carboxylates, so that their charges are effectively neutralized, is sufficient to activate the olefin and to allow the nucleophile to attack. The result implies that if coordinated OH^- were the nucleophile, malic acid could be generated by this route. It thus provides some justification for the Glusker proposals (111) for the mode of action of aconitase.

Superficially, it also creates some difficulty for the argument raised earlier concerning the exclusive formation of five-membered chelates in these olefin additions. However, the nucleophile and chelate orientations shown in Figure 7 go some way to countering these difficulties. The λ conformation (133) of the nucleophile chelate is clearly the most favorable for reaction; the δ conformation does not put the lone pair of electrons of the aminate ion anywhere near the olefin center.

Orienting the maleate chelate with one carboxyl and the olefin roughly coplanar (Fig. 8) indicates how the specificity might be achieved. Models indicate that the conformational arrangement to give the two five-membered rings *trans* to each other is slightly more favorable than that which gives a five- and a six-membered ring *trans* to each other. Since the energy differences required may be quite small (2–3 kcal mol^{-1}), the conformational and chemical effects may be quite subtle.

7 INTER- AND INTRAMOLECULAR REACTIONS OF COORDINATED IMINES—SERINE SULFOHYDROLASE

7.1 Transamination and Deamination of Amino Acids

The common catalyst in biology for both transamination and deamination is pyridoxal phosphate. There has been much work done in this area both with enzymes and with metal ion model systems (134–137). However, it is not clear if metal ions are also required in the enzymes. Reviews have presented both arguments (136, 137). Some metal ions catalyze the reactions to a modest extent ($\sim 10^2$) and the nature of the β-substituent appears to govern whether transamination or deamination is observed (Scheme 29). Since the subject has been well covered in the literature, this chapter merely directs the reader to salient references and addresses a more recent problem concerning a type of deamination which is pyridoxal independent.

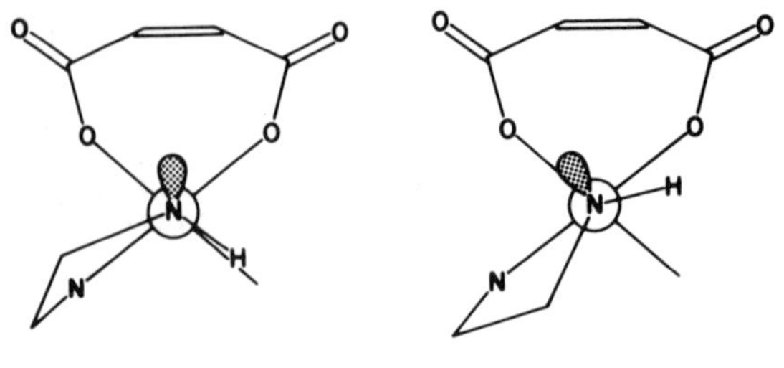

Figure 7 Orientations of the lone pair of electrons on a deprotonated ethylenediamine (en) in the $[Co(en)_2$ maleato$]^+$ ion for the δ and λ conformations of the en chelate.

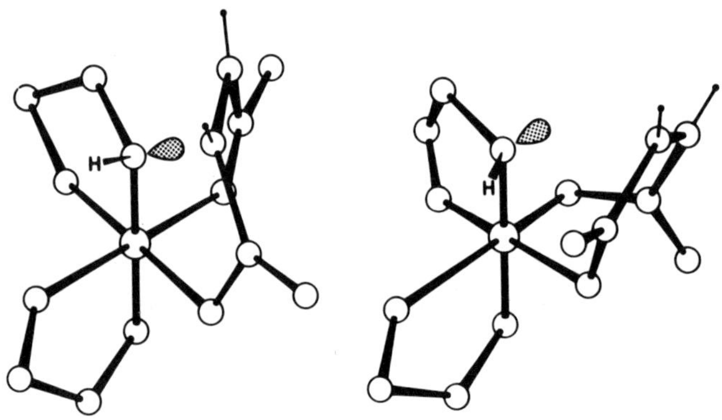

Figure 8 Orientations for the addition of the en aminate ion to chelated maleate in the $[Co(en)_2$ maleato$]^+$ ion.

An enzyme capable of catalyzing the deamination of (*S*)-serine-O-sulfate and (*S*)-β-chloroalanine was first described by Lloyd et al. (138). Since then it has been shown to be widely distributed in nature (139), even though the natural substrate still remains unknown. The enzyme does function *in vivo* on serine-O-sulfate, however, and the products of degradation are SO_4^{2-}, NH_3, and pyruvate in equimolar amounts (140). The enzyme is not a sulfatase since free serine is not a product nor is serine degraded by the enzyme. A β-elimination reaction which is independent of pyridoxal phosphate is implicated. Since labeling studies with $^{18}OH_2$ have shown that the SO_4^{2-} ion eliminated is free of label (141), the C—O rather than the S—O bond is ruptured. Addition of BH_4^- to the culture medium resulted in the isolation of racemic alanine and this implied the formation of aminoacrylate (dehydroalanine) or the tautomeric imine as an intermediate. Also, liberation of nitroform from tetranitromethane by the enzyme system suggests the involvement of a carbanion in the

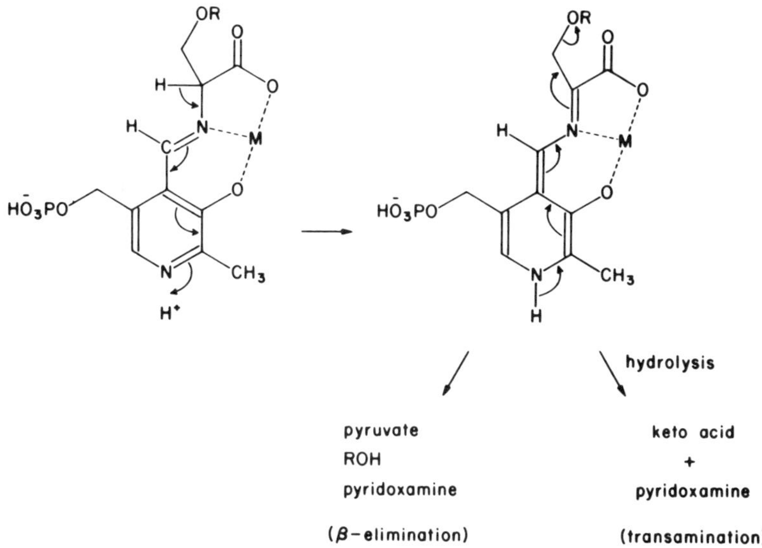

Scheme 29

mechanism (142). Presumably an enzyme base B (Scheme 30) reacts with serine sulfate to produce a carbanion which then reacts rapidly with tetranitromethane to produce nitroform and the nitrated amino acid. It is difficult to envision the enzyme base removing a proton from the anionic serine sulfate without some form of additional activation. In the absence of pyridoxal phosphate a metal ion is a logical prospect for such a role, and the following chemistry explores the effect of metal ions on such β-elimination reactions.

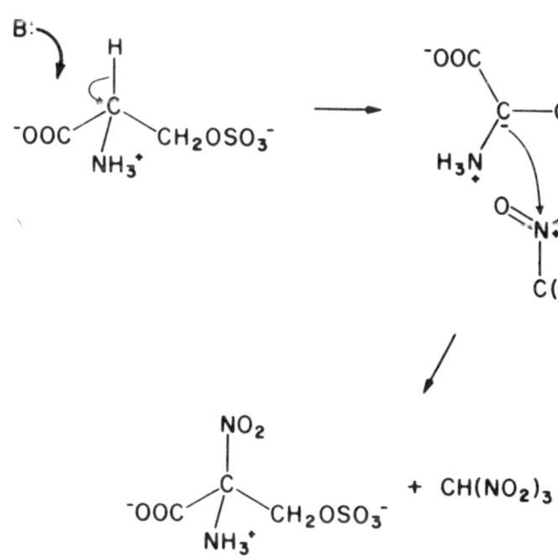

Scheme 30

Three complexes were synthesized (120): the bis(1,2-ethanediamine)-Co(III) complexes of (*S*)-serine sulfate and (*S*)-serine acetate and the tris(2-aminoethyl)amine-Co(III) complex of (*R*)-S-methyl cysteine (Fig. 9). There was little difference in reactivity between the Λ and Δ forms of the $(en)_2$ complexes.

The serine derivatives reacted readily with OH^- at 25°C to eliminate sulfate ($k_{OH}=22M^{-1}\ sec^{-1}$) and acetate ions ($15M^{-1}\ sec^{-1}$) and to generate the chelated pyruvate imine complex. The rate-determining step was the removal of the proton. Each such act led quantitatively to imine product. No serine chelate was produced and the sulfate and acetate ion therefore arose by way of C—O bond rupture. An E_1CB mechanism (Scheme 31) is proposed which leads in the first instance to the aminoacrylate intermediate. This rearranges rapidly to the tautomeric imine product. So far the aminoacrylate intermediate has not been trapped but presumably such a species conjugated with a coordinated carboxylate should be susceptible to attack by nucleophiles. More will be said of this chemistry later.

The rate of the elimination is about 10^7-fold more rapid for the coordinated chelates than the uncoordinated molecules. The reactivity derives at least in part from the enhanced acidity of the α-methyne proton in the chelated reactants. This property has been known for some time and arises from coordination of both $—COO^-$ and $—NH_2$ groups. The metal ion presumably neutralizes the negative charge on the carboxylate ion and at the same time confers some ammonium ion character on the bound $—NH_2$ group. Both effects would act to increase the acidity of the methyne proton (143).

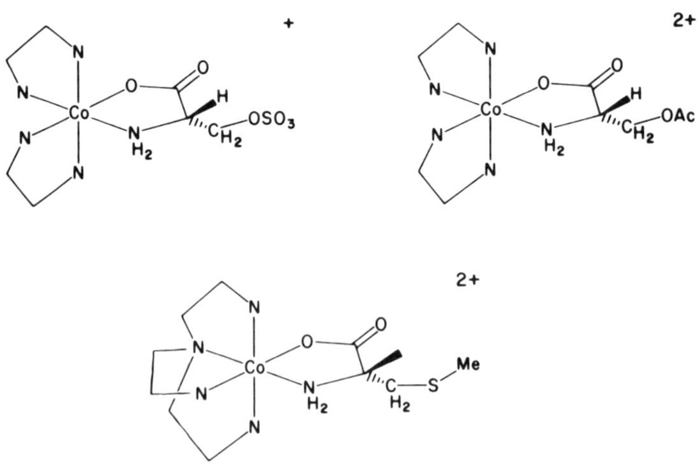

Figure 9 Co(III) chelate complexes containing the serine sulfate, serine acetate, and S-methyl cysteine ions, used to observe β-elimination reactions.

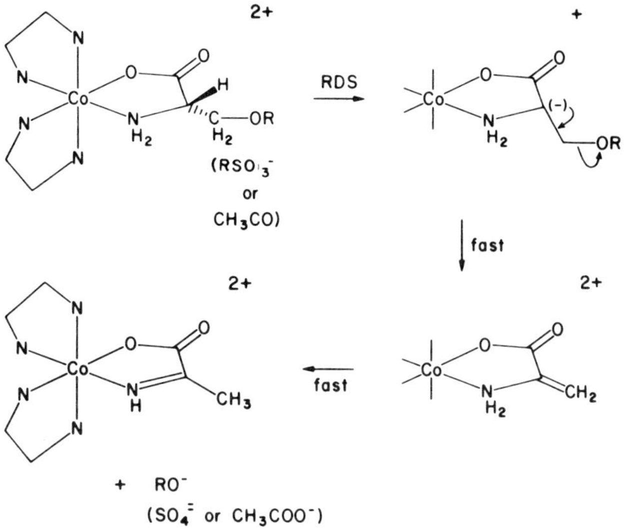

Scheme 31

The question of a concerted E_2 process does arise, and with the two former systems it cannot be excluded. However, with the S-methyl cysteine chelate in the $[Co(tren)(S\text{-}Me\text{-}cysteine)]^{2+}$ ion as a substrate in the same circumstances, a slightly different pattern of reactivity emerges. The β-elimination still occurs in base, but more slowly ($\sim$100-fold), and the chelated imine is the product after elimination of the methyl mercaptide ion. The difference between the two systems emerges in the ratio of the rate of proton exchange to β-elimination. The former is $\sim$25-fold faster than the latter, so that the ^2H label exchanges into the methyne position before elimination. The results support an E_1CB process and not a concerted path (E_2). The elimination of such a good nucleophile as CH_3S^- from the carbanion also indicates why the aminoacrylate was not trapped. Clearly, the rearrangement to the imine must be fast and very favorable in order to expel the mercaptide ion. Although it is not surprising that the imine is stabilized relative to the enamine, its inherent stability is unusual by comparison with imines not bound to metal centers.

In $1M$ OH^- the serine acetate and sulfate substrates react with rate constants comparable to the turnover number of the enzyme acting on serine sulfate at neutral pH ($\sim$40 sec^{-1} at 37°C). In the enzyme, an intramolecular base could be responsible for removal of the methyne proton and it is conceivable that an as yet undetected metal ion could assist the proton abstraction in the manner described above.

There have been identified in plants and bacteria biosynthetic pathways

which involve O-acetylserine as a substrate for the production of a variety of amino acids, for example, mimosine (144), cysteine (145, 146), S-methyl cysteine (146), pyrazolyl alanine (144), and so forth. Some, at least, of these systems require pyridoxal phosphate, which effectively stabilizes the dehydroalanine (aminoacrylate) moiety against rearrangement to the imine and allows the nucleophile to add via a Michael-type addition. Metalion-Schiff-base complexes could also stabilize this type of enamine and allow the synthesis of these unusual amino acids (Scheme 32). This chemistry is currently being explored.

8 PHOSPHATE ESTER CHEMISTRY—ALKALINE AND ACID PHOSPHATASES, PHOSPHORYL TRANSFERASES

Most of the phosphate chemistry in biology (147) appears to be catalyzed by metalloenzymes and the role of the metal ions in such systems clearly needs to be understood. Often the analogous chemistry in the laboratory is very slow or inaccessible and the attempts to catalyze it by adding metal ions like Cu^{2+}, Ni^{2+}, Ca^{2+}, Mn^{2+}, and Zn^{2+} have on the whole been disappointing (148–152). The problem resides in the multiplicity of coordinated species when the experiments are conducted in this manner. Equilibria between the metal and the substrates and between coordinated water and hydroxide ion bound to the metal ion introduce complexities into the analysis of the catalysis. Coordination of more than one molecule of substrate can sometimes negate a catalysis that occurs when just a single molecule of substrate is coordinated. Poly-

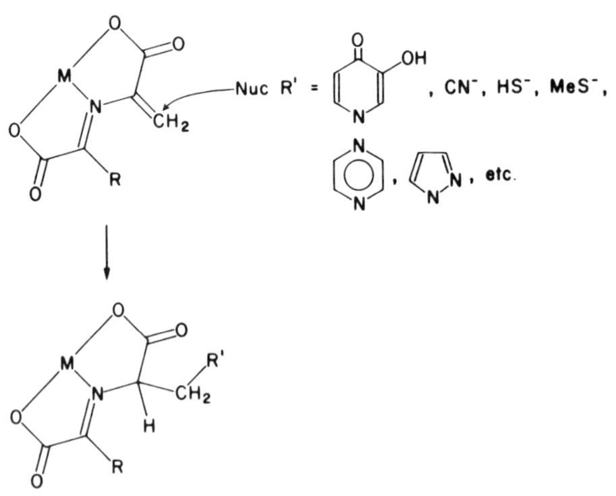

Scheme 32

merization of the coordination complexes can divert the assembly from a catalytic path. Some or all of these problems are difficult to control or evaluate. Even when catalysis is observed, it is often difficult to be sure which intermediate is the effective one and what its concentration is. These comments pertain to phosphate chemistry probably more than to any other metal ion-catalyzed chemistry related to biology. Phosphates bind to metals readily, polymeric structures are common, and much of the simple phosphate chelate chemistry is not well defined (148).

All of these factors make it necessary to approach the problem with well-defined complexes, preferably where the phosphate residue remains bound to the metal in a mode that is known and where its reaction path and fate can be determined by a variety of techniques.

With these notions in mind, the hydrolysis of coordinated phosphate esters in basic media has been examined using kinetically "inert" Co(III) amine complexes in which the phosphate residue at least reacts competitively with loss of the ligand (151,153–157). The immediate stimuli for the work in this laboratory were twofold. One was the role of the metal ion in *Escherichia coli* alkaline phosphatase (158), an enzyme that hydrolyzes phosphate monoesters rapidly and which contains Zn^{2+} ions apparently binding to the substrate. The other was the rapid ester hydrolysis observed for the ethylene methyl phosphate triester in both acid and base, an effect attributed to the steric strain engendered by the ethylene dioxy-phosphorus ring (159). It seemed that the two might be linked if the phosphate ester was bound to the Zn^{2+} ion as a four-membered chelate ring (153,154). Under these circumstances the strain would be considerable and an activation toward attack by nucleophiles in the enzymatic reactions could be expected. Independently, Farrell et al. (153b), and Harrowfield (154b) had shown that methylphosphate bound to either a bis(ethylenediamine)Co(III) or a triethylenetetramine-Co(III) moiety was hydrolyzed considerably more rapidly than in the absence of the Co(III) complex. Unfortunately, the studies did not allow the two obvious mechanisms to be discriminated:

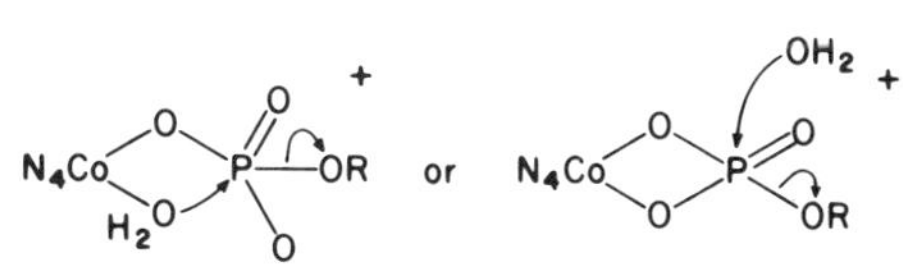

but both processes involve four-membered rings and the results looked sufficiently promising to pursue the ideas involved. In these particular instances, however, it now appears that dimeric ions are the active components (see later).

Although some detail of the chemistry associated with the enzyme will be

dealt with later, the main features of the organic chemistry associated with ester hydrolysis appropriate to this chapter need to be treated now in order to set the scene for the hydrolyses observed in the complexes.

The ethylene phosphate anion $\overline{OCH_2CH_2OP}O_2^-$ hydrolyzes $\sim 10^6$-fold faster than dimethylphosphate ion in basic conditions to give exclusively the ring-opened monoester (159). The rate acceleration observed for the cyclic versus the acyclic species is attributed to the relief of steric strain experienced by the five-membered ring in the ground state as the pentaoxyphosphane transition state forms (159):

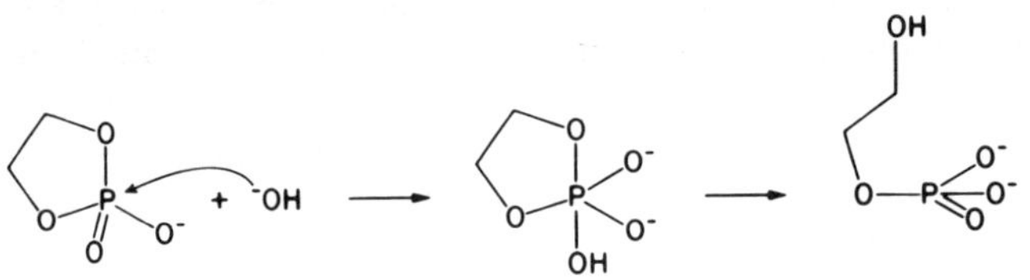

The same acceleratory effect is not observed when the ring size is expanded and the strain in the cyclic ester ring is reduced. It is presumed that the O-P-O angle of 99° in the distorted tetrahedral ground state is relieved to a strain-free 90° angle in a trigonal bipyramidal transition state provided the ring is oriented over axial-equatorial positions. Such an orientation also places the leaving group in the required position for ring opening. In this system no exchange between the solvent and phosphorus oxygen atoms occurs, and the entry of nucleophile and ring opening could be concerted without the intermediacy of a pentaoxyphosphorane. These diester systems have been chosen to outline the organic chemistry as a prelude to an appropriate comparison with a coordinated phosphate monoester. It can be argued that a coordinated phosphate has some ester quality and that a coordinated monoester ought to be intermediate between a phosphate diester and monoester in its behavior.

The other possibility that needs to be considered is the efficacy of an intramolecular nucleophile in such hydrolysis reactions. Molecules such as

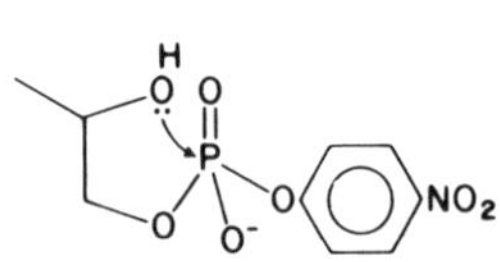

have been examined to determine how effective —OH and —O⁻ are as nucleophiles at the P atom (160). In the enzyme a coordinated OH^- ion could be involved in an intramolecular role during the hydrolysis. Alternatively, coordinated ions such as RHN^-, RO^-, or RS^- could be involved as nucleophiles in phosphoryl transfer reactions.

With these prospects in mind, several complexes have been prepared where a coordinated nucleophile can be generated *cis* to a reactive coordinated phosphate derivative. The first of these is the simple ion $[(NH_3)_5CoO_3POC_6H_4NO_2]^+$ (161) (Scheme 33).

Deprotonation of an ammonia *cis* to the ester leads to rapid attack of the coordinated aminate ion at the P center (estimated $k_3 = 0.4\ sec^{-1}$ at 25°C). It is not clear if the process is concerted or if the aminophosphorane has a lifetime sufficient for its intermediacy to be probed. However, if it is an intermediate, it does not live long enough to exchange its oxygen atoms with the solvent H_2O. What it achieves relatively rapidly is elimination of nitrophenol with the production of phosphoramidate ion. The phosphoramidate is presumably bound in the first instance as a chelate which rapidly undergoes ring opening (unobserved) to give the hydroxo-N-bound phosphoramidate complex.

In addition to hydrolysis to produce nitrophenolate ion, this reaction involves rapid transfer of a phosphate residue from O to N. If the aminate concentration is taken into account, then the ester hydrolysis and O-to-N phosphoryl transfer are accelerated at least 10^8-fold relative to the uncoordinated ester reacting with NH_3 or OH^-. (The pK_a of the coordinated ammonia is estimated as ~ 17.) If the concentration of the intramolecular nucleophile is ignored ($k_{OH} = 3.6 \times 10^{-4} M^{-1}\ sec^{-1}$, 25°C, $\mu = 1.0M$), then the acceleratory factor is still about 10^5-fold. The hydrolysis is complicated somewhat by a parallel path leading to 4-nitrophenylphosphate loss from the Co center by the dissociative conjugate base path characteristic of Co(III) amine chemistry. The kinetics of both paths are first order in OH^- ion as anticipated from the pK_a of the bound ammonia, and first order in reactant. The ^{18}O-tracer experiments showed no incorporation of solvent water in the product phosphoramidate ion; nor is there any incorporation of ^{18}O in the 4-nitrophenol product nor in the cleaved 4-nitrophenylphosphate. All these results indicate the lack of a competing path for addition of $^{18}OH^-$ at the P atom or exchange with solvent in the five-coordinate aminophosphorane.

The attack of the aminate ion may be concerted with entry of NH_2^- apically and loss of phenol or phenolate ion from the *trans*-apical position. There is no special evidence to support this proposal other than the failure to find any indication of a reasonable lifetime for the aminophosphorane. However, with a suitable unsymmetrical diester, inversion or pseudorotation might be observed, provided that the chelated phosphoramidate ester product could be observed by using a sufficiently good leaving group. Such an experiment could reveal more of the mechanistic detail.

The rate enhancement for the intramolecular reaction is surprisingly large when it is borne in mind that a strained four-membered chelate is formed. The strain in a trigonal bipyramid about the P atom is, however, not as great as that expected in the tetrahedral ground state provided the chelate spans ax-

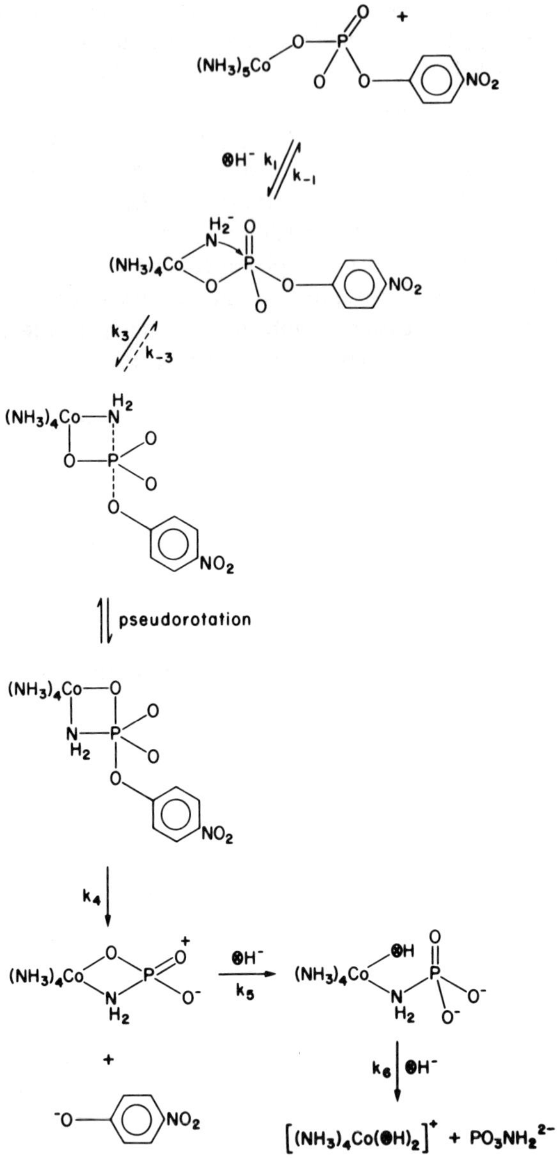

(NH3)5Co
NO2
k1
k-1
(NH3)4Co
NO2
k3
k-3
pseudorotation
k4
k5
k6
[(NH3)4Co(OH)2]+ + PO3NH2 2-

Scheme 33

ial–equatorial positions. Structurally it should not be much different from the dimensions of chelated PO_4^{3-} in an analogous octahedral ion $[Co(en)_2PO_4]$ where O–P–O subtends 76° at the metal ion and O–Co–O subtends 98.7° at the P atom (154). It is presumed that part of the efficiency arises because of the proximity of the P atom to the nucleophile generated. There are minimal losses from vibrational and rotational degrees of freedom, and the rate gain from loss of translational entropy is maximized (162).

The other factor that could contribute to the rate enhancement is the neutralization of charge on the phosphate residue by the metal ion. Even though it is clear that most metal ions are not as efficient in this role as H^+, they still should make it easier to add an electron-rich reagent to what is basically an anionic residue. Some estimate of this effect may be gauged from results communicated later in this chapter.

The metal ion chemistry may be compared with related intramolecular organic substitutions using the substrate:

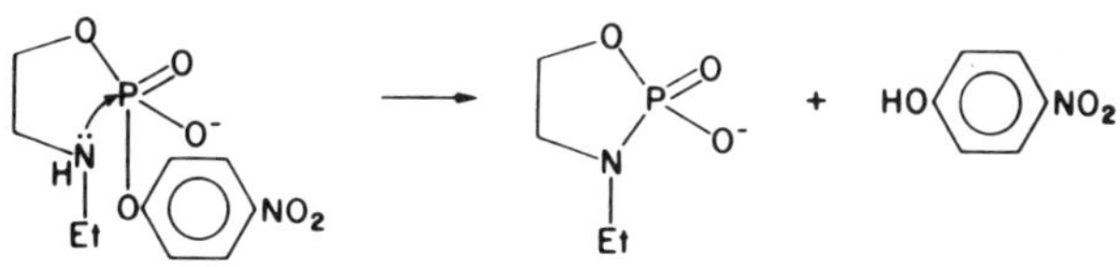

This process is also first order in OH^- up to the pK_a of the amine moiety, after which a limiting rate is reached ($k = 2.7 \times 10^{-3}$ sec^{-1}, 35°C, $\mu = 1.0M$) (163). Clearly the process is much slower than that estimated for the complex ($k \sim 0.4$ sec^{-1}, 25°C, $\mu = 1.0M$). Correspondingly, the basicity of the ethanolamine moiety (pK_a 9.4) is less than that of the coordinated aminate ion ($pK_a \sim 17$) (60). The different reactivities reflect in part this pK_a difference. However, the strain for the five-membered organic cyclization should also be significantly less than for the four-membered chelation, and these factors are yet to be separated.

Analogous chemistry has also been observed for the coordinated fluorophosphate ion in the cation $[(NH_3)_5CoO_3PF]^+$. In the presence of OH^-, rapid F^- loss is observed (164) following attack of coordinated NH_2^- to form the chelated phosphoramidate complex ($k = 3.5 \times 10^{-3}M^{-1}$ sec^{-1}, 25°C), which then undergoes rapid ring opening to give the hydroxo-phosphoramidate complex.

The reaction is related in principle to a reaction catalyzed by pyruvate kinase (165):

$$FPO_3^{2-} + ADP^{3-} \xrightarrow{Mn^{2+}} F^- + ATP^{4-}$$

The model involves the cleavage of fluorophosphate by addition of an aminate ion at the P center and the formation of a phosphoramide, that is, transfer of a

PO_3^- residue from F^- to the nitrogen base NH_2^-. The enzymatic reaction involves the transfer of a PO_3^- residue from F^- to an anionic oxygen base, adenosine diphosphate (ADP). Both systems involve metal ions and it is tempting to propose a requirement for an intramolecular process in the enzyme analogous to that observed in the simple chemistry:

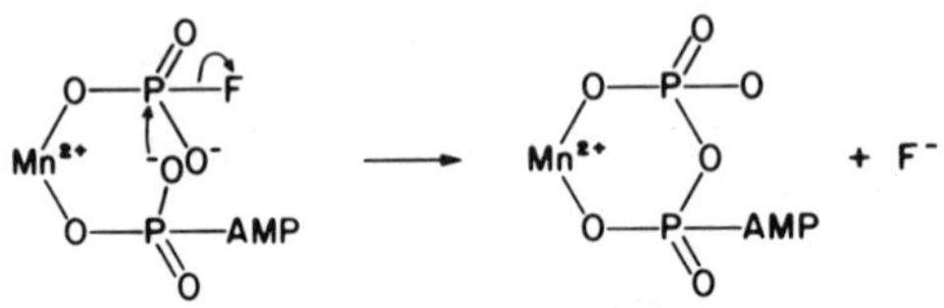

Admittedly, the ring sizes are different in the two instances, but the six-membered chelated product would be more than acceptable as a binding mode for such a pyrophosphate residue. The effect of ring size in such a reaction needs to be probed.

Alternatively, a process like the following:

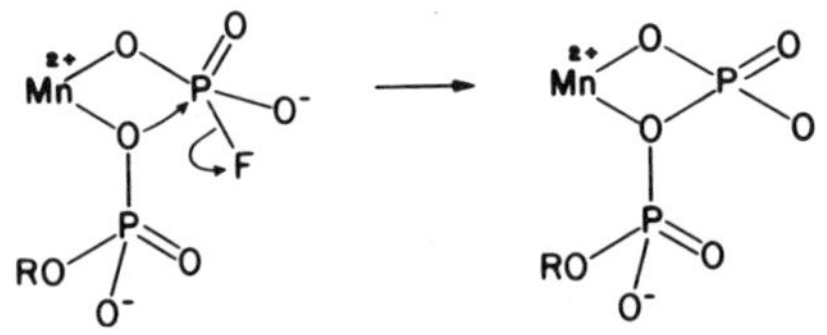

may obtain. This approaches more closely the model chemistry and leaves the pyrophosphate residue in a strained and relatively unfavorable condition. Under these circumstances chelate ring opening could be expected to be fast and the monodentate ATP produced could be expected to dissociate readily.

In this way, exchange of the reactants and products at the metal center could be achieved along with acceptable reaction geometries and rate constants. The latter arise from the intramolecularity of the reaction; a rate increase derives from the loss of translational entropy of one reacting partner (162). It is important then that the rate gain is not cancelled by a loss from increased vibrational and rotational degrees of freedom. Although a four-membered ring may be more favorable than a six-membered ring in this situation, the question has yet to be tested in model systems.

The transfer of a PO_3^- residue from oxygen to nitrogen is a relatively general problem in biochemistry and is typified by the enzyme creatine kinase. It requires Mg^{2+} to effect catalysis of a phosphoryl transfer between ATP and creatine to produce creatine phosphate and ADP (Scheme 34). It is tempting to view the enzymology in the light of the complex chemistry just described where one nitrogen of the basic guanidine residue is bound to Mg^{2+} as shown

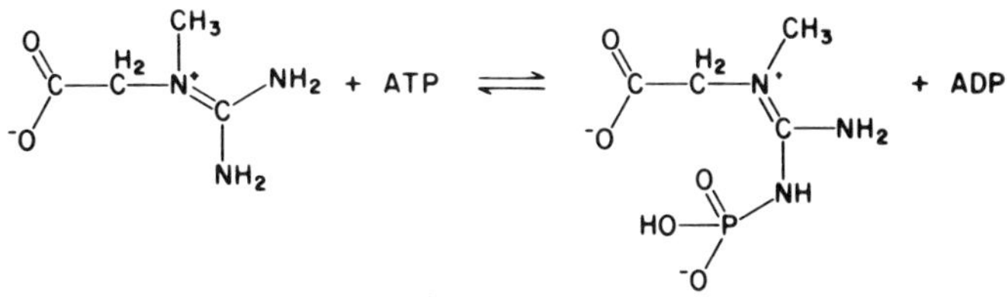

Scheme 34

below and the terminal phosphate residue of ATP is also bound in a position *cis* to the potential N nucleophile.

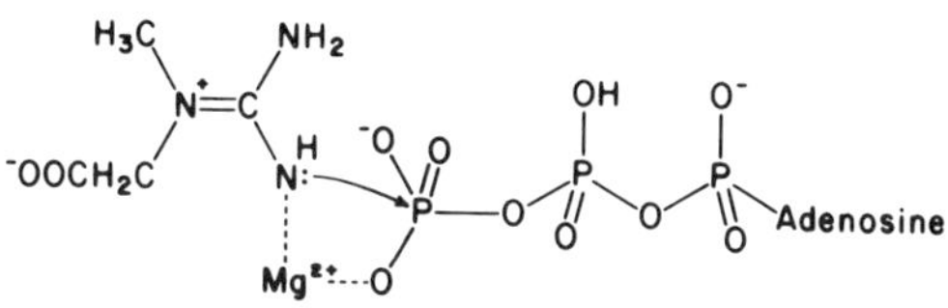

Intramolecular deprotonation of the bound guanidine group would allow attack at the terminal P center to eliminate ADP and transfer of a PO_3^- residue from an O to a N nucleophile. Clearly one of the consequences of the proposal is that a $[(NH_3)_5CoATP]^-$ complex could undergo such an O-to-N transfer and generate ADP in base.

A system that also merits close examination is that involving attack of coordinated OH^- at a phosphate ester bound *cis* to the nucleophile, for example:

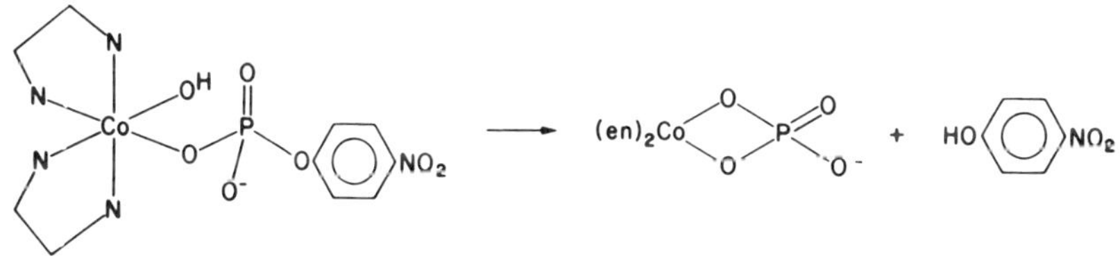

This reaction proceeds relatively rapidly at rates independent of the concentrations of general bases and OH^- in the pH range 9–12 ($k = 7.6 \times 10^{-4}$ sec^{-1} at 25°C) (166).

An initial rate increase over the pH range 7–9 coincides with deprotonation of a coordinated water molecule to give the coordinated OH^- ion. The pH-independent region (9–12) corresponds to the hydroxo complex reacting exclusively and yields 86% 4-nitrophenol and 14% nitrophenylphosphate as the products. Tracer studies (^{18}O) show that the former derives solely from attack of the coordinated nucleophile and the latter by Co—O rupture of isomerized *trans*-$[(en)_2Co(OH_2)(O_3POC_6H_4NO_2)]^+$ via the usual dissociative conjugate base path (Scheme 35). It is only the ester hydrolysis path which is of interest here.

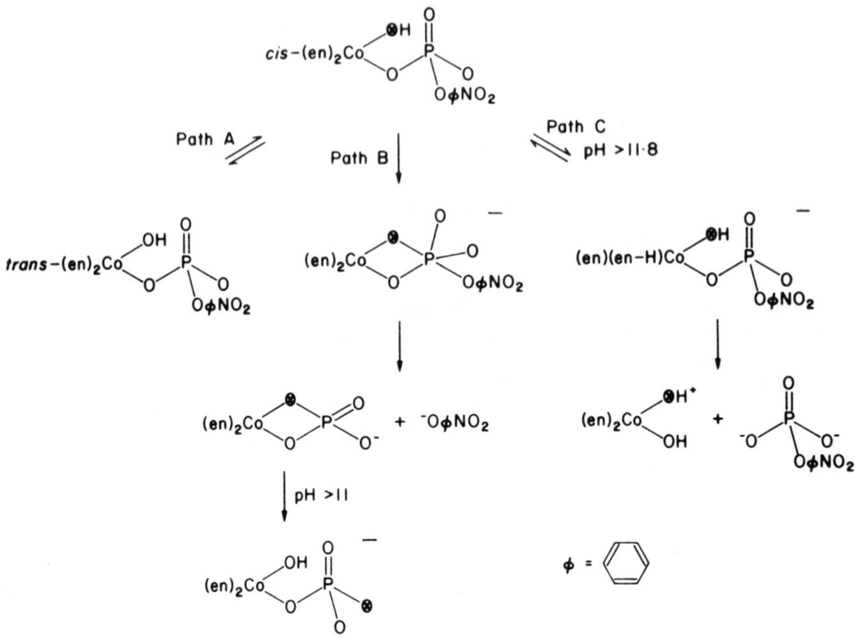

Scheme 35

Chelated phosphate is the sole phosphorus-containing product up to pH 12 apart from the product of the isomerization of the *cis* reactant. When the hydroxyl group is labeled, the label appears in the phosphato product, so the intramolecularity of the process is established. Above pH 12, the reaction changes character again. A path first order in OH^- becomes significant and in 1*M* OH^- it dominates the reaction. The yield of 4-nitrophenol falls sharply to 17% and 4-nitrophenylphosphate (83%) is the major product. No label was observed in the nitrophenylphosphate nor in the nitrophenol, either when the coordinated OH^- was labeled or when the experiment was conducted in $^{18}OH_2$. However, at pH 10, in $^{18}OH_2$ using unlabeled complex, some label (~15% of one oxygen atom) was observed in the phosphate of the $[Co(en)_2PO_4]$ product. Additional experiments showed that this arose during the hydrolysis process and not after nitrophenol was lost. Clearly, the label had to reside in the oxygen atoms *exo* to the chelating oxygen atoms.

The reaction scheme (paths A, B, C) takes account of these results. Up to pH 12, the significant path is attack of coordinated OH^- at the P center to give a five-coordinate oxyphosphorane as a transition state or intermediate. This decomposes to give chelated phosphate and nitrophenol. The question that arises immediately is whether the process is concerted or whether the oxy-

phosphorane has a reasonable lifetime. Clearly, in this path it exchanges oxygen atoms with the solvent and we can therefore be sure of its intermediacy. A hexacoordinate phosphorus species is implicated by addition of H_2O to the pentacoordinate oxyphosphorane. Decay of the hexacoordinate species leads to some label in the noncoordinating oxygen atoms of the oxyphosphorane and ultimately in the chelated phosphate product. The same type of intermediate cannot be involved in the high base path since it would decompose to give labeled nitrophenylphosphate. Presumably, there we are observing a conjugate base path which eliminates intact nitrophenylphosphate. Pseudorotation may also occur in the base-independent region, although it is not observable in these experiments in the absence of stereospecific labeling.

The essential feature is a relatively rapid hydrolysis of the ester moiety by the intramolecular nucleophile at biologically accessible pH. The reaction occurs 10^5–10^6-fold more rapidly at neutral pH than hydrolysis of the uncoordinated ester. This acceleration occurs despite the formation of a strained four-membered ring and despite the reduction in basicity of OH^- on coordination to the metal ion. Coordinated OH^- in this system is 10^7–10^8-fold less basic than free OH^-, yet it is a more efficient nucleophile by a factor of at least 10^5–10^6. The comparison clearly attests to the efficiency of doing the chemistry by the intramolecular route.

It is relevant to compare this metal complex chemistry with analogous intramolecular organic reactions (160) such as:

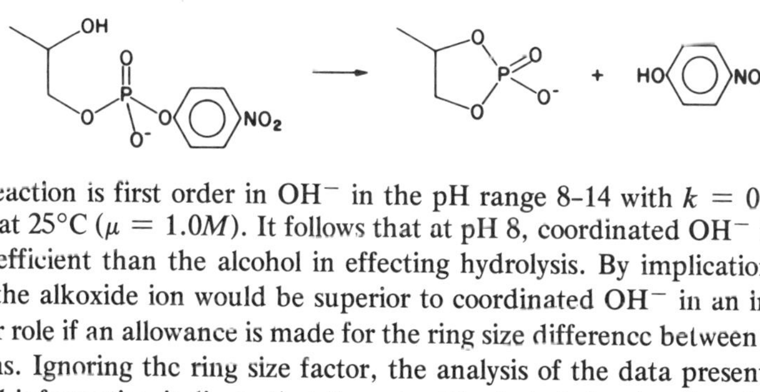

The reaction is first order in OH^- in the pH range 8–14 with $k = 0.21M^{-1}$ sec^{-1} at 25°C ($\mu = 1.0M$). It follows that at pH 8, coordinated OH^- is much more efficient than the alcohol in effecting hydrolysis. By implication, however, the alkoxide ion would be superior to coordinated OH^- in an intramolecular role if an allowance is made for the ring size difference between the two systems. Ignoring the ring size factor, the analysis of the data presented and related information indicate that the rate of hydrolysis is related to the basicity of the intramolecular nucleophile. In this context it is clear that the metal ion can provide a ready source of coordinated OH^- as an effective nucleophile over a relatively wide range of pH suitable for reaction in biological conditions, whereas an alcohol group would be in difficulty on this count. An intramolecular general base to remove the proton as the nucleophile attacked would seem to be required to make such an alcohol group effective.

A phosphate monoester, chelated as a four-membered ring to the metal

ion, is the inorganic analog of ethylenemethylphosphate. Correspondingly, there would be an expectation for rapid ester hydrolysis because of the excessive strain in the four-membered chelate, and it was thought that such a proposal might accommodate the rapid ester hydrolyses observed with the metalloenzyme systems. In the literature there is a report (154) of this kind which proposes the chelation of 4-nitrophenylphosphate to a Co(III)–bis(1,3-propanediamine) moiety to give an anhydrous perchlorate salt, $[Co(tn)_2(O_3POC_6H_4NO_2)]ClO_4$. On treatment with OH^-, this system produced 4-nitrophenol $\sim 10^9$-fold faster than the uncoordinated ester under the same conditions. The large acceleratory effect provided presumptive evidence for the reactivity of the chelate. However, it has now been discovered (168) that this system and an analogous ethylenediamine complex do not contain chelated phosphate esters (**I**) but are unusual dimeric complexes of two Co(III) ions bound to two molecules of phosphate ester in a surprisingly stable eight-membered chelate (**II**):

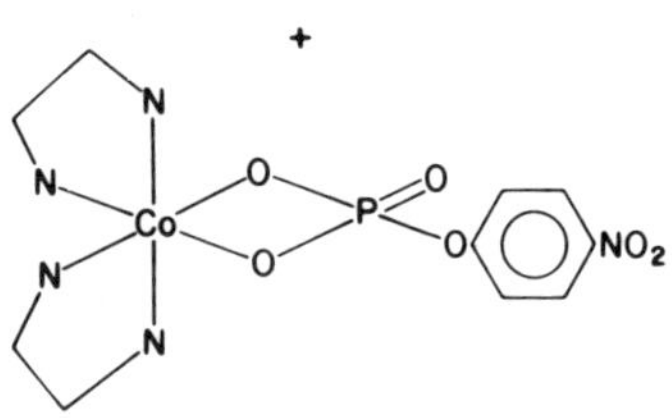

I

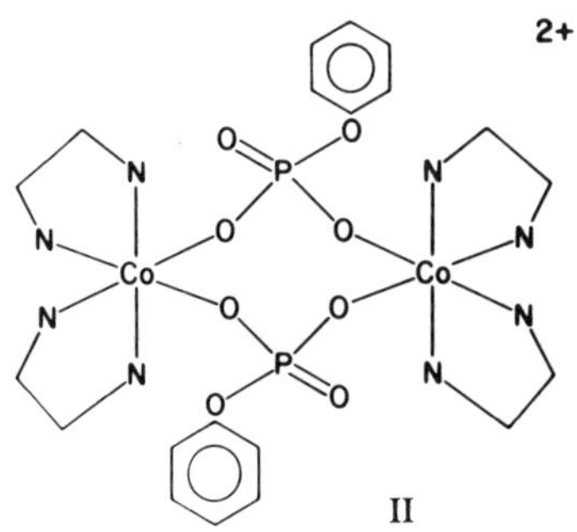

II

An X-ray crystallographic study on the *meso*-isomer of the analogous phenylphosphato-ethylenediamine complex shows its dimeric nature (168). The unusually stable large ring system even survives nitration of the phenyl groups. Conductivity and molecular-weight determinations of the trimethylenediamine and ethylenediamine-4-nitrophenylphosphate complexes clearly establish their dimeric nature.

Since the reactivity in these systems therefore does not derive from chelation of the phosphate ester, they obviously needed reexamination. ^{31}P NMR spectrometry, product analysis, ^{18}O-labeling studies, and kinetics have helped unravel a complex set of reaction paths (169).

The results of a ^{31}P NMR experiment using the $[\{Co(en)_2(\mu\text{-}O_2P(O)OC_6H_4NO_2)\}_2]^{2+}$ ion in the presence of excess OH^- are displayed in Figure 10. The reaction sheme which accommodates these results, the product distribution, the ^{18}O-tracer, and the kinetic results, is given in Scheme 36. In a rapid initial step, a Co–O bond is ruptured by the dissociative conjugate

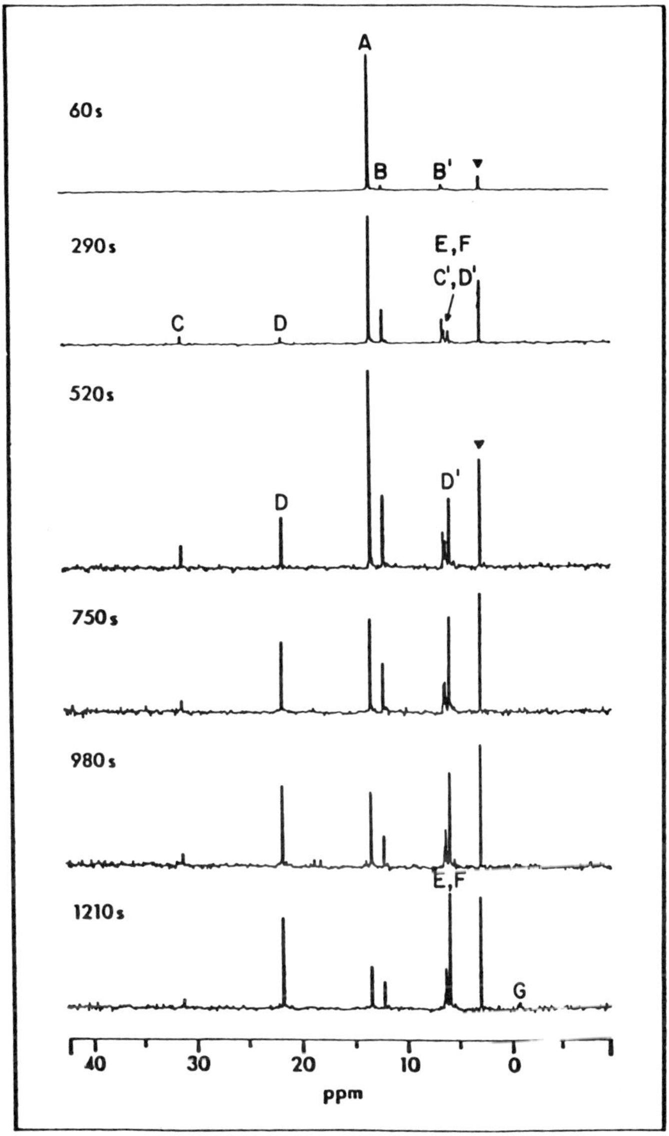

Figure 10 ^{31}P NMR spectra for the base hydrolysis ($[OH^-] = 0.2M$) of *meso*-$\{[Co(en)_2\mu - O_3POC_6H_4NO_2)]_2\}^{2+}$ at 5 ± 1°C. Trimethyl phosphate (▼) was used as an internal reference; chemical shifts (ppm) are relative to 85% H_3PO_4 as external reference. Selected peak assignments (see Scheme 36) are: A (13.31 ppm); C (31.15) and C′ (6.30); D (21.69) and D′ (5.95) (bridging and nonbridging P atoms, respectively); E (6.35); F (6.00); G (−0.73).

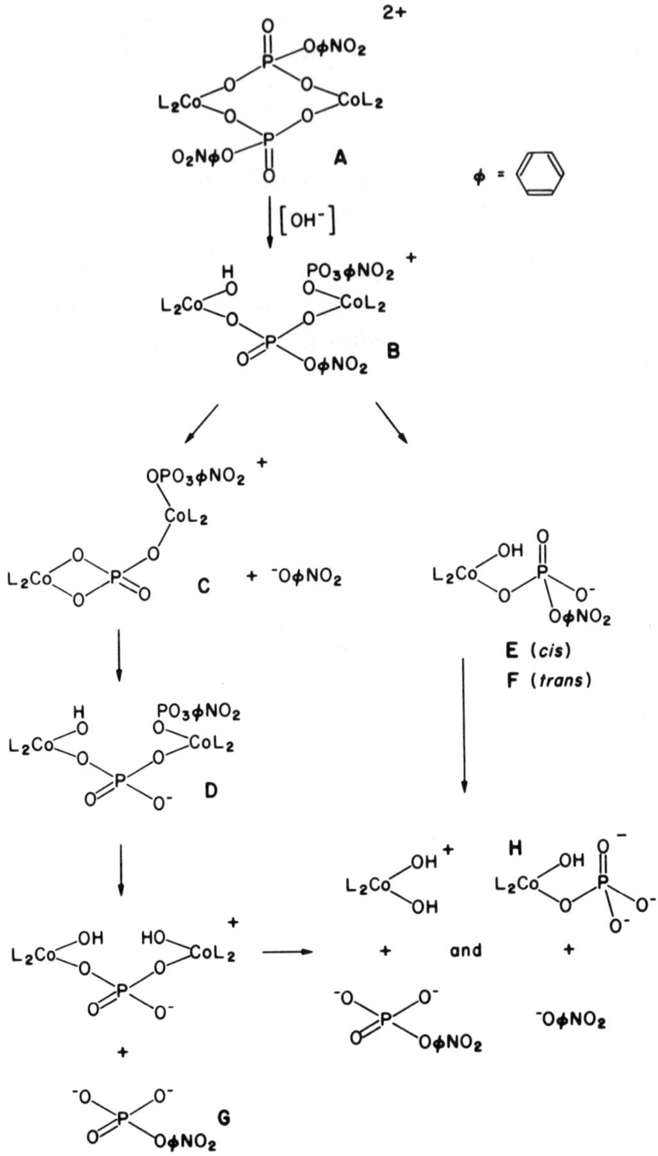

Scheme 36

base mechanism (60) to yield the ring-opened species B. This species undergoes a fast intramolecular reaction of coordinated OH^- at the bridging phosphorus atom to give an oxyphosphorane intermediate which decays rapidly to the chelated phosphate entity C with release of 4-nitrophenolate ion. The chelated phosphate species is not observed in strongly basic conditions ($>0.2M$) and undergoes ring opening readily to give species D with a bridging phosphate group and a nitrophenylphosphate ion attached to one of the Co(III) residues monodentate. The nitrophenyl ester is slowly eliminated from the complex by Co–O rupture.

Along with the initial intramolecular hydrolysis, rupture of the dimer link also occurs to give monomeric *cis*- and *trans*-$[Co(en)_2(OH)O_3POC_6H_4NO_2]$ (E and F). The *cis* isomer then undergoes hydrolysis to yield nitrophenol in the manner described previously.

These NMR experiments are supported by characterizations of products and intermediates, by Co-to-P ratios for various intermediates, and by tracer and kinetic studies. For example, when the kinetics of nitrophenol release were followed, an induction period was observed which corresponded to the opening of the dimer ring. This was followed by a rapid burst and then by a slower release of nitrophenolate ion. The two ester hydrolysis paths can be related to a reactive dimer and to the somewhat less reactive monomer *cis*-$[Co(en)_2(OH)O_2P(O)OC_6H_4NO_2]$. Experiments in $^{18}OH_2$ showed that no label was incorporated in the 4-nitrophenol produced nor in the nitrophenylphosphate released. Clearly P–O cleavage is exclusive in the first instance as is Co–O cleavage in the second. Further, product analyses using chromatography showed the presence of *cis*- and *trans*-$[Co(en)_2(OH)O_3POC_6H_4NO_2]$ early in the reaction and the *cis*- and *trans*-$[Co(en)_2(OH)PO_4]^-$ products at the end. Cobalt and phosphorus analyses of products during the course of the reaction identified the bridging phosphate dimers *cis*- and *trans*-$[en_2Co(OH)OP(O)_2O(OH)Co(en)_2]^+$.

In essence, the ester hydrolysis observed here is of the same type as that recorded in the other examples; namely, intramolecular attack by coordinated hydroxide ion, for example:

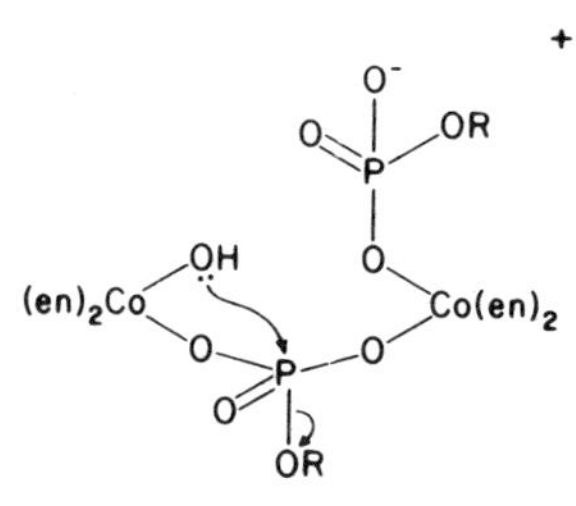

The increased efficiency of the hydrolysis presumably arises from two main sources. Attack of coordinated OH^- at the P center is akin to an intramolecular reaction of a monodentate diester that would be expected to be somewhat more reactive than a monoester with the same leaving group. The other facet of having the Co(III) ion bound to the reactive P center is that it acts to neutralize the charge on the phosphate unit and assist the attack of the nucleophile. These two factors are not unrelated but it might be possible to gauge their effectiveness by adding suitable metal ions to a reactive monomeric system of the type $[en_2Co(OH)(O_3POR)]$ in the region where the reaction is pH independent. The other factor that emerges is that coordination of the phosphate ester to the Co(III) ion does not alone generate a large increase in reactivity. Most of the rate enhancement comes from the intramolecularity of the reaction and the ability of the metal ion to group the reactants.

So far, a genuine example of a chelated phosphate ester reacting with a nucleophile has not been found. The implications in the previous chemistry are that such systems should be reactive since the phosphorane:

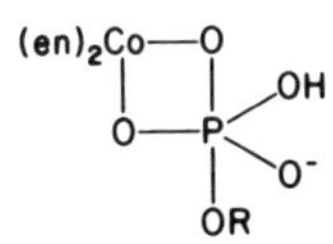

is generated by the intramolecular route and rapidly eliminates nitrophenol. The same intermediate should be achievable from the chelated ester with free OH^- as the nucleophile. This chemistry is generally so attractive from kinetic and synthetic viewpoints that it should be pursued.

Although the chemistry of *E. coli* alkaline phosphatase has been rather well studied (147), it does have obscure features. The native enzyme is dimeric and consists of two identical subunits (~43,000 daltons). There are normally four Zn(II) ions and one or two Mg(II) ions associated with each dimer. Two of the Zn(II) ions (one per monomer unit and separated by 32 Å) are essential for catalytic activity, and an X-ray crystallographic analysis (170) of the enzyme places them close to clefts in the protein which would allow a substrate molecule to come within 3 Å of the metal ion. Recent NMR studies (158) with the ^{113}Cd-substituted enzyme also indicate that the phosphate esters bind to the metal ion which is in turn bound to three histidine residues at the active site.

The catalysis of phosphate ester hydrolysis by the enzyme occurs in two steps: a phosphorylated enzyme intermediate is generated by attack of the hydroxyl group of Ser-99 at the P atom of the bound phosphate ester, leading to loss of the alcohol and the formation of a seryl ester intermediate. Subse-

quently, the phosphorylated enzyme hydrolyzes to generate free enzyme and phosphate ion. Elegant work (171) with a chiral (^{16}O ^{17}O ^{18}O) phosphate ester has shown the enzyme reaction to occur with overall retention of configuration, presumably via consecutive inversion processes.

The dephosphorylation is largely rate determining and this accommodates the lack of sensitivity in overall rate to the substrate phosphate ester. Both substrate binding and dephosphorylation of the enzyme intermediate have an absolute requirement for Zn(II). The pH-activity profile of the enzyme also reflects the pH dependence of the dephosphorylation reaction. The sigmoidal increase in rate between pH 7 and 9 has been attributed to the dissociation of an acidic group in the enzyme with a pK_a of 7.4; thereafter the rate changes little. It has been suggested that the acidic group is a water molecule coordinated to Zn(II) in the active site. At least two questions immediately arise. Can the pK_a of a water molecule bound to Zn(II) be as low as 7.4 and can a mechanism using coordinated OH^- be conceived as a mechanism for the enzymatic phosphate ester hydrolysis?

Most simple complexes of Zn(II) containing coordinated H_2O have pK_a values in the range 9–10. However, five-coordinate complexes are known that have apparently lower values. For example, in aqueous solution the distorted square pyramidal complex (172):

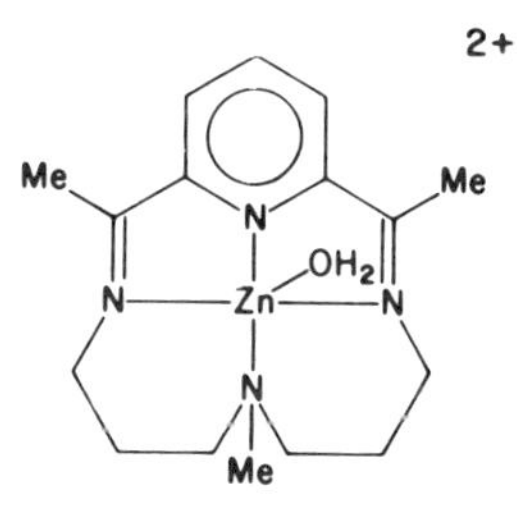

has a pK_a of 8.12, a value that approaches that observed for the enzyme. The coordinating atoms are not dramatically different from those encountered in metalloenzymes, for example, imidazole groups.

A feasible mechanism involving coordinated OH^- is outlined in Scheme 37. It requires, as two preliminary events, coordination of the phosphate ester and deprotonation of a coordinated water molecule. Rapid attack by the serine OH group at the P center, presumably assisted by an intramolecular base, leads to the phosphorylated enzyme. The general base would seem to be required since the organic models are unreactive unless the alkoxide ion is generated. Intramolecular attack by a coordinated hydroxyl group then cleaves the seryl ester to give the chelated phosphate which subsequently leaves the metal ion and regenerates the active enzyme. Such a mechanism accommodates the essential pH independence of the hydrolysis after deprotona-

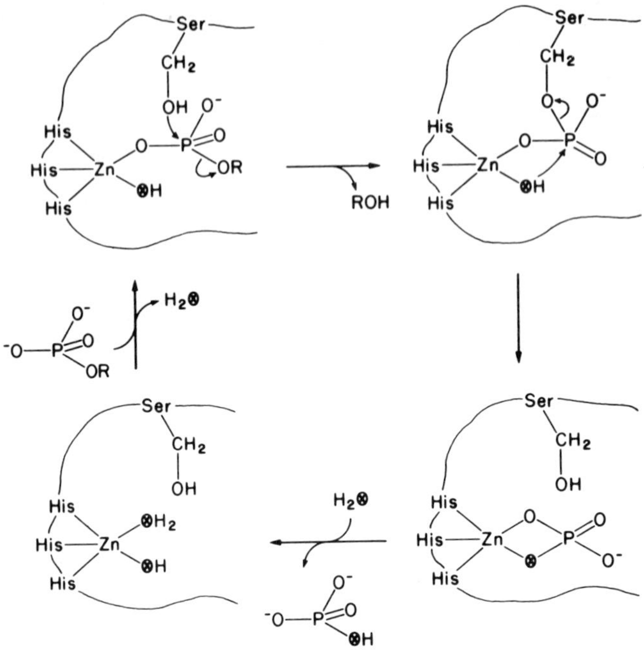

Scheme 37

tion of the coordinated water molecule. It is conceivable that the coordinated OH, acting as a general base, could also assist the serine alcohol group to attack the P center. The rate constant for dephosphorylation has been evaluated as 100 sec^{-1}, which is about 10^5–10^6-fold greater than those observed for the model complexes. The difference could be accommodated by the several prospects for assistance by intramolecular bases on the enzyme.

Alternatively, the serine alcohol could coordinate to the Zn^{2+} ion. The coordinated alkoxide ion could readily attack the bound phosphate ester by an intramolecular route, to produce a four-membered chelated seryl ester intermediate (Scheme 38). The metal ion would serve two functions:

1. To produce an appreciable concentration of a good intramolecular nucleophile (i.e., Zn-O-R) at physiological pH (cf. coordinated OH^-, Secs. 3 and 5); and,

2. To produce a superb leaving group for the subsequent general base assisted dephosphorylation step (cf. Sec. 5).

Although the efficacy of coordinated alcohols as nucleophiles has yet to be evaluated in model systems, it is notable that the pK_a of coordinated methanol

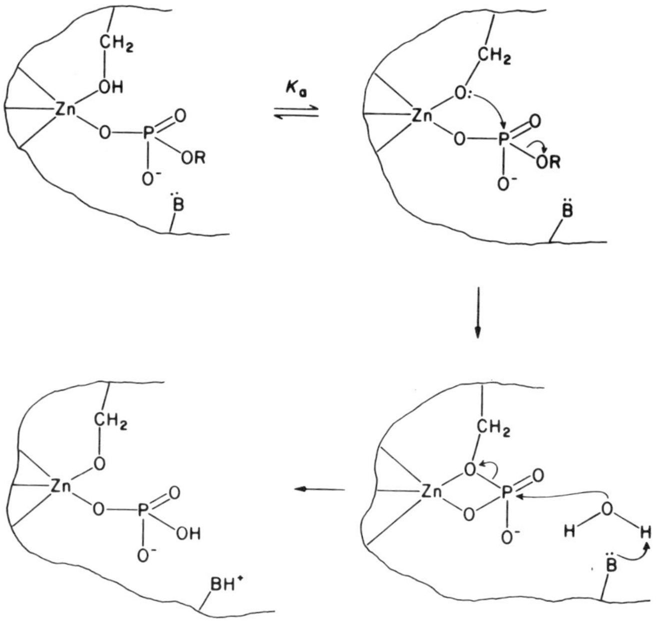

Scheme 38

in the $[(NH_3)_5CoOHCH_3]^{3+}$ ion (5.58) (173) is comparable to that of water coordinated to Co(III) (Table 2). As discussed earlier (Sec. 5), it has not yet been possible to produce complexes of esters with coordinated alkoxide ions as potential leaving groups. However, the concept of the role of metal ions as proton substitutes at neutral pH predicts that such an ester (of a carboxylic or phosphoric acid) ought to be extremely sensitive to attack by an external nucleophile. In such a mechanism at the active site of alkaline phosphatase, the pK_a governing dephosphorylation could logically be that of the conjugate acid of an intramolecular general base (e.g., histidine).

A further mechanism, outlined in Scheme 39, involves general base-assisted attack of the seryl OH group on a four-membered chelate. The rapid initial ester cleavage is readily accommodated both by the anticipated reactivity of the strained chelate ring and the phosphotriester character of the bound substrate in which both anionic oxygen atoms are largely neutralized by the metal ion. The addition of the intramolecular nucleophile could lead directly to the phosphorylated seryl ester which may be reactive enough to be attacked by water. In these circumstances the chelated phosphate would be generated and the serine alcohol restored.

Alternatively, ring opening of the seryl phosphate ester could occur fol-

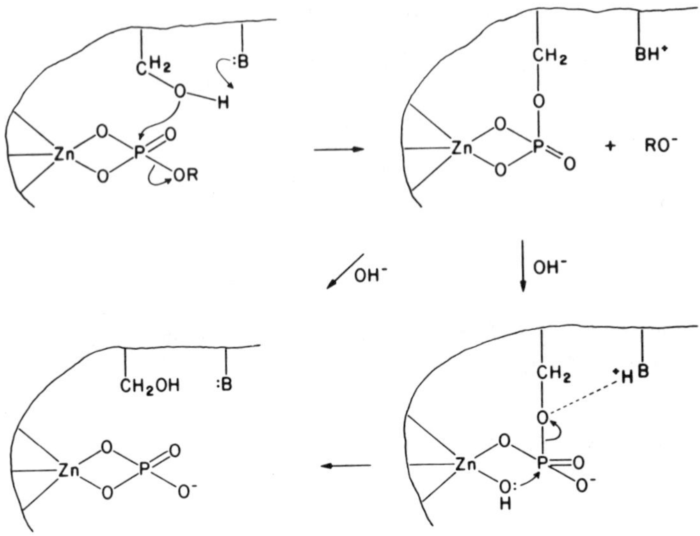

Scheme 39

lowed by attack of a coordinated OH^- at the P center as shown in the latter part of Scheme 39.

Clearly, there are factors here still to be probed. So far the models have used esters with good leaving groups. It is important to observe high reactivity induced in simple monoalkylphosphates, and four-membered chelates are a possible source for such enhanced reactivity.

Purple phosphomonoesterases, active at acid pH, that contain stoichiometric amounts of iron have recently been isolated from beef spleen and porcine allantoic fluid (174,175), and an apparently similar manganese-containing enzyme is present in sweet potato (176). The porcine enzyme contains two atoms of iron per molecule, one of which can be replaced by Zn^{2+} with substantial retention of enzymatic activity (175). Little is yet known of the mechanisms of action of any of these enzymes, but the possibility that both metal ions act concertedly in a manner analogous to the "dimeric" model system (Scheme 36, B → C) is an attractive speculation.

There have been some elegant studies concerning the role of metal ions in enzymes that use adenosine triphosphate (ATP). The coordination of ATP to a metal ion as shown in Figure 11 generates a chiral center at the β-P center and complexes of this type have now been synthesized and resolved enzymatically into catoptric forms (157, 177). Following cleavage of the adenosine moiety with periodate and aniline, the resulting triphosphate complex $[Co(NH_3)_4P_3H_2O_{10}]$ was crystallized. The absolute configuration of one of

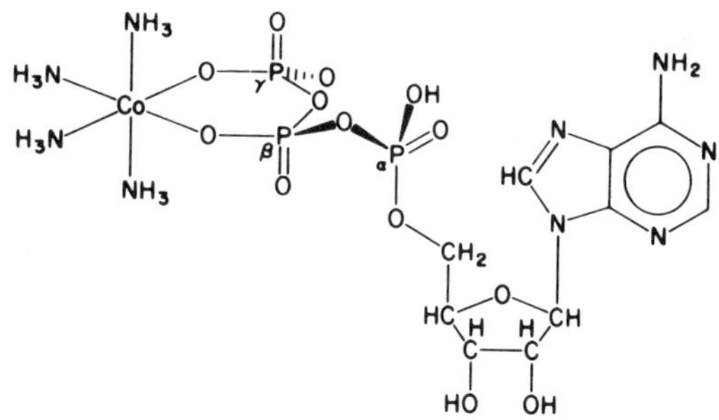

Figure 11 Stereochemical relationship of chelated ATP to the tetraammine-Co(III) moiety indicating the chiral β-P center.

the isomers ($\dot{\Delta}$) was then established by an X-ray crystallographic examination (177).

The especially interesting feature of these studies is the ability of the well-defined Co(III) tetraammine-ATP complex to substitute for a metal ion–ATP complex (frequently Mg^{2+}–ATP) in enzymes (178, 179). One such example involves yeast hexokinase, which catalyzes the reaction:

$$[(NH_3)_4CoATP] + \text{glucose} \rightleftharpoons [(NH_3)_4Co(\text{glucose-6-phosphate})ADP]$$

The Co(III) complex apparently replaces the Mg^{2+}–ATP complex in the enzyme, although the rate is considerably less (0.06%) than that in the Mg^{2+}-catalyzed system. Confidence that the Co(III) and Mg^{2+} complexes react analogously derives from at least two facts. Only one isomer is consumed in the reaction and the Co(III) complex competitively inhibits reaction with the MgATP complex. The enzyme clearly recognizes the stereochemistry at the chiral P center in the Co(III) complex, and this implies a similar binding mode and stereochemistry for the labile MgATP complex. With the advantage of hindsight it is not so surprising that the Co(III) ammine complex mimics the MgATP complex; the latter presumably chelates the β,γ-phosphate residues in the same way, and four H_2O molecules are coordinated in the remaining octahedral positions around the Mg^{2+} ion. Bound NH_3 would be essentially isomorphous with coordinated H_2O in this context. The other implication is that the Mg^{2+} ion does not tie the ATP to the enzyme. Presumably that association is between the phosphate groups and/or the nucleoside and the enzyme.

The association with glucose is also obscure and the origin of the reactivity in the system is still a puzzle. The β,γ-chelation of the pyrophosphate moiety does not seem to lead to any special reactivity at the P centers other than that which might arise from neutralization of the charge. On the contrary, some

diminution in reactivity might be expected since rupture of the stable chelate ring must be energetically unfavorable:

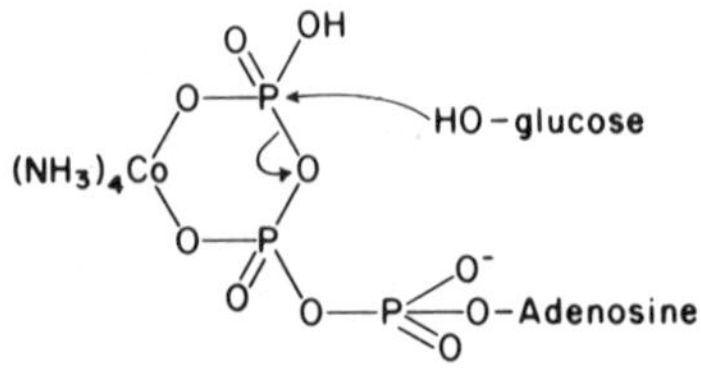

Also attack by a glucose–OH group at the γ-P center should not be a facile process unless it is organized to be essentially intramolecular and is deprotonated by an enzyme base. Sugar hydroxyl groups are not noted nucleophilic reagents. Elimination of the nucleoside or the nucleotide would both be expected to be more facile processes than rupture of the terminal pyrophosphate.

A more-detailed mechanistic picture may not become available until the structural detail of the active site of hexokinase is known, but at least these experiments appear to define important stereochemical aspects of the process. Similar studies have been carried out with Cr(III) analogs, and the stereochemical requirements of a variety of other enzyme systems have since been probed using these model compounds (179).

However, reactions of bis(diamine)-Co(III) and related complexes with pyrophosphate (155) do indicate roles for metal ions in the hydrolysis of polyphosphates. With complexes containing 1,3-propanediamine (tn), there is evidence that three cobalt ions bind to pyrophosphate, leading to hydrolysis of $P_2O_7^{4-}$ to two HPO_4^{2-} ions. At pH 7 the reaction is promoted $\sim 10^5$-fold relative to the uncoordinated substrate. A mechanism proposed to account for the handsome increase in rate is given in Scheme 40. The $[(tn)_2Co(OH_2)(OH)]^{2+}$ rapidly adds to the chelated $P_2O_7^{4-}$ ion and the coordinated OH^- ion then attacks the P atom eliminating a phosphate residue. The latter is assisted to leave by having another Co residue bound to it and the neutralization of the negative charge on the phosphate groups makes the addition of the nucleophile much more facile. A more kinetically inert complex $[(en)_2CoP_2O_7]^-$, has also been used as the starting point. Here the $P_2O_7^{2-}$ ion is chelated (180) and kinetically inert to ring opening in basic solution. Similar results are obtained upon addition of $[(tn)_2Co(OH)(OH_2)]^{2+}$ and the reaction is even more rapid (181).

These results are pertinent to the way by which a metal ion could activate ATP to hydrolysis, and in this respect the vanadium(V)-catalyzed hydrolysis of triphosphate ion and ATP is also relevant (182). Here, a detailed study of a relatively complex system indicated a $[(V(V))_2(P_3O_{10})]$ complex to be the most active species. The $[V(V)(P_3O_{10})_2]$ complex did not hydrolyze. In this

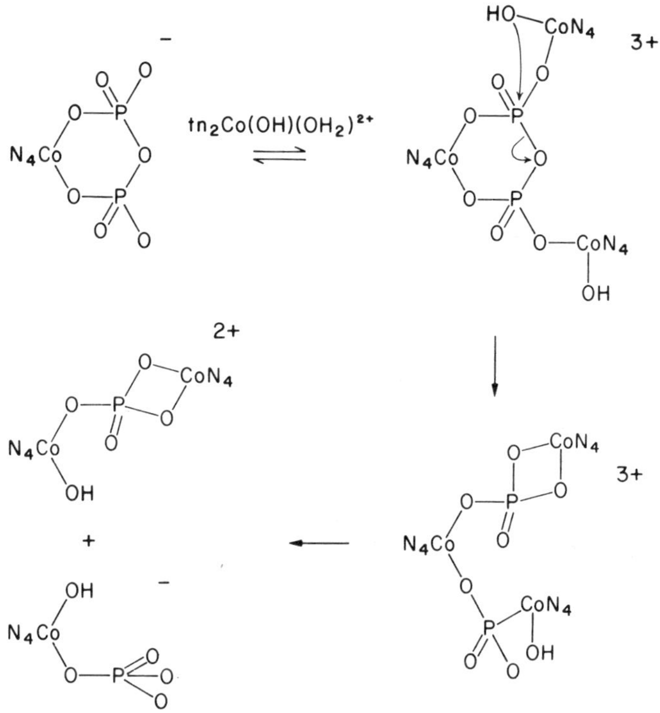

Scheme 40

respect the results are similar to those observed by Hübner and Milburn (155). There are indications that the active complexes are not equilibrating with the medium and all potential ligands for vanadium (e.g., acetate, peroxide, and the triphosphates themselves) inhibit catalysis by V(V). It is presumed therefore that OH^- or O^{2-} bound to vanadium(V) is involved in the hydrolysis and that ligands such as those above displace the active nucleophile.

Hydrolysis is presumed to occur predominantly through intermediates of the type:

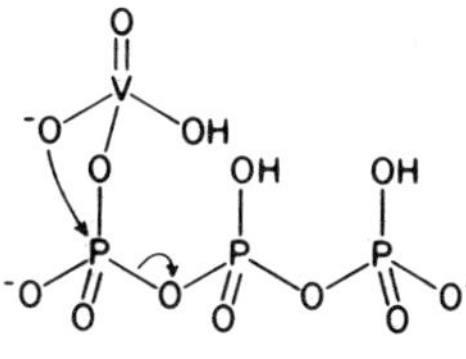

to give $P_2O_7^{4-}$ and PO_4^{3-}, while an entity such as:

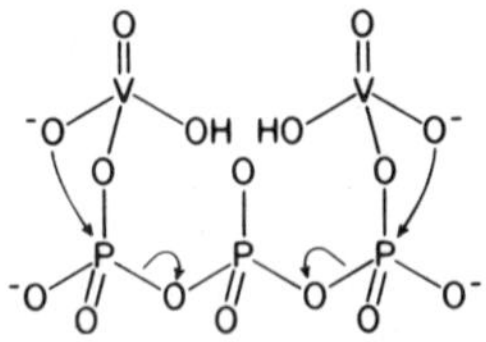

could be envisioned as a reactive intermediate acting in a sequential manner to give PO_4^{3-} ion directly. One vanadium(V) ion is about half as effective as two and the experiments indicate that in the presence of multiple vanadium(V) ions, $P_2O_7^{4-}$ ion is not an intermediate product. Similar conclusions have been reached using Ni^{2+}, Cu^{2+}, and Zn^{2+} in the presence of ATP (183). Vanadium(IV) is about 1000-fold less effective than vanadium(V), and it catalyzes the hydrolysis at about the level of divalent metal ions. Although vanadium(V) would be expected to withdraw electron density from the P center more efficiently and thereby enhance attack of the nucleophile, the coordinated nucleophile must necessarily be weaker (Secs. 1 and 3).

At this stage it is clear our understanding of how reactivity is induced in the enzymology of phosphate and polyphosphates is primitive. Much more work needs to be done on these model systems, and some of the experiments described above do indicate fruitful research directions.

The chemistry organized by the metal ions and discussed in the previous pages go some way to removing the mystery about the rapid reactions of those metalloenzymes that are not redox systems. The efficacy of intramolecular coordinated nucleophiles and the activation of substrates bound to metal centers, coupled with intramolecular nucleophiles or acids and bases on the enzyme, can often account for rate increases of greater than 10^{12} relative to the uncoordinated molecules. The manner in which metal ions activate or deactivate the substrates is revealing, and the way in which a metal ion can provide a nucleophile over a wide range in pH is a power that should not be ignored. In total, these effects are significant pathways for rapid organic chemistry. They are also capable of revealing mechanistic information that is often difficult or impossible to find in catalytic processes because of the rapid turnover. We have therefore attempted to take the understanding they give and apply it to some biological systems in the hope that it will stimulate more probing experiments with the enzymes and promote another view of the chemistry of the nonredox metalloenzymes.

ACKNOWLEDGMENTS

The authors would like to acknowledge the proofreading and critical comments of G. P. Haight and the artwork of Mrs. C. Jacobs.

REFERENCES

1. F. H. Westheimer, *Trans. N.Y. Acad. Sci.*, **18,** 15 (1955).
2. N. E. Dixon, C. Gazzola, R. L. Blakeley, and B. Zerner, *Science*, **191,** 1144 (1976), and references therein.
3. N. F. Curtis, *Coord. Chem. Rev.*, **3,** 3 (1968).
4. A. M. Sargeson, *Chem. Brit.* **15,** 23 (1979).
5. (a) J. Sen and H. Taube, *Acta. Chem. Scand. Ser. A.*, **33,** 125 (1979), and references therein. (b) I. P. Evans, G. W. Everett, and A. M. Sargeson, *J. Chem. Soc. Chem. Comm.*, **139** (1975); *J. Am. Chem. Soc.*, **98,** 8041 (1976).
6. M. M. Jones, *Ligand Reactivity and Catalysis*, Academic, New York, 1968.
7. N. E. Dixon, P. W. Riddles, C. Gazzola, R. L. Blakeley, and B. Zerner, *Can. J. Biochem.*, **58,** 1335 (1980).
8. M. L. Tobe, *Sci. Progr.*, **48,** 483 (1960).
9. C. J. Hawkins, A. M. Sargeson, and G. H. Searle, *Aust. J. Chem.*, **17,** 598 (1964).
10. J. MacB. Harrowfield, V. Norris, and A. M. Sargeson, *J. Am. Chem. Soc.*, **98,** 7282 (1976).
11. S. F. Lincoln, D. R. Stranks, *Aust. J. Chem.* **21,** 37 (1968).
12. D. A. Buckingham, C. E. Davis, D. M. Foster, and A. M. Sargeson, *J. Am. Chem. Soc.*, **92,** 5571 (1970).
13. D. A. Buckingham, F. R. Keene, and A. M. Sargeson, *J. Am. Chem. Soc.*, **95,** 5649 (1973).
14. N. E. Dixon, W. G. Jackson, W. Marty, and A. M. Sargeson, *Inorg. Chem.*, **21,** 688 (1982), and references therein.
15. N. J. Curtis, N. E. Dixon, and A. M. Sargeson, to be published.
16. W. P. Jencks and J. Regenstein in *Handbook of Biochemistry and Molecular Biology—Physical and Chemical Data*, Vol. 1, 3rd ed., G. D. Fasman, Ed., Chemical Rubber Company, Cleveland, 1976, p. 305.
17. R. Breslow, R. Fairweather, and J. Keana, *J. Am. Chem. Soc.*, **89,** 2135 (1967).
18. Early papers include the following: K. Sakai, T. Ito, K. Watanabe, *Bull. Chem. Soc. Japan* **40,** 1660 (1967); K. Watanabe, S. Komiya, S. Suzuki, *Bull. Chem. Soc. Japan* **46,** 2792 (1973); P. F. B. Barnard, *J. Chem. Soc. A* 2140 (1969). A recent example in which carbon nucleophiles are readily condensed at the nitrile carbon of EtOC(O)$-$C$\equiv$N in the presence of Zn^{2+} is given in: T. Iimori, Y. Nii, T. Izawa, S. Kobayashi, M. Ohno, *Tetrahedron Lett.* 2525 (1979).
19. D. Pinnell, G. B. Wright, and R. B. Jordan, *J. Am. Chem. Soc.*, **94,** 6104 (1972).
20. I. I. Creaser, J. MacB. Harrowfield, F. R. Keene, and A. M. Sargeson, *J. Am. Chem. Soc.*, **103,** 3559 (1981).
21. A. W. Zanella, P. C. Ford, *Inorg. Chem.*, **14,** 42 (1975).
22. N. E. Dixon, D. P. Fairlie, W. G. Jackson, and A. M. Sargeson, to be published.
23. D. A. Buckingham, P. Morris, A. M. Sargeson, and A. Zanella, *Inorg. Chem.*, **16,** 1910 (1977).
24. See reference 20, and R. J. Balahura, P. Cock, W. L. Purcell, *J. Am. Chem. Soc.* **96,** 2739 (1974).

25. I. I. Creaser, A. M. Sargeson, *J. Chem. Soc. Chem. Comm.*, **974** (1975).

26. I. I. Creaser, S. F. Dyke, A. M. Sargeson, and P. A. Tucker, *J. Chem. Soc. Chem. Comm.*, **289** (1978).

27. D. G. Butler, I. I. Creaser, S. F. Dyke, and A. M. Sargeson, *Acta Chem. Scand. Ser. A*, **32,** 789 (1978).

28. R. E. Clarke and P. C. Ford, *Inorg. Chem.*, **9,** 227, 495 (1970); P. C. Ford, *J. Chem. Soc. Chem. Comm.*, 7 (1971).

29. M. Eigen, *Pure Appl. Chem.*, **6,** 97 (1963).

30. D. A. Buckingham, A. M. Sargeson, and A. Zanella, *J. Am. Chem. Soc.*, **94,** 8246 (1972).

31. D. A. Buckingham in *Biological Aspects of Inorganic Chemistry*, A. W. Addison, W. R. Cullen, D. Dolphin, B. R. James, Eds., Wiley, New York, 1977, p. 141, and references therein.

32. R. K. Murmann and H. Taube, *J. Am. Chem. Soc.*, **78,** 4886 (1956). D. E. Klimek, B. Grossman, and A. Haim, *Inorg. Chem.*, **11,** 2382 (1972).

33. A. M. Sargeson and H. Taube, *Inorg. Chem.*, **5,** 1094 (1966).

34. F. A. Posey and H. Taube, *J. Am. Chem. Soc.*, **75,** 4099 (1953). E. Chaffee, T. P. Dasgupta, and G. M. Harris, *J. Am. Chem. Soc.*, **95,** 4169 (1973).

35. R. van Eldik and G. M. Harris, *Inorg. Chem.*, **19,** 880 (1980).

36. D. A. Buckingham, J. MacB. Harrowfield, and A. M. Sargeson, *J. Am. Chem. Soc.*, **95,** 7281 (1973).

37. A. D. Fowless and D. R. Stranks, *Inorg. Chem.*, **16,** 1271, 1276, 1282 (1977).

38. H. Gamsjager, G. A. K. Thompson, W. Sagmuller, and A. G. Sykes, *Inorg. Chem.*, **19,** 997 (1980).

39. R. S. Taylor, *Inorg. Chem.*, **16,** 116 (1977).

40. T. A. Beech, N. C. Lawrence, and S. F. Lincoln, *Aust. J. Chem.*, **26,** 1877 (1973).

41. R. K. Wharton, R. S. Taylor, and A. G. Sykes, *Inorg. Chem.*, **14,** 33 (1975).

42. (a) M. F. Dunn, *Struct. Bonding (Berlin)*, **23,** 61 (1975). (b) R. H. Prince, *Advan. Inorg. Chem. Radiochem.*, **22,** 349 (1979). (c) B. T. Golding, G. J. Leigh, *Inorg. Biochem. (Chem. Soc. London)*, **1,** 35 (1979).

43. S. Lindskog, L. E. Henderson, K. K. Kannan, A. Liljas, P. O. Nyman, and B. Strandberg, *Enzymes*, **5,** 587 (1971).

44. (a) D. W. Appleton and B. Sarkar, *Proc. Nat. Acad. Sci. U.S.A.*, **71,** 1686 (1974). (b) D. W. Appleton and B. Sarkar, *Bioinorg. Chem.*, **4,** 309 (1975). (c) J. M. Pesando, *Biochemistry*, **14,** 675 (1975). (d) R. K. Gupta and J. M. Pesando, *J. Biol. Chem.*, **250,** 2630 (1975).

45. M. W. Makinen, L. C. Kuo, J. J. Dymowski, and S. Jaffer, *J. Biol. Chem.*, **254,** 356 (1979).

46. M. Stiles, *J. Am. Chem. Soc.*, **81,** 2598 (1959).

47. A. A. Frost and R. G. Pearson, *Kinetics and Mechanism*, 2nd ed., Wiley, New York, 1963, pp. 307–316.

48. (a) N. E. Dixon, C. Gazzola, R. L. Blakeley, and B. Zerner, *J. Am. Chem. Soc.*, **97,** 4131 (1975). (b) N. E. Dixon, R. L. Blakeley, and B. Zerner, *Can. J. Biochem.*, **58,** 469, 481 (1980). (c) N. E. Dixon, C. Gazzola, C. J. Asher, D. S. W. Lee, R. L. Blakeley, and B. Zerner, *Can. J. Biochem.*, **58,** 474 (1980). (d) N. E. Dixon, J. A. Hinds, A. K. Fihelly,

C. Gazzola, D. J. Winzor, R. L. Blakeley, and B. Zerner, *Can. J. Biochem.*, **58,** 1323 (1980).

49. R. L. Blakeley, J. A. Hinds, H. E. Kunze, E. C. Webb, and B. Zerner, *Biochemistry*, **8,** 1991 (1969), and references therein.
50. W. G. Jackson and A. M. Sargeson in *Rearrangements in Ground and Excited States*, Vol. 2, P. de Mayo, Ed., Academic, New York, 1980, pp. 273-378, and references therein.
51. R. J. Balahura and R. B. Jordan, *Inorg. Chem.*, **9,** 1567 (1970).
52. P. C. Ford, *Inorg. Chem.*, **10,** 2153 (1971).
53. D. A. Buckingham, D. J. Francis, and A. M. Sargeson, *Inorg. Chem.*, **13,** 2630 (1974).
54. M. B. Jensen, *Acta. Chem. Scand.*, **12,** 1657 (1958).
55. F. J. W. Roughton, *J. Am. Chem. Soc.*, **63,** 2930 (1941).
56. M. Linhard and H. Flygare, *Z. Anorg. Chem.*, **251,** 25 (1943).
57. D. A. Buckingham, I. I. Olsen, A. M. Sargeson, and H. Satrapa, *Inorg. Chem.*, **6,** 1027 (1967).
58. N. E. Dixon, W. G. Jackson, M. J. Lancaster, G. A. Lawrance, and A. M. Sargeson, *Inorg. Chem.*, **20,** 470 (1981).
59. D. A. Buckingham, J. MacB. Harrowfield, and A. M. Sargeson, *J. Am. Chem. Soc.*, **96,** 1726 (1974).
60. A. M. Sargeson, *Pure Appl. Chem.*, **33,** 527 (1973).
61. W. P. Jencks, *Catalysis in Chemistry and Enzymology*, McGraw-Hill, New York, 1969.
62. D. P. Fairlie and W. G. Jackson, unpublished results.
63. R. L. Blakeley, A. Treston, R. K. Andrews, and B. Zerner, *J. Am. Chem. Soc.*, **104,** 612 (1982).
64. H. Kroll, *J. Am. Chem. Soc.*, **74,** 2036 (1952).
65. (a) N. C. Li and R. A. Manning, *J. Am. Chem. Soc.*, **77,** 5225 (1955). (b) J. M. White, R. A. Manning, and N. C. Li, *J. Am. Chem. Soc.*, **78,** 2367 (1956). (c) N. C. Li, F. Doody, and J. M. White, *J. Am. Chem. Soc.*, **79,** 5859 (1957). (d) R. Mathur and N. C. Li, *J. Am. Chem. Soc.*, **86,** 1289 (1964).
66. M. L. Bender and B. W. Turnquest, *J. Am. Chem. Soc.*, **79,** 1889 (1957).
67. H. L. Conley, Jr. and R. B. Martin, *J. Phys. Chem.*, **69,** 2914, 2923 (1965).
68. (a) W. A. Connor, M. M. Jones, and D. L. Tuleen, *Inorg. Chem.*, **4,** 1129 (1965). (b) J. E. Hix, Jr. and M. M. Jones, *Inorg. Chem.*, **5,** 1863 (1966). (c) J. E. Hix, Jr. and M. M. Jones, *J. Am. Chem. Soc.*, **90,** 1723 (1968).
69. (a) R. W. Hay and P. J. Morris, *J. Chem. Soc. Chem. Comm.*, **23** (1967). (b) R. W. Hay and L. J. Porter, *J. Chem. Soc. A.* 127 (1969). (c) R. W. Hay and P. J. Morris, *J. Chem. Soc. A.* 1524 (1971). (d) R. W. Hay and P. J. Morris, in *Metal Ions in Biological Systems*, Vol. 5, H. Sigel, Ed., Dekker, New York, 1976, pp. 173–243.
70. (a) R. J. Angelici and B. E. Leach, *J. Am. Chem. Soc.*, **89,** 4605 (1967). (b) R. J. Angelici and B. E. Leach, *J. Am. Chem. Soc.*, **90,** 2499 (1968). (c) B. E. Leach and R. J. Angelici, *J. Am. Chem. Soc.*, **90,** 2504 (1968). (d) D. Hopgood and R. J. Angelici, *J. Am. Chem. Soc.*, **90,** 2508 (1968). (e) R. J. Angelici and D. Hopgood, *J. Am. Chem. Soc.*, **90,** 2514 (1968). (f) B. E. Leach and R. J. Angelici, *Inorg. Chem.*, **8,** 907 (1969).
71. L. Meriwether and F. H. Westheimer, *J. Am. Chem. Soc.*, **78,** 5119 (1956).
72. (a) J. I. Hoppé and J. E. Prue, *J. Chem. Soc.*, 1775 (1957). (b) G. L. Johnson, R. J. Angelici, *J. Am. Chem. Soc.*, **93,** 1106 (1971).

73. D. S. Sigman, C. T. Jorgensen, *J. Am. Chem. Soc.*, **94,** 1724 (1972).
74. M. A. Wells and T. C. Bruice, *J. Am. Chem. Soc.*, **99,** 5341 (1977).
75. J. T. Groves and R. M. Dias, *J. Am. Chem. Soc.*, **101,** 1033 (1979).
76. W. E. Stewart and T. H. Siddall, III, *Chem. Rev.*, **70,** 517 (1970).
77. C. R. Smith and K. Yates, *J. Am. Chem. Soc.*, **94,** 8811 (1972).
78. J. P. Collman and D. A. Buckingham, *J. Am. Chem. Soc.*, **85,** 3039 (1963).
79. D. A. Buckingham, J. P. Collman, D. A. R. Happer, and L. G. Marzilli, *J. Am. Chem. Soc.*, **89,** 1082 (1967).
80. D. A. Buckingham and J. P. Collman, *Inorg. Chem.*, **6,** 1803 (1967).
81. D. A. Buckingham, L. G. Marzilli, and A. M. Sargeson, *J. Am. Chem. Soc.*, **89,** 2772 (1967).
82. J. P. Collman and E. Kimura, *J. Am. Chem. Soc.*, **89,** 6096 (1967).
83. D. A. Buckingham, L. G. Marzilli, and A. M. Sargeson, *J. Am. Chem. Soc.*, **89,** 4539 (1967).
84. E. Kimura, S. Young, and J. P. Collman, *Inorg. Chem.*, **9,** 1183 (1970); S. K. Oh and C. B. Storm, *Bioinorg. Chem.*, **3,** 89 (1973); S. K. Oh and C. B. Storm, *Biochemistry*, **13,** 3250 (1974).
85. M. D. Alexander and D. H. Busch, *J. Am. Chem. Soc.*, **88,** 1130 (1966).
86. D. A. Buckingham, P. A. Marzilli, I. E. Maxwell, A. M. Sargeson, M. Fehlmann, and H. C. Freeman, *J. Chem. Soc. Chem. Commun.*, 488 (1968).
87. D. A. Buckingham, D. M. Foster, and A. M. Sargeson, *J. Am. Chem. Soc.*, **90,** 6032 (1968).
88. D. A. Buckingham, D. M. Foster, and A. M. Sargeson, *J. Am. Chem. Soc.*, **92,** 5701 (1970).
89. R. B. Martin, *J. Am. Chem. Soc.*, **89,** 2501 (1967).
90. D. A. Buckingham, J. Dekkers, A. M. Sargeson, and M. Wein, *J. Am. Chem. Soc.*, **94,** 4032 (1972).
91. W. P. Jencks, *Acc. Chem. Res.*, **9,** 425 (1976).
92. D. A. Buckingham, J. Dekkers, and A. M. Sargeson, *J. Am. Chem. Soc.*, **95,** 4173 (1973).
93. R. W. Hay and L. J. Porter, *J. Chem. Soc. B*, 1261 (1967).
94. D. A. Buckingham, D. M. Foster, and A. M. Sargeson, *J. Am. Chem. Soc.*, **91,** 4102 (1969).
95. D. A. Buckingham, D. M. Foster, L. G. Marzilli, and A. M. Sargeson, *Inorg. Chem.*, **9,** 11 (1970).
96. D. A. Buckingham, D. M. Foster, and A. M. Sargeson, *J. Am. Chem. Soc.*, **92,** 6151 (1970).
97. C. J. Boreham, D. A. Buckingham, and F. R. Keene, *Inorg. Chem.*, **18,** 28 (1979).
98. D. A. Buckingham, F. R. Keene, and A. M. Sargeson, *J. Am. Chem. Soc.*, **96,** 4981 (1974).
99. C. J. Boreham, D. A. Buckingham, and F. R. Keene, *J. Am. Chem. Soc.*, **101,** 1409 (1979).
100. D. A. Buckingham, D. M. Foster, and A. M. Sargeson, *J. Am. Chem. Soc.*, **91,** 3451 (1969).
101. C. J. Boreham, D. A. Buckingham, D. J. Francis, A. M. Sargeson, and L. G. Warner, *J. Am. Chem. Soc.*, **103,** 1975 (1981).

102. T. H. Fife and V. L. Squillacote, *J. Am. Chem. Soc.*, **99,** 3762 (1977).
103. T. H. Fife and V. L. Squillacote, *J. Am. Chem. Soc.*, **100,** 4787 (1978).
104. T. H. Fife, T. J. Przystas, and V. L. Squillacote, *J. Am. Chem. Soc.*, **101,** 3017 (1979).
105. T. J. Przystas and T. H. Fife, *J. Am. Chem. Soc.*, **102,** 4391 (1980).
106. T. H. Fife and T. J. Przystas, *J. Am. Chem. Soc.*, **102,** 7297 (1980).
107. M. W. Makinen, K. Yamamura, and E. T. Kaiser, *Proc. Natl. Acad. Sci. U.S.A.*, **73,** 3882 (1976).
108. R. Breslow and D. L. Wernick, *Proc. Natl. Acad. Sci. U.S.A.*, **74,** 1303 (1977).
109. D. A. Buckingham and L. M. Englehardt, *J. Am. Chem. Soc.*, **97,** 5915 (1975).
110. J. March, *Advanced Organic Chemistry*, 2nd ed., McGraw-Hill, New York, 1977, p. 693.
111. (a) J. P. Glusker, *Enzymes*, **5,** 413 (1971). (b) J. P. Glusker, *J. Mol. Biol.*, **38,** 149 (1968).
112. D. M. Kurtz, R. H. Holm, F. J. Ruzicka, H. Beinert, C. J. Coles, and T. P. Singer, *J. Biol. Chem.*, **254,** 4967 (1979), and references therein.
113. J. J. Villafranca and A. S. Mildvan, *J. Biol. Chem.*, **247,** 3454 (1972), and references therein.
114. S. Englard, *J. Biol. Chem.*, **235,** 1510 (1960).
115. K. R. Hanson and I. A. Rose, *Proc. Natl. Acad. Sci. USA*, **50,** 981 (1963).
116. I. A. Rose and E. L. O'Connell, *J. Biol. Chem.*, **242,** 1870 (1967).
117. L. R. Gahan, J. MacB. Harrowfield, A. J. Herlt, L. F. Lindoy, A. M. Sargeson, and P. O. Whimp, to be published.
118. P. O. Whimp, to be published.
119. (a) A. J. Herlt, to be published. (b) N. E. Dixon and A. M. Sargeson, *J. Am. Chem. Soc.*, **104,** 6716 (1982).
120. E. K. Chong, Ph.D. Thesis, Australian National University (1979).
121. K. R. Hanson and E. A. Havir, *Enzymes*, **7,** 75 (1972), and references therein.
122. H. Bright, *J. Biol. Chem.*, **239,** 2307 (1964), and references therein.
123. C. B. Klee, *J. Biol. Chem.*, **247,** 1398 (1972), and references therein.
124. M. M. Rechler, *J. Biol. Chem.*, **244,** 551 (1969).
125. D. S. Hodgins, *J. Biol. Chem.*, **246,** 2977 (1971).
126. I. L. Givot, T. A. Smith, and R. H. Abeles, *J. Biol. Chem.*, **244,** 6341 (1969).
127. K. R. Hanson and E. A. Havir, *Arch. Biochem. Biophys.*, **141,** 1 (1970).
128. R. B. Wickner, *J. Biol. Chem.*, **244,** 6550 (1969).
129. J. MacB. Harrowfield and A. M. Sargeson, *J. Am. Chem. Soc.*, **96,** 2634 (1974).
130. J. C. Duff, *J. Chem. Soc.*, **119,** 385 (1921).
131. W. L. Steffen, to be published.
132. A. Hammershøi, Ph.D. Thesis, Australian National University (1981).
133. The conformational and configurational nomenclature for these systems are defined in *Inorg. Chem.*, **9,** 1 (1970).
134. D. E. Metzler, M. Ikawa, and E. E. Snell, *J. Am. Chem. Soc.*, **76,** 648 (1954).
135. K. Tatsumoto and A. E. Martell, *J. Am. Chem. Soc.*, **100,** 5549 (1978), and references therein.
136. T. C. Bruice and S. J. Benkovic, *Bioorganic Mechanisms*, Vol. 2, W. A. Benjamin, New York, 1966, pp. 181–300, and references therein.

137. R. H. Holm in *Inorganic Biochemistry*, Vol. 2, G. L. Eichhorn, Ed., Elsevier, New York 1973, pp. 1137–1167, and references therein.

138. A. G. Lloyd, N. Tudball, and K. S. Dodgson, *Biochim. Biophys. Acta*, **52,** 413 (1961).

139. Y. Noda, *J. Biochem.* (*Tokyo*), **57,** 709 (1965).

140. J. H. Thomas and N. Tudball, *Biochem. J.*, **105,** 467 (1967).

141. N. Tudball and P. Thomas, *Biochem. J.*, **126,** 187 (1972).

142. (a) N. Tudball, personal communication. (b) N. Tudball and J. G. O'Neill, *Biochim. Biophys. Acta*, **429,** 616 (1976), and references therein.

143. K. R. Turnbull, Ph.D. Thesis, Australian National University (1965); D. H. Williams and D. H. Busch, *J. Am. Chem. Soc.*, **87,** 4644 (1965).

144. I. Murakoshi, H. Kuramoto, J. Haginiwa, and L. Fowden. *Phytochemistry*, **11,** 177 (1972).

145. N. M. Kredich and G. M. Tomkins, *J. Biol. Chem.*, **241,** 4955 (1966).

146. I. K. Smith and J. F. Thompson, *Biochim. Biophys. Acta*, **227,** 288 (1971).

147. Some enzymic phosphate chemistry is reviewed by J. E. Coleman and J. F. Chlebowski (pp. 1–66), and W. W. Cleland and A. S. Mildvan (pp. 163–191), in *Advances in Inorganic Biochemistry*, Vol. 1, G. L. Eichhorn and L. G. Marzilli, Eds., Elsevier/North-Holland, New York, 1979.

148. R. K. Osterheld in *Topics in Phosphorus Chemistry*, Vol. 7, E. J. Griffith and M. Grayson, Eds., Wiley-Interscience, 1972, pp. 103–254.

149. M. Dixon and E. C. Webb, *Enzymes*, 2nd ed., Longmans, New York, 1971, pp. 422–424.

150. J. F. Morrison and E. Heyde, *Ann. Rev. Biochem.*, **41,** 29 (1972).

151. T. G. Spiro, W. A. Kjellstrom, M. Zeydel, and R. A. Butow, *Biochemistry*, **7,** 859 (1968).

152. M. Tetas and J. M. Lowenstein, *Biochemistry*, **2,** 350 (1963).

153. (a) T. G. Spiro, in *Inorganic Biochemistry*, Vol. 1, G. L. Eichhorn, Ed., Elsevier, New York, 1973, pp. 549–581, and references therein. (b) F. J. Farrell, W. A. Kjellstrom, and T. G. Spiro, *Science*, **164,** 320 (1969).

154. (a) B. Anderson, R. M. Milburn, J. MacB. Harrowfield, G. B. Robertson, A. M. Sargeson, *J. Am. Chem. Soc.*, **99,** 2652 (1977). (b) J. MacB. Harrowfield, unpublished work.

155. P. W. A. Hübner and R. M. Milburn, *Inorg. Chem.*, **19,** 1267 (1980).

156. D. R. Jones, M. R. Snow, L. F. Lindoy, R. M. Milburn and A. M. Sargeson, to be published.

157. R. D. Cornelius and W. W. Cleland, *Biochemistry*, **17,** 3279 (1978), and references therein.

158. J. D. Otvos, J. R. Alger, J. E. Coleman, and I. M. Armitage, *J. Biol. Chem.*, **254,** 1778 (1979), and references therein.

159. F. H. Westheimer, in *Rearrangements in Ground and Excited States*, Vol. 2, P. de Mayo, Ed., Academic, New York, 1980, pp. 229-271; F. H. Westheimer, *Accounts Chem. Research*, **1,** 70 (1968), and references therein.

160. D. M. Brown and D. A. Usher, *J. Chem. Soc.*, 6558 (1965).

161. J. MacB. Harrowfield, D. R. Jones, L. F. Lindoy, and A. M. Sargeson, *J. Am. Chem. Soc.*, **102,** 7733 (1980).

162. M. I. Page and W. P. Jencks, *Proc. Natl. Acad. Sci. U.S.A.*, **68,** 1678 (1971).

163. R. A. Lazarus, P. A. Benkovic, and S. J. Benkovic, *J. Chem. Soc. Perkin Trans. 2*, 373 (1980).

164. I. I. Creaser, R. V. Dubs, and A. M. Sargeson, to be published.

165. A. S. Mildvan, J. S. Leigh, and M. Cohn, *Biochemistry*, **6,** 1805 (1967).
166. D. R. Jones, L. F. Lindoy, and A. M. Sargeson, to be published.
167. A. J. Kirby and W. P. Jencks, *J. Am. Chem. Soc.*, **87,** 3209 (1965).
168. D. R. Jones, L. F. Lindoy, A. M. Sargeson, and M. R. Snow, *Inorg. Chem.*, **21,** 4155, (1982).
169. D. R. Jones, L. F. Lindoy, R. M. Milburn, and A. M. Sargeson, to be published.
170. J. R. Knox and H. W. Wyckoff, *J. Mol. Biol.*, **74,** 533 (1973).
171. S. R. Jones, L. A. Kindman, and J. R. Knowles, *Nature*, **275,** 564 (1978).
172. P. Woolley, *Nature*, **258,** 677 (1975).
173. R. B. Jordan, A. M. Sargeson, and H. Taube, *Inorg. Chem.*, **5,** 1091 (1966).
174. H. D. Campbell and B. Zerner, *Biochem. Biophys. Res. Commun.*, **54,** 1498 (1973); H. D. Campbell, D. A. Dionysius, D. T. Keough, B. E. Wilson, J. de Jersey, and B. Zerner, *Biochem. Biophys. Res. Commun.*, **82,** 615 (1978).
175. D. T. Keough, D. A. Dionysius, J. de Jersey, and B. Zerner, *Biochem. Biophys. Res. Commun.*, **94,** 600 (1980).
176. Y. Sigiura, H. Kawabe, H. Tanaka, S. Fujimoto, and A. Ohara, *J. Biol. Chem.*, **256,** 10664 (1981).
177. E. A. Merritt, M. Sundaralingam, R. D. Cornelius, and W. W. Cleland, *Biochemistry*, **17,** 3274 (1978).
178. D. Dunaway-Mariano and W. W. Cleland, *Biochemistry*, **19,** 1506 (1980), and references therein.
179. J. Granot, A. S. Mildvan, H. N. Bramson, N. Thomas, and E. T. Kaiser, *Biochemistry*, **20,** 602 (1981), and references therein.
180. W. B. Robinson, D. A. Buckingham, and A. M. Sargeson, unpublished work.
181. G. P. Haight, Jr. and R. Peachey, unpublished work.
182. T. Imamura, D. M. Hinton, R. L. Belford, R. L. Gumport, and G. P. Haight, Jr., *J. Inorg. Biochem.*, **11,** 241 (1979).
183. H. Sigel and P. E. Amsler, *J. Am. Chem. Soc.*, **98,** 7390 (1976).

Index

Acetazolamide, 105–107
Aconitase, 304
  reaction mechanism, 311–313
  substrate reactions, 305
Aconitic acid hydratase, *see* Aconitase
Acylamino acids, hydrolysis of, 54
Acyl enzymes, hydrolysis of, 271
Adenosine triphosphatase, 341
  metal ion activation, 343
    by vanadium, 344, 345
  phosphorothioate analog, 227
  stereochemistry, 342
Alberti, G., 108
Albumins, 9
Alcohol dehydrogenase, 14, 15, 123–152
  catalytic mechanism, aspects, 144–147
  coenzyme binding, 134
  electron paramagnetic resonance, 144
  hydride transfer, 147
  inhibitor binding, 137–140
  ligand binding crystallography of, 134–140
  metal substitution, 141
  molecular structures, 125
    conformational change, 131–133
    dynamic role, 132–134
    open and closed forms, 131
  NMR studies, 142
  oxidation properties, 15
  proton abstraction, 144
  resonance Raman spectroscopy, 143
  schematic drawings of, 126
  stereodiagram of, 127
  substrate binding, 136
  substrate-binding pocket, 129–130
  and zinc, 127
    location, 129
Alcohols, 15. *See also individual alcohols*
Aldolases, 13
Alexander, M.D., 291
Alkaline phosphatase, 9, 14, 15, 153–217
  activation by magnesium, 174
  active centers, 178
  activity of, 16
  chemistry of, 323
  crystal structure, 175
  enzymatic reactions, 160–166
  kinetics, 160–166
  metal binding properties, 173
  metal binding sites, 177–182
    electron spin resonance, 182–185
    optical spectroscopy, 180–182
  metal ion binding, 166–174
    competition for, 170
  metal ion substitutions, 164
  molecular boundaries, 176
  NMR studies, 178, 185–188, 194–203
  primary structure of, 157
  requirement for zinc, 166
  sources:
    mammalian, 157–159
    microorganisms, 157–159
  substrate specificity, 160
  synthesis of, 156
  zinc binding, 169
Amides:
  amino acid hydrolysis, 288
  coordinated, 287

Amination, 313
Aminoacetonitrile, 266
Amino acid amides, hydrolysis, 288
Amino acid esters:
  chelated derivatives, 291
    intramolecular hydrolysis, 292
  hydrolysis, 288
    mechanism, 288–291
Amino acid methyl esters, 288
Amino acids:
  chelation, 315
  deamination of, 318
  transamination, 318
Amino acid sequences, of alkaline phosphatase, 157
Aminopropionitrile, 266
Ammonia lyases, 304
Anencephaly, 10
Aniline, bromination, 258
Anions:
  binding with carbonic anhydrase, 101–105
  equilibrium, 104
Anomalous dispersion, 178
Apoalkaline phosphatase, 167–170
  cadmium binding, 167–169
  zinc binding, to, 169–170
Apoenzymes, 88
  of carbonic anhydrase, 89, 96
Arginine, 20
  modification of, 109
Arsanilazo carboxypeptidase A, 45
  ligand binding sites, 48
  spectra, 48
Arsanilic acid, 45
Aspartic acid, carboxyl group, 11
Aspergillus niger, 3, 7
Auld, D. S., 25

Bacillus subtilis, RNA polymerase, 232
  zinc content, 232
Bacteriophage:
  DNA, 233
  genome, 233
  RNA polymerase, 233–236
    enzymatic properties, 233
    inhibition, 234
    physiochemical properties, 234
    preparation, 233
    zinc role, 233
Bauer, R., 88
Bertini, I., 92
Blake, C. C. F., 54
Bollum, E. J., 17
Bonds, chemical, *see* Chemical bonds, around zinc atoms
Boreham, C. J., 300
Bradshaw, R. A., 157
Branden, C., 123
Bromobenzyl alcohol, 136
Brown, R. D., 100
Burgen S. V., 106
Busch, D. H., 291

Cadmium alkaline phosphatase, 185–188, 203
  phosphorylation of, 188–194
    equilibrium, 192–194
Carbamate complexes, 274
Carbanions, protonation, 314
Carbon dioxide, elimination of, 16
Carbonic anhydrase, 3, 9, 13, 15, 77–121
  active site:
    aldehyde hydration, 270
    inhibition of, 79–80
    metal ion centers in, 86
    modification of, 108–109
    structure of, 83–86
  anion complexes, 101–105
  evolution of, 82
  and hydroxide ion, 270
  inhibition, 97–108
  isoenzymes of active site residues in, 87
  kinetic mechanisms, 97–108
  kinetic properties of, 79–80
  metal ion binding, rates and equilibria, 95–97
  molecular mechanism, 108–113
  structure of 82–83, 85
  zinc coordinated histidine, 270
  zinc hydroxide mechanism, 271
Carbon NMR, 194–196
Carbonyl group, polarization of, 281
Carboxyaspartic acid, 314
Carboxylation, 274
Carboxylato-olefin complexes, 306
Carboxypeptidase, 13, 15, 16
  catalysis of peptides, 54
  histidine modification in, 45
  and proteolysis, 16
Carboxypeptidase A, 25–75
  acetylation, 44
  active sites, 35
  activity, 35

and amino acid residues, 44
amino acid sequence, 28
catalysis of, esters, 54
catalytic intermediates, 63
spectroscopy, 64, 65
catalytic properties, 54
characteristics, 27, 28
chemical modification, 44
complexes, 32
crystals;
kinetic studies, 53
spectra, 51
derivatives, 51
ester binding sites, 68
ester substrates, hydrolysis, 67
hydrolysis mechanism, 56, 57
role of zinc, 60
inhibitors, 35
mechanism:
histidine inactivation, 69
tyrosine role, 69
metal substitution, 34
peptide substrates, 67
in solution:
kinetic studies, 52
spectra, 51
sources, 27
structure, 28
substrates of, 64
spectroscopic properties, 35
x-ray crystallography, 29
Carpy, S., 96
Cell biology, 17, 18
Cell division, zinc requirment, 239
Chang, L. M. S., 17
Chelating agent, 221
Chemical bonds, around zinc atoms, 128
Chlorine NMR, 198
Chromosomal replication, 243
Chymotrypsin, 28
Citirc acid, isomerization of, 304, 305
Cleft palate, 10
Cleland, W. W., 64
Cobalt, in model chemistry, 265
Cobalt alkaline phosphatase, 180–185
Cobalt carboxypeptidase, 36
catalytic activity, 44
ERR spectra, 43
inactivation, 44
spectral perturbations, 38
Cobalt complexes, 257, 274, 275
and peptide hydrolysis, 291
reactions, 275–287
with pyrophosphate, 343
solvent exchange, 280
Cobalt nitrile complexes, 264
Cobalt procarboxypeptidase, 37
Cobalt urea complexes, 286
Coleman, J. E., 153, 219
Collier, E. S., 8
Conformational changes, in alcohol dehydrogenase, 131–134
Coordinated hydroxide, as intramolecular nucleophile, 298
Coordinated ligands, as nucleophiles, 266
Coordinated nucleophiles, hydrolysis by, 296
Coordination geometry, 14
Copper alkaline phosphatase, 182–185
Copper complex, as DNA polymerase inhibitor, 243–244
Coupling, 208
Cryokinetics, 69
Cryospectroscopy, 69
Crystallography, of alkaline phosphatase, 175–176
Cyanate complexes, 274

Deacylation, 273
Deamination, 313
Dephosphorylation, 273, 338
pH dependence, 338
zinc requirement, 338
Dimethylaminocinnamaldehyde, 136, 142
Dimethylphosphate ion, 325
Dimethylsulfoxide, 137, 271
hydrolysis, 279
Dissociation, 210–211
Dixon, N. E., 253, 264
DNA polymerase, 17, 243–245
inhibition, 221, 243–244
molecular properties, 243
NMR studies, 246
subunits, 243
zinc content, 221, 244
zinc coordination sites, 245
zinc role in, 219–252
DNA transcription, zinc role, 239–240
Drinker, K. P., 8
Dwarfism, 10

Eklund, H., 123

Electron:
magnetic resonance, of alcohol dehydrogenase, 144
spin resonance, of alkaline phosphatase, 182–185
Endonuclease, 243
Enzymatic reactions, modeling, 265
Enzyme catalysis, intermediate species, 63
Equilibrium, and NMR spectra, 203–205, 209–210
Erythrocytes, 9, 15
Escherichia coli, RNA polymerase, 221, 229–232
amino acid sequence, 229
genetic mapping, 229
ESR, *see* Electron spin resonance, of alkaline phosphatase
Esters:
carboxyl group, 68
chelated, 291, 293
reaction kinetics, 294–295
hydrolysis mechanism, 56
Ethylene glycol, 16
Ethylene phosphate anion, 325
Euglena gracilis:
RNA polymerase, 236
inhibition, 237
zinc content, 236, 237
zinc deficiency, 239
Eukaryotes, RNA polymerase, 236
zinc content, 236

Fabry, M. E., 93
Fife, T. H., 304
Fluorine, NMR, 196–198

Galdes, A., 25
Gene expression, 20, 21, 228
Genetic transcription, 21
Geoghegan, K., 43
Gettins, P., 153
Glusker, J. P., 311
Glutamic acid, carboxyl group, 11
Glycine amides, chelated, alkaline hydrolysis, 295
Glycine derivatives, intramolecular hydrolysis, 301
Glycine esters, chelated, 292
Glycolysis, 305
Glycyglycine, hydrolysis, 300

Haffner, P. H., 94
Hexokinase, active sites, 343
Histidine, modification in carbonic anhydrase, 108–109
Histones, 20
Holmes, R. S., 82
Hubner, P. W. A., 344
Hydration, 261–264
Hydrocephalus, 10
Hydrogen bond, 62
Hydroxide ion:
coordination to cobalt, 267
as nucleophile, 267
Hydroxo complex:
acylation of, 268
incomplete ionization, 269

Imidazole, 107–108, 137–138, 271
Imines, intramolecular reactions, 318–323
Isocitric acid, 305
Isocyanato complexes, 286

Jendrisak, J., 237

Keilin, D., 3
Ketones, carboxylation, 273
Khalifah, R. G., 98
King, R. W., 106
Koenig, S. H., 100, 103
Kroll, H., 288

Leucine aminopeptidase, 14
Leukocytes, 9
alkaline phosphatase, 16
Lindskog, S., 77
Lipscomb, W. N., 67
Lomofungin, 237

Macroglobulin, 13
Magnesium:
effect on alkaline phosphatase, 174
functional effects of, 174–176
and zinc binding, 169–172
Makinen, M. W., 67
Maleate:
chelate, ring opening, 316
ester complex, 313
Maleic acid:
complexes, 307
reactivity, 316

Manganese alkaline phosphatase, 182–185
Mann, T., 3
Martinelli, R., 43
Meldrum, N. U., 79
Metal extraction techniques, 239
"Metal fume fever," 8
Metal hydroxide mechanism, 60, 67
Metal ions:
  charge density, 257
  comparison to proton, 256, 257
  and coordinate substrates, 253–352
  coordination sites, 256
  Lewis acidity, 303
  as Lewis acids, 256
  in olefin reactions, 304
  oxidation states, 255
Metalloalkaline phosphatases, 166, 207
Metallocarbonic anhydrases, 86–95
  cadmium complexes, 88–89
  cobalt complexes, 91–94
  copper complexes, 90
  inhibitor complexes, 85–95
  manganese complexes, 94–95
  mercury complexes, 89
  nickel complexes, 90–91
  spectroscopic properties, 88–95
  vanadium complexes, 95
  zinc ion removal, 86, 88
Metallocarboxypeptidases, 37
Metalloenzymes:
  catalytic, 13
  crystal structure, 12
  and metal ions, 253–352
  metal ions roles, 255
  metal ligands, 11
  noncatalytic, 13
  nucleophilic substitution, 255
  regulatory, 13
  structural, 13
  and zinc, 13, 141
Metallothionein, 10, 13
Metals:
  biological functions, 3
  in the body, 3
Methanol, 16
Methylation, 21
Michaelis-Menten kinetics, 64
Milburn, R. M., 344

Nakagawa, S., 67
Nitriles:
  hydration of, 261–263
  intramolecular reactions, 261
Nitrophenyl acetate, reaction with nucleophiles, 272
Nitrophenyl phosphate, hydrolysis of, 161
Nuclear magnetic resonance:
  of alcohol dehydrogenase, 142
  of alkaline phosphatase, 178, 185, 194–203
  of cadmium alkaline phosphatase, 185–188
  of phosphoenzyme intermediates, 199–213
Nucleic acid:
  metabolism, 13, 17
    zinc role, 239–241
  phosphate group, 21
Nucleophiles, 264, 270. *See also individual compounds*
Nucleophilic substitition, 255
Nucleoside triphosphate, 227

O'Connel, E. L., 305
Olefins:
  biological reactions, 304
  cyclization reactions, 307, 308
    electronic effects, 310
  hydration, 307–309
  nucleophilic additions, 304
  reactions, metal ion promoted, 304–311
O-Phospharothioates, hydrolysis of, 160, 163
Optical spectroscopy, of active site metal ions, 180–182
Oxidation, of alcohol, 144
Oxyphosphorane, 331

Pancreas, 27
Pancreatic juice, 16
Peptide bond hydrolysis, by peptidases, 288–304
Peptides:
  hydrolysis, 16, 54, 56
  synthesis, 19
Pesano, J. M., 110
pH:
  and alkaline phosphatase, 161
  and zinc binding, 169–172
Phenanthroline, 137–221
Phenylazophenylphosphate, hydrolysis of, 161
Phosphatase, alkaline, *see* Alkaline phosphatase

Phosphate esters, hydrolysis of, 324
  catalysis by, 337
  efficiency, 337
  pH effect, 332
Phosphates, and zinc binding, 169–172
Phosphodiesterase, 227
Phosphoenzyme intermediates, 161
  chemical shifts, 205–207
  dissociation of enzyme–phorphorus complex, 210
  mechanistic implications, 212–213
  negative cooperativity, 211
  NMR studies, 199–213
Phosphorus NMR, 199–202
Phosphorylation, 21, 174–175
  of cadmium alkaline phosphatase, 188–194
  rate, 160
Phosphoryl transferase, 323
Phytic acid, 10
Pocker, Y., 79
Polymerases, molecular properties, 221–228
Pregnancy, and zinc concentration, 9, 10
Propionic anhydride, 268
Protein:
  structural stability, 13
  in synthesis, 17, 18
    zinc role, 240
Proteolysis, 28
Proton:
  coordination number, 257
  transfer, 144–177
Protonation, 257
Purple phosphomonoesterases, 341
Pyrazoles, 137–140
Pyridoxal phosphate, as deamination catalyst, 318
Pyruvate kinase, 328

Raulin, J., 3, 7
Resonance Raman spectroscopy, of alcohol dehydrogenase, 143
Retinene reductase, 15
Reverse transcriptases, 241
  inhibition by phenanthroline, 241
  zinc content, 242
Rhodium complexes, 274, 284
  linkage isomerization, 287
  reaction kinetics, 285, 286
Ribonucleoside triphosphate, 226
Ribosomal proteins, 230
Riordan, J. F., 25
RNA-DNA hybrid, 226
RNA polymerase, 17
  catalytic activity, 234
  denaturation, 230
  inhibition, 221
  manganese binding, 246
  metal ion contents, 239
  NMR studies, 246
  nondialyzable zinc, 221
    RNA chain elongation, 227
    RNA chain termination, 227
  renaturation, 230
  zinc analysis, 237
  zinc binding sites, 230
  zinc content, 222–225
  zinc role in, 219–252
Romans, A. Y., 95
Rose, I. A., 305
Roughton, F. J. W., 79

Sargeson, A. M., 253
Sarkanen, S., 79
Scheiner, S., 67
Scrutton, M. C., 221
Serine alcohol, 339
Serine chelate, 321
Serine sulfate, deamination, 319
Serine sulfohydrolase, intermolecular reactions, 318–323
Seryl phosphate ester, 340
Shellfish, 10
Silverman, D. N., 99
Speckhard, D. C., 230
Spectroscopy, and role of catalytic zinc, 141–144
Spina bifida, 10
Stoichiometry:
  of phosphate binding, 208–209
  of zinc binding, to alkaline dehydrogenase, 164
Sulfonamides, binding with carbonic anhydrase, 105–107
Superacid catalysis, 256

Taylor, P. W., 104
Tetrazole, 108
Thermolysin, 13
Thiophosphate, hydrolysis of, 163
Thompson, J., 51

Tibell, L., 88
Torriani, A., 156
Transition metal complexes, 257
Transphosphorylation, 163–164
Triazole, 108
Trypsin, 28

Urea:
  coordination to nickel, 283
  synthesis of, 274
Urea carbonyl, polarization of, 282
Urea complexes:
  and carbamate complexes, 279
  cobalt-oxygen bond, 282

Vallee, B. L., 25

Water, molecular splitting, 109–113
Weber, G., 54
Westheime, F. H., 256
Wheat germ, RNA polymerase from, 237
  zinc content, 237

X-ray crystallography, 11

Yeast, RNA polymerase from, 238
Yee, D., 227

Zinc:
  alcohol dehydrogenase, role in, 125–149
  in alcohol dehydrogenase, 127–129
  binding:
    to alkaline phosphatase, 169
    to carbonic anhydrase, 95–97
  binding sites, 11, 64
    in alcohol dehydrogenase, 133–144
  biological reactions, 11
  biological role, 3, 6, 7
  biology of, 1–24
  in the blood, 9
  in the body, 9
  catalytic mechanism, 14
  catalytic role, 13
    in enzymes, 11
    spectroscopy, 141–144
  and cell biology, 17
  and coenzyme binding, 134–135
  and DNA, 17
  and enzymatic reactions, 12
  genetic expression role, 7, 20
  as industrial hazard, 8
  and nucleic acid metabolism, 17, 19
  pathology of, 10
  physiology of, 9
  and plant growth, 7
  protein complexes, 12
  and protein synthesis, 17, 18
  and RNA, 17
  tissue accumulation, 9
  toxic effects, 7, 8, 9
Zinc absorption, 10
Zinc alkaline phosphatase, NMR studies, 200–202
Zinc biochemistry, 1–24
Zinc complexes, 11, 257
  and carbonic anyhdrase, 11–113
  dissociation, 96
Zinc deficiency, 19
  congenital diseases, 7, 9
  diseases, 7, 9
  and nucleic acid synthesis, 221
Zinc enzymes, 3
  coordination geometry, 14
  genetic material transcription, 7
  metal exchange, 14
  see also specific enzymes
Zinc function, in polymerases, molecular aspects, 245
Zinc-hydroxide model, 111–113
Zinc as inhibitor, 14
Zinc ion, location in carbonic anhydrase, 85
Zinc metabolism, abnormal, 7
Zinc metalloproteases, 69
Zinc proteins, 7
Zwitterions, 277
Zymogen, 27